ALLELOPATHY

ALLELOPATHY

Basic and applied aspects

Edited by

S. J. H. Rizvi

and

V. Rizvi

Department of Botany and Plant Physiology
Rajendra Agricultural University
India

CHAPMAN & HALL
London · New York · Tokyo · Melbourne · Madras

Published by Chapman & Hall, 2–6 Boundary Row, London SE1 8HN

Chapman & Hall, 2–6 Boundary Row, London SE1 8HN, UK

Van Nostrand Reinhold Inc., 115 5th Avenue, New York NY10003, USA

Chapman & Hall Japan, Thomson Publishing Japan, Hirakawacho Nemoto Building, 6F, 1-7-11 Hirakawa-cho, Chiyoda-ku, Tokyo 102, Japan

Chapman & Hall Australia, Thomas Nelson Australia, 102 Dodds Street, South Melbourne, Victoria 3205, Australia

Chapman & Hall India, R. Seshadri, 32 Second Main Road, CIT East, Madras 600 035, India

First edition 1992

Typeset in 10/12 pt Times by Excel Typesetters Company, Hong Kong
Printed in Great Britain by St Edmundsbury Press, Bury St Edmunds, Suffolk

ISBN 0 412 39400 6 0 442 31465 5 (USA)

A catalogue record for this book is available from the British Library

Library of Congress Cataloging-in-Publication data available

Contents

List of Contributors

H. Aguilar-Erazo
Department of Crop Protection
Department of Chemistry
University of Puerto Rico
Mayaguez
Puerto Rico

H. Alborn
USDA/ARS Insect Attractants
Behaviour and Basic Biology Research Laboratory
PO Box 14565 Gainesville
Florida
USA

I. S. Alsaadawi
Department of Botany
Nuclear Research Centre
PO Box 765
Baghdad
Iraq

A. L. Anaya
Instituto de Fisiologia Celular
Universidad Nacional Autonoma de Mexico
Apartado Postal 70-600
04510 Mexico, DF
Mexico

I. C. Anderson
Department of Agronomy
Iowa State University
Ames
Iowa 50010
USA

M. E. A. Aquila
Departamento de Botanica
Universidade Federal do Rio Grande do Sul
Av Paulo Gama #40
CEP 90049
Porto Alegre, RS
Brazil

V. H. Argandoña
Departamento de Biologia
Faculdad de Cienceas
Universidad de Chile
Casilla 653
Santiago
Chile

H. H. Cheng
Department of Soil Science
University of Minnesota
St Paul
Minnesota 55108
USA

Chang-Hung Chou
Institute of Botany
Academia Sinica
Taipei
Taiwan 11529
Republic of China

L. J. Corcuera
Departamento de Biologia
Faculdad de Cienceas
Universidad de Chile
Casilla 653
Santiago
Chile

M. Diaz
Department of Crop Protection
Department of Chemistry
University of Puerto Rico
Mayaguez
Puerto Rico

A. G. Ferreira
Departamento de Botanica
Universidade Federal do Rio Grande do Sul
Av Paulo Gama #40
CEP 90049
Porto Alegre, RS
Brazil

N. H. Fischer
Department of Chemistry
Louisiana State University
Baton Rouge
Louisiana
USA

P. M. Górski
Department of Biochemistry
Institute of Soil Science and Plant Cultivation
24-100 Pulawy
Poland

A. M. Grodzinsky
Central Republican Botanical Garden
Academy of Sciences of the Ukrainian SSR
Kiev
USSR

H. Haque
Department of Biochemistry
Faculty of Basic Sciences and Humanities
Rajendra Agricultural University
Pusa
Samastipur 848125
India

A. Hasan
Department of Nematology
Narendra Dev University of Agriculture and Technology
Kumarganj
Faizabad
224 229
India

P. Hepperly
Tropical Agricultural Research Station
USDA-ARS
Box 70
Mayaguez
Puerto Rico 00709

U. S. Jacobi
Departamento de Biologia
Pontifice Universidade Catolica
Porto Alegre, RS
Brazil

R. Jobidon
Service de l'Amélioration des Arbres
Ministère de l'Énergie et des Resources
2700 rue Einstein
Sainte-Foy
Quebéc
Canada
G1P 3W8

M. Jurzysta
Department of Biochemistry
Institute of Soil Science and Plant Cultivation
24-100 Pulawy
Poland

Bong-Seop Kil
Department of Biology Education
Wonkwang University
Iri 570-749
Republic of Korea

O. Koul
Regional Research Laboratory
Canal Road
Jammu Tawi 180001
India

K. Leuschner
SADCC/ICRISAT
PO Box 776
Bulawayo
Zimbabwe

J. Lovett
Department of Agronomy and Soil Science
University of New England
Armidale
NSW 2351
Australia

A. U. Mallik
Department of Biology
Lakehead University
Thunder Bay
Ontario
Canada
P7B 5E1

N. P. Melkania
Faculty of Ecosystem Management
Indian Institute of Forest Management
PO Box 357
Bhopal 462 003
India

D. A. Miller
Department of Agronomy
University of Illinois
Urbana-Champaign
Illinois
USA

W. Oleszek
Department of Biochemistry
Institute of Soil Science and Plant Cultivation
24-100 Pulawy
Poland

R. C. Ortega
Instituto de Fisiologia Celular
Universidad Nacional Autonoma de Mexico
Apartado Postal 70-600
04510 Mexico, DF
Mexico

R. Perez
Department of Crop Protection
Department of Chemistry
University of Puerto Rico
Mayaguez
Puerto Rico

C. Reyes
Department of Crop Protection
Department of Chemistry
University of Puerto Rico
Mayaguez
Puerto Rico

E. L. Rice
Department of Botany and Microbiology
University of Oklahoma
Norman
Oklahoma 73019
USA

D. R. Richardson
Department of Botany
Louisiana State University
Baton Rouge
Louisiana
USA

S. J. H. Rizvi
Laboratory of Allelopathy and Natural Products
Department of Botany and Plant Physiology
Faculty of Basic Sciences and Humanities
Rajendra Agricultural University
Pusa
Samastipur 848125
India

V. Rizvi
Laboratory of Allelopathy and Natural Products
Department of Botany and Plant Physiology
Faculty of Basic Sciences and Humanities
Rajendra Agricultural University
Pusa
Samastipur 848125
India

V. N. Rodriguez
Instituto de Fisiologia Celular
Universidad Nacional Autonoma de Mexico
Apartado Postal 70-600
04510 Mexico, DF
Mexico

M. Ryuntyu
Department of Agronomy and Soil Science
University of New England
Armidale
NSW 2351
Australia

E. Sarobol
Department of Agronomy
Kasetsart University
Bangkhen
Bangkok 10900
Thailand

V. K. Singh
Department of Microbiology
Faculty of Basic Sciences and Humanities
Rajendra Agricultural University
Pusa
Samastipur 848125
India

G. Stenhagen
Department of Medical and Physiological Chemistry
Box 33 031
400 33 Göteborg
Sweden

T. Twardowski
Institute of Bioorganic Chemistry
Polish Academy of Science
Noskoskiego 12
P-61-704 Poznan
Poland

D. Vokou
Department of Ecology
School of Biology
University of Thessaloniki
UPB 119
540 06 Thessaloniki
Greece

B. Williamson
Department of Botany
Louisiana State University
Baton Rouge
Louisiana
USA

M. Wink
Institut für Pharmazeutische Biologie
Universität Heidelberg
Im Neuenheimer Feld 364
D-6900 Heidelberg
Germany

G. E. Zúñiga
Departamento de Quimica
Faculdad de Ciencia
Universidad de Santiago
Casilla 5659-2
Santiago
Chile

Preface

Science is essentially a descriptive and experimental device. It observes nature, constructs hypotheses, plans experiments and proposes theories. The theory is never contemplated as the 'final truth', but remains ever subject to modifications, changes and rejections. The science of allelopathy in a similar way has emerged, and exists on a similar footing; our endeavour should be to keep it fresh and innovative with addition of newer information and concepts with the rejection of older ideas and antiquated techniques. During the past few decades encouraging results have been obtained in various aspects of allelopathic researches. However, in addition to continuing efforts in all these directions, constant attempts are to be made to describe the mechanics of allelopathic activity in molecular terms and to discover ways and means to exploit it for the welfare of mankind. We feel that multidisciplinary efforts are the only tool to achieve this goal. It is the hope of the editors that this book will serve as a document which identifies an integrated approach, through which research both to understand and exploit allelopathy can be conducted.

The present volume arose out of an attempt to bring together eminent scientists in allelopathy to describe their work, of a highly diverse nature, under one title. Basic, as well as applied, aspects of allelopathic research have been given equal importance and an attempt has been made to provide an overall picture of allelopathic research conducted in different parts of the world. As authors of various chapters hail from different ethnic/cultural backgrounds, in the course of editing we have attempted as much as possible to preserve the linguistic flavour of the many cultural groups represented.

Only Chapter 1 endeavours to give an overview of the various aspects of the discipline which are tackled in the subsequent chapters comprehensively. General, theoretical and conceptual topics have been addressed in Chapters 2–5. Chapters 6–7 describe the specific plant(s) studies, and 8–11 cover research done with isolated allelochemicals. An overview of allelopathic researches in various global regions has been provided in Chapters 12–22. These informations are expected to serve as key references for future studies on a particular biological system of a region.

Chapters 23–25 review the state of knowledge on some specific allelopathic aspects and their direct applications.

We are indeed thankful to the contributors for their enthusiastic support throughout the preparation of the book. We deeply mourn the sad and untimely demise of Professor A. M. Grodzinsky, who was doyen in the field of allelopathy and remained a constant source of inspiration to us. We would specially like to thank Professor Elroy L. Rice of the University of Oklahoma at Norman, for his constant support and suggestions throughout the preparation of the manuscript and to Dr H. Haque of the Biochemistry Department of this University for his help and constructive suggestions. We consider it our sacred duty to show our deep sense of gratitude for our respected teacher, the late Dr D. Mukerji, who introduced us to scientific research.

We acknowledge with thanks the permission granted by numerous publishers and authors to use previously published materials. We would specially like to thank Mr Nigel J. Balmforth, Senior Commissioning Editor, and the staff of Chapman & Hall, London, for their cooperation during preparation of this volume. Support of the Rajendra Agricultural University also deserve a thankful mention.

We would be failing in our duties if we did not express our deep sense of gratitude to our parents, children and other family members who suffered inexcusable neglect during the completion of this volume.

S. J. H. Rizvi
V. Rizvi

CHAPTER 1

A discipline called allelopathy

S. J. H. Rizvi, H. Haque, V. K. Singh and V. Rizvi

1.1 INTRODUCTION AND HISTORY

Allelopathy (rootwords: *allelon* and *pathos*) is derived from the Greek *allelon*, 'of each other', and *pathos*, 'to suffer'; hence it means: the injurious effect of one upon another. The term denotes that body of scientific knowledge which concerns the production of biomolecules by one plant, mostly secondary metabolites, that can induce suffering in, or give benefit to, another plant. The phenomenon could also be considered as a biochemical interaction among plants. The concept suggests that biomolecules (specifically termed *allelochemicals*) produced by a plant escape into the environment and subsequently influence the growth and development of other neighbouring plants. The subject not only deals with the gross biochemical interactions and their effects on physiological processes but also with the mechanism of action of allelochemicals at specific sites of action at the molecular level.

The term 'allelopathy' was coined by Molisch in 1937 and his definition referred to both the detrimental and beneficial biochemical interactions among all classes of plants, including microorganisms. This has led Rice (1984) to give the following definition of allelopathy: 'any direct or indirect harmful or beneficial effect by one plant (including microorganisms) on another through production of chemical compounds that escape into the environment'.

Allelopathy, if studied closely, is a very old component of agricultural sciences. Agriculture was developed by man about 10 000 years ago, though the circumstances and the time of original domestication of plants appears to have occurred in southwestern Asia about a million years back. During the long development of agriculture the earliest reference concerning phytotoxicity appears in a volume called *Enquiry into Plants*, written by Theophrastus (*c.* 300 B.C.). He also reported that chickpea does not re-invigorate soil but rather exhausts it. Earlier than that, the Greek philosopher Democritus had reported the use of naturally occurring plant products as a practical method of controlling weeds and that

Allelopathy: Basic and applied aspects Edited by S. J. H. Rizvi and V. Rizvi
Published in 1992 by Chapman & Hall, London ISBN 0 412 39400 6

the trees could be killed by treating their roots with a mixture of lupine flowers soaked in hemlock juice.

In his encyclopaedic work, *Natural History*, Plinius (first century A.D.) gives numerous examples of apparent allelopathic interactions. Plants, such as chickpea, barley, fenugreek and retch, were reported to scorch-up cornlands.

In comparatively recent times, DeCandolle (1832) has pointed out that interactions occur in many crops, but, in spite of the suggestions concerning apparent biochemical interactions, it was Molisch in 1937 who described such biological processes specifically as allelopathy. During the last 20 odd years this subject has developed into an effective branch of science which, through interdisciplinary endeavours, has come into sharp prominence in plant sciences.

1.2 BACKGROUND INFORMATION

1.2.1 Role of allelopathy in natural and manipulated ecosystems

There is convincing evidence that allelopathic interactions between plants play a crucial role in natural as well as in manipulated ecosystems. Studies of these interactions, especially those in natural ecosystems, provided the basic data for the science of allelopathy. These data were eventually applied to understand the problems of plant–plant, plant–microbe and plant–insect interactions and to exploit these in improving the production of manipulated ecosystems. During the past decade a number of fine books have addressed allelopathy in greater detail (Rice, 1984; Thompson, 1985; Putnam and Tang, 1986; Waller, 1987; Chou and Waller, 1989). Here we intend to mention only the major aspects of allelopathy which are under investigation and those which need greater attention.

(a) Patterning of vegetation and succession

The credit for a specific vegetational pattern has mostly been given to the competition. However, in recent times evidence is accumulating that, apart from competition, allelopathy does play an important role. Research findings have demonstrated that all types of plants, viz. herbs, shrubs and trees, allelopathically affect the patterning of vegetation, largely in their immediate vicinity.

Succession is another ecological phenomenon that has been a topic of interest for the last few years. The majority of workers have tried to explain this on the basis of changes in physical factors in the habitat, availability of essential minerals and competition, etc., but in 1911 Cowels suggested that toxins produced by plants may act as an important factor in plant succession. Rice (1984) has very nicely reviewed some of the case studies in old fields, algal and urban successions, and may be consulted for detailed information on these aspects.

(b) Allelopathy and agriculture

One of the most worked out aspects of allelopathy in manipulated ecosystem is the role of allelopathy in agriculture. In this, the effect of weeds on crops, crops on weeds and crops on crops have been invariably emphasized. Several attempts have been made to separate the competition and allelopathic effects while working with such interferences. The results obtained so far clearly demonstrate that some of the findings on allelopathic control of weeds, elimination of deleterious allelopathic effects of crops on crops, or exploitation of beneficial interactions in a rotation or mixed cropping system, have a direct bearing on crop production (Duke, 1986; Cutler, 1988; Rizvi *et al.*, 1988, 1989a, b). Not much work has been done, but convincing results are available to demonstrate the important role of allelochemicals/allelopathy in the promotion of pathogen infestation/developing resistance to diseases/insects in plants. Further, the role of allelochemicals in providing protection to the seeds prior to germination, and the implication of allelopathy in nitrification/denitrification, are some of the aspects that have been recognized but have yet to receive the attention they deserve. In addition, the possibility of using allelochemicals as growth regulators and natural pesticides has also been explored. As a result, a number of allelochemicals are either commercially available or are in the process of large-scale manufacture. The greatest advantage of their use lies in the fact that, being easily biodegradable, most of them are safer and environmentally cleaner than the synthetic pesticides in current use.

(c) Allelopathy and forestry

Allelopathic interactions have been demonstrated to play a crucial role in natural as well as man-made forests. Such interactions are pivotal in determining the composition of the vegetation growing as understorey and in understanding the forest regeneration problems. Results obtained so far have shown that almost all types of plants (viz. angiosperms, gymnosperms, lower plants like ferns and microorganisms, including mycorrhizae) present in forests indulge in allelopathic interactions (Waller, 1987).

(d) Allelopathy and agroforestry

Agroforestry is a relatively younger area of research and is defined as 'a suitable land management system which increases the yield of a particular piece of land, combines the production of crops (including tree crops) and forest plant and/or any animal, simultaneously or sequentially on the same unit of land and applies management practices that are compatible with the local cultural practices' (King and Chandler, 1978). Since, in agroforestry, trees are grown in association with the crop and fodder plants, there is a good chance that allelochemicals produced by the intercropped trees and shrubs will affect the food and fodder crops. Therefore, it seems essential that the allelopathic compatibility of crops with trees should be checked before being introduced to an agroforestry system (Rizvi *et al.*, 1990). Some of the recent findings have demonstrated that tree–crop interactions may have significant bearings on the total productivity of an agroforestry system.

(e) Allelopathy and horticulture
Some of the finest examples of allelopathy in field conditions have come from horticultural researches (Rice, 1984). Allelochemical-based soil sickness leading to replant problems in apple, peach and citrus provides tremendous impetus for furthering such studies with other horticultural plants. Considering the detailed and exhaustive works on some of the horticultural plants, there has been surprisingly little research on allelopathic phenomenon in ornamental plants.

The above-quoted examples are some of the major aspects of allelopathic interaction in natural and manipulated ecosystems. Research is continuing in all these areas and many few ideas have been floated in an attempt to understand the phenomenon of allelopathy more deeply and to exploit it more fully to boost the production of manipulated ecosystems.

1.2.2 Chemical nature of allelochemicals

Various types of chemicals implicated in allelopathy have been discussed in detail by Rice (1974, 1979, 1984), Thompson (1985) and Putnam and Tang (1986). Most of these chemicals are secondary metabolites and are produced as offshoots of primary metabolic pathways. These secondary products could be classified into five major categories: phenyl propanes, acetogenins, terpenoids, steroids and alkaloids (Whittaker and Feeny, 1971). It is almost impossible to enumerate each and every chemical identified as an allelochemical. However, to classify these into various major chemical groups would be a viable approach. Rice (1984) has also adopted the same approach and has classified allelochemicals produced by higher plants and microorganisms into the following major categories:

- simple water soluble organic acids, straight chain alcohols, aliphatic aldehydes and ketones
- simple unsaturated lactones
- long-chain fatty acids and polyacetylenes
- naphthoquinones, anthroquinones and complex quinones
- simple phenols, benzoic acid and derivatives
- cinnamic acid and derivatives

- flavonoids
- tannins
- terpenoids and steroids
- amino acids and polypeptides
- alkaloids and cyanohydrins
- sulphides and glucosides
- purines and nucleotides.

The chemical properties of the allelochemicals, classified as above, can be followed in the above references.

1.2.3 Mode of action of allelochemicals

As noted by Winter (1961), the visible effects of allelochemicals on plant processes are only secondary signs of primary changes. Therefore, studies on the effects of allelochemicals on germination and/or growth are only the manifestation of primary effects occurring at the molecular level. Although a strong tendency is being developed to look into the actual mechanism of action, the experimental work is in its infancy.

The mode of action of allelochemicals can broadly be divided into indirect and direct action. Indirect action may include effects through alteration of soil property, its nutritional status and an altered population and/or activity of harmful/beneficial organisms like microorganisms, insects, nematodes, etc. This is relatively a less studied aspect. On the other hand, the direct mode of action, which includes effects of allelochemicals on various aspects of plant growth and metabolism, has received fairly wide attention.

The following are some important sites and processes known to be attacked or influenced by allelochemicals:

- cytology and ultrastructure
- phytohormones and their balance
- membrane and its permeability
- germination of pollens/spores
- mineral uptake
- stomatal movement, pigment synthesis and photosynthesis
- respiration
- protein synthesis
- leghaemoglobin synthesis and nitrogen fixation
- specific enzyme activity
- conducting tissue
- water relations of plants
- genetic material.

Under natural conditions the action of allelochemicals seems to revolve round a fine-tuned regulatory process in which, perhaps, many such compounds act together on one or more than one of the above processes in a simultaneous or sequential manner. What is observed solely in terms of one compound – one aspect study – is perhaps some sort of artefact and unless a much more holistic approach is adopted, it would be rather unfair to form an opinion regarding the mode of action of a specific allelochemical.

Apart from the above, factors affecting the production of allelochemicals and their release into the environment, their absorption and translocation in the receptor organism, concentration at the site of action and factors determining the effectiveness of allelochemicals after their release from the producing organism, are some of the important factors which should be considered if the action of allelochemicals is to be understood in its entirety.

1.3 FUTURE DIRECTIONS FOR ALLELOPATHIC RESEARCHES

Most of the results obtained in allelopathic studies cannot be explained in terms of a single disciplinary approach. To understand and exploit allelopathy fully, different disciplines should come together to work out allelopathy, both in its minute and detailed terms. Such an approach needs the cooperation of scientists of diverse disciplines like botany, microbiology, plant chemistry, ecology, etc. Allelopathy is a relatively younger science and therefore research activities in all its aspects should continue. These aspects involve isolation and characterization of allelochemicals, their movement and threshold concentrations in the substrate (soil, air or water), factors affecting the production of allelochemicals and their activities, impact of allelochemicals on yield of crops grown under various cultural practices in different agroclimatic zones, exploitation of allelopathy/allelochemicals in plant protection, etc. While research in all the above aspects is unavoidable, the exploration of new ideas is also essential. Recently, Rice (1987) and Putnam (1987) have suggested some problems which are to be solved by scientists from various disciplines; and these are noteworthy and worth contemplating. We also suggest that the following approaches should receive more attention.

1.3.1 Biotechnology

Biotechnology promises to play a very significant role in increasing plant productivity in the coming years. The improvement in plants using DNA technologies will require the development of transformation vectors and regeneration techniques, the use of engineered Ti plasmids, plant tissue culture propagation and identification of genes to be transferred.

However, the introduction of a particular gene for a specific character may cause unexpected and undesirable effects. L-canavinine, a non-protein amino acid present in some of the legumes, exerts adverse effects on insects, when ingested. Thus, the genes for L-canavinine might be inserted genetically into the crop plant which, in turn, may provide protection to engineered plant against some insects. But plants with L-canavinine would also prove harmful for mammals; hence the purpose would be defeated. Therefore, before selecting any gene (for a specific allelochemical) to be inserted into plants, the non-target toxicity and agronomic fitness, etc., of the transformed plants, will have to be evaluated. However, there are enormous possibilities to induce pest-resistance in useful plants through regulating their capacity to produce specific allelochemicals.

1.3.2 Marine biology

Studies of allelopathy are mostly concerned with terrestrial plants. Nevertheless, several chemicals from marine plants have been found to exert adverse or beneficial effects on organisms living in the same environment. Some marine dinoflagellates are known to produce allelochemicals which exert adverse effects on phytoplankton, fish and other marine animals. Considering the large area covered by the oceans and about 500 000 species living in/on the water, the progress of allelopathic studies is too slow in comparison to other ecosystems. The thought that the allelochemical concentration would be too small to act owing to the huge dilution capacity of the sea was prohibitory. However, the activity of ectocarpene and other brown algal pheromones at concentrations as low as 1×10^{-12} M leaves no room for any doubt that the small quantity of allelochemical will not be effective in marine ecosystems.

It has been shown that allelochemicals may play an important role in habitat recognition by carabid beetles. It is very interesting to note that *Thalassotrechus barbarae*, an intertidal carabid beetles, never go beyond the confines of the zone they normally occupy and they almost always get back to the appropriate crevice's opening they inhabit. By the process of elimination, all other factors such as visual recognition, idiothetic information, track memory, change in temperature, etc., can be discarded as an aid in habitat recognition, and chemo-orientation remains as the only' consistently reliable explanation for correct habitat recognition. There is convincing evidence that allelochemicals produced by microalgae or cyanobacteria growing in and around crevices may, at least partially, be responsible for the chemo-orientation of carabid beetles leading to recognition of their habitat. Recently some attention has been paid to these aspects (Evans, 1986; Chou and Su, 1989), but such problems should be looked into in a more comprehensive manner by marine biologists. It is quite possible that studies of vast marine ecosystems will provide information and material to extend further the scope for exploitation of allelopathy.

1.3.3 Industry and environment

Several hundred chemicals have been used to manufacture agrochemicals, food preservatives, additives, cosmetics, medicines, etc. However, the use of many of these chemicals is a cause of concern from both environmental and health points of view. These hazards are largely attributed to their non-biodegradable nature as most of them are artificially synthesized. On the other hand, owing to their biodegradability, most of the plant products (allelochemicals) are considered to be safer for the consumers and their environment. It is therefore essential for industrial technologists to work effectively to exploit such compounds directly and/or to explore the possibility of using these chemicals as molecular models to create more active and safer products. Also, a critical scientific evaluation of the generally held view on the environmental cleanliness of plant products as future industrial chemicals, is a challenging task for environmental scientists.

1.3.4 Human and animal nutrition

It is well known that certain allelochemicals – e.g. mimosine – not only adversely affect plants and microorganisms, but also cause various serious health disorders in domestic animals. Such allelochemicals, through accumulation in the food chain, may also equally affect man. Further, during growth and storage of food commodities, various kinds of microbial infestation occur. As a result, highly toxic chemicals are produced which, apart from causing damage to the hosts, could pose danger to consumers as well. Recent findings have shown that the production of such toxic substances can be checked through the use of allelochemicals.

In this chapter we have attempted to give a broad picture of the advances made in allelopathy in the last few decades, though this overview does not provide a complete scenario of facts and conclusions. Nevertheless, we hope that this will inspire experts in many disciplines to look at this fast-emerging subject more closely and explore ways and means to exploit allelopathy from their viewpoint.

REFERENCES

Chou, C. H. and Waller, G. R. (eds) (1989) *Phytochemical Ecology: Allelochemicals, Mycotoxins, and Insect Pheromones and Allomones*, Academia Sinica Monograph Ser. No. 9, Acad. Sinica, Taipei, ROC.

Chou, H. N. and Su, H. M. (1989) Allelochemicals in marine ecosystems, in

Phytochemical Ecology: Allelochemicals, Mycotoxins, and Insect Pheromones and Allomones (eds C. H. Chou and G. R. Waller), Academia Sinica Monograph Ser. No. 9, Acad. Sinica, Taipei, ROC, pp. 119–28.

Cowells, H. C. (1911) The cause of vegetative cycles. *Bot. Gaz.* (Chicago), **51**, 161–83.

Cutler, Horace G. (ed.) (1988) *Biologically Active Natural Products, Potential Use in Agriculture*, ACS Symp. Ser. 380, Amer. Chem. Soc. Washington, DC.

DeCandolle, M. A. P. (1832) *Physiologie Végétale*, Tome III, Béchet Jeune, Lib. Fac., Méd., Paris, pp. 1474–5.

Duke, Stephen O. (1986) Naturally occurring chemical compounds as herbicides. *Rev. Weed Sci.*, **2**, 15–44.

Evans, W. G. (1986) Edaphic and allelochemic aspects of intertidal crevice sediments in relation to habitat recognition by *Thallassotrechus barbarae* (Horn) (Coleoptera: carabidae). *J. Exp. Mar. Biol. Ecol.*, **95**, 57–66.

King, K. F. S. and Chandler, M. T. (1978) *The Waste Lands*, Internatl. Council for Research in Agroforestry, Nairobi, Kenya.

Molisch, H. (1937) *Der Einfluss einer Pflanze auf die andere-Allelopathie*, Fischer, Jena.

Plinius Secundus, C. (A.D. 1) *Natural History*, 10 vols (Engl. transl. by H. Rackam, W. H. S. Jones and D. E. Eichholz), Harvard Univ. Press, Cambridge, Massachusetts, 1938–63.

Putnam, A. R. (1987) Introduction, in *Allelochemicals, Role in Agriculture and Forestry*, (ed. G. R. Waller), ACS Symp. Ser. 330, Amer. Chem. Soc., Washington, DC.

Putnam, A. R. and Tang, C. S. (eds) (1986) *The Science of Allelopathy*, Wiley, New York.

Rice, E. L. (1974) *Allelopathy*, Academic Press, New York.

Rice, E. L. (1979) Allelopathy – an update. *Bot. Rev.*, **45**, 15–109.

Rice, E. L. (1984) *Allelopathy*, Academic Press, New York.

Rice, E. L. (1987) Allelopathy: an overview, in *Allelochemicals, Role in Agriculture and Forestry* (ed. G. R. Waller), ACS Symp. Ser. 330, Amer. Chem. Soc., Washington, DC, pp. 8–22.

Rizvi, S. J. H., Mishra, G. P. and Rizvi, V. (1989a) Allelopathic effects of nicotine on maize. I. Its possible importance in crop rotation. *Plant and Soil*, **116**, 289–91.

Rizvi, S. J. H., Mishra, G. P. and Rizvi, V. (1989b) Allelopathic effects of nicotine on maize. II. Some aspects of its mechanism of action. *Plant and Soil*, **116**, 292–4.

Rizvi, S. J. H., Singh, V. K., Rizvi, V. and Waller, G. R. (1988) Geraniol, an allelochemical of possible use in integrated pest management. *Plant Protect.*, **3**, 112–14.

Rizvi, S. J. H., Sinha, R. C. and Rizvi, V. (1990) Implications of mimosine allelopathy in Agroforestry. *Proc. 19th IUFRO World Congress Forestry, Montreal, Canada*, **2**, 22–7.

Theophrastus (*c.* 300 B.C.) *Enquiry into Plants and Minor Works on Odours and Weather Signs*, 2 vols (transl. into English by A. Hort), Heinemann, London, 1916.

Thompson, A. C. (ed.) (1985) *The Chemistry of Allelopathy, Biochemical*

Interactions Among Plants, ACS Symp. Ser. 268, Amer. Chem. Soc., Washington, DC.

Waller, G. R. (ed.) (1987) *Allelochemicals, Role in Agriculture and Forestry*, ACS Symp. Ser. 330, Amer. Chem. Soc., Washington, DC.

Whittaker, R. H. and Feeny, P. P. (1971) Allelochemics: chemical interactions between species. *Science*, **171**, 757–70.

Winter, A. G. (1961), New physiological and biological aspects in the interrelationships between higher plants. *Symp. Soc. Exp. Biol.*, **15**, 229–44.

CHAPTER 2

Allelopathy: broadening the context

John Lovett and Matthew Ryuntyu

2.1 INTRODUCTION

The recent upsurge of interest in allelopathy, with major volumes of collected papers regularly published (Rice, 1984; Thompson, 1985; Putnam and Tang, 1986; Waller, 1987; Chou and Waller, 1989), has established the topic as one of biological significance, although the ecological impact of allelopathy remains a subject of debate.

Trans-disciplinary studies suggest that the significance of allelopathy may be extended even further. Entomologists, for example, write of 'allelochemicals' in a context much wider than plant scientists concerned with allelopathy, *per se*. Thus, Reese (1979) defines allelochemicals as 'nonnutritional chemicals produced by one organism that affect the growth, health, behaviour or population biology of other species'. Behaviour-controlling chemicals (semiochemicals) are beginning to find a place in integrated pest management systems (Pickett 1988), realizing a potential which has frequently been discussed in the literature.

We perceive allelopathy as one of the many stresses with which plants must cope in their environment. Recent data suggest that the responses to other stress factors, for example, invasion by viruses (Bassi *et al.*, 1986), or stress by heavy metals (Wierzbicka, 1987) are similar. To extend this concept further, there is evidence that chemicals identified with allelopathy may also affect other organisms, and that the responses of these organisms follow a common pattern (Lovett *et al.*, 1989). Some of these data will be discussed below.

It seems likely, given the present state of knowledge, that allelopathy might best be regarded as part of a complex network of chemical communication between organisms (Harborne, 1987) in which groups of chemical compounds elicit similar, quantifiable, responses from disparate organisms (Lovett, 1982a). Strictly in the context of plants, allelopathy may represent a chemical contribution to defensive adaptations, which include also physical characteristics.

Allelopathy: Basic and applied aspects Edited by S. J. H. Rizvi and V. Rizvi
Published in 1992 by Chapman & Hall, London ISBN 0 412 39400 6

2.2 ALLELOPATHIC PHENOMENA

2.2.1 Secondary effects

Reports of allelopathic phenomena most frequently identify effects which are readily observed in the field or under controlled conditions. Delayed or inhibited germination and the stimulation or inhibition of root and shoot growth are often reported. Yet, the visible effects of allelopathy are merely secondary expressions of primary effects upon metabolic processes (Winter, 1961).

2.2.2 Primary effects

Many possible primary effects on plant metabolism, affecting the majority of vital processes, have been suggested (Rice, 1984) but few have been rigorously investigated. Most attention has probably been paid to effects of allelochemicals on cell elongation and ultrastructure of root tips; for example, Lorber and Muller (1976). Koch and Wilson (1977) reported on the effects of allelochemicals on mitochondria but the total volume of work on primary effects remains small. Thus, in the five recent collections of papers previously cited, few focus mainly on this topic, although Waller (1989) notes that studies of allelopathy are moving 'quite rapidly' from practical considerations to molecular biology. Among recent contributions are a hypothetical sequence of action for the effects of phenolics, one of the most frequently reported groups of allelochemicals, proposed by Einhellig (1986). Membrane function and interaction with hormones are included. However, supporting evidence is not abundant and there is a paucity of data on the primary effects of other important groups of allelochemicals, for example, alkaloids. An examination of mono- and sesqui-terpenes as plant germination and growth regulators, again, points to the lack of evidence for elucidating primary effects (Fischer, 1986).

Most studies of primary effects have focused on early plant growth, a time of high metabolic activity but great susceptibility to environmental stress. Some weeds, for example, *Datura stramonium* L. (thornapple) (Lovett *et al.*, 1981) release from the seed coat, compounds which have the ability to inhibit early growth of competing seedlings in their vicinity. Some crop plants, for example, *Hordeum vulgare* L. (Liu and Lovett, 1990), have a similar facility. The chemicals involved in both these examples are alkaloids, compounds widely used in medicine and veterinary science and highly active, biologically. In thornapple and barley they appear to contribute to plant defence by causing disruption at the level of the cell (the primary effect) which is observed as impaired germination or reduced early seedling growth (the secondary effects).

2.3 SOME TECHNIQUES FOR THE STUDY OF ALLELOPATHY

2.3.1 Bioassays

Bioassays, in the context of allelopathy, have been reviewed by Leather and Einhellig (1986). Probably the simplest forms of bioassay used in studies of allelopathy have been to quantify germination and/or emergence of seedlings, and to measure the length of the radicle or its equivalent. As defined by Winter (1961), although useful, such gross morphological data define only the secondary, readily observable effects of allelopathy.

The main tool used in the small number of critical investigations of primary effects of allelochemicals has also been the bioassay, but in more refined forms. Thus, the inhibition of mineral uptake in excised plant roots treated with phenolic acids are reported as being a consequence of the alteration of cellular membrane function (Balke, 1985), while phenolic acids, coumarins and flavonoids inhibit carbon dioxide dependent oxygen evolution in intact chloroplasts of spinach (*Spinacia oleracea* L.) and inhibit electron transport with mitochondria prepared from mung bean (*Vigna radiata* Roxb.) hypocotyls (Moreland and Novitzky, 1987). Few published reports have combined bioassays with microscopy to elucidate primary allelopathic effects.

2.3.2 Electron microscopy

Radicle elongation in linseed (*Linum usitatissimum* L.) is affected by benzylamine, an allelochemical produced from leaf washings of the weed *Camelina sativa* (L.) Crantz, and by leaf washings themselves. Light microscopy indicates interference with development of linseed root tips (Lovett, 1982b). Examination of root tips treated with $1000\,\mu g\,ml^{-1}$ benzylamine by transmission electron microscopy shows arrested development (Fig. 2.1). Organelles cannot be distinguished, in contrast to controls (Fig. 2.2). Lipid bodies are abundant in the cell contents and amyloplasts may also be present. These features suggest that metabolism of food reserves by affected cells has been disrupted.

While present in controls, root tip cells affected by allelochemicals show greatly increased evidence of vacuolation (Fig. 2.3). Vacuoles may be important in sequestering phytotoxic secondary metabolites such as alkaloids within the cell (Kutchan *et al.*, 1986). Detoxification may take place during sequestration (Matile, 1984). Increased vacuolation is also observed when root cells are exposed to other forms of stress, for example, invasion by plant parasitic nematodes in *Apium graveolens* L. (celery) (Bleve-Zacheo *et al.*, 1979) and exposure of *Allium cepa* L. (onion) to heavy metals such as lead (Wierzbicka, 1987). Our observa-

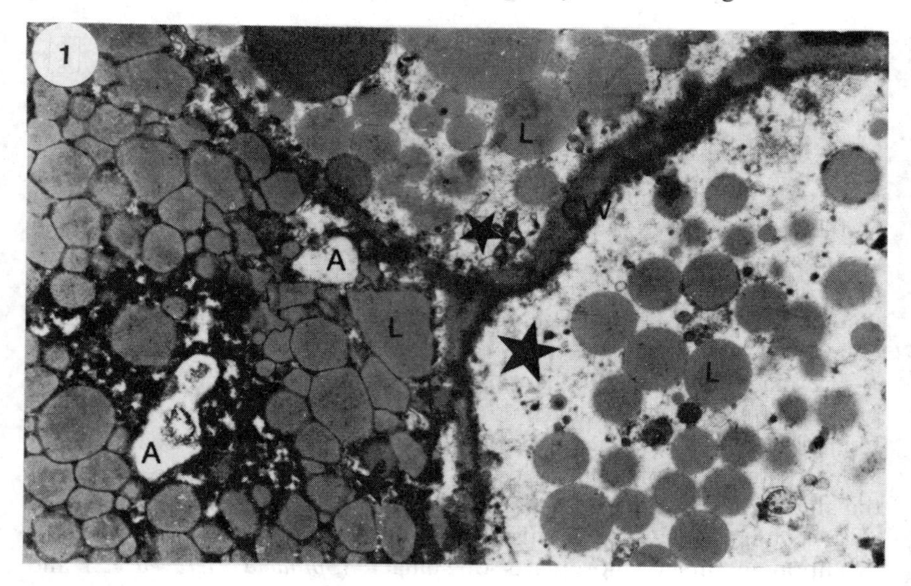

Fig. 2.1 1000 μg l^{-1} benzylamine applied to *Linum* root tip cells. A, amyloplast; L, lipid body. (Scale bar = 1 μm.)

tions support the view that, in the context of allelopathy, vacuoles play an autophagic role.

2.4 RESPONSES COMMON TO ALLELOCHEMICALS AND OTHER BIOLOGICALLY ACTIVE CHEMICALS

On the basis of the above observations, the effects of allelochemicals on plant cells may be seen to have similarities with damaged animal cells. In this respect, the observations of apparent 'phagocytosis', paralleling similar data for mammals such as rat (Ahlberg *et al.*, 1982), are of particular interest. Our data are not confined to one species. Transmission electron microscopy applied to root tip cells of sunflower (*Helianthus annuus* L.) following treatment with the alkaloids of thornapple (Levitt *et al.*, 1984) showed accumulations of amyloplasts together with increase in size and number of microbodies (Lovett *et al.*, 1987) which perform oxidation processes. These observations again indicate interference with metabolism of food reserves in germinating seedlings. Most recently, responses by *Sinapis alba* L. (white mustard) to barley allelochemicals show close similarities to the work with *Camelina* and linseed (Lovett *et al.*, 1989).

Whether observations of retarded metabolism of energy reserves is an adequate explanation of primary effects of allelochemicals is debatable

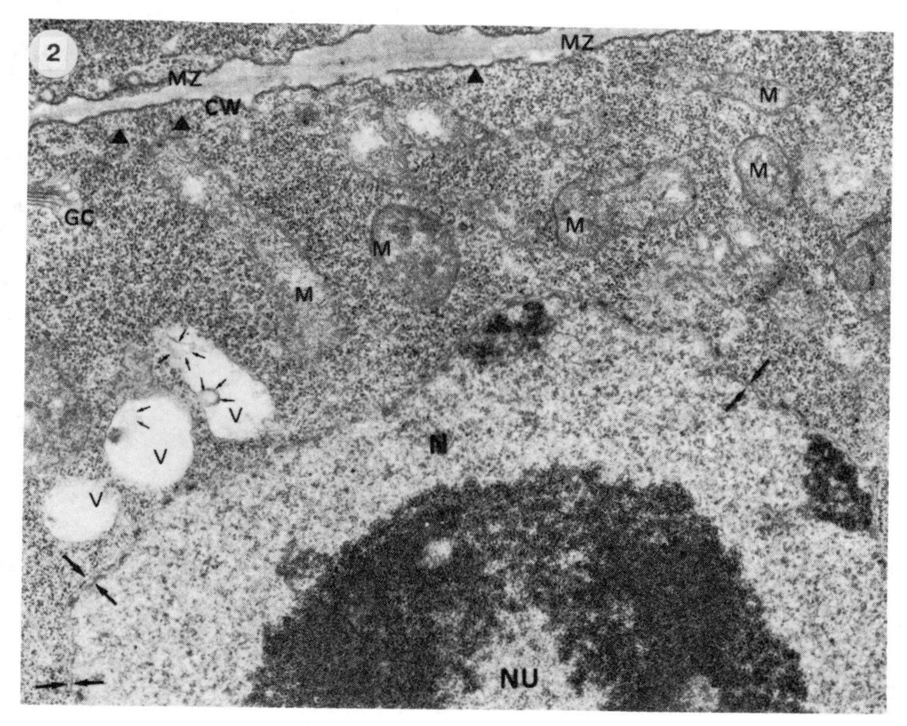

Fig. 2.2 Sterile water (control) applied to *Linum* root tip cells. GC, Golgi complex; M, mitochondrion; V, vacuole; N, nucleus; NU, nucleolus. Small arrows indicate apparent engulfing of cellular components and membrane invaginations. (Scale bar = 1 μm.)

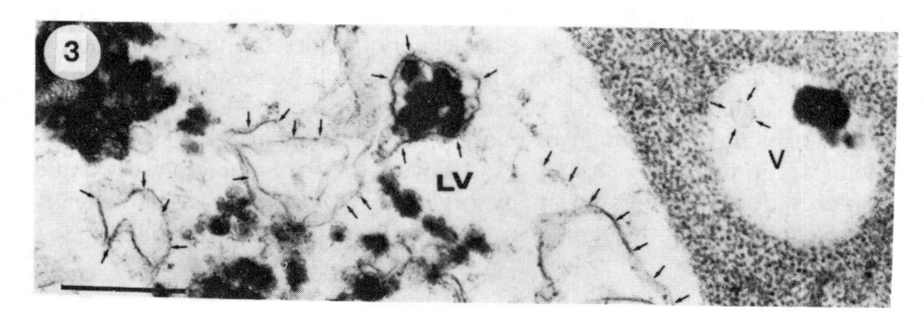

Fig. 2.3 Leaf washings of *Camelina* applied to *Linum* root tip cells. V, vacuole; LV, large vacuole; small arrows indicate apparent engulfing of damaged cellular components. (Scale bar = 1 μm.)

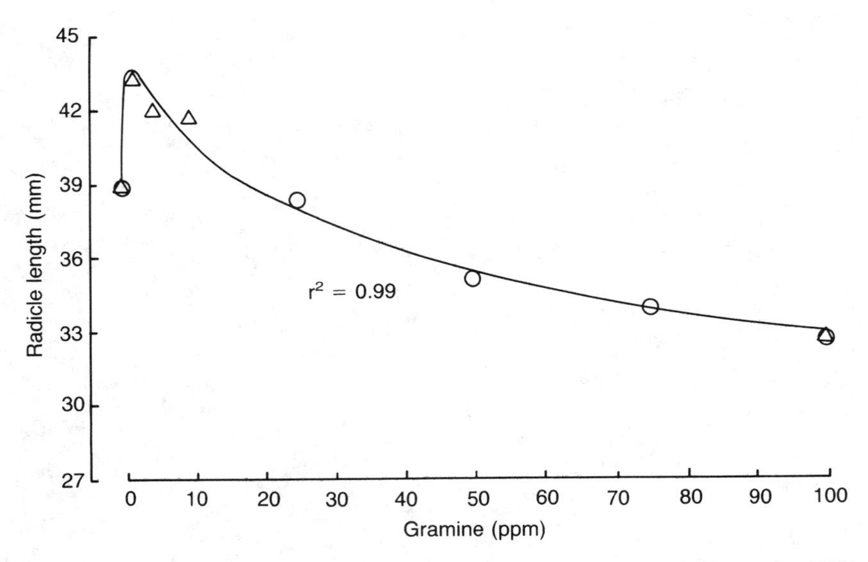

Fig. 2.4 Response in radicle length of white mustard to gramine, after 72 h. (After Liu and Lovett, 1990. Reproduced with permission, Lovett, J. and Forest Research Institute, New Zealand.)

(Lovett and Potts, 1987). Bioassays and microscopy may need to be combined with biochemical studies if these effects are to be satisfactorily elucidated. Impaired enzyme activity, identified by Rice (1984) as a primary target for allelopathic activity, may offer an explanation for reduced ability to metabolize reserve materials such as starch. Working with α-amylase, an enzyme involved in starch breakdown, we have shown its susceptibility to the allelochemical scopolamine, found in thornapple (Lovett *et al.*, 1989). The curve illustrated for this, postulated, primary effect is similar to that for secondary responses, for example, radicle length of white mustard in response to application of the barley allelochemical, gramine, Fig. 2.4 (Liu and Lovett, 1990). Similar curves can be produced for responses by many organisms to a range of biologically active chemicals, both organic and inorganic (Lovett *et al.*, 1989)

2.5 CONCLUSIONS AND PROSPECTS

We have drawn attention to a number of intriguing similarities in the nature of compounds involved in allelopathy and in wider chemical activity by plants; in the readily apparent morphological responses to allelochemicals (secondary effects), and in primary effects sensu Winter (1961). Cytological evidence reinforces our suggestion that a common response to varied damage, or stress, may be found in plant and, perhaps, animal cells (Lovett *et al.*, 1989). The observation of Guern (1987) that there are similarities between plants and animals in respect of hormonal activity complements this postulate.

Further elucidation of allelopathy as chemical communication between plants and other organisms, and the possible existence of a far-reaching network of chemical effect and response, will depend on an increase in morphological, biochemical and cytological studies across disciplinary boundaries.

ACKNOWLEDGEMENTS

We thank Mr P. R. Garlick and Mrs B. H. Ward, Electron Microscope Unit, University of New England, for their assistance, and the Australian Research Grants Scheme for financial support.

REFERENCES

Ahlberg, J., Henell, F. and Glaumann, H. (1982) Proteolysis in isolated autophagic vacuoles from rat liver. Effect of pH and of proteolytic inhibitors. *Exp. Cell Res.* **142**, 373–83.

Balke, N. E. (1985) Effects of allelochemicals on mineral uptake and associated physiological processes, in *The Chemistry of Allelopathy* (ed. A. C. Thompson), ACS Symp. Ser. 268, Amer. Chem. Soc., Washington, DC, pp. 161–78.

Bassi, M., Barbieri, N., Appian, A. and D'Agostina, G. (1986) Origin and function of tomato bushy stunt virus-induced inclusion bodies. *J. Ultrastruct. Mol. Struct. Res.*, **96**, 194–203.

Bleve-Zacheo, T., Zacheo, G., Lamberti, F. and Arrigoni, O. (1979) Cell wall protrusions and associated membranes in roots parasitized by *Longidorus apulus*. *Nematologica*, **25**, 62–6.

Chou, C. H. and Waller, G. R. (eds) (1989) *Phytochemical Ecology: Allelochemicals, Mycotoxins, and Insect Pheromones and Allomones*, Academia Sinica Monograph Ser. No. 9, Acad. Sinica, Taipei, ROC.

Einhellig, F. A. (1986) Mechanisms and modes of action of allelochemicals, in *The Science of Allelopathy* (eds A. R. Putnam and C. S. Tang), Wiley, New York, pp. 171–88.

Fischer, N. H. (1986) The function of mono and sesquiterpenes as plant germination and growth regulators, in *The Science of Allelopathy* (eds A. R. Putnam and C. S. Tang), Wiley, New York, pp. 203–18.

Guern, J. (1987) Regulation from within: The hormone dilemma. *Ann. Bot.* **60** (Supplement) 4, 75–102.

Harborne, J. B. (1987) Chemical signals in the ecosystem. *Ann. Bot.*, **60** (Supplement) 4, 39–57.

Koch, S. J. and Wilson, R. H. (1977) Effects of phenolic acids on hypocotyl growth and mitochondrial respiration in mung bean (*Phaseolus aureus*). *Ann. Bot.*, **41**, 1091–2.

Kutchan, T. M., Rush, M. and Coscia, C. J. (1986) Subcellular localization of alkaloids and dopamine in different vacuolar compartments of *Papaver bracteum*. *Plant Physiol.*, **81**, 161–6.

Leather, G. R. and Einhellig, F. A. (1986) Bioassays in the study of allelopathy, in *The Science of Allelopathy* (eds A. R. Putnam and C. S. Tang), Wiley, New York, pp. 133–45.

Levitt, J., Lovett, J. V. and Garlick, P. R. (1984) *Datura stramonium* allelochemcals: longevity in soil and ultrastructural effects on root tip cells of *Helianthus annuus* L. *New Phytol.*, **97**, 213–18.

Liu, D. L. and Lovett, J. V. (1990) Allelopathy in barley: potential for biological suppression of weeds, in *Alternatives to the Chemical Control of Weeds*, Forest Research Institute, Rotorua, pp. 85–92.

Lorber, P. and Muller, W. H. (1976) Volatile growth inhibitors produced by *Salvia leucophylla*: effects on seedling root tip ultrastructure. *Amer. J. Bot.*, **63**, 196–200.

Lovett, J. V. (1982a) Allelopathy and self-defence in plants. *Australian Weeds*, **2**, 33–6.

Lovett, J. V. (1982b) The effects of allelochemicals on crop growth and development, in *Chemical Manipulation of Crop Growth and Development* (ed. J. S. McLaren), Butterworths, London, pp. 93–110.

Lovett, J. V. and Potts, W. C. (1987) Primary effects of allelochemicals of *Datura stramonium* L. *Plant and Soil*, **98**, 137–44.

Lovett, J. V., Ryuntyu, M. Y. and Garlick, P. R. (1987) Allelopathic effects of Thorn-apple (*Datura stramonium* L.). *Proc. 8th Aust. Weeds Conf., Sydney*, pp. 179–81.

Lovett, J. V., Ryuntyu, M. Y. and Liu, D. L. (1989) Allelopathy, chemical communication and plant defence. *J. Chem. Ecol.*, **15**, 1193–202.

Lovett, J. V., Levitt, J., Duffield, A. M. and Smith, N. G. (1981) Allelopathic potential of *Datura stramonium* L. (Thorn-apple). *Weed Res.*, **21**, 165–70.

Matile, P. (1984) Das toxische Kompartiment der Pflanzencelle. *Naturwiss.*, **71**, 18–24.

Moreland, D.E. and Novitzky, W. P. (1987) Effects of phenolic acids, coumarins, and flavonoids on isolated chloroplasts and mitochondria, in *Allelochemicals: Role in Agriculture and Forestry* (ed. G. R. Waller), ACS Symp. Ser. 330, Amer. Chem. Soc., Washington, DC, pp. 247–61.

Pickett, J. A. (1988) The future of semiochemicals in pest control. *Aspects Appl. Biol.*, **17**, 397–406.

Putnam, A. R. and Tang, C. S. (eds) (1986) *The Science of Allelopathy*, Wiley, New York.

Reese, J. C. (1979) Interactions of allelochemicals with nutrients in herbivore food, in *Herbivores: their Interaction with Secondary Plant Metabolites* (eds G. A. Rosenthal and D. H. Janzen), Academic Press, New York, pp. 309–30.

Rice, E. L. (1984) *Allelopathy*. Academic Press, New York.

Thompson, A. C. (ed.) (1985) *The Chemistry of Allelopathy*, ACS Symp. Ser. 268, Amer. Chem. Soc., Washington, DC.

Waller, G. R. (ed.) (1987) *Allelochemicals: Role in Agriculture and Forestry*, ACS

Symp. Ser. 330, Amer. Chem. Soc., Washington, DC.

Waller, G. R. (1989), Allelochemical action of some natural products, in *Phytochemical Ecology: Allelochemicals, Mycotoxins, and Insect Pheromones and Allomones* (eds C. H. Chou and G. R. Waller), Academia Sinica Monograph Ser. No. 9, Acad. Sinica, Taipei, ROC, pp. 129–54.

Wierzbicka, M. (1987) Lead accumulation and its translocation barriers in roots of *Allium cepa* L. – autoradiographic and ultrastructural studies. *Plant Cell Environ.*, **10**, 17–26.

Winter, A. G. (1961) New physiological and biological aspects in the interrelations between higher plants. *Soc. Exp. Biol. (Cambridge) Symp.*, **15**, pp. 229–44.

CHAPTER 3

A conceptual framework for assessing allelochemicals in the soil environment

H. H. Cheng

3.1 INTRODUCTION

There has been a surge of interests in recent years on the phenomenon of allelopathy, the chemical agents involved in allelopathy, and the role of allelopathy in crop production (Chou and Waller, 1983; Rice, 1984; Thompson, 1985; Waller, 1987). Studies have generally concentrated either on the symptoms and severity of the adverse effects on affected plants or on the production and identification of the allelochemicals from producing organisms. Few reports have considered the mechanisms and processes involved in the overall relationship from the production of the allelochemicals to the actual occurrence of allelopathy. To be sure, a number of chemicals have been implicated in allelopathic relationships, the phytotoxic nature of these chemicals has been demonstrated, and these chemicals have been isolated from the surroundings of affected plants. Invariably the relationship has been established by inference rather than by direct evidence showing that the isolated chemicals actually caused the allelopathic effects. As Fisher (1979) remarked: 'It seems unlikely that the allelochemicals that may be extracted from plant material are actually those that reach the host plant, yet all our information on allelopathic compounds is derived from extracts that have never been exposed to the soil.' Similar concerns were expressed by Thompson (1985) who echoed the calls by a number of others for more attention on the effects of soil and microbial flora on understanding the allelopathic activities in the natural environment. In fact, in spite of the abundance of recent literature on allelopathy, few reports have addressed the fate of the allelochemicals in the soil environment.

Allelopathy: Basic and applied aspects Edited by S. J. H. Rizvi and V. Rizvi
Published in 1992 by Chapman & Hall, London ISBN 0 412 39400 6

3.2 DEFINING THE QUESTION

The basic question that must ultimately be addressed is what constitutes proof of an allelopathic effect. Commonly, the evidence for allelopathy includes: (1) symptoms of plant damage, such as reduced germination, growth, or development; (2) presence of substances or organisms (plants or microbes) which contain or are capable of producing phytotoxic chemicals in the vicinity of affected plants; and (3) presence of the phytotoxic chemicals in the extracts of plants or soils in the vicinity of affected plants. Most evidence on allelopathy has not gone beyond the presentation of such observations. However, it may be argued that the cause-and-effect relationship cannot be established merely by observing the appearance of phytotoxic symptoms on the one hand and finding the presence of a certain chemical of demonstrated toxicity in the vicinity of an affected plant on the other. Surely for every phytotoxic chemical identified in a plant or soil extract, there are likely numerous others also present but not identified (Rice, 1984; Mandava, 1985). Einhellig (1987) has also raised the question whether the biochemical agents alleged to be involved were in sufficient concentration in the environment to influence germination or growth. In addition, many of these chemicals are reactive and may be transformed during the course of extraction and analysis. How can one be certain that the chemical identified in an extract is in the form that actually causes the allelopathic effect? Furthermore, most of the reported allelopathic symptoms are observations of gross effects and not based on measurement of actual damage. The allelopathic symptom may not be manifested at the time or the site where plant damage has actually occurred. By the time the symptoms are observed, the culprit may no longer be present. The possibility that the observed effects are unrelated to the chemical in question cannot be ignored (Elliott and Cheng, 1987).

3.3 ESTABLISHING THE PROOF

To establish the cause-and-effect relationship in allelopathy, the following events must occur in sequence: (1) a phytotoxic chemical is produced; (2) the chemical is transported from the producing organism to the target plant; and (3) the target plant is exposed to the chemical in sufficient quantity and for sufficient time to cause damage. Most reports on allelopathy have given attention to the first and third events. However, the critical link between the production and the exposure has not been well established (Rice, 1987). Much could happen to a chemical in the soil during its transit from the producing organism to the target plant. The complexity of the processes and interactions involved that affect the presence of a chemical in the soil is not readily appreciated. Knowledge in soil science and from studies on environmental chemicals can provide the background needed to understand the behaviour and fate of allelochemicals in the soil environment. The objective of this chapter is to develop a conceptual framework useful for reaching an understanding of the time-course fate of allelochemicals under natural conditions. The approach will be to characterize the processes and factors that affect the fate and transport of the chemical and to integrate their interactions from the point of entry of the chemical into the environment to the site of its exposure to the target plant.

Once a chemical enters into the environment, a number of interacting processes will take place. These processes can be classified broadly as the retention, the transformation, and the transport processes. The retention processes will retard the movement of a chemical from one location to another through the media of soil, water and air. The transformation processes will change the form or structure of the chemical leading to a partial alteration or total decomposition of the molecule. The transport processes will define how a chemical moves in the environment. The nature of the chemical and the organisms present, the properties of the soil, and the environmental conditions are key factors that can influence each of the processes. Thus, the fate of a chemical in the environment will depend upon the kinetics and interactions of the individual processes in the course of time at a particular site under a set of natural conditions.

3.4 THE SETTING

Although not all the processes are at present clearly understood, nor all the influencing factors readily characterized, it is nonetheless possible even with a limited knowledge to start building a conceptual model that includes the major processes and factors useful for evaluating the potential of a chemical to become an allelochemical. How each of the processes operates and how the processes interact will depend upon the setting in which the various factors can influence the processes and contribute to the fate of the chemical in a specific environment. The governing factors are mainly those related to the nature of the allelochemical produced, the nature of the producing organisms and affected plants, the properties of the surrounding soil, and the climatic and other environmental conditions under which the transport of the allelochemical takes place.

The allelochemicals are by nature reactive compounds. How effectively they cause allelopathy will depend on how they interact with the surrounding environment. Their solubility will affect their mobility in soil water, their vapour pressure will affect their volatility in the air, and their structure will affect their affinity to the soil surface and their degradability by microorganisms. Such soil properties as the soil mineral and organic matter contents, particle size distribution, pH and ion exchange characteristics and oxidation state, will play a prominent role in influencing the behaviour of a chemical in the soil. In general, the higher the clay and organic matter contents, the higher is the soil adsorptive power. Soil pH can affect the ionization state of the allelochemical and, in turn, its mobility. The soil oxidation state will govern the nature and activity of soil microorganisms present. After all, soil is a matrix in which water and heat move and soil microorganisms reside; water and heat will affect the movement and soil microorganisms will affect the decomposition of the allelochemicals in the soil. Just as soil microorganisms can be both producers and degraders of chemicals, so can plants affect the amount of allelochemicals present in the environment. While a target plant may be damaged by the presence of an allelochemical, it may also be able to defend itself by exuding enzymes to degrade harmful chemicals in its surroundings. The role of affected plants is not always appreciated in studying allelopathy.

In each setting, the chemical, the plant and the soil are the main components. At the interface between the soil and plant roots, known as the rhizosphere, the microorganisms will be most active and the various interactions will be most intense. Both the macro-environmental conditions such as temperature, moisture, air movement and radiation, and the micro-environment of the rhizosphere are variables that determine which of the processes and interactions would be most influential on the fate of the chemical in each setting.

3.5 THE PROCESSES

While the kinetics of transport of a chemical in the environment must be assessed in the setting in which the transport occurs, the mechanisms of transport must be elucidated in relation to the retention and transformation processes that compete with the transport processes.

3.5.1 Retention

Strictly speaking, the retention process implies an attraction or sorption of a chemical to the soil surface leading to a restriction of the mobility of the chemical in the soil. It is a physical process and a number of intermolecular forces may be involved. There is an abundance of literature on the adsorption of environmental chemicals in the soil. By one type of force or another, all chemicals have a certain affinity to the medium, whether it be the soil particle surface or the soil solution. Thus, measurements on retention provides information on any changes in the distribution of a chemical between the adsorbed phase and the solution phase over a period of time. Methods for measuring retention can either be on the dynamics of the change of distribution or, more commonly, the resulting change at equilibrium. Unfortunately, the commonly used methodology for characterizing sorption is often of limited value (Cheng and Koskinen, 1986). Many indices of retention do not take into consideration the varying influence of soil properties on the adsorbability of the chemical. Even when the soil influence is accounted for, the test may not differentiate the physical process from chemical or biochemical transformation processes. The frequently used batch equilibration method and the Freundlich isotherm for expressing adsorption are often used without any considerations whether the basic assumptions of equilibrium conditions and reversibility of the sorption process are applicable as required by the methodology. When different sorption forces are involved, the Freundlich-type of adsorption isotherm may not always be appropriate in depicting the retention relationship. However, in spite of these experimental inadequacies, retention is recognized as an important influential factor in the transport process.

3.5.2 Transformation

The transformation processes may involve chemical, biochemical and photochemical means with the net result in a reduction of the amount of the original allelochemical available for transport. The term 'transformation' is frequently used synonymously with 'degradation' in the literature on environmental chemicals. Whereas degradation implies a breakdown of the original compound to simpler components with concurrent loss of toxicity, transformation is used here in the broadest sense, implying any change to the chemical structure of the original molecule. It is possible that the transformed product may be more toxic or more complex in structure, as well as simpler in structure and less toxic. Photochemical transformation can be a major process when the chemical is exposed to sunlight, but not when it is underneath the soil surface. Biochemical transformation involves the participation of plant roots and microorganisms and is the predominant transformation process in soil. The organisms present working in concert can provide a wide array of enzymes to degrade the chemical through various pathways into the simplest forms such as H_2O, CO_2 and salts. The kinetics of transformation is most sensitive to environmental variables. Too frequently the biochemical reaction is assumed to follow first-order kinetics, without conducting the necessary tests to verify the true reaction order. Such a simplistic approach for characterizing the transformations does not take into account either the multistage nature of the degradation process or the environmental influence on plant and microbial activities. Not only temperature level but also the daily and seasonal temperature fluctuations will affect the rate of biotransformation. Both plant and microbial activities are dependent upon moisture availability. When the soil is too dry, biological activities will be drastically reduced; but when the soil becomes too wet and oxygen diffusion becomes limited, anaerobic conditions will set in and the mechanisms for transformation can totally change. Many allelochemicals which are readily transformed under aerobic conditions are stable under anaerobic conditions (Chou *et al.*, 1981). The kinetics of biotransformation of a chemical is governed by the kinetics of biological activities, which is in turn affected by not only the environmental conditions but by such factors as soil properties and nutrient availability.

Insufficient attention has been given to the role of strict chemical reactions on the transformation and fate of allelochemicals in the soil. Depending on soil conditions and the presence of specific soil components, chemical transformation processes, such as oxidation, reduction, hydrolysis, substitution, complexation and polymerization, can play a significant role in reducing the allelopathic potential of certain chemicals. For instance, phenolic acids react readily with soil manganese oxides, resulting in the occurrence of reduced Mn^{2+} and oxidized products of phenolic acids (Lehmann *et al.*, 1987). Moreover, these oxidized products appear to be bound on the metal oxide surfaces in polymerized forms (Lehmann and Cheng, 1988). This oxidation–reduction reaction can occur readily in soil; within a few hours of contact much of the free phenolic acids can be oxidized chemically and be unavailable for biological transformation. Thus, results from any long-term biochemical degradation study on these phenolic acids could be easily misinterpreted, if the chemical reactions taking place at the onset of the experiment were not considered. It is imperative, therefore, that methods for assessing the transformation of allelochemicals in the soil include characterization of all modes of transformation and their interactions as well as the rates of transformation under each setting.

3.5.3 Transport

Whether a chemical is transported in a soil environment will depend on how the competing processes and all the factors interact in a specific setting. Different means are available to transport a chemical from one site to another. Transport can be either through the air as a vapour or in the soil solution. A chemical can move by means of mass flow of air or water, or by diffusion in the media. The movement follows a water potential gradient or a concentration gradient. Diffusion can be a significant mechanism of dispersal, especially at short distances as the case of allelopathy may often be. Plant roots can be a sink for chemicals as they absorb water from the surrounding soil, although probably only a small portion of the allelochemical produced will actually reach the target plant. The extent that the target plant is exposed to the chemical and the resulting damage incurred will be a function of the concentration of the

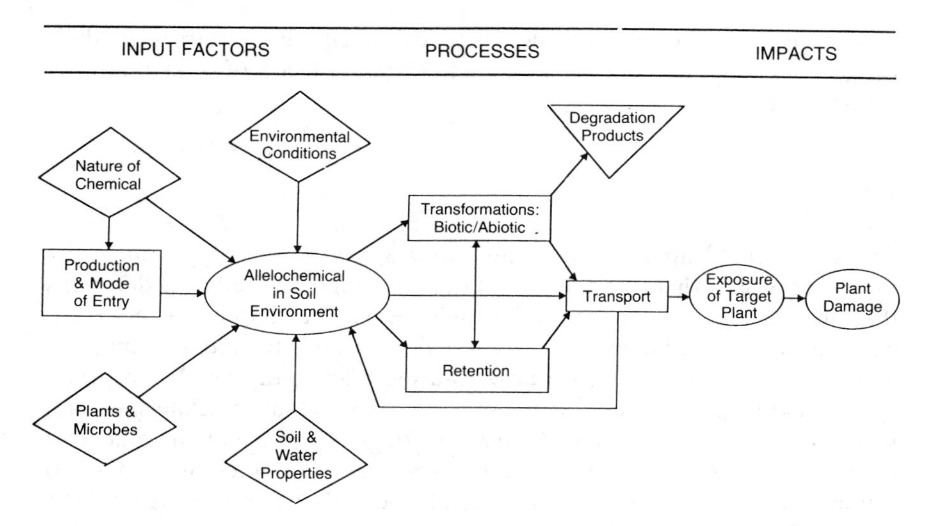

Fig. 3.1 A conceptual framework depicting the relationship of the factors and processes affecting the transport of allelochemicals from the site of production to the target plant. (After Cheng, 1989. Reproduced with permission, Cheng, H. H. and Academia Sinica, Taipei, ROC.)

chemical at the exposure site and the duration of the exposure. Thus, exposure can be assessed from the kinetics of transport.

3.6 THE CONCEPTUAL FRAMEWORK

When all the competing processes and all the influencing factors are considered together, their relationship can be diagrammatically depicted in a flow chart (Fig. 3.1). This flow chart provides a conceptual framework useful in assessing how a chemical can be transported from the site of production to the site at which the target plant is exposed. The conceptual framework can serve as the basis for assessing the processes quantitatively and establishing the interrelationship among the processes. Useful models have been developed for simulation and prediction of transport of many chemicals in the soil environment. Similar models could be developed to depict the transport of allelochemicals in the root environment. The conceptual framework can provide the foundation for such a mechanistic model. The input factors will define the setting, the processes will govern the transport of the chemical and the impact assessment should reflect the specific effect of allelopathy. These models should be tested with actual data for their applicability to characterize the fate and transport of allelochemicals in the root environment. Only when the process of transport of allelochemicals from their sources to their targets is known can the cause-and-effect relationship of allelopathy be ascertained.

3.7 FUTURE NEEDS

The conceptual framework presented in this chapter has only defined the overall approach that one needs to undertake and described considerations that one must give to any study on allelopathy. The chapter has not dealt with the approaches needed to devise valid methods for the identification and quantification of the chemicals causing allelopathy and appropriate experimental procedures for elucidating the actual mechanisms and kinetics of the processes under specific settings. Valid data will be needed to characterize the interactions of the processes on a time-course basis to estimate the nature and quantity of allelochemicals transported to the target organisms.

ACKNOWLEDGEMENT

The author was formerly Professor of Soils, Department of Agronomy and Soils, Washington State University, Pullman, Washington, USA. This contribution was from Project 1858, as Scientific Paper No. 7932, College of Agriculture and Home Economics Research Center, Washington State University.

REFERENCES

Cheng, H. H. (1989) Assessment of the fate and transport of allelochemicals in the soil, in *Phytochemical Ecology: Allelochemicals, Mycotoxins, and Insect Pheromones and Allomones* (eds C. H. Chou, and G. R. Waller), Academia Sinica Monograph Ser. No. 9, Acad. Sinica, Taipei, ROC, pp. 209–16.

Cheng, H. H. and Koskinen, W. C. (1986) Processes and factors affecting transport of pesticides to ground water, in *Evaluation of Pesticides in Ground Water* (eds W. Y. Garner, R. C. Honeycutt, and H. N. Nigg), ACS Symp. Ser. 315, Amer. Chem. Soc., Washington, DC, pp. 2–13.

Chou, C. H., Chiang, Y. C. and Cheng, H. H. (1981) Autointoxication mechanisms of *Oryza sativa* III. Effect of temperature on phytotoxin production during rice straw decomposition in soil. *J. Chem. Ecol.*, **7**, 741–52.

Chou, C. H. and Waller, G. R. (eds) (1983) *Allelochemicals and Pheromones*, Academia Sinica Monograph Ser. No. 5, Acad. Sinica, Taipei, ROC.

Einhellig, F. A. (1987) Interactions among allelochemicals and other stress factors of the plant environment, in *Allelochemicals: Role in Agriculture and Forestry* (ed. G. R. Waller), ACS Symp. Ser. 330, Amer. Chem. Soc., Washington, DC, pp. 343–57.

Elliott, L. F. and Cheng, H. H. (1987) Assessment of allelopathy among microbes and plants, in *Allelochemicals: Role in Agriculture and Forestry* (ed. G. R. Waller), ACS Symp. Ser. 330, Amer. Chem. Soc., Washington, DC, pp. 504–15.

Fisher, R. F. (1979) Allelopathy, in *Plant Disease: An Advanced Treatise*. Vol. IV (eds J. G. Horsfall and E. B. Cowling), Academic Press, New York, pp. 313–30.

Lehmann, R. G. and Cheng, H. H. (1988) Reactivity of phenolic acids in soils and formation of oxidation products. *Soil Sci. Soc. Amer. J.*, **52**, 1304–9.

Lehmann, R. G., Cheng, H. H. and Harsh, J. B. (1987) Oxidation of phenolic acids by soil iron and manganese oxides. *Soil Sci. Soc. Amer. J.*, **51**, 352–6.

Mandava, N. B. (1985) Chemistry and biology of allelopathic agents, in *The Chemistry of Allelopathy* (ed. A. C. Thompson), ACS Symp. Ser. 268, Amer. Chem. Soc., Washington, DC, pp. 33–54.

Rice, E. L. (1984) *Allelopathy*, 2nd edn, Academic Press, Orlando, Florida.

Rice, E. L. (1987) Allelopathy: An Overview, in *Allelochemicals: Role in Agriculture and Forestry* (ed. G. R. Waller), ACS Symp. Ser. 330, Amer. Chem. Soc., Washington, DC, pp. 8–22.

Thompson, A. C. (ed.) (1985) *The Chemistry of Allelopathy: Biochemical Interactions Among Plants*, ACS Symp. Ser. 268, Amer. Chem. Soc., Washington, DC.

Waller, G. R. (ed.) (1987) *Allelochemicals: Role in Agriculture and Forestry*, ACS Symp. Ser. 330, Amer. Chem. Soc., Washington, DC.

CHAPTER 4

Allelopathic effects on nitrogen cycling

E. L. Rice

4.1 INTRODUCTION

Molisch (1937) coined the term allelopathy to refer to inhibitory or stimulatory biochemical interactions between plants, including the microorganisms traditionally placed in the plant kingdom. Thus, allelopathy could conceivably affect all phases of the nitrogen cycle which involve plants and microorganisms. This obviously includes most phases of the cycle.

There is considerable evidence in the literature concerning allelopathic effects on biological N_2-fixation. Virtually no research has been done on possible allelopathic effects on aminization or ammonification. It is doubtful, however, that there are significant allelopathic effects on these processes because they are carried out by a very heterogeneous group of soil organisms (Harmsen and Kolenbrander, 1965; Brady, 1974). On the other hand, nitrification is apparently carried out chiefly by two genera of bacteria, *Nitrosomonas* which oxidizes ammonium to nitrite and *Nitrobacter* which oxidizes nitrite to nitrate (Brady, 1974; Alexander, 1977). It is not surprising, therefore, that much evidence exists that allelopathy strongly inhibits nitrification in many ecosystems.

It appears likely that allelopathy could affect losses of nitrogen from soil due to plant removal, leaching and volatilization. Allelopathy has definitely been shown to affect mineral uptake by plants and productivity (Rice, 1984), and thus loss of minerals due to plant removal by cropping or removal of animals could be altered. There is evidence that concentrations of nitrate are lower in some climax ecosystems than in successional stages or crop lands, and that losses of nitrogen due to leaching and runoff are therefore lower in such climax areas. Moreover, there is much evidence that the low nitrate concentrations in some climax areas are due to allelopathic inhibition of nitrification.

There is considerable loss of nitrogen from some soils due to denitrification, resulting in the volatilization of the nitrogen (Klubek *et al.*,

Allelopathy: Basic and applied aspects Edited by S. J. H. Rizvi and V. Rizvi
Published in 1992 by Chapman & Hall, London ISBN 0 412 39400 6

1978; Westerman and Tucker, 1978). It appears likely that allelopathy could affect denitrification directly through an effect on denitrifying bacteria and indirectly through its effect on nitrification, which would determine the amount of substrate available for the denitrifying organisms. Unfortunately, there is no information available on allelopathic effects on denitrifying bacteria.

4.2 ALLELOPATHIC EFFECTS ON NITROGEN FIXATION

4.2.1 Asymbiotic nitrogen fixers

Many bacteria, actinomycetes and fungi isolated from soil have been found to inhibit growth of free-living N_2-fixing bacteria (Iuzhina, 1958; Chan *et al.*, 1970; Leuck and Rice, 1976). This chapter will be concerned chiefly, however, with inhibition of symbiotic N_2-fixation by plants.

In the eastern part of Oklahoma and throughout the eastern United States broomsedge, *Andropogon virginicus*, invades old fields 3–5 years after abandonment and remains for many years, sometimes in almost pure stands (Rice, 1972). These fields are generally very low in nitrogen so it was hypothesized that broomsedge might inhibit nitrogen-fixing bacteria. Sterile aqueous extracts of roots and shoots inhibited growth of two species of *Azotobacter*.

Booth (1941) found that plant succession in fields abandoned from cultivation in Oklahoma and southeast Kansas included four stages: (1) pioneer weed, (2) annual grass, (3) perennial bunch grass and (4) true prairie. The weed stage lasted only 2–3 years. The annual grass stage lasted from 9–13 years and was dominated by *Aristida oligantha*. The perennial bunch grass stage was dominated by *Schizachyrium scoparium* (little bluestem) and this stage was still present 30 years after abandonment. The true prairie in central Oklahoma is dominated by little bluestem, *Andropogon gerardii* (big bluestem), *Panicum virgatum* and *Sorghastrum nutans*. Two problems concerning old-field succession in this area have been of major concern: (1) why the weed stage is replaced so rapidly by a small depauperate species like *A. oligantha* and (2) why the *A. oligantha* and perennial bunch grass stages remain so long.

Kapustka and Rice (1976) found asymbiotic N_2-fixation in the soils of the first two successional stages to be very low compared with the climax (Table 4.1). Seven phenolic acids, previously identified as allelochemicals from pioneer weeds, inhibited growth of three free-living N_2-fixers, *Azotobacter chroococcum*, *Enterobacter aerogenes* and *Clostridium* sp. Aqueous soil extracts from the pioneer weed stage, the annual grass stage and the climax inhibited growth of *E. aerogenes*. Rice (1964, 1965) reported that water extracts of several pioneer weeds and of *A. oligantha*, the dominant of the second stage, inhibited growth of *A. chroococcum*

TABLE 4.1 *Mean values of C_2H_2-reduction in successional stages in old fields*[a]

	1974														1975				
	Apr 13	May 15	May 25	Jun 2	Jun 9	Jun 16	Jul 1	Jul 15	Aug 16	Sep 14	Sep 28	Oct 19	Nov 16	Dec 17	Jan 15	Feb 15	Mar 1	Mar 15	Yearly mean
	p moles C_2H_2-reduction. g $soil^{-1}$ -h^{-1}																		
P_1	20a	0	117b	4b	59	9ab	0	0	25ab	16a	13a	19ab	28b	32ab	17ab	5b	13	16	23ab
P_2	7c	0	121c	15c	36	5c	0	0	13	7	5c	8	18	10	11	4c	5	12	15c
P_3	24	0	820	21	35	19	0	0	17	10	11	8	13	11	9	466	7	15	83

[a] Data from Kapustka and Rice (1976). Differences significant at 0.05 level or better: a, between P_1 and P_2; b, between P_1 and P_3; c, between P_2 and P_3.

TABLE 4.2 *Effect of desert shrub canopy upon acetylene reduction of soil cores with intact crust collected from three desert plant communities*[a]

Site	Ethylene production C_2H_4/soil core per 24 hours[b] Interspace (nmol)	Under canopy (nmol)
Atriplex confertifolia	78.7	20.7
Eurotia lanata	4.5	2.3
Artemisia tridentata	79.5	10.8

[a] After Rychert and Skujins (1974).
[b] Means of triplicate assays.

TABLE 4.3 *Inhibition of acetylene reduction of soil cores with intact bluegreen algae–lichen crust collected from the* Atriplex confertifolia *site*[a]

Desert shrub extract	Ethylene production C_2H_4/soil core per 24 hours[b] (nmol)
H_2O	24.8
Artemisia tridentata	0
Eurotia lanata	5.2
Atriplex confertifolia	8.6

[a] From Rychert and Skujins (1974).
[b] Means of triplicate assays.

and *A. vinelandii*. Several pioneer weeds and *A. oligantha* inhibited growth of some N_2-fixing bluegreen algae, whereas little bluestem from stage 3 and the climax did not in most tests (Parks and Rice, 1969). Another finding of significance here is that the order in which species invade old fields during succession is the same as the order based on increasing nitrogen requirements (Rice *et al.*, 1960). It seems logical therefore that the slowing of the rate of N_2-fixation by the pioneer species and *A. oligantha* would favour the species with lower nitrogen requirements and cause them to remain longer.

Nitrogen fixation by desert algal crusts probably constitutes a major input of nitrogen into desert ecosystems according to Skujins and coworkers (Rychert and Skujins, 1974; Rychert *et al.*, 1978). These investigators determined that bluegreen algae–lichen crusts from *Atriplex*, *Eurotia* (*Ceratoides*), and *Artemisia*-dominated sites in the Great Basin Desert had a laboratory potential of fixing nitrogen at rates up to 84 g of nitrogen per hectare per hour. They found that the rates were reduced under canopies of the desert shrubs (Table 4.2), and demonstrated that water extracts of leaves of *Atriplex*, *Eurotia* and *Artemisia* markedly inhibited N_2-fixation (Table 4.3). Moreover, volatile chemicals from the leaves of these shrubs inhibited this process (Skujins and West, 1974). Thus it appears that allelopathy plays an important role in the nitrogen economy of desert ecosystems.

Rice *et al.* (1980) found that four of five phenolic compounds present in decomposing rice straw markedly inhibited growth of *Anabaena cylindrica*, a N_2-fixing bluegreen alga important in rice fields (Table 4.4). All except two inhibited N_2-fixation (acetylene reduction) by this organism also. A combination of all five compounds was particularly effective in inhibiting growth and nitrogen-fixation. This suggests a synergistic action and seems particularly significant because the five compounds always occur together in decomposing rice residues in rice paddies (Chou and Lin, 1976).

TABLE 4.4 *Effects of phenolics produced by decomposing rice straw on growth and acetylene reduction (N_2-fixation) in* Anabaena cylindrica *(mean ±SE), DF = 6*[a]

	Treatment	Dry weight of algae (mg)	C_2H_4 ($\mu mol\, g^{-1} h^{-1}$)
p-Coumaric acid	Control	120.0 ± 12.4	1.36 ± 0.59
	10^{-4} M	142.4 ± 7.9	1.03 ± 0.17
	10^{3} M	62.3 ± 1.6[bc]	0.61 ± 0.09
Ferulic acid	Control	70.6 ± 0.9	0.13 ± 0.01
	10^{4} M	60.8 ± 0.8[b]	0.15 ± 0.01
	10^{-3} M	51.2 ± 2.5[bc]	0.07 ± 0.02[bc]
p-Hydroxybenzoic acid	Control	76.5 ± 3.8	6.65 ± 0.49
	10^{4} M	136.5 ± 7.6[b]	1.64 ± 0.36[b]
	10^{-3} M	77.0 ± 5.8	0.75 ± 0.31[b]
o-Hydroxyphenylacetic acid	Control	130.4 ± 5.1	0.18 ± 0.07
	10^{-4} M	141.6 ± 3.5	0.14 ± 0.01
	10^{-3} M	79.6 ± 9.9[bc]	1.21 ± 0.43[c]
Vanillic acid	Control	142.3 ± 21.3	0.35 ± 0.06
	10^{4} M	134.9 ± 5.1	0.36 ± 0.02
	10^{3} M	89.9 ± 7.3[bc]	1.49 ± 0.53[b]
All of above[d]	Control	123.0 ± 8.2	2.28 ± 0.51
	10^{-4} M	118.1 ± 7.9	5.36 ± 2.88
	10^{-3} M	63.1[bc]	0.00 ± 0.00[b]

[a] After Rice *et al.* (1980).
[b] Significantly different from respective control at 0.05 level or better.
[c] Significantly different from 10^{-4} M at 0.05 level or better.
[d] The indicated molarity is for each of the phenolics in the mixture.

4.2.2 Symbiotic nitrogen fixers

(a) Legumes

The recognized importance of symbiotic nitrogen fixation by *Rhizobium* has caused many workers to research possible allelopathic effects against this organism. Thorne and Brown (1937) found that most legume nodule bacteria investigated by them were able to grow in freshly expressed juices of their host plants, but such juices were bactericidal to other species of root-nodule bacteria. Elkan (1961) found that a non-nodulating, near isogenic soybean strain decreased significantly the number of nodules produced on its normally nodulating sister strain when inoculated plants of the two types were grown together in nutrient solution. Nodulation in ladino clover was inhibited also by the mutant soybean. Both results suggested the exudation by the roots of the non-nodulating strain of substances inhibitory to *Rhizobium* or to the nodulating process, or to both.

Beggs (1964) reported that oversown, inoculated white clover failed to nodulate properly and become established in large areas of *Nasella* and *Danthonia* grasslands in New Zealand. Subsequent research caused him to conclude that the failure to nodulate was due to some toxin in the soil.

Duff underneath *Rhus copallina* contained 6000–46 000 ppm of tannic acid throughout the year in Oklahoma, and the top 5 cm of soil under the duff contained 600–800 ppm (Blum and Rice, 1969). The addition of 30 ppm of tannic acid to soil that was previously free of it significantly reduced nodulation of red kidney beans (*Phaseolus vulgaris*) growing in that soil. It is not surprising, therefore, that soil from underneath

TABLE 4.5 *Effects of field soil from underneath* Rhus copallina *on nodulation and haemoglobin content of bean plants*[a]

Date of soil sample	Control soil: Mean nodule number[b]	Control soil: Haemoglobin, (μg) mean total per plant[c]	Test soil: Mean nodule number[b]	Test soil: Haemoglobin, (μg) mean total per plant[c]
23 Sept. 1966	139.2 ± 7.5	nd[e]	106.6 ± 6.9[d]	nd
26 April 1967	163.4 ± 10.7	nd	125.5 ± 6.7[d]	nd
26 June 1967	130.9 ± 8.0	90.9 ± 7.68	107.1 ± 8.2[d]	57.6 ± 9.48[d]
6 Oct. 1967	130.9 ± 8.0	90.9 ± 7.68	102.3 ± 7.6[d]	45.7 ± 8.1[d]

[a] Data from Blum and Rice (1969).
[b] Each figure represents mean of 20 plants with SE.
[c] Each figure represents mean of 15 plants with SE.
[d] Difference from control significant at 0.05 level or below.
[e] nd – not determined.

R. copallina significantly reduced nodulation and haemoglobin formation in bean plants (Table 4.5).

Twelve of 24 species of plants important in old-field succession in Oklahoma inhibited growth of *Rhizobium* and nodulation and haemoglobin formation in several legumes (Rice, 1964, 1965, 1968, 1971; Blum and Rice, 1969). The two most important legumes in old-field succession in central Oklahoma, *Lespedeza stipulacea* and *Trifolium repens*, were markedly affected. Root exudates of allelopathic species (Table 4.6), leaf leachates (artificial rain), and decomposing residues were all found to be effective in reducing nodule numbers and haemoglobin concentration. It has been demonstrated, of course, that the amount of nitrogen fixed is directly proportional to the haemoglobin content of the nodules (Stewart, 1966; Virtanen *et al.*, 1947). The allelopathic effects against symbiotic N_2-fixation would supplement the effects against asymbiotic N_2-fixation and favour the low nitrogen requiring early successional species, thus prolonging the *A. oligantha* and perennial bunch grass stages.

It was mentioned previously that broomsedge extracts inhibited some free-living N_2-fixers. Sterile aqueous extracts of roots and shoots of broomsedge also inhibited growth of two species of *Rhizobium* (Rice, 1972). Small amounts of decaying shoots of broomsedge (1 g per 454 g of soil) inhibited growth and nodulation of the two most important species of legumes in old-field succession in eastern Oklahoma, *Lespedeza stipulacea* and *Trifolium repens* (Table 4.7). Broomsedge competes vigorously and grows well on soils of low fertility; thus, inhibition of nodulation of legumes could help keep the nitrogen supply low and give broomsedge a selective advantage in competition over species that have higher nitrogen requirements. This could explain why it remains so long in almost pure stands.

The numbers of nodules on *Indigofera cordifolia* growing in association with *Aristida adscensionis* in the field in India are significantly lower than on *I. cordifolia* growing where *Aristida* is absent (Murthy and Ravindra, 1974; Murthy and Nagodra, 1977; Sarma, 1983). Moreover, the nodules on control plants are pinkish-red in colour whereas they are black on plants growing with *Aristida* (Sarma, 1983). Shoot and root extracts of *A. adscensionis* and leachates of its residues inhibited growth of *Rhizobium* isolated from nodules of *I. cordifolia*. Moreover, aqueous shoot and root extracts of *Aristida* inhibited nodulation of *I. cordifolia*.

Five phenolic compounds present in decomposing rice straw and sterile extracts of decomposing rice straw in soil were very inhibitory to growth of three strains of *Rhizobium* (Rice *et al.*, 1981). The effects were additive and in several instances synergistic. The phenolic compounds also reduced nodule numbers and haemoglobin content of the nodules in two bean varieties. Extracts of decomposing rice straw in soil significantly reduced N_2-fixation in bush black seeded beans (Table 4.8). This may help

TABLE 4.6 *Effects of root exudates of inhibitor species on nodulation of legumes*[a]

Inhibitor species	Average nodule number with standard error					
	Red kidney bean		Korean lespedeza		White clover	
	Control	Test	Control	Test	Control	Test
FORBS						
Ambrosia psilostachya	180.6 ± 10.0	62.1 ± 3.8[b]	8.1 ± 0.3	3.9 ± 0.2[b]	8.2 ± 0.6	4.3 ± 0.4[b]
Euphorbia supina	198.4 ± 11.2	181.4 ± 16.4	14.0 ± 0.5	6.5 ± 0.5[b]	5.5 ± 0.3	4.2 ± 0.4[b]
Helianthus annuus	298.6 ± 17.4	22.5 ± 2.8[b]	17.0 ± 1.5	2.3 ± 0.6[b]	6.5 ± 0.6	0.4 ± 0.1[b]
GRASSES						
Aristida oligantha	214.3 ± 12.3	145.5 ± 7.7[b]	6.4 ± 0.2	5.5 ± 0.3[b]	11.1 ± 0.7	4.3 ± 0.4[b]
Bromus japonicus	129.8 ± 5.1	127.7 ± 5.5	9.9 ± 0.3	6.5 ± 0.3[b]	5.9 ± 0.8	2.3 ± 0.4[b]
Digitaria sanguinalis	174.1 ± 7.7	109.2 ± 6.3[b]	14.1 ± 0.6	8.3 ± 0.5[b]	19.5 ± 1.0	3.5 ± 0.4[b]

[a] From Rice (1968).
[b] Difference from corresponding control significant at the 0.01 level or better.

TABLE 4.7 *Effects of decaying* Andropogon virginicus *on nodulation and fresh weight of legumes*[a]

Test legume	Average nodule number with standard error		Average fresh weight (g) with standard error	
	Control	Test	Control	Test
Lespedeza stipulacea	13.3 ± 0.8	7.3 ± 0.4[b]	1.0 ± 0.04	0.2 ± 0.01[b]
Trifolium repens	31.6 ± 2.4	6.2 ± 0.7[b]	1.2 ± 0.07	0.2 ± 0.08[b]

[a] After Rice *et al.* (1972).
[b] Difference from control significant at better than 0.001 level.

TABLE 4.8 *Effects on N_2-fixation (acetylene reduction) by bush black seeded beans of extracts of decaying rice straw in soil (mean ±SE) (DF = 6)*[a]

	Treatment	C_2H_4 (μmol plant^{-1} h^{-1})
Rice straw in soil	0 g per 3 kg soil (control)	8.56 ± 1.64
	25 g per 3 kg soil	3.71 ± 0.77[b]
	100 g per 3 kg soil	2.45 ± 0.35[b]

[a] Data from Rice *et al.* (1981).
[b] Significantly different from control at 0.05 level or better.

explain the great reduction in soybean yields in Taiwan following rice crops when the rice stubble is left in the field.

Yields of grain legumes decrease in Taiwan when grown more frequently than once a year (Young and Chao, 1983). Soil previously planted in *Vigna radiata* (mungbeans) and root residues of mungbeans reduced growth of that species, nodule number, nodule weight and N_2-fixation.

Mowed *Agropyron repens* (quackgrass) sod significantly reduced nodule number, nodule weight and N_2-fixation in three legumes in field and greenhouse tests (Weston and Putnam, 1985 – Tables 4.9, 4.10). In many cases, legume nodulation was decreased in glyphosate-treated quackgrass sod as compared with screened quackgrass soil or control soil.

(b) Non-legumes

Jobidon and Thibault (1981) found that aqueous extracts of leaf litter and buds, and fresh leaf leachates of *Populus balsamifera* (balsam poplar) severely inhibited root hair development in *Alnus crispa* (green alder). Moreover, necrosis of the radicle meristems occurred. Subsequent investigation demonstrated that the average number of nodules per green alder plant, after treatment with balsam poplar extracts, was only 51% that of

TABLE 4.9 *Influence of various quackgrass regimens on nodule number and nodule fresh weight of three legume indicators grown in the field*[a]

Regimen	Snapbean	Navybean	Soybean
		Nodule no.[b]	
Living	2c	5b	61ab
Glyphosate	3b	5b	54b
Removed	44a	15a	79a
		Nodule wt[b] (g)	
Living	0.03	0.03b	0.60b
Glyphosate	0.10	0.16a	1.10a
Removed	0.14	0.07b	1.20a

[a] After Weston and Putnam (1985).
[b] Figures are means of 40 plants harvested per treatment. Means within columns followed by different letters are different at 0.05 level.

TABLE 4.10 *Influence of various quackgrass regimens on N_2-fixing ability of three legume indicators grown in the field*[a]

	Fixed N_2^b (nmol plant^{-1} h^{-1})		
Regimen	Snapbean	Navybean	Soybean
Living	28c	212c	1654
Glyphosate	158ab	191b	1702
Removed	198a	399a	2871

[a] After Weston and Putnam (1985).
[b] Figures are means of 16 plants harvested per treatment. Means within columns followed by different letters are different at 0.05 level.

the controls (Jobidon and Thibault, 1982). Moreover, there was a 62% decrease in N_2-fixation (acetylene reduction) by alder seedlings treated with the most concentrated bud and leaf litter extracts employed (Table 4.11).

Numerous phenolic acids were identified as allelochemicals produced by balsam poplar (Perradin, 1982), and several of these compounds markedly inhibited growth and development of several *Frankia* strains (Perradin *et al.*, 1983).

Alnus glutinosa (black alder) was interplanted with *Juglans nigra* (black walnut) at four locations in Illinois and Missouri (USA). The alders suddenly declined and died after 8–13 years in all locations (Rietveld *et al.*, 1982). Alternative causes were investigated including competition, adverse soil properties, frost, insects, diseases, unsuitable seed source

TABLE 4.11 *Effects of different plant parts of* Populus balsamifera *on nitrogenase activity*[ab]

Treatments (%, w/v)		Mol C_2H_4 per 3 h per root dry weight × 10^{-11} (±SE)	% of inhibition
Bud extracts	2	4.36 ± 0.59a	61.72
	1	4.12 ± 0.49a	63.88
	0.5	5.26 ± 0.26ab	53.89
	0.1	6.54 ± 0.66b	42.67
	Control	11.41 ±0.99c	0.00
Leaf litter extracts	2	4.25 ± 0.96a	62.74
	1	6.19 ± 0.78a	45.74
	0.5	4.14 ± 0.54a	63.71
	0.1	5.20 ± 0.55a	54.62
	Control	11.41 ± 0.99b	0.00

[a] After Jobidon and Thibault (1982).
[b] Within each treatment, means followed by the same letter are not different at 0.05 level, Duncan's new multiple range test.

and allelopathy. Only allelopathy could be substantiated. It was suggested that the allelopathy results from (1) sufficient walnut biomass to contribute substantial amounts of juglone to soil, and (2) wet soil that greatly restricts the aerobic metabolism of juglone by soil microorganisms, allowing it to build up to toxic concentrations. Dawson and Seymour (1983) reported that a 10^{-4} M concentration of juglone completely inhibited growth of a *Frankia* isolate from nodules of *Alnus rubra*. It appears, therefore, that juglone could inhibit growth of alder directly, and also indirectly through inhibition of N_2-fixation.

4.3 ALLELOPATHIC EFFECTS ON NITRIFICATION

Leather (1911) reported low nitrification under perennial grass in India, indicating an inhibition of the nitrifying bacteria by certain perennial grass species. Russell (1914) reported that cropped soil had a much lower total nitrate content than uncropped similar soil even when the amount taken up by plants was included, and suggested that the lower amount was due to a diminished production in the presence of plants. Lyon *et al.* (1923) found that maize, wheat and oats markedly depressed nitrate production. They suggested that the reason for the recovery of smaller amounts of nitrates from the soil and plants was that plants liberate carbonaceous matter into the soil which favours the development of nitrate-consuming organisms.

Richardson (1935, 1938) studied the nitrogen cycle in grassland soils at Rothamsted Experimental Station in the UK. He found that the concentration of ammonium nitrogen was several times greater than that of nitrate nitrogen in the soil, and the ratio of ammonium to nitrate nitrogen increased with the age of the sward. He also reported that the grasses and other plants absorbed the ammonium nitrogen as readily as nitrate, and that much of the nitrogen was taken up as ammonium.

Theron (1951) investigated problems which arose in South Africa when grasses were used as a means of rehabilitating wornout soils. Normally, a luxuriant growth of grass took place during the first season after the sward was established but growth soon deteriorated, and by the third and fourth season the sward was so poor that it hardly afforded any grazing and appeared to be valueless as a rebuilder of soil.

Theron ran numerous lysimeter experiments in which some lysimeters contained crop plants, some contained a local perennial grass species, and some were left fallow. An example of his usual results were those obtained using annual millet as the crop plant and *Hyparrhenia* sp. as the perennial grass. Large quantities of nitrogen were mineralized during the first year in all three types of lysimeters. Under the millet, virtually a steady state was reached with respect to both yield and the nitrogen lost to the crop and to the percolate that year. The nitrate in the percolate of the cropped soil had cyclic changes associated with the growth and maturation of the millet. By the time the second crop had matured, the nitrate content had fallen to a very low concentration. However, it again increased to a high value during the ensuing winter. Virtually no nitrates were found, however, in the percolate of the perennial grass plots at any time after the first year of growth. In the fallow soil, nitrogen was freely mineralized consistently. These results illustrate the very important point that nitrification took place actively from the second season on throughout the entire winter in the cropped and fallow soil but not in the soil under perennial grass, even though the grass was dormant from May to September. Although little or no nitrate occurred under perennial grass, ammonium-N was generally present in greater concentrations than were usually present in cultivated soils. According to Theron, the soil remained equally moist in all plots and other external conditions were similar.

Theron presented several strong arguments against the suggestion of Lyon *et al.* (1923) that the continued low concentration of nitrate in the percolate under perennial grass was caused by the consumption of nitrate by microorganisms due to their stimulation by carbonaceous matter from the roots of the grass. He concluded that perennial grasses and other actively growing plants interfere only with nitrification and not with ammonification. He concluded further that the inhibition of nitrification is probably due to bacteriostatic excretions by living roots, and suggested that only very minute quantities of the inhibitors would probably be necessary to inhibit the rather sensitive nitrifying bacteria.

Greenland (1958) found almost no nitrate throughout the year in permanent grassland plots in Ghana, Africa. On the other hand, he found that amounts of nitrate under temporary or successional grass plots were greater than under climax grasses, but lower than in cropland. He pointed out that the cause of low nitrate concentrations was not likely to be nitrate absorption, because he found that little microbial nitrate absorption took place under crops. In addition, soil samples taken from the permanent grassland plots and incubated in the laboratory showed a high rate of nitrification in spite of a very high carbon/nitrogen ratio of over 20. Greenland concluded, therefore, that the low level of nitrate under grassland was caused by suppression of mineralization due to an excretion of the plant roots which was toxic to the nitrification process.

Boughey *et al.* (1964) reported that two species of *Hyparrhenia*, grasses abundant in Zimbabwe (previously Rhodesia) high-veld savanna, secrete a toxin which suppresses the growth of nitrifying bacteria. Warren (1965) found that the populations of nitrifiers in the climax purple-veld of South Africa were much lower than in successional stages. During the greater part of the year the nitrite oxidizers were almost completely absent in the purple-veld soil and were only slightly higher in *Hyparrhenia* soils. Moreover, there was a gradual decrease in nitrite and nitrate with succession. Munro (1966a, b) found that root extracts of several climax species from the Zimbabwean (previously Rhodesian) high-veld were more inhibitory to nitrification than were those from several seral species investigated. The evidence suggests, therefore, that inhibition of nitrification increases with succession in grassland areas in Africa, and that it is pronounced in the climax grasslands.

Moore and Waid (1971) tested the suggestion made by Theron (1951) and others that roots of some plants produce compounds that inhibit nitrification in soil. Root exudates (washings) were obtained from *Lolium perenne* (ryegrass), *Triticum aestivum* (wheat), *Lactuca sativa* (Cos lettuce), *Brassica napus* var. *arvensis* (salad rape) and *Allium cepa* (onion) by leaching columns of quartz chips in which the plants were growing. The exudates and ammonium solution were then added to columns of clay loam in which ammonium-ion was being converted to nitrate-ion at a known rate. The exudates of all plants investigated reduced the rate of nitrification (Table 4.12), but the effects of rape and lettuce were temporary. Ryegrass exudates had the most pronounced and persistent effects, reducing the rate of nitrification up to 84%. As neither microbial immobilization of inorganic nitrogen nor denitrification appeared to be taking place, it was concluded that the exudates contained inhibitors which retarded nitrification.

Dommergues (1954) did a microbiological study of five forest soil types from central and eastern Madagascar. He found that ammonification was higher and nitrification was lower in the forest soils than in cultivated soils. In a subsequent study of dry tropical forest soils in Senegal,

TABLE 4.12 *Effects of washings of roots of ryegrass, onion and wheat on the rate of nitrification in soil columns after 45 days*[a]

Treatment in stage 2	Control	Ryegrass	Onion	Wheat
Rate (mg N per kg soil per day) of:				
ammonium disappearance	375	125	195	210
nitrate formation	370	135	200	220
Ratio (NH_4^+/NO_3^-)	1.01	0.93	0.97	0.95
Decrease in rate (% control) of:				
ammonium disappearance	–	67	48	44
nitrate formation	–	64	46	41
Total N recovered from both stages of the experiment (% applied)	103.3	99.4	100.9	101.4

[a] After Moore and Waid (1971).

Dommergues (1956) stated that nitrification in dry tropical forest soils is more active than in dense, humid forest soils. Nevertheless, nitrification increases greatly in the dry forest soils on clearing and cultivating.

Jaquemin and Berlier (1956) reported that the nitrifying power in forest-covered soils of the lower Ivory Coast of Africa was low, and that it increased on clearing. There was an 18-fold increase in *Nitrosomonas* and a 34-fold increase in *Nitrobacter* after clear-cutting a forest ecosystem in Connecticut (Smith *et al.*, 1968). The evidence indicated that the pronounced increase in numbers of nitrifiers was not due to changes in physical conditions, but to the elimination of uptake of nitrate by the vegetation, or to a reduced production of substances inhibitory to the autotrophic nitrifying population, or both. Likens *et al.* (1969) reported a 100-fold increase in nitrate loss in the same ecosystems after cutting.

Concentrations of ammonium and nitrate nitrogen and numbers of nitrifiers were determined throughout the year in soils from two old-field successional stages and the climax in three vegetation types in Oklahoma, the tall grass prairie (TGP), the post oak–blackjack oak forest and the oak–pine forest (Rice and Pancholy, 1972). The inverse correlation between the concentration of nitrate nitrogen and the concentration of ammonium nitrogen in all plots was striking. The concentration of ammonium increased from a low in the first successional stage to a high in the climax, whereas the concentration of nitrate decreased from a high in the first successional stage to a low value in the climax (Table 4.13 – only data for the TGP are shown). Moreover, the numbers of nitrifiers were high in the first successional stage and low in the climax (Table 4.14). Thus, some factor or factors reduced the populations of nitrifiers during succession, resulting in an apparent reduction in the rate of oxidation of ammonium to nitrate. It was obvious from the general soil data that the

TABLE 4.13 *Concentrations of ammonium and nitrate nitrogen in 0–15 cm level of research plots*[ab]

	Tall grass prairie					
	ppm NH_4^+			ppm NO_3^-		
Date	P_1	P_2	P_3	P_1	P_2	P_3
March 1971	3.28	3.98	4.60[de]	4.42	1.42[c]	1.50[e]
May 1971	2.87	3.71[c]	4.86[de]	3.11	1.98[c]	1.39[de]
July 1971	0.94	1.45[c]	2.21[de]	4.12	2.70[c]	1.77[de]
September 1971	1.68	5.63[c]	6.68[e]	4.27	2.78[c]	1.78[de]
November 1971	2.82	6.69[c]	7.83[de]	2.42	0.91[c]	0.20[de]
January 1972	2.08	3.10	4.35[e]	2.49	1.31[c]	0.57[de]

[a] Data from Rice and Pancholy (1972).
[b] Each number is average of ten analyses.
[c] Difference between P_1 and P_2 significant at 0.05 level or better.
[d] Difference between P_2 and P_3 significant at 0.05 level or better.
[e] Difference between P_1 and P_3 significant at 0.05 level or better.

TABLE 4.14 *Numbers (MPN) of nitrifiers in 0–15 cm level of research plots*[ab]

	Tall grass prairie					
	Nitrosomonas per g soil			*Nitrobacter* per g soil		
Date	P_1	P_2	P_3	P_1	P_2	P_3
April 1971	111	42[c]	140	25	25	25
June 1971	3012	525[c]	147[de]	347	417	280
August 1971	334	158	50[e]	62	72	3[de]
October 1971	817	51	37	24	24	24
December 1971	177	26[c]	32[e]	23	23	18
February 1972	127	51	51	110	270[c]	43[e]

[a] Data from Rice and Pancholy (1972).
[b] Each number is average of four determinations at different locations.
[c] Difference between P_1 and P_2 significant at 0.05 level or better.
[d] Difference between P_2 and P_3 significant at 0.05 level or better.
[e] Difference between P_1 and P_3 significant at 0.05 level or better.

low rates of nitrification in the climax plots were not due to pH or textural differences. Moreover, the lack of definite trends in amounts of organic carbon in relation to succession in the oak–pine and post oak–blackjack areas indicated that the quantity of organic carbon was not responsible for the low rate of nitrification in the climax plots. These facts, along with the previous discovery (Rice, 1964) that the climax species investigated were very inhibitory to nitrification, led to the inference that the climax plants reduced the rates of nitrification in the three ecosystems involved. It appears that inhibition of nitrification started during succession and increased in intensity as succession proceeded towards the climax.

All the dominant herbaceous and tree species from the intermediate and climax areas produced considerable amounts of condensed tannins, and the concentration of these tannins in the top 15 cm of soil was always higher in the climax stand than in the intermediate successional stage in each vegetation type (Rice and Pancholy, 1973). Gallic and ellagic acids, which result from the digestion of hydrolysable tannins in oak species, were present in the climax oak–pine forest soil also. Condensed tannins, hydrolysable tannins, ellagic acid, gallic acid, digallic acid and commercial tannic acid (a hydrolysable tannin), in very small concentrations, completely inhibited nitrification by *Nitrosomonas* in soil suspensions for 3 weeks, the duration of the tests. Moreover, the concentrations of tannins, gallic acid and ellagic acid found in the soil of the climax research plots were several times higher than the minimum concentrations necessary completely to inhibit nitrification.

Woodwell (1974) reported a relatively high concentration of nitrate in a shallow water table under a cultivated field (4.48 mg ml^{-1}) on Long Island, New York, whereas the concentration was only 0.93 mg l^{-1} in the water table under an abandoned field, 0.42 mg l^{-1} under a pine forest, and 0.02 mg l^{-1} under an oak–pine forest. He stated that 'mature systems are tight; disturbance causes leakage of nutrients'.

Todd *et al.* (1975) measured the nitrate content of stream water and the nitrifying bacterial population of the upper 40 cm of soil of three Appalachian watersheds over a 22-month period. The watersheds were a fescue grass catchment, a 15-year-old white pine plantation, and a mature undisturbed hardwood forest. Monthly averages of nitrate nitrogen in stream water were 730, 190 and 3 ppb respectively for the three ecosystems. The respective nitrifying populations averaged 16 000, 175 and 22 per gram dry weight of soil. There was an obvious correlation between numbers of nitrifiers and nitrate contents of the streams. Thus, the authors concluded that 'nitrifying activity appears to be dependent on vegetation type and successional stage'. It apears that inhibition of nitrification decreased as succession progressed towards the climax hardwood forest of the region.

The ratio of ammonium nitrogen to nitrate nitrogen increased with succession on mine-spoils in North Dakota, and numbers of *Nitrosomonas* and *Nitrobacter* decreased with succession (Lodhi, 1979). It was concluded that the vegetation inhibits nitrification and the inhibition increases with succession.

Low levels of nitrate nitrogen relative to ammonium nitrogen and

TABLE 4.15 *Numbers of* Nitrosomonas *(MPN method) in control and treatment cultures. Secondary plant chemicals were extracted from ponderosa pine needles and bark, and from soils for the treatment cultures*[a]

Control	Extract treatment		% inhibition
12 270	(needles-water)	875[b]	93
12 270	(needles-ether)	4050[b]	68
12 720	(bark-acetone)	1100[b]	91
12 720	(soil-acetone)	1485[b]	88
12 720	(soil-hydrolysis)	950[b]	93

[a] After Lodhi and Killingbeck (1980).
[b] Statistical t value indicates significant difference between numbers of *Nitrosomonas* in treatment and control groups at $p \leq 0.05$.

low numbers of *Nitrosomonas* and *Nitrobacter* were found in the soils of climax *Pinus ponderosa* stands in western North Dakota (Lodhi and Killingbeck, 1980). This suggested a low nitrification rate, and the low rate could not have been due to low pH because the soils were alkaline (pH 7.25–7.75). Evidence suggested that the reduction in nitrate synthesis was due to the production and subsequent transfer to the soil of allelochemicals toxic to *Nitrosomonas*. Several compounds inhibitory to nitrification were found in extracts from ponderosa pine needles, bark and A-horizon soils (Table 4.15).

Data presented to this point support the hypothesis of Rice and Pancholy (1972) that the rate of nitrification is increasingly slowed as succession proceeds towards the climax in many ecosystems. This makes good biological sense because ammonium ions are positively charged and are therefore adsorbed on the negatively charged micelles in the soil, thus preventing leaching below the depth of rooting due to percolating water. On the other hand, nitrate ions are negatively charged and are repelled by the micelles in the soil. Thus, they readily leach below the depth of rooting or are easily carried away in surface drainage. It would appear from these facts that inhibition of nitrification would help to conserve nitrogen.

If plants take up nitrate ions, they generally have to reduce these ions to nitrite and then to ammonium ions before the nitrogen can become involved in reactions leading to the formation of amino acids, and subsequently to other nitrogenous organic compounds. The reduction of nitrate ions to ammonium ions requires energy; thus, inhibition of nitrification would conserve energy.

The conservation of energy and of nitrogen resulting from the inhibition of nitrification would appear to be a strong force in the selection of plants inhibitory to nitrification. If nitrification is inhibited, this would mean that ammonium nitrogen would be the chief form of available nitrogen in later successional stages and in climax ecosystems. The next item of importance, therefore, concerns the ability of plants to use ammonium nitrogen, especially by non-crop plants. There is growing evidence that many plant species can use ammonium nitrogen as effectively or more so than nitrate nitrogen in ecologically meaningful concentrations (Rice, 1984).

The evolution of only two genera of bacteria primarily responsible for nitrification also suggests that this process is not a necessary link in the N-cycle in many ecosystems. Considerable additional evidence from other types of investigations supports the hypothesis that nitrification is increasingly inhibited during some types of succession. If the ratio of ammonium to nitrate does increase with succession, there should be a selection of species along the sere in favour of those which selectively absorb and grow better on ammonium nitrogen. Wiltshire (1973) found that climax grasses in Rhodesia showed a greater preference for ammonium than did earlier seral species or crop cultivars (Table 4.16). Haines (1977) investigated, in the field, the relative uptake of nitrate and ammonium by plants from different stages in a South Carolina old field to forest sere. He found that nitrate uptake decreased whereas ammonium uptake increased with succession.

If the hypothesis is correct, species that are adapted to a particular successional stage must be adapted to the form of nitrogen available

TABLE 4.16 *Effect of nitrogen source on yield of highveld grasses*[a]

Species	Yield from nitrate as % yield from ammonium nitrogen	p[b]
Ruderal		
Eragrostis racemosa	94	n.s.
Setaria pallidifusca	84	n.s.
Pennisetum setaceum	83	n.s.
Pioneer		
Pogonarthria squarrosa	76	0.001
Sporobolus pyramidalis	76	0.001
Aristida adscensionis	52	0.001
Climax		
Heteropogon contortus	48	0.05
Schizachyrium jeffreysii	42	0.001
Andropogon schirensis	39	0.001
A. gayanus	30	0.001
Hyparrhenia filipendula	30	0.01
Danthoniopsis intermedia	22	0.05

[a] After Wiltshire (1973).
[b] P based on SE of yields.

TABLE 4.17 *Nitrate content and nitrate reductase activity of plants grown under different nitrogen status*[a]

Measurement	Veld grown[b]		Nitrate grown[b]	
	H.f.	*S.p.*	*H.f.*	*S.p.*
Enzyme activity in early growing season (μM NO_2^- $g^{-1}h^{-1}$)	0.43	1.60	–	0.77
Enzyme activity in late growing season (μM NO_2^- $g^{-1}h^{-1}$)	0.05	0.23	0.14	0.42
Leaf NO_3^- content (μM NO_3^- g^{-1})	0.09	0.37	3.05	0.78

[a] After Bate and Heelas (1975).
[b] *H.f.* = *Hyparrhenia filipendula*; *S.p.* = *Sporobolus pyramidalis*.

to them. Thus, the enzymatic systems for absorbing and incorporating nitrogen should reflect the form of nitrogen used. Nitrate reductase is the enzyme involved in the first step of the reduction of nitrate into a form the plant can use. This enzyme thus controls the ability of a plant to use nitrate; therefore, species that use mostly nitrate as their nitrogen source should have relatively high levels of nitrate reductase activity (NRA). Conversely, species that use mostly ammonium as their nitrogen source should have relatively low levels of NRA. Bate and Heelas (1975) reported that *Sporobolus pyramidalis*, a pioneer species in the Rhodesian grasslands, had a NRA four times that of the climax species, *Hyparrhenia filipendula* (Table 4.17). Franz and Haines (1977) found that the NRA of vascular plants decreased during an old field to forest succession in South Carolina. Smith and Rice (1983) measured the NRA of leaves and roots of characteristic species from different stages of an old field to prairie sere in Oklahoma and found that the pioneer species had high NRA concentrations whereas the four climax species had low NRA levels (Table 4.18).

Overall, there is considerable indirect evidence that supports the hypothesis that nitrification is increasingly inhibited during succession in some seres. Nevertheless, there are numerous ecologists who disagree with this hypothesis, and there is increasing evidence that the generalized hypothesis of Rice and Pancholy is not valid.

Vitousek and Reiners (1975) hypothesized that increases in biomass cause decreasing losses of nutrients from ecosystems during succession because the added biomass ties up more of the nutrients. They suggested that this probably accounts for a lowering in concentration of nitrate in soil during succession instead of an inhibition of nitrification. Robertson and Vitousek (1981) questioned the use of pool sizes of ammonium and

TABLE 4.18 *Nitrate reductase activity of species grown on different nitrogen treatments. Comparisons are made horizontally in the table. Numbers that share the same letters are not significantly different at the 0.01 level*[a]

	μM NO_2 per g dry wt per hour – antilog of log mean with s.d.			
Species	NO_3	2:1	1:2	NH_4
PIONEER SPECIES				
Helianthus annuus				
leaves	22.4 ± 1.3a	22.5 ± 1.3a	22.4 ± 1.1a	13.0 ± 1.4b
roots	8.6 ± 1.4a	6.7 ± 1.8b	7.6 ± 1.3ab	2.4 ± 1.6c
Ambrosia trifida				
leaves	53.9 ± 1.2a	45.0 ± 1.3a	45.5 ± 1.2a	31.9 ± 1.7a
roots	19.2 ± 1.3a	15.8 ± 1.2b	14.6 ± 1.4b	4.6 ± 1.6c
Aristida oligantha				
leaves	45.1 ± 1.2a	43.1 ± 1.3a	41.9 ± 1.2a	28.5 ± 1.3b
roots	12.6 ± 2.5a	11.9 ± 3.0a	5.3 ± 2.6a	5.4 ± 2.3a
Bromus japonicus				
leaves	27.7 ± 1.5a	32.4 ± 1.3a	27.4 ± 1.6a	19.5 ± 1.7b
roots	61.1 ± 3.5a	82.3 ± 1.8a	88.9 ± 1.9a	79.8 ± 1.8a
CLIMAX SPECIES				
Schizachyrium scoparium				
leaves	2.0 ± 1.6a	1.8 ± 1.7a	1.5 ± 1.6a	1.5 ± 1.7a
roots	4.8 ± 1.7a	4.2 ± 1.8a	3.1 ± 2.8b	1.5 ± 1.8c
Sorghastrum nutans				
leaves	4.6 ± 3.0a	4.7 ± 3.7a	6.5 ± 2.2a	4.4 ± 2.7a
roots	12.1 ± 2.2a	12.4 ± 1.2a	7.8 ± 1.5b	5.8 ± 1.7c
Panicum virgatum				
leaves	7.4 ± 1.4a	5.5 ± 1.7a	5.2 ± 1.7a	4.8 ± 1.7a
roots	3.3 ± 2.2a	2.1 ± 3.0b	1.9 ± 3.2b	1.0 ± 3.5c
Andropogon gerardii				
leaves	10.5 ± 1.7a	8.1 ± 1.7b	7.7 ± 1.4b	5.6 ± 1.6c
roots	4.3 ± 2.2a	3.8 ± 2.1a	3.4 ± 1.9a	3.0 ± 1.6a

[a] After Smith and Rice (1983).

nitrate and numbers of nitrifiers to indicate relative rates of nitrification. This is a valid criticism because it would obviously be much more closely related to the real world if rates of nitrification could be measured in the field. They chose to measure nitrification by incubating soil samples in the laboratory. Their method completely prevented the continued addition of allelochemicals, which could occur in the field, and would also speed the dilution and breakdown of any allelochemicals present in the soil at the time of collection. This could cause the population of nitrifiers to build rapidly and cause nitrification in all samples.

Robertson and Vitousek (1981) investigated nitrification from a primary sere in the Indiana Dunes and a secondary sere in the Hutcheson Memorial Forest in New Jersey, using the laboratory incubation technique. They concluded that their results did not support the hypothesis that nitrification is progressively inhibited during the course of ecological succession. Robertson (1982) investigated nitrification in the same seres as Robertson and Vitousek, but his method differed in that he monitored nitrate production in laboratory-incubated soils treated with different combinations of nutrients. He also performed experiments to try to determine if allelochemicals were involved in regulating nitrification. He prepared whole-leaf washings and extracts, forest-floor washings and extracts, and soil washings. These extracts were amended with NH_4^+–N to equalize available NH_4^+–N among treatments and were added to soil to be incubated as in other experiments. Robertson inferred from his investigation that nitrification appeared to be controlled by ammonium availability in at least the first four sites in the primary sere, and perhaps in the fifth. He inferred that allelochemicals might be important also in the fifth site. In the secondary sere, he inferred that ammonium availability appeared to control nitrification in all sites. Several other investigators who used laboratory or field incubations to measure nitrification either failed to find a successional trend or found the reverse from that hypothesized by Rice and Pancholy (Robertson, 1982).

Vitousek *et al.* (1989) did a very thorough literature search of publications concerning changes in rates of nitrification during primary, secondary (pure) and old-field succession. They restricted their use of the term secondary succession to situations where disturbance kills or removes vegetation but leaves the soil intact, and in which regrowth of vegetation can begin immediately following disturbance. They reported that primary succession invariably involves a low nitrogen availability and potential nitrification early in succession and this increases in rate with succession. In contrast, destructive disturbance followed by immediate regrowth (pure secondary succession) invariably increases nitrogen availability, and generally potential nitrification, in recently disturbed sites and nitrification declines during later stages of succession. They concluded that succession following a period of chronic disturbance (agriculture) does not follow any clear pattern. The duration and intensity of disturbance may control whether potential nitrification increases or decreases early in such seres.

The problem of a reliable technique to determine the actual rate of nitrification in soil is still with us. Woldendorp and Laanbroek (1989) tested several techniques and concluded that no reliable estimate of the rate of nitrification can be obtained from the number of nitrifying bacteria, and estimates made using field-incubation and ^{15}N–NH_4^+ tech-

TABLE 4.19 *Results of soil analyses*[a]

		μg per g soil		(MPN) Nitrifiers per g soil	
Soil taken	pH	NH_4–N[b]	NO_3–N[b]	*Nitrosomonas*	*Nitrobacter*
Elm	7.20	6.20	2.76	26	8
Hackberry	6.75	7.15	3.17	20	6
Red oak	5.38	14.28	2.83	14	3
Sycamore	6.95	7.50	2.10	21	7
White oak	4.80	8.93	1.80	16	3
Pooled mean	6.22	8.81	2.53	19.4	5.4

[a] After Lodhi (1978).
[b] Significantly different at 0.05 level.

niques do not yield reliable data. They stated that possibly the best results can be obtained using Schimel's method to estimate the actual nitrification rate using $^{15}N-NO_3$, but this method has still not been tested under different sets of soil conditions.

The nitrification potential of soils under three desert shrubs (*Artemisia tridentata*, *Ceratoides lanata*, *Atriplex confertifolia*) was determined by the perfusion technique over the period January to October (Skujins and Trujillo y Fulgham, 1978). Considerable nitrite accumulation was evident in all three soils before conversion to nitrate occurred, and the nitrite production rates appeared similar. The nitrite oxidation rate decreased however from a high in the *C. lanata* soil to a low in the *A. confertifolia* soil. When the soil columns were amended with litter from the shrubs, there was a build-up of nitrite and a very slow rate of nitrate accumulation, indicating a suppression of the second stage of nitrification.

Lodhi (1977, 1978) investigated soil properties in relation to the distance from the tree trunks of several forest species. He found that nitrate nitrogen had the greatest overall variability at different distances from the trunks. Ammonium nitrogen was always considerably higher than nitrate nitrogen when compared at each distance from the trunk of each species, and he stated that the low nitrate could not have been due to its uptake by the intact vegetation because the soil samples were taken before active growth (Table 4.19). Additionally, low numbers of *Nitrosomonas* and *Nitrobacter* were associated with large amounts of ammonium nitrogen in most samples. Lodhi concluded that the high ratio of $NH_4:NO_3$ was due to an inhibition of nitrification and that the differences under various species were caused by variations in the type of litter.

Mineral soil (Al horizon) under *Thuja plicata* (western redcedar) had higher total microbial counts and populations of ammonium oxidizing bacteria than soil under *Tsuga heterophylla* (western hemlock) (Turner

TABLE 4.20 *Mineralization and nitrification potentials of soils in three week incubations. Nitrification % = nitrate production per total nitrogen mineralized*[a]

	Final ($\mu g\ g^{-1}$)[b]		Production ($\mu g\ g^{-1}\ d^{-1}$)[b]			
Treatment	NO_3–N	NH_4–N	NO_3–N	NH_4–N	Sum	Nitrification (%)
Field soil brought to field capacity						
Cedar	1.27a	12.7a	0.01a	0.17a	0.18a	5.5
Hemlock	1.11a	91.9b	0.02a	4.02b	4.04b	0.5
Air-dried soil brought to field capacity						
Cedar	15.3a	26.1a	0.65a	0.66a	1.31a	49.6
Hemlock	0.82b	157.1b	0.00b	6.54b	6.54b	0.0
Air-dried soil brought to field capacity +20 $\mu g\ g^{-1}$ NH_4–N						
Cedar	34.0a	53.3a	1.56a	1.28a	2.84a	54.9
Hemlock	0.67b	211.3b	0.05b	8.21b	8.26b	0.0

[a] After Turner and Franz (1985).
[b] Different letters within comparisons of cedar and hemlock indicate significantly different means ($p < 0.05$/t-test).

and Franz, 1985). Nitrogen mineralization rates were greater in laboratory incubations of hemlock soil, but nitrification was observed only in incubations of redcedar soil (Table 4.20). The low nitrification rate in hemlock soil was not due to ammonium limitation or to low phosphorus, thus indicating a possible allelopathic effect of western hemlock on nitrification.

Root exudates and residues of four cultivars of *Sorghum bicolor* were tested against nitrification using a soil incubation method (Alsaadawi *et al.*, 1986). All four cultivars significantly reduced nitrification, but there was a genetic effect on the level of inhibition.

4.4 CONCLUSIONS AND PROSPECTS

Considerable progress has been made in determining the effects of allelopathy on the nitrogen cycle but much remains to be done. As more knowledge is gained, farmers and others who grow plants will be able to help prevent detrimental effects on the cycle and to use beneficial effects to improve growth, reproduction and yield.

REFERENCES

Alexander, M. (1977) *Introduction to Soil Microbiology*, 2nd edn, Wiley, New York.

Alsaadawi, I. S., Al-Uqaili, J. K., AlRubeaa, A. J. *et al.* (1986) Allelopathic suppression of weed and nitrification by selected cultivars of *Sorghum bicolor* (L.) Moench. *J. Chem. Ecol.*, **12**, 209–19.

Bate, G. C. and Heelas, B. V. (1975) Studies on the nitrate nutrition of two indigenous Rhodesian grasses. *J. Appl. Ecol.*, **12**, 941–52.

Beggs, J. P. (1964) Spectacular clover establishment with formalin treatment suggests growth inhibitor in soil. *New Zealand J. Agric.*, **108**, 529–35.

Blum, U. And Rice, E. L. (1969) Inhibition of symbiotic nitrogen-fixation by gallic and tannic acid, and possible roles in old-field succession. *Bull. Torrey Bot. Club*, **96**, 531–44.

Booth, W. E. (1941) Revegetation of abandoned fields in Kansas and Oklahoma. *Amer. J. Bot.*, **28**, 415–22.

Boughey, A. S., Munro, P. E., Meiklejohn, J. *et al.* (1964) Antibiotic reactions between African savanna species. *Nature*, **203**, 1302–3.

Brady, N. C. (1974) *The Nature and Properties of Soils*, 8th edn, Macmillan, New York.

Chan, E. C. S., Basavanand, P. and Liivak, T. (1970) The growth inhibition of *Azotobacter chroococcum* by *Pseudomonas* sp. *Can. J. Microbiol.*, **16**, 9–16.

Chou, C. H. and Lin, H. J. (1976) Autointoxication mechanisms of *Oryza sativa*. I. Phytotoxic effects of decomposing rice residues in soil. *J. Chem. Ecol.*, **2**, 353–67.

Dawson, J. O. and Seymour, P. E. (1983) Effects of juglone concentration on growth *in vitro* of *Frankia* Ar13 and *Rhizobium japonicum* Strain 71. *J. Chem. Ecol.*, **9**, 1175–83.

Dommergues, Y. (1954) Biology of forest soils of central and eastern Madagascar. *Trans. 5th Int. Congr. Soil Sci.*, **3**, 24–8.

Dommergues, Y. (1956) Study of the biology of soils of dry tropical forests and their evolution after clearing. *Trans. 6th Int. Congr. Soil Sci.*, **E**, 605–10.

Elkan, G. H. (1961) A nodulation-inhibiting root excretion from a non-nodulating soybean strain. *Can. J. Microbiol.*, **7**, 851–6.

Franz, E. H. and Haines, B. L. (1977) Nitrate reductase activities of vascular plants in a terrestrial sere: relationship of nitrate to uptake and the cybernetics of biogeochemical cycles. *Bull. Ecol. Soc. Amer.*, **58**, 62.

Greenland, D. J. (1958) Nitrate fluctuations in tropical soils. *J. Agric. Sci.*, **50**, 82–92.

Haines, B. L. (1977) Nitrogen uptake: apparent pattern during old-field succession in southeastern United States. *Oecologia*, **26**, 295–303.

Harmsen, G. W. and Kolenbrander, G. J. (1965) Soil inorganic nitrogen, in *Soil Nitrogen* (eds W. V. Bartholomew and F. E. Clark), American Society of Agronomy, Madison, Wisconsin, pp. 43–92.

Iuzhina, Z. I. (1958) Relationship between toxic properties of soil of Kola Peninsula and number of bacterial antagonists of *Azotobacter*. *Microbiology*, **27**, 452–6 (transl. from Russian).

Jacquemin, H. and Berlier, Y. (1956) Evolution du pouvoir nitrifiant d'un sol de basse Côte d'Ivoire sous l'action du climat et de la végétation. *Trans. 6th Int.*

Congr. Soil Sci., **C**, 343–7.

Jobidon, R. and Thibault, J. R. (1981) Allelopathic effects of balsam poplar on green alder germination. *Bull. Torrey Bot. Club*, **108**, 413–18.

Jobidon, R. and Thibault, J. R. (1982) Allelopathic growth inhibition of nodulated and un-nodulated *Alnus crispa* seedlings by *Populus balsamifera*. *Amer. J. Bot.*, **69**, 1213–23.

Kapustka, L. A. and Rice, E. L. (1976) Acetylene reduction (N_2-fixation) in soil and old field succession in central Oklahoma. *Soil Biol. Biochem.*, **8**, 497–503.

Klubek, B., Eberhardt, P. J. and Skujins, J. (1978) Ammonia volatilization from Great Basin Desert soils, in *Nitrogen in Desert Ecosystems* (eds N. E. West and J. Skujins), US/IBP Synthesis Series 9, Dowden, Hutchinson & Ross, Stroudsberg. PA., pp. 107–29.

Leather, J. W. (1911) Records of drainage in India. *Mem. Dep. Agric., India, Chem.*, **2**, 63–140.

Leuck, E. E. II and Rice, E. L. (1976) Inhibition of *Rhizobium* and *Azotobacter* by rhizosphere bacteria of *Aristida oligantha*. *Bot. Gaz.*, **137**, 160–4.

Likens, G. E., Bormann, F. H. and Johnson, N. M. (1969) Nitrification: Importance to nutrient losses from a cutover forested ecosystem. *Science*, **163**, 1205–6.

Lodhi, M. A. K. (1977), The influence and comparison of individual forest trees on soil properties and possible inhibition of nitrification due to intact vegetation. *Amer. J. Bot.*, **64**, 260–4.

Lodhi, M. A. K. (1978) Comparative inhibition of nitrifiers and nitrification in a forest community as a result of the allelopathic nature of various tree species. *Amer. J. Bot.*, **65**, 1135–7.

Lodhi, M. A. K. (1979) Inhibition of nitrifying bacteria, nitrification and mineralization in spoil soils as related to their successional stages. *Bull. Torrey Bot. Club*, **106**, 284–9.

Lodhi, M. A. K. and Killingbeck, K. T. (1980) Allelopathic inhibition of nitrification and nitrifying bacteria in a ponderosa pine (*Pinus ponderosa* Dougl.) community. *Amer. J. Bot.*, **67**, 1423–9.

Lyon, T. L., Bizzell, J. A. and Wilson, B. D. (1923) Depressive influence of certain higher plants on the accumulation of nitrates in the soil. *J. Amer. Soc. Agron.*, **15**, 457–67.

Molisch, H. (1937) *Der Einfluss einer Pflanze auf die andere-Allelopathie*, Gustav Fischer, Jena.

Moore, D. R. E. and Waid, J. S. (1971) The influence of washings of living roots on nitrification. *Soil Biol. Biochem.*, **3**, 69–83.

Munro, P. E. (1966a) Inhibition of nitrite-oxidizers by roots of grass. *J. Appl. Ecol.*, **3**, 227–9.

Munro, P. E. (1966b) Inhibition of nitrifiers by grass root extracts. *J. Appl. Ecol.*, **3**, 231–8.

Murthy, M. S. and Nagodra, T. (1977) Allelopathic effects of *Aristida adscensionis* on *Rhizobium*. *J. Appl. Ecol.*, **14**, 279–82.

Murthy, M. S. and Ravindra, R. (1974) Inhibition of nodulation of *Indigofera cordifolia* by *Aristida adscensionis*. *Oecologia*, **16**, 257–8.

Parks, J. M. and Rice, E. L. (1969) Effects of certain plants of old-field succession on the growth of blue-green algae. *Bull. Torrey Bot. Club*, **96**,

345–60.

Perradin, Y. (1982) Etude des Acides Phenoliques de *Populus balsamifera* L. et d'*Alnus crispa* var. mollis Fern. et de Leur Influence sur *Frankia* ACN1^{AG}. Thèse pour le Doctorat de Troisième Cycle, Biologie et Physiologie Végétales, L'Université Claude-Bernard, Lyon 1, France.

Perradin, Y., Mottet, M. J. and Lalonde, M. (1983) Influence of phenolics on *in vitro* growth of *Frankia* strains. *Can. J. Bot.*, **61**, 2807–14.

Rice, E. L. (1964) Inhibition of nitrogen-fixing and nitrifying bacteria by seed plants. I. *Ecology*, **45**, 824–37.

Rice, E. L. (1965) Inhibition of nitrogen-fixing and nitrifying bacteria by seed plants. III. Comparison of three species of *Euphorbia*. *Proc. Oklahoma Acad. Sci.*, **45**, 43–4.

Rice, E. L. (1968) Inhibition of nodulation of inoculated legumes by pioneer plant species from abandoned fields. *Bull. Torrey Bot. Club*, **95**, 346–58.

Rice, E. L. (1971) Inhibition of nodulation of inoculated legumes by leaf leachates from pioneer plant species from abandoned fields. *Amer. J. Bot.*, **58**, 368–71.

Rice, E. L. (1972) Allelopathic effects of *Andropogon virginicus* and its persistence in old fields. *Amer. J. Bot.*, **59**, 752–5.

Rice, E. L. (1984) *Allelopathy*, 2nd edn, Academic Press, Orlando, Florida.

Rice, E. L. and Pancholy, S. K. (1972) Inhibition of nitrification by climax ecosystems. *Amer. J. Bot.*, **59**, 1033–40.

Rice, E. L. and Pancholy, S. K. (1973) Inhibition of nitrification by climax ecosystems II. Additional evidence and possible role of tannins. *Amer. J. Bot.*, **60**, 691–702.

Rice, E. L., Lin, C. Y. and Huang, C. Y. (1980) Effects of decaying rice straw on growth and nitrogen fixation of a bluegreen alga. *Bot. Bull. Academia Sinica*, **21**, 111–17.

Rice, E. L., Lin, C. Y. and Huang, C. Y. (1981) Effects of decomposing rice straw on growth of and nitrogen fixation by *Rhizobium*. *J. Chem. Ecol.*, **7**, 333–44.

Rice, E. L., Penfound, W. T. and Rohrbaugh, L. M. (1960) Seed dispersal and mineral nutrition in succession in abandoned fields in central Oklahoma. *Ecology*, **41**, 224–8.

Richardson, H. L. (1935) The nitrogen cycle in grassland soils. *Trans. 3rd Int. Congr. Soil Sci.*, 1, 219–21.

Richardson, H. L. (1938) Nitrification in grassland soils: with special reference to the Rothamsted Park Grass experiment. *J. Agric. Sci.*, **28**, 73–121.

Rietveld, W. J., Schlesinger, R. C. and Kessler, K. J. (1982) Allelopathic effects of black walnut on European black alder coplanted as a nurse species. *J. Chem. Ecol.*, **9**, 1119–33.

Robertson, G. P. (1982) Factors regulating nitrification in primary and secondary succession. *Ecology*, **63**, 1561–73.

Robertson, G. P. and Vitousek, P. M. (1981) Nitrification potentials in primary and secondary succession. *Ecology*, **62**, 376–86.

Russell, E. J. (1914) The nature and amount of the fluctuations in nitrate contents of arable soils. *J. Agric. Sci.*, **6**, 50–3.

Rychert, R. C. and Skujins, J. (1974) Nitrogen fixation by blue-green algae-lichen crusts in the Great Basin Desert. *Proc. Soil Sci. Soc. Amer.* **38**, 768–71.

Rychert, R., Skujins, J., Sorensen, D. and Porcella, D. (1978) Nitrogen fixation by lichens and free-living microorganisms in deserts, in *Nitrogen in Desert Ecosystems* (eds N. E. West and J. Skujins), US/IBP Synthesis Series 9, Dowden, Hutchinson & Ross, Stroudsberg, PA, pp. 20–30.

Sarma, K. K. V. (1983) Allelopathic potential of the phytoextracts of *Aristida adscensionis*. Linn. *Trop. Ecol.*, **24**, 19–21.

Skujins, J. and Trujillo y Fulgham, P. (1978) Nitrification in Great Basin Desert soils, in *Nitrogen in Desert Ecosystems* (eds N.E. West and J. Skujins), US/IBP Synthesis Series 9, Dowden, Hutchinson & Ross, Stroudsberg, PA, pp. 60–74.

Skujins, J. J. and West, N. E. (1974) Nitrogen Dynamics in Stands Dominated by Some Cool Desert Shrubs. *US/IBP Desert Biome Res. Memo.*, 74,42 Ecology Center, Utah State Univ., Logan, Utah.

Smith, J. L. and Rice, E. L. (1983) Differences in nitrate reductase activity between species of different stages in old field succession. *Oecologia*, **57**, 43–8.

Smith, W., Bormann, F. H. and Likens, G. E. (1968) Response of chemo-autotrophic nitrifiers to forest cutting. *Soil Sci.*, **106**, 471–3.

Stewart, W. D. (1966) *Nitrogen Fixation in Plants*, Oxford Univ. Press, London and New York.

Theron, J. J. (1951) The influence of plants on the mineralization of nitrogen and the maintenance of organic matter in the soil. *J. Agric. Sci.*, **41**, 289–96.

Thorne, D. W. and Brown, P. E. (1937) The growth and respiration of some soil bacteria in juices of leguminous and non-leguminous plants. *J. Bacteriol.*, **34**, 567–80.

Todd, R. L., Swank, W. T., Douglass, J. E. *et al.* (1975) The relationship between nitrate concentration in the southern Appalachian mountain streams and terrestrial nitrifiers. *Agro-Ecosystems*, **2**, 127–32.

Turner, D. P. and Franz, E. H. (1985) The influence of western hemlock and western redcedar on microbial numbers, nitrogen mineralization, and nitrification. *Plant and Soil*, **88**, 259–67.

Virtanen, A. I., Erkama, J. and Linkola, H. (1947) On the relation between nitrogen fixation and leghaemoglobin content of leguminous root nodules. II. *Acta Chem. Scand.*, **1**, 861–70.

Vitousek, P. M. and Reiners, W. A. (1975) Ecosystem succession and nutrient retention: a hypothesis. *Bioscience*, **25**, 376–81.

Vitousek, P. M., Matson, P. A. and Van Cleve, K. (1989) Nitrogen availability and nitrification during succession: Primary, secondary, and old-field seres. *Plant and Soil*, **115**, 229–39.

Warren, M. (1965) A study of soil – nutritional and other factors operating in secondary succession in highveld grassland in the neighbourhood of Johannesburg. PhD Thesis, Univ. of Witwatersrand, Johannesburg, South Africa.

Westerman, R. L. and Tucker, T. C. (1978) Denitrification in desert soils, in *Nitrogen in Desert Ecosystems* (eds N. E. West and J. Skujins), US/IBP Synthesis Series 9, Dowden, Hutchinson & Ross, Stroudsberg, PA, pp. 75–100.

Weston, L. A. and Putnam, A. R. (1985) Inhibition of growth, nodulation, and nitrogen fixation of legumes by quackgrass (*Agropyron repens*). *Crop Sci.*,

25, 561–5.
Wiltshire, G. H. (1973) Response of grasses to nitrogen source. *J. Appl. Ecol.*, **10**, 429–35.
Woldendorp, J. W. and Laanbroek, H. J. (1989) Activity of nitrifiers in relation to nitrogen nutrition of plants in natural ecosystems. *Plant and Soil*, **115**, 217–28.
Woodwell, G. M. (1974) Success, succession, and Adam Smith. *Bioscience*, **24**, 81–7.
Young, C. C. and Chao, C. C. (1983) Legume production and application of *Rhizobium* in Taiwan, in *Proc. ROC-Japan Seminar on Promoting Nitrogen Fixation in Agriculture*, National Science Council, Taipei, Taiwan, pp. 39–44.

CHAPTER 5

Allelopathic mechanism in fire-prone communities

G. B. Williamson, D. R. Richardson and N. H. Fischer

5.1 INTRODUCTION

The Southeastern Coastal Plain of the USA supports two different communities on its upland, sandy soils. The communities share a common subtropical climate with heavy rains, seasonally concentrated in the summer. Both are exposed to fires on a regular basis, and both are dominated by pines (*Pinus* spp) and oaks (*Quercus* spp). However, the species composition of the two communities is almost entirely different.

The prevalent vegetation is the sandhill, an open woodland of longleaf pine (*Pinus palustris* Mill.) or slash pine (*P. elliottii* Engelm.) and turkey oak (*Quercus laevis* Walt.), sand live oak (*Q. geminata* Small) and bluejack oak (*Q. incana* Bartr.) with a dense graminoid ground cover of wiregrass (*Aristida stricta* Michx.) and beard grasses (*Andropogon* and *Schizachyrium* spp.). Shrubs are uncommon with the exception of clumps of saw palmetto (*Serenoa repens* (Bartr.) Small). Surface fires, fuelled by pine needles and the deciduous grasses, sweep through the sandhills on a regular basis every 3–8 years. However, community composition is virtually unchanged by the fires because plant reproduction and development are stimulated by post-fire conditions.

The second community, the sand pine scrub, is much less common, being interspersed as islands and strands among the sandhills. At maturity the scrub is a closed canopy of sand pine (*Pinus clausa* Vasey ex Sarg.) with a solid understorey of scrub oaks (*Q. myrtifolia* Willd., *Q. chapmannii* Sarg. and *Q. geminata*). In its early seral stages, the scrub is characterized by a diverse array of endemic shrubs. At no time does the scrub community have any significant herbaceous ground cover. Periodic crown fires consume all the standing biomass in the scrub every 20–50 years, but afterwards the oak species resprout and sand pine releases seed from serotinous cones.

Despite the differences in vegetative composition, the existence of two communities appears unwarranted on the basis of abiotic factors.

Allelopathy: Basic and applied aspects Edited by S. J. H. Rizvi and V. Rizvi
Published in 1992 by Chapman & Hall, London ISBN 0 412 39400 6

TABLE 5.1 *Comparison of scrub and sandhill community traits*

	Scrub	Sandhill
Physiognomy		
Ground cover	None	Complete
Shrub cover	Very dense	Very sparse
Fuel traits		
Surface litter, quantity	Low	High
Surface litter, quality	Compact	Loose, aerated
Crown litter, quantity	High	None
Foliage phenology	Evergreen	Deciduous
Fire traits		
Frequency	20–50 yr	3–8 yr
Periodicity	High	High
Intensity	Crown fire	Surface fire
Allelochemical release	Yes	No

Attempts to find soil differences have been more popular than fruitful (Harper, 1914; Webber, 1935; Kurz, 1942; Laessle, 1958, 1968). More recently, Kalisz and Stone (1984) found differences in soil profiles but then attributed them to differences in vegetation. Alternatively, the occurrence of the scrub community on coastal dunes hints that it may be more tolerant of salt spray than the sandhill vegetation (Laessle, 1968), but its persistence at numerous inland locations, such as Ocala National Forest, precludes salt tolerance as the decisive abiotic factor.

Our studies have led us to believe that biotic, not abiotic, factors, are responsible for the different communities (Richardson, 1977, 1988; Richardson and Williamson, 1988; Williamson and Black, 1981; Fischer *et al.*, 1987). The vegetation types may regulate the factors that in turn determine community composition. Principal among these factors are variables related to fire and allelopathy (Table 5.1)

5.2 FIRE AND ALLELOPATHY: A HYPOTHESIS

Woody plants in fire-prone communities have developed various adaptations for survival and reproduction in response to the abrupt, but severe selection of wildfires. Adults have three mechanisms for escaping genetic death (mortality without reproduction). If the adults have insulating bark and are tall enough to preclude scorch damage to shoots and buds, then often they survive surface fires. If the adults have large storage organs underground together with subterranean buds, then consuming crown fires become periodic interludes followed by resprouting. Lastly, if adults fuel the crown fires in which they die but release seeds from fruits, serotinous cones or the seed bank after the fire, then they colonize the post-fire soils. These three adult response types – live, resprout and die – represent spatial escape above the fire, spatial escape below the fire and temporal escape through keying the reproductive cycle around the fire.

The fire variables that select for the various response types in plants are fire frequency, periodicity and intensity. These variables are dependent on abiotic factors such as incidence of lightning strikes and seasonal temperature and precipitation patterns, as well as on certain biotic factors – namely, the plants themselves (Mutch, 1970; Platt *et al.*, 1988; Rebertus *et al.*, 1989). That is, coexisting species through their response types and associated patterns of fuel deposition will regulate the fire variables in a community. For example, retention of dead branches and leaves in the crown fosters the less frequent but more intense crown fires, whereas frequent shedding of dead leaves and branches promotes the frequent but less intense surface fires. Likewise, a dense shrub layer promotes crown fires, whereas a dense ground cover promotes surface fires.

In the sandhill, adult oaks and pines survive fires by being above the heat produced by surface fuels (Williamson and Black, 1981). Litter deposition by sandhill species is frequent, at least annually, and is concentrated in the dry season, December through May. As the pines and deciduous oaks shed their lower branches as well as leaves, their boles remain free of fuel that could otherwise carry flames into their crowns. Thus, the sandhill fires are frequent, every 3–8 years, but confined to surface fuels (Table 5.1).

In contrast, the scrub community suffers a high intensity crown fire every 20–50 years (Table 5.1). Species in the scrub exhibit the adult die response, exemplified by sand pine and wild rosemary (*Ceratiola ericoides* Michx.) or the resprout response, typified by the scrub oaks (Burns, 1968, 1972; Johnson *et al.*, 1986). Dead branches are retained on the trees, and although dead leaves and needles are shed, their small sizes result in a compact, moist litter layer. More importantly, the absence of a graminoid ground cover in the scrub pre-empts an aerated, fine fuel accumulation. The scrub oaks and other shrubs are evergreen, another trait that hinders the ignition of the scrub fuels. Fires are difficult to start in scrub vegetation and only occur after several decades of fuel accumulation in the crowns; even then, fires spread only after a cold front has lowered the moisture content of the live foliage and a strong wind is present (Doren *et al.*, 1988).

The system of two communities with different sets of species and fuel production patterns that stimulate different fire frequencies and intensities would not be unusual were it not for the fact that they occur on the same substrate across noticeably abrupt and narrow ecotones (Nash, 1895; Whitney, 1898; Mulvania, 1931; Webber, 1935; Laessle, 1958;

Abrahamson, 1984; Myers, 1985). Given the short distances, seed dispersal across the ecotones must be frequent.

Scrub plants are found frequently colonizing sandhill communities, and under a management programme of fire suppression, they will become reproductive adults (Veno, 1976). Eventually, the maturing forest loses its woodland openness that permits growth of the graminoids that provide the fuel for the frequent surface fires. Then, any subsequent fire is likely to spread into the tree crowns and further speed the replacement of sandhill oaks and pines by their scrub counterparts. However, under the natural sandhill fire regimen, the frequent surface fires kill the young scrub colonizers before they become adults. Therefore, fire appears to be the factor that periodically purges the sandhills of invading scrub plants.

In contrast, sandhill species generally do not colonize scrub communities. In the mature scrub community, the closed canopy of sand pine and the dense understorey of shrubs appear to prevent colonization by the sandhill species, most of which require full sunlight. However, young scrub sites are very open, and here we suspect that allelopathy has evolved as a mechanism to prevent the invasion of grasses and pines that provide fuel for surface fires so hazardous to the scrub shrubs. In essence each shrub creates a miniature fuel break around itself; together, a group of shrubs becomes an island devoid of surface fuels – a barrier against sandhill surface fires that are extinguished at the scrub ecotone.

5.3 TESTS FOR ALLELOPATHY IN THE SCRUB

5.3.1 Bioassays and seasonal activities

Foliage of three shrub species common in immature scrub communities was collected monthly from plants in central Florida: *Ceratiola ericoides* (A. Gray) Heller (Empetraceae), *Conradina canescens* (Torr. & Gray) (Labiatae), and *Calamintha ashei* (Weatherby) Shinners (Labiatae). In addition, litter under one species, *Ceratiola*, was collected monthly because it accumulates visibly in the field. Test solutions for bioassays were prepared by soaking 50 g of foliage or litter in 500 ml of distilled water for 24 h under refrigeration at 8°C, and then filtering through No. 1 Whatman paper.

Bioassays were performed by adding 5 ml of test solution from each of the four sources to 30 target seeds placed on a sheet of Whatman No. 1 filter paper in a 10 cm diameter Petri dish (Richardson, 1985). Seeds of three native sandhill grasses, *Schizachyrium scoparium* (Michx.) Nash, *Andropogon gyrans* Ashe and *Leptochloa dubia* (HBK) Nees, were used as target species, and each species was treated in four replicate dishes kept in the dark at 26°C, for 8 to 24 days until the control seeds had germinated. Control dishes were given distilled water which was unadjusted because the test solutions exhibited osmolalities (2–9 mOsm) and pH values (5.2–6.9) that normally do not affect seedling germination and growth (Stout and Tolman, 1941; Bell, 1974; Reynolds, 1975a, b).

Bioassay results were tabulated as percent germination and as radicle lengths of germinated seedlings. Germination and radicle length means were computed for each of the four replicate dishes for each treatment and the control for each month of bioassays. The magnitude of inhibition or stimulation was compared through a response index (RI), determined as follows for the treatment mean (T) and the control mean (C):

$$\text{If } T \geq C \text{ then RI} = 1 - (C/T)$$
$$\text{If } T < C \text{ then RI} = (T/C) - 1$$

RI corresponds to the proportional increase of the treatment versus the control means: it is positive if the control mean is less than the treatment mean (stimulation) and negative if the treatment mean is less than the control mean (inhibition). RI values are useful because empirically they approximate normal distributions (Williamson and Richardson, 1988).

The three leaf washes and one litter wash from the scrub species significantly inhibited germination of the three sandhill grasses in five of the 12 source–target combinations (Table 5.2). The tests produced inhibition in all 12 cases, although significant monthly variation tended to mask statistical significance in seven of the source–target means. The source means and target means with their larger sample sizes, 24 and 36, respectively, revealed significant inhibition in every case (Table 5.2).

Overall, the average RI for germination was −0.20, indicating a 20% reduction in germination relative to the controls. However, this average effect masks much specific variation. For example, *Leptochloa* was twice as sensitive as *Andropogon* to the scrub washes. Among the washes, *Calamintha* proved to be twice as inhibitory as *Ceratiola*.

TABLE 5.2 *Effects of source leaf washes and one litter wash on germination of target species*[a]

	Target species			
Source wash	*Andropogon*	*Leptochloa*	*Schizachyrium*	Source mean
Calamintha	−0.10	−0.44**	−0.27*	−0.27**
Conradina	−0.16	−0.34**	−0.17	−0.22**
Ceratiola	−0.09	−0.20**	−0.09	−0.12**
Ceratiola litter	−0.18	−0.09	−0.26**	−0.18**
Target mean	−0.13**	−0.27**	−0.20**	

[a] Shown as means of 8 monthly RI values, with overall source means (3 targets by 8 months) and overall target means (4 sources by 8 months). Means significantly different from 0 are indicated by '*' for $p \leq 0.05$ and by '**' for $p \leq 0.01$.

TABLE 5.3 *Effects of source leaf washes and one litter wash on radicle length of target species*[a]

Source wash mean	Target species			Source
	Andropogon	*Leptochloa*	*Schizachyrium*	
Calamintha	+0.12	−0.23**	−0.18*	−0.10
Conradina	−0.01	−0.13*	−0.10	−0.08
Ceratiola	+0.03	−0.01	−0.02	−0.00
Ceratiola litter	−0.25	−0.01	−0.16	−0.13*
Target mean	−0.03	−0.09**	−0.12**	

[a] Shown as means of 8 monthly RI values, with overall source means (3 targets by 8 months) and overall target means (4 sources by 8 months). Means significantly different from 0 are indicated by '*' for $p \leq 0.05$ and by '**' for $p \leq 0.01$.

In contrast to the strong inhibition of germination, the effects of the scrub washes on radicle growth were mild (Table 5.3). Statistically significant inhibition was found in only 3 of the 12 source–target combinations, and the overall mean for all combinations was −0.08, an 8% reduction relative to the controls. Among the source washes, only the litter of *Ceratiola* showed significant inhibition; however, among the target species, both *Leptochloa* and *Schizachyrium* suffered significant inhibition. Again the magnitude of the effects varied greatly among source–target combinations; however, in all but a few cases the effects were mild relative to the effects of germination. There was no correlation between inhibition of germination and inhibition of radicle growth over the source–target combinations for all months (Pearson's $r = 0.14$, $p = 0.15$, $N = 100$). However, the source–target means for germination (Table 5.1) and for radicle length (Table 5.3) were correlated strongly ($r = 0.72$, $p = 0.008$, $N = 12$). Even within a target species, the source means for germination and those for radicle lengths were correlated ($N = 4$: *Andropogon* $r = 0.83$, $p = 0.17$; *Leptochloa* $r = 0.95$, $p = 0.05$; *Schizachyrium* $r = 0.99$, $p = 0.01$).

The absence of a correlation in the monthly effects on germination and on radicle growth reflected different seasonal activities. Inhibition of germination was greater than 20% from May through October with a severe effect (50%) in June (Fig. 5.1). Radicle growth showed strong inhibition in April and May, little or no effect during the summer, and some stimulation in the winter (Fig. 5.1). The monthly effect (RI) of the washes on germination was highly correlated with 30-year averages for precipitation ($r = -0.47$, $p = 0.0001$) and temperature ($r = -0.45$, $p = 0.0001$) – a result indicating that inhibition was concentrated during the growing season (Fig. 5.1). The monthly effect (RI) on radicle growth

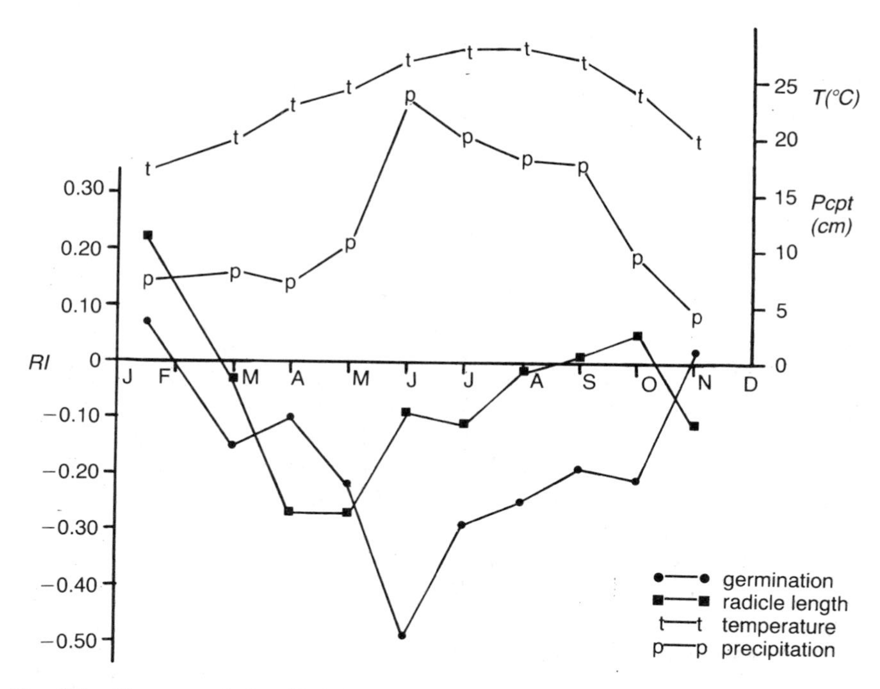

Fig. 5.1 Response index (RI) for germination and radicle length of three grasses in monthly bioassays, with comparative temperature and precipitation data.

showed no such relationships (temperature $r = -0.15$, $p = 0.13$; precipitation $r = 0.01$, $p = 0.91$).

5.3.2 Transplant experiments

A field growth experiment was designed to test for allelopathic interactions of plants at scrub sites versus plants from sandhill sites (Richardson, 1985). Plugs of wiregrass (*Aristida stricta*) were removed from the sandhill with a 10 cm diameter golf green cupcutter. The plugs were inserted immediately into sections of PVC pipes, 10 cm diameter by 20 cm long, but open at both ends. The grass was pruned to the soil surface and monitored in a greenhouse until new growth emerged. Then, 20 pipes with healthy grass shoots were transplanted into each of eight field plots, four scrub sites and four sandhill sites. In the field the plants remained in the sandhill soil within the pipes but were open to soil and gas exchange through the ends of the pipes.

A number of provisions were made to preclude resource competition between the native vegetation and the transplants. Plots were 1 by 20 m with the long axis in an east–west orientation to ensure daily exposure to

TABLE 5.4 *Growth responses of transplanted* Aristida stricta *plugs after 16 months at scrub and sandhill communities*[a]

	Survival (%)	Wet wt (g)	Dry wt (g)
Sandhill	95 (74/78)	26.0 ± 3.0	7.87 ± 0.76
Scrub	76 (53/70)	12.7 ± 5.8	4.48 ± 1.71

[a] All three variables are significantly different ($p \leq 0.05$) between the two communities.

sunlight. Additionally, all overhanging vegetation was pruned. Vegetation in the plots was pruned to the soil surface, and the plot perimeter was root pruned by inserting a spade to a depth of 30 cm. All pruning was repeated every two weeks. Finally, the pipes were rotated 180° every two weeks to ensure that there was no root growth into or out of the pipes. During periods of drought, all pipes were watered with 500 ml, but only once in any two-week period.

After 16 months, which included two growing seasons, the pipes of *Aristida* were removed and each plant's biomass recorded. Tests for differences through analysis of variance compared plant variables (aboveground, belowground and total biomass) between scrub and sandhill communities. For the scrub sites only 76% (53/70) of the plugs survived, whereas 95% (74/78) survived in the sandhill sites (Table 5.4). Of the survivors, *Aristida* in the sandhill weighed 93% (fresh wt) and 76% (dry wt) more than in the scrub (Table 5.4).

5.4 CHEMICAL MECHANISMS OF ALLELOPATHY BY SCRUB PLANTS

Investigation of the active compounds in the scrub plants was conducted through a series of extractions and fractional separations, followed by bioassays with seeds of *Schizachyrium scoparium*. Then, further separations of the active fractions and more bioassays were performed until single compounds were isolated and identified. Preliminary bioassays showed that *Schizachyrium* seeds were unaffected by pH and osmolality in the ranges of the extracts used in the bioassays. Bioassays were conducted with 30 seeds in 5 ml of solution in petri dishes as described above for the field foliar washes.

5.4.1 Active constituents of scrub plants

The leaves of *Ceratiola* contained large quantities of triterpenes (ursolic acid and erythrodiol) that were not active and small quantities of monoter-

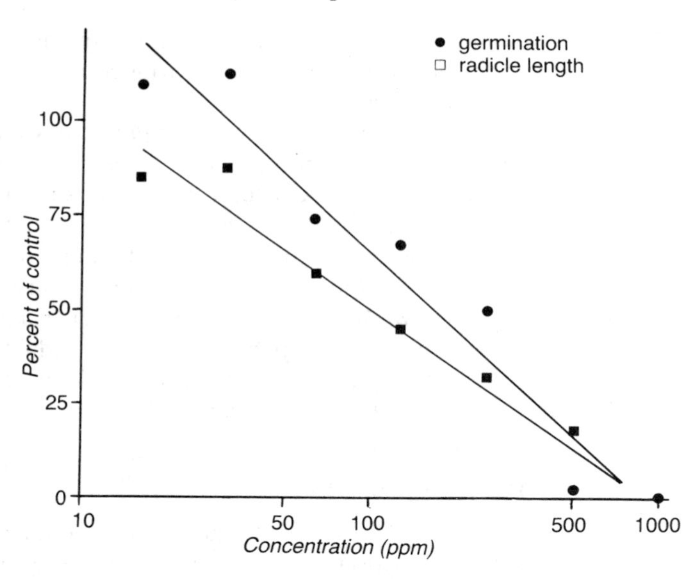

Fig. 5.2 Response of *Schizachyrium* seeds to different concentrations of hydrocinnamic acid, expressed as a percent of the distilled water control. Both regression lines are statistically significant ($p = 0.003$).

penes, many of which are known to be active from other studies (α-pinene, camphor, salicylic acid methyl ester, eugenol, γ-ionone, β-ionone, geraniol propanoate, farnesol, γ-amorphene, humulene and caryophyllene). However, biological activity was concentrated not in the non-polar extracts where these terpenoids were found, but in the most polar fraction. Furthermore, simply dipping fresh foliage in water produced a nearly pure solution of the novel flavonoid, ceratiolin, which appears to be responsible for the activity of the plant (Tanrisever *et al.*, 1987). Ceratiolin itself has little effect on *Schizachyrium* seeds, but its degradation product hydrocinnamic acid causes significant inhibition of germination and radicle growth. The effects are detectable to less than 30 ppm in single dose applications, and the degree of inhibition increases with concentration (Fig. 5.2).

The breakdown of ceratiolin is facilitated by light, heat and acid. Thus, the compound can be leached from the plants by the frequent rains (Fig. 5.1) into the sandy soils of the scrub where it is exposed to intense light and high soil temperatures (Fig. 5.1). Hydrocinnamic acid is quite water soluble, so it leaches through the soil with the rainfall. Samples of soil and litter under *Ceratiola* contain hydrocinnamic acid in extremely low concentrations (2–10 ppm), but due to its constant release with rainfall, it is continually bathing seeds and seedlings at the time of water availability. These soils are very well-drained with virtually no clay and only 7–9% organic matter (Richardson, 1985), so there is very little absorptive material to hold allelopathic compounds in the soil. Therefore, in this environment continual release in precipitation may be more effective than production of lipophilic chemicals that adhere to soil particles.

In contrast to *Ceratiola*, the two mints *Calamintha* and *Conradina* showed inhibitory activity from the non-polar fractions of their leaf extracts. Both species showed very high concentrations of ursolic acid and a large number of monoterpenes, many of which are known to have inhibitory action on plants. Over 20 monoterpenes, including 1,8-cineole, camphor, pinocarveol, borneol, pinocamphone, myrtenol, carveol and carvone, were isolated from *Conradina* (de la Peña, 1985). *Calamintha* contained a different set of monoterpenes, many of which were menthofuran derivatives, including evodone, the novel calaminthone and menthofuran itself (Tanrisever, 1986; Tanrisever *et al.*, 1988).

5.4.2 Micelles as a transport mechanism

The presence of a variety of active monoterpenes raised the question of how these compounds might function as allelotoxins. In pure form, many monoterpenes are volatile at ambient temperatures, and Muller *et al.* (1964) suggested that such compounds simply volatilize from the leaf surfaces and drift onto the soil where they are adsorbed by lipophilic soil particles, seeds or seedlings. Such a mechanism seems inefficient in the scrub for several reasons: (1) winds are a regular feature in all the coastal plain habitats, so a plant's monoterpenes are unlikely to settle near its neighbours; (2) in contrast to clay particles, the sandy soil provides little lipophilic surface to trap the monoterpenes, and (3) the high temperatures at the soil surface are likely to revolatilize any condensing monoterpene molecules.

An alternative transport mechanism may be the formation of tensides between the monoterpenes and other plant compounds. Other molecules in the leaf with surfactant properties could form micelles with the non-polar monoterpenes – a result with the following effects: (1) volatilization of the monoterpenes would be reduced, (2) solubilization of the non-polar monoterpenes in the aqueous foliar leachates would be increased and (3) entry into target seed or seedling cells would be facilitated by creating leaks in their membranes.

One test for micelle formation employs an aqueous solution of the fluorescent dye, acridine (Wolff, 1981). As the suspected surfactant is added, the dye's fluorescence decreases when a micelle is formed because acridine's fluorescence is much more intense in protic solvents than in a lipophilic environment. If a micelle is not formed the fluorescence remains unchanged. The two different patterns are demonstrated with the non-tenside forming ethyl acetate and two known biological detergents, SDS and emulphogen (Fig. 5.3) (Tanrisever, 1986). Aqueous leachates of fresh foliage were prepared by soaking 360 g of fresh leaves in 3 litres of

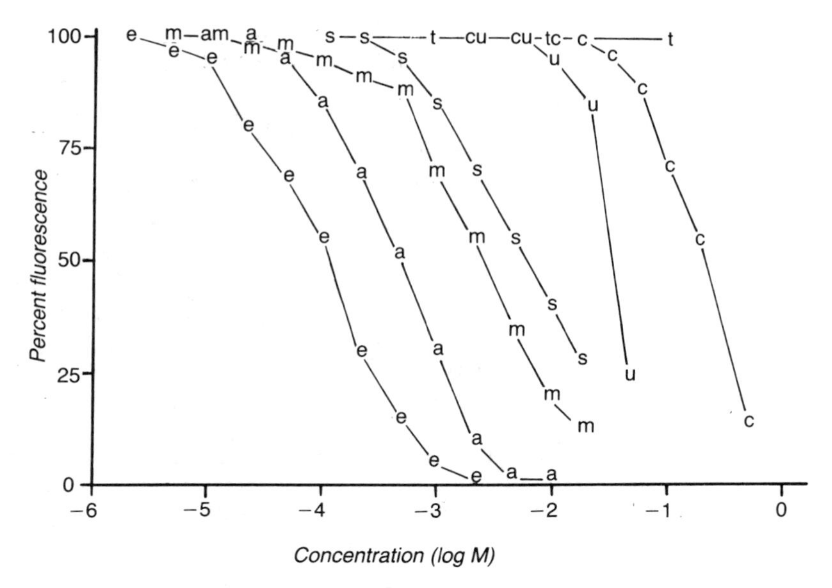

Fig. 5.3 Percent fluorescence of acridine in different concentrations of solutions: the non-micelle forming ethyl acetate (t), two known surfactants, emulphagene (m) and SDS (s), three aqueous plant extracts, *Ceratiola* (e), *Calamintha* (a) and *Conradina* (c), and ursolic acid (u).

water for 3 days at 8°C; the leaves were neither ground nor treated with organic solvents. The leachates were then filtered before testing in 1.2×10^{-5} M acridine solution. The results showed that leachates of both mints formed micelles, although *Calamintha* formed a tenside at a lower concentration than *Conradina* (Fig. 5.3) (Tanrisever, 1986).

The presence of significant quantities of the triterpene, ursolic acid, in both mints (as well as in *Ceratiola*) suggested that this compound, available commercially as a biological detergent, might be involved in micelle formation. Investigation with pure ursolic acid in acridine revealed micelle formation at high concentrations, similar to those of the *Conradina* extract, but not at lower concentrations, similar to the *Ceratiola* and *Calamintha* extracts (Fig. 5.3).

A second test of ursolic acid interaction with monoterpenes employed bioassays with *Schizachyrium* where solutions were prepared as follows: (1) ursolic acid and the monoterpene saturated simultaneously in water, (2) the monoterpene saturated in water, (3) ursolic acid saturated in water and (4) water. The three monoterpenes, β-pinene, camphor and borneol, produced quite different results (Table 5.5). Borneol was much more inhibitory when dissolved in the ursolic acid solution. Camphor and pinene showed little activity with or without ursolic acid, although pinene was generally less inhibitory with ursolic acid than alone. Ursolic acid

TABLE 5.5 *Percent germination and mean radicle length of germinating seedlings of* Schizachyrium scoparium *in saturated solutions of ursolic acid, boreol, camphor and β-pinene, alone and in mixtures*[a]

Treatment	Germination (%)	Radicle length (cm)
Water	17.0 ± 2.8	4.1 ± 0.0
Ursolic acid (UA)	20.0 ± 5.7	5.3 ± 0.2*
Borneol + UA	4.0 ± 0.0*	2.2 ± 1.8*
Borneol	8.0 ± 4.2*	2.8 ± 0.2*
Camphor + UA	14.5 ± 0.7	6.0 ± 1.6
Camphor	20.0 ± 1.4	5.2 ± 1.7
Pinene + UA	19.5 ± 2.1	4.6 ± 0.1*
Pinene	19.5 ± 3.5	3.7 ± 0.1*

[a] The ursolic acid control is tested against the water control; all other solutions are tested against the ursolic acid control. Significant differences are indicated by '*' for $p \leq 0.05$ and by '**' for $p \leq 0.01$.

alone had little effect, being only slightly stimulatory to radical growth but not to germination, in comparison to distilled water.

5.5 SCRUB–SANDHILL ANALOGUES

The best documented case of allelopathy occurring among plants as a community phenomenon is the chaparral of southern California (Muller, 1966, 1969; Muller *et al.*, 1964, 1968; Rice, 1974). Despite numerous challenges to Muller's interpretation of allelopathy (Bartholomew, 1970; Halligan, 1973; Harper, 1975; Kaminsky, 1981), alternative explanations have been shown plausible in only a few instances. It is somewhat remarkable that all the traits of the Florida scrub community, listed in Table 5.1, are shared by the chaparral despite Florida and California differing drastically in soil types, topography and climate. Furthermore, the chaparral occurs adjacent to grasslands and forb communities, just as the scrub is contiguous with sandhill, and fire frequencies of 1–5 years convert scrub to sandhill (Richardson, 1977; Christensen, 1981; Kalisz and Stone, 1984) or chaparral to grassland (Kay, 1960; Biswell, 1974; Zedler *et al.*, 1983).

However, the analogy is not perfect. In Florida, the herbaceous community (sandhill) is more abundant than the shrubland (scrub), whereas in California the shrubland (chaparral) is dominant. In Florida the graminoid species are native; in California they are European introductions. Historically, the ignition of fires in Florida has a natural origin, whereas in California it has an anthropogenic one (Komarek, 1964, 1974).

A second scrub–sandhill analogue is found in the high elevations of the tropics. Vegetation and ecological processes are even less well documented for these communities, but it is known that neotropical paramo is dominated by fire-sensitive shrubs (Janzen, 1973; Smith, 1981, 1984). Repeated burning of the paramo produces a herbaceous community dominated by grasses and sedges (Williamson *et al.*, 1986). Furthermore, several lines of indirect evidence suggest that allelopathy is likely to be involved as well. Annual rings indicate that a fire every 5–10 years is sufficient to maintain the grassland community, but fire-free intervals greater than 15 years will cause reversion to shrub dominance.

The common process in these three pairs of vegetation types is a herbaceous community that burns much more frequently than an adjacent shrub community, and shrubs that are eliminated by the more frequent fires. In such a selective regimen any factor that increases a shrub's ability to retard fires will provide an advantage. Allelopathy is only one mechanism of preventing herbaceous growth around the shrubs. Shading and nutrient competition are others. However, allelopathy offers at least two advantages as a mechanism of fuel reduction; first, it may pre-empt rather than reduce the growth of the graminoids; and second, it may operate beyond the dripline of the shrub canopy or beyond the roots, if allelochemicals can diffuse through the soil or spread as litter.

Our interpretation of the allelopathy–fire interactions for the scrub and sandhill in Florida implies a community level phenomenon that differs somewhat from that postulated by Muller *et al.* (1968) for chaparral. They argued for a 'fire cycle' on a local site where a fire in the shrub crowns was followed by germination of annuals, and later, the gradual suppression of annuals as the shrubs recovered and produced toxins again. While such cycles are evident in the chaparral, the herb community is merely transitional during periods of shrub dominance. It is common for extremely large areas to be occupied by one vegetation type or the other – be it scrub or sandhill, chaparral or grassland, paramo or grassland – for decades, centuries or even millennia. During periods of vegetational stability the two ongoing interactive processes are easily overlooked: shrub species colonizing the grasslands and usually being eliminated in surface fires, and fires in grasslands burning to shrublands, often to be extinguished at the ecotone. Once established, the pattern of the communities would be maintained although ecotones might expand and contract episodically with natural changes in fire frequency or habitat disruption, as has been suggested for Florida (Kalisz and Stone, 1984) and for California (Zedler *et al.*, 1983).

While we have described the allelopathy–fire interactions at a community level, none of our interpretations requires abandoning individual selection. The potential advantage of frequent fires to sandhill individuals and the potential disadvantage of light fuels around scrub individuals are clear. However, the spread of fires and the heat release is such that individuals may be highly influenced by their immediate neighbours

(Mutch, 1974); temperatures and mortality rates may vary over distances of only a few metres when fuel loads vary spatially (Williamson and Black, 1981; Rebertus *et al.*, 1989). Obviously, one large allelopathic shrub has a higher probability of surviving a surface fire in the sandhill if it is part of a patch of alleopathic shrubs than if it is alone. Likewise, a bunch grass, dependent on fire to eliminate woody competitors, is more likely to be burned if it occurs in a large continuous stand of grasses than if it is growing alone. Therefore, while individual selection is not negated, there are group or community level effects that influence the selection coefficient of an individual, based on its neighbours. This is apparent not only in regard to the effects of fire, but also to the non-additive effects of allelochemicals (Table 5.5).

5.6 BIOTIC CONTROL AND ALTERNATIVE EQUILIBRIA

Through time some patches of sandhill, surrounded by scrub, have become scrub in the absence of fire (Kalisz and Stone, 1984), and likewise, patches of scrub have suffered successive fires and been converted to sandhill. Because the dominant species of each community have evolved traits to perpetuate themselves, transitions from scrub to sandhill and vice versa may be rare. Nevertheless, persistence at a site, sensu Connell and Sousa (1983), is not guaranteed.

In this regard, it is noteworthy that both Laessle (1968) and Harper (1914) concluded that the current distribution of scrub vegetation is associated with past and present coastal shorelines. The prevalence of scrub along present shorelines indicates that it is tolerant of salt spray and, therefore, occurs on some sites not suitable for sandhill vegetation. However, scrub communities persisting on the central ridge of Florida have not been shorelines for at least 5000–20 000 years (Kurz, 1942; Laessle, 1958, 1968; Alt and Brooks, 1965; Brooks, 1972; Long, 1974). Isolated from salt spray, these plants must have another mechanism to persist among the sandhill communities, and the only plausible mechanism at present is allelopathy. Some circumstantial evidence suggests that the rate of conversion of scrub to sandhill may be greater than conversion of sandhill to scrub. But even if the rates are different, the net replacement of scrub by sandhill is so slow that climatic shifts are equally likely to occur before non-coastal scrub sites are converted to sandhill vegetation. Therefore, we support the view that scrub and sandhill communities represent alternative configurations that are as stable as the climate in which they occur, and the stability results from biotic interactions of fire and allelopathy.

ACKNOWLEDGEMENTS

This material is based upon work supported by the US Department of Agriculture, Competitive Grants Program for Forest and Rangeland Renewable Resources under Agreement No. 88-33520-4077.

REFERENCES

Abrahamson, W. G. (1984) Post-fire recovery of Florida Lake Wales Ridge vegetation. *Amer. J. Bot.*, **7**, 9–21.

Alt, D. and Brooks, H. K. (1965) Age of Florida marine terraces. *J. Geol.*, **73**, 406–11.

Bartholomew, B. (1970) Bare zone between California shrub and grassland communities: the role of animals. *Science*, **170**, 1210–12.

Bell, D. T. (1974) The influence of osmotic pressure in tests for allelopathy. *Trans. Ill. State Acad. Sci.*, **67**, 312–17.

Biswell, H. H. (1974) The effects of fire on chaparral, in *Fire and Ecosystems* (eds T. T. Kozlowski and C. E. Ahlgren), Academic Press, New York, pp. 321–64.

Brooks, H. K. (1972) The geology of the Ocala National Forest, in *Ecology of the Ocala National Forest* (eds S. C. Snedaker and E. A. Lugo), USDA For. Serv., Atlanta, Georgia, pp. 81–92.

Burns, R. M. (1968) Sand pine: a tree for west Florida's sandhills. *J. For.*, **66**, 561–2.

Burns, R. M. (1972) Sand pine: distinguishing characteristics and distribution, in *USDA Forest Service, Sand Pine Symp. Proc.*, USDA For. Serv. Gen. Tech. Rep. SE-2, pp. 13–27.

Christensen, N. L. (1981) Fire regimes in southeastern ecosystems, in *Fire Regimes and Ecosystem Properties* (eds H. A. Mooney, T. M. Bonnickson, N. L. Christensen, J. E. Lotan and W. A. Reiners), USDA For. Serv. Gen. Tech. Rep. WO-26, Washington, DC, pp. 112–36.

Connell, J. H. and Sousa, W. P. (1983) On the evidence needed to judge ecological stability or persistence. *Amer. Nat.*, **121**, 789–924.

de la Peña, A. C. (1985) Terpenoids from *Conradina canescens* (Labiatae) with possible allelopathic activity. MS Thesis, Louisiana State Univ., Baton Rouge.

Doren, R. F., Richardson, D. R. and Roberts, R. E. (1988) Prescribed burning of the sand pine scrub community. *Fla. Sci.*, **50**, 182–92.

Fischer, N. H., Tanrisever, H., de la Peña, A. and Williamson, G. B. (1987) The chemistry and allelopathic mechanisms in the Florida scrub community. *Proc. Plant Growth Regul. Soc. Amer.*, **14**, 192–208.

Halligan, J. P. (1973) Bare areas associated with shrub stands in grassland: the case of *Artemisia californica*. *Bioscience*, **23**, 429–32.

Harper, R. M. (1914) Geography and vegetation of northern Florida. *Ann. Rep. Fla. State Geol. Surv.*, **6**, 163–391.

Harper, J. L. (1975) Allelopathy. *Q. Rev. Biol.*, **50**, 493–5.

Janzen, D. H. (1973) Rate of regeneration after a tropical high elevation fire. *Biotropica*, **5**, 117–22.

Johnson, A. F., Abrahamson, W. G. and McCrea, K. D. (1986) Comparison of biomass recovery after fire of a seeder (*Ceratiola ericoides*) and a sprouter (*Quercus inopina*) species from south-central Florida. *Amer. Midl. Nat.*, **116**, 423–8.

Kalisz, P. J. and Stone, E. L. (1984) The longleaf pine islands of the Ocala National Forest, Florida: a soil study. *Ecology*, **65**, 1743–54.

Kaminsky, R. (1981) The microbial origin of the allelopathic potential of *Adenostoma fasciculatum* H and A. *Ecol. Monogr.*, **51**, 365–82.

Kay, B. L. (1960) Effect of fire on seeded forage species. *J. Range Manage.*, **13**, 31–3.

Komarek, E. V. (1964) The natural history of lightning. *Proc. Tall Timbers Fire Ecol. Conf.*, **3**, 139–83.

Komarek, E. V. (1974) Effects of fire on temperate forests and related ecosystems: southeastern United States, in *Fire and Ecosystems* (eds C. E. Ahlgren and T. T. Kozlowski), Academic Press, New York, pp. 251–77.

Kurz, H. (1942) Florida sand dunes and scrub, vegetation and geology. *Fla. Geol. Surv., Geol. Bull.*, **23**, 1–154.

Laessle, A. M. (1958) The origin and successional relationship of sandhill vegetation and sand-pine scrub. *Ecol. Monogr.*, **28**, 361–87.

Laessle, A. M. (1968) Relationships of sand pine scrub to former shorelines. *Q. J. Fla. Acad. Sci.*, **30**, 269–86.

Long, R. W. Jr (1974) Origin of the vascular flora of south Florida, in *Environments of South Florida: Past and Present* (ed. J. Gleason), *Mem. Miami Geol. Soc.* **2**, 28–36.

Muller, C. H. (1966) The role of chemical inhibition (allelopathy) in vegetational composition. *Bull. Torrey Bot. Club*, **93**, 332–51.

Muller, C. H. (1969) Allelopathy as a factor in ecological process. *Vegetacio*, **18**, 348–57.

Muller, C. H., Hanawalt, R. B. and McPherson, J. K. (1968) Allelopathic control of herb growth in the fire cycle of California chaparral. *Bull. Torrey Bot. Club*, **95**, 225–31.

Muller, C. H., Muller, W. H. and Haines, B. L. (1964) Volatile growth inhibitors produced by aromatic shrubs. *Science*, **143**, 471–3.

Mulvania, M. (1931) Ecological survey of a Florida scrub. *Ecology*, **12**, 528–40.

Mutch, R. W. (1970) Wildland fires and ecosystems – a hypothesis. *Ecology*, **51**, 1046–51.

Myers, R. L. (1985) Fire and the dynamic relationship between Florida sandhill and sand pine scrub vegetation. *Bull. Torrey Bot. Club*, **112**, 241–52.

Nash, G. V. (1895) Notes on some Florida plants. *Bull. Torrey Bot. Club*, **22**, 141–61.

Platt, W. J., Evans, G. W. and Rathbun, S. J. (1988) The population dynamics of a long-lived conifer (*Pinus palustris*). *Amer. Nat.*, **131**, 491–525.

Rebertus, A. J., Williamson, G. B. and Moser, E. B. (1989) Longleaf pine and turkey oak mortality in Florida xeric sandhills. *Ecology*, **70**, 60–70.

Reynolds, T. (1975a) Characterization of osmotic restraints on lettuce fruit germination. *Ann. Bot.* (*London*), **39**, 791–6.

Reynolds, T. (1975b) pH restraints on lettuce fruit germination. *Ann. Bot.* (*London*), **39**, 797–805.

Rice, E. L. (1974) *Allelopathy*. Academic Press, New York.

Richardson, D. R. (1985) Allelopathic effects of species in the sand pine scrub of Florida. PhD Dissertation, Univ. of South Florida, Tampa.

Richardson, D. R. (1977) Vegetation of the Atlantic Coastal Ridge of Palm Beach Country. *Florida. Fla. Sci.*, **40**, 281–330.

Richardson, D. R. (1988) Sand pine: an annotated bibiliography. *Fla Sci.*, **52**, 65–93.

Richardson, D. R. and Williamson, G. B. (1988) Allelopathic effects of shrubs of the sand pine scrub on pines and grasses of the sandhills. *For. Sci.*, **34**, 592–605.

Smith, A. P. (1981) Growth and population dynamics of *Espletia* (Compositae) in the Venezuelan Andes. *Smithsonian Cont. Bot.*, **48**, 1–45.

Smith, A. P. (1984) Postdispersal parent–offspring conflict in plants: antecedent and hypothesis from the Andes. *Amer. Nat.*, **123**, 354–70.

Stout, M. and Tolman, B. (1941) Factors affecting the germination of sugar-beet and other seeds, with special reference to the toxic effects of ammonia. *J. Agric. Res.*, **63**, 687–713.

Tanrisever, N. (1986) Plant germination and growth inhibitors from *Ceratiola ericoides* and *Calamintha ashei*. PhD Dissertation, Louisiana State Univ., Baton Rouge.

Tanrisever, N., Fischer, N. H. and Williamson, G. B. (1988) Menthofurans from *Calamintha ashei*: effects on *Schizachyrium scoparium* and *Lactuca sativa. Phytochemistry*, **27**, 2523–6.

Tanrisever, N., Fronczek, F. R., Fischer, N. H. and Williamson, G. B. (1987) Ceratiolin and other flavonoids from *Ceratiola ericoides. Phytochemistry*, **26**, 175–9.

Veno, P. A. (1976) Successional relationships of five Florida plant communities. *Ecology*, **57**, 498–508.

Webber, H. J. (1935) The Florida scrub, a fire-fighting association. *Amer. J. Bot.*, **22**, 344–61.

Whitney, M. (1898) The soil of Florida. *US Dept. Agric. Bull.*, **13**, 14–27.

Williamson, G. B. and Black, E. M. (1981) High temperatures of forest fires under pines as a selective advantage over oaks. *Nature*, **293**, 643–4.

Williamson, G. B. and Richardson, D. R. (1988) Bioassays for allelopathy: measuring treatment responses with independent controls. *J. Chem. Ecol.*, **14**, 181–7.

Williamson, G. B., Schatz, G. E., Alvarado, A. *et al.* (1986) Tropical high elevation fire: effects of repeated burning. *Trop. Ecol.*, **27**, 62–9.

Wolff, T. (1981) The solvent dependent fluorescence quantum yield of acridine as a probe for critical micelle concentrations. *J. Colloid Interface Sci.*, **83**, 658–60.

Zedler, P. H., Gautier, C. R. and McMaster, G. S. (1983) Vegetation change in response to extreme events: the effect of short interval between fires in California chaparral and coastal scrub. *Ecology*, **64**, 809–18.

CHAPTER 6

Allelopathic effects of cruciferous plants in crop rotation

A. M. Grodzinsky[†]

6.1 INTRODUCTION

In recent years, cruciferous crops have become widespread in temperate zone agriculture due to their fodder and oil potential (Persanowski, 1970; Gonet, 1970; Makowski and Schutt, 1973; Uteush, 1979). The popularity of their cultivation is mainly based on economic considerations, since they are grown as additive or intermediate crops and, therefore, require no additional investment. Also, the cultivation of crucifers is quite convenient with the agricultural implements available. Further, they also have a high coefficient of seed germination and tolerance towards various unfavourable environmental conditions. Some of the commonly grown species in the Ukraine are winter rape, *Brassica napus* (var. *Oleifera* D. C.), spring colza, *B. campestris* L., winter colza, *B. rapa* (var. *Oleifera biennis*) and oilradish, *Raphanus sativus* (var. *Oleifera metzg*). Cruciferous plants represent about 4.23% of the total plant species (Malyshev, 1972). These are well represented in most Arctic regions, south of the European territory of the USSR, the republics of Central Asia and partially along the Okhotsk seashore. However, their numbers are comparatively fewer in the humid environment of the north-west USSR, the Far East, as well as in the Caucasus and mountainous regions of central Asia. Thus, their versatility in various environmental conditions is quite evident.

When introduced in rotation with crops, the yield has been found to be 17–20% higher than monoculture (Uteush, 1979). Winter rape and other crucifers have been found to suppress weeds, leading to a reduction in the use of herbicides (Grummer, 1955; Jiménez-Osornio and Gliessman, 1987). They are also known to support the development of soil microflora and assist the survival of useful insects. Many cruciferous plants have a high content of sulphur-containing compounds, oils, glycosides, glucosinolates as well as indole and other physiologically active components. Besides providing a specific odour and taste, their presence may also play

[†] Sadly, A. M. Grodzinsky is now deceased.

Allelopathy: Basic and applied aspects Edited by S. J. H. Rizvi and V. Rizvi
Published in 1992 by Chapman & Hall, London ISBN 0 412 39400 6

an important role in the crop rotation system. In order to verify this possibility, various aspects of cruciferal allelopathy were studied.

6.2 ALLELOPATHIC ACTIVITY OF FORAGE CRUCIFERS

Some cruciferous species are known to show a very high allelopathic activity. We noted a high activity of *Crambe tataria* Sebeok, which enabled it to invade dense steppe grassland very easily, but owing to the same reason it cannot form monospecial stands since its own plants begin to suffer from severe soil sickness. Similarly, *Brassica nigra* has also been found to exhibit high allelopathic activity (Bell and Muller, 1973). However, investigations conducted in our laboratory (Yurchak *et al.*, 1977) with some cultivars of winter rape and colza showed a relatively low allelopathic activity. For this, the effects of aqueous extracts of various plant parts were studied on seed germination and seedling growth of winter wheat, maize and radish. Among all the extracts tested, the water extract of influorescence inhibited seed germination between 26.5 to 79.5% as compared with the control. Leaf and stalk extracts showed almost similar allelopathic activities and suppressed seed germination by 60%. A moderate inhibition of radish seed germination, ranging from 20 to 30%, was caused by rape root extract. However, seed germination and subsequent seedling growth was slightly stimulated in wheat and was significantly promoted in maize. There have also been earlier reports that rape and related species do not exhibit any negative allelopathic effect. Martin and Rademacher (1960) observed that when grounded rape roots were added to soil, these caused an increase in the growth of wheat root. The extracts from the soil in which rape residues had been composted, firstly caused a slight growth retardation, followed by an induction.

In the rape roots, mustard oil glycosides (mainly, of β-phenylmustard oil) have been found. However, they are rather rapidly destroyed in soil, producing phenylethylamine. Rape is sometimes proposed as an effective means of biological control against such a strong weed as couch grass, *Agropyrum repens* (L.) PB. However, in this instance it is probable that some mechanism other than the direct influence of allelopathically active rape substances is involved.

Some authors found that 'Brassins', or the 'Brassin Complex' extracted from rape pollen, produced a powerful growth-accelerating effect when applied to young pinto bean plants, and thus was found a new class of plant growth regulators. As many as 14 native Brassinosteroids have since been found in many kinds of plants, specially in crucifers (Mandava *et al.*, 1978). However, their role in agriculture is yet to be thoroughly investigated.

As forage crucifers do not contribute to the accumulation of poisonous and physiologically active substances in soil, and do not lead to soil sickness, they may therefore be recommended as good predecessors to any agricultural crops, provided that there is a sufficient water supply.

These tests, as well as field observations, have indicated that allelopathic activity of forage crucifers range from moderate to strong; and before they are recommended for any rotation system, attention must be paid to this aspect.

6.3 INTERACTION OF FORAGE CRUCIFEROUS SPECIES WITH MICROORGANISMS

A study of various groups of microorganisms colonizing the cruciferous rhizosphere has provided some very interesting data which are summarized in Table 6.1. It was observed that the highest quantity of bacteria is present during summer time, when there is an initiation of flowering, i.e. during the period of highest root secretion, after which their numbers considerably decrease. This tendency was found throughout the years of the experiment, although the absolute quantity of microorganisms notably varied, depending on the climatic conditions. Of all the microorganisms found in rape and colza rhizospheres, 90% were Gram-negative and non-sporous bacilli, while coccal and bacillary forms occurred only randomly. Particularly widespread were the representatives of *Pseudomonas* and mycobacteria. *Pseudomonas* seems to play a major role in the establishment of a specific microbiological balance under cruciferous crops. It further seems that bacteria are particularly active during the initial stages of the destruction of rape and colza organic remains, because their population has been found to be enormous during this period. According to recent reports, *Pseudomonas* actively produce a fluorescent pigment, siderophore, which binds iron necessary for phytopathogenic fungi. Siderophore transforms iron into an accessible form for the fungi. This is an interesting aspect of indirect allelopathy of cruciferae, which has yet to receive its deserved attention.

The number of fungi in the cruciferae rhizosphere was gradually increased by the end of vegetative growth phase and their number was always found to be significantly higher than in the soil without crucifers (Table 6.1). This clearly indicates that fungi also take an active part in the decomposition of residues and substances produced by the crucifers. The predominant fungus in this experimental series was identified as belonging to *Mucor* species.

Another test series dealt with the microbial population during decomposition of cruciferous residues in soil. Crucifers are mainly used as fodder while the remains are ploughed in to enrich the soil with organic substances. The decomposition of plant residues in soil is accompanied by the development of saprophytic microflora as this provides a good substrate for these microorganisms. However, their quantity and group ident-

TABLE 6.1 *Quantitative composition of rhizospheric microflora of winter rape and winter colza determined at different growth phases*

		Bacteria (million g^{-1})			Fungi (thousands g^{-1})			Actinomycetes (thousands g^{-1})		
Crops	Number of cultivars	May	June	July	May	June	July	May	June	July
Winter rape	2	4.1	11.5	7.3	21.5	83.0	95.0	14.5	5.0	5.0
	6	7.1	12.9	16.8	23.0	50.0	61.0	0.0	18.3	30.0
	10	8.8	27.8	26.4	14.5	39.0	81.2	9.0	1.0	1.0
	12	27.8	78.9	26.0	18.2	120.0	141.5	11.0	16.8	32.0
	14	4.0	85.3	40.3	8.0	65.8	85.0	2.5	13.2	64.0
	18	4.1	112.5	11.8	7.0	15.0	33.0	1.0	40.5	57.0
	21	1.3	8.9	40.4	41.0	16.8	72.0	7.5	17.3	20.0
	24	12.3	47.7	15.6	12.0	19.4	138.0	14.5	5.1	10.0
	27	1.6	28.4	18.2	7.5	35.1	412.2	1.5	11.3	20.0
	29	8.5	33.8	12.1	9.5	14.0	105.0	4.0	18.7	31.0
Winter colza	30	8.4	41.3	44.3	8.0	115.0	534.3	10.0	20.3	30.0
Bare fallow		1.33	12.1	11.8	3.8	5.0	12.0	1.5	1.1	3.5

May: phase of renewed growth. June: budding and flowering phases. July: wax rapes phase.

ity were shown to vary, depending upon different external and internal factors.

6.4 EFFECT OF CRUCIFEROUS PLANTS ON PHYTOPATHOGENIC ORGANISMS

An attempt was made to find out any possible direct effect of forage cruciferous species on pathogenic organism affecting principal crops. As a test object we selected *Fusarium oxysporum* f. *vasinfectum* which causes wilt in thin-fibre cotton plants, cultivated in Turkmenia. From the very beginning of cotton cultivation in these regions, there was an indication that, owing to the pest attack, a prolonged cultivation was not possible. To overcome such problems, a cotton–alfalfa rotation was adopted. For the last seven decades, cotton–alfalfa rotation has been in practice which involves 3–4 years of cotton cultivation followed by 2–3 years of alfalfa plantation. However, this practice could not effectively counteract the pest infestation. Therefore, to search for a suitable replacement of alfalfa, we studied the effect of extracts from various crop residues, root exudates and soils collected from under the canopy of these crops, on the growth of *F. oxysporum* (*in vitro*). The findings are summarized in Tables 6.2, 6.3 and 6.4.

Alfalfa showed extremely low efficiency with regard to various forms of pathogens and their reproduction. Surprisingly, earlier reports on the strong inhibitory allelopathic potential of alfalfa (Naumova, 1959; Nielsen *et al.*, 1960; Grant and Sallans, 1964; Pederson, 1965; Stepanova, 1971; Marchain *et al.*, 1972; Stepanova and Prutenskaya, 1975; Miller, 1982, 1983) proved to be ineffective in relation to *F. oxysporum*. A comparison of the effects of alfalfa and cotton showed that spore germination

TABLE 6.2 *Effect of root exudates from various crop plants on* Fusarium oxysporum *f.* vasinfectum

	In vitro tests (% of control)		*In vivo* tests (% of control)	
Crops	Number of germinated spores	Germ tube length	Number of germinated spores	Germ tube length
Alfalfa	119.6	63.4	91.0	62.9
Barley	22.7	60.5	58.4	44.4
Colza	32.7	34.2	77.2	39.6
Cotton ASh-25	84.1	120.0	85.5	89.5
Rape	23.5	42.4	59.1	42.2

TABLE 6.3 *Fungicidal effect of aqueous extracts of crop plant residues on* Fusarium oxysporum *f.* vasinfectum

Crops	*In vitro* tests (% of control)		*In vivo* tests (% of control)	
	Number of germinated spores	Germ tube length	Number of germinated spores	Germ tube length
Alfalfa	137.0	89.7	90.1	72.5
Barley	30.5	32.2	39.4	38.4
Colza	49.2	29.7	59.1	40.5
Cotton ASh-25	85.7	79.3	78.7	86.0
Rape	55.6	39.9	53.2	39.2

TABLE 6.4 *Effect of extract of soils collected from various crop fields on* Fusarium oxysporum *f.* vasinfectum

Crops	*In vitro* tests (% of control)		*In vivo* tests (% of control)	
	Number of germinated spores	Germ tube length	Number of germinated spores	Germ tube length
Rape	65.5	101.7	30.8	36.9
Colza	116.3	55.9	25.2	42.0
Barley	55.7	55.5	62.7	63.3
Cotton ASh-25	74.5	51.9	90.2	96.9

and tube growth is favoured by alfalfa, and is greater than cotton. Hence, it may be suggested that during the course of systematic application of alfalfa against *Fusarium* sp., the latter adapted itself to the allelochemicals produced by alfalfa. Such resistance developed by *Fusarium* in the course of time has thus completely jeopardized the prospects of using alfalfa as an aid to pest control in cotton cultivation. As an alternative to alfalfa, the inclusion of cruciferous plants in rotation has already been started for wilt suppression. Besides being ploughed in as green manure, their cultivation as one component of a rotation system has achieved 25–49% wilt control in cotton fields (Parishkura, 1968, 1970; Avazov, 1970; Marupov, 1986; Urupov *et al.*, 1986).

Pursianova and Prutenskaya (1987) have reported that spore germination and tube growth of *Fusarium* was suppressed by 33 and 41% by root exudates from colza and rape, respectively. Regressive characteristic changes of pathogen response to unfavourable conditions were also observed. In particular, the content of the macrospore cells was compressed, leading eventually to cell destruction. Even if the spores germinated, the germ tubes failed to develop and finally became lysed. When treated with rape and colza decay products, the spore germination, as well as the length of *F. oxysporum* germ tubes, decreased more than 50%. Extracts of soils collected from under the canopy of these species also suppressed *Fusarium* through inhibiting the spore germination and conidial growth. Interestingly, the addition of residues of these cruciferous plants did not exhibit any toxic effect on agronomically valuable groups of microorganisms; instead, the growth of cellulose-destroying actinomycetes was found to be enhanced. Thus, the cruciferous plants are a potential factor in influencing the microflora and reducing the harmful microorganisms in the soil. The results presented here clearly demonstrate that the inclusion of plants like rape and colza increases the usefulness of the rotational system. Since a wide range of soils had been investigated – from northern podzol to southern serozem and takyr soils – the above crops may be recommended as an aid to *Fusarium* control.

6.5 WEED CONTROL EMPLOYING CRUCIFEROUS PLANTS

There have been frequent reports concerning the reduction of weed population when winter and spring crucifers are employed as intermediate crops. Uteush (1979), in long-term experiments, established that weeds were reduced by 40.0% when winter rape was used as an intermediate crop. Similarly, a 48.7% reduction in weed population was recorded when rape was grown in spring. Bell (1970) and Bell and Muller (1973) have also recommended post-harvest cropping of cruciferous plants in corn fields, which leads to a reduction of weeds by 90–96%.

6.6 CONCLUSIONS AND PROSPECTS

It can be concluded that crucifers are effective intermediate crops that provide sanitation to fields from pests and unwanted microflora, which otherwise compete with higher plants. However, their overcropping may lead to an increase in pests and heterotrophic microflora population adapted to these species. Therefore, systems of incorporating cruciferous species into rotation systems should be fully examined before making any definite recommendation to this effect. Further, comprehensive studies about allelopathic compatibility of all components of a rotation system and allelopathic interactions between higher plants and microorganisms would be a significant leap forward to exploit allelopathy in optimizing the production of rotation systems.

REFERENCES

Avazov, I. (1970) Siderates and wilt disease in cotton. *Khlopkovodstvo*, **9**, 15.

Bell, D. T. (1970) Allelopathic effects of *Brassica nigra* on annual grassland. PhD Thesis, Univ. of California, Santa Barbara.

Bell, D. T. and Muller, C. H. (1973) Dominance of California Annual grasslands by *Brassica nigra*. *Amer. Mid. Nat.*, **90**, 227–99.

Gonet, L. (1970) Intensyfikacja produkcji pash zielonych i kryteria jej oceny. *Nowe rol.*, **4**, 17–19.

Grant, E. A. and Sallans, W. G. (1964) Influence of plant extracts on germination and growth of eight forage species. *J. Brit. Grass Soc.*, **19**, 191–7.

Grummer, G. (1955) *Die gegenseitige Beeinflussung höherer Pflanzen-Allelopathie*, Fischer, Jena.

Jiménez-Osornio, I. I. and Gliessman, S. R. (1987) Allelopathic interference in wild mustard (*Brassica campestris* L.) and broccoli (*Brassica oleracea* L. var. *italica*) intercrop agroecosystems, in *Allelochemicals: Role in Agriculture and Forestry* (ed. G. R. Waller), ACS Symp. Ser. 330, Amer. Chem. Soc., Washington, DC, pp. 262–4.

Makowski, N. and Schutt, J. (1973) Der Raphanbau auf Leichten Böden. *Feldwirtschaft.*, **6**, 226–32.

Malyshev, L. I. (1972) Floristic spectra of the Soviet Union, in *History of Flora and Vegetation of Eurasia*, Nauka, Leningrad, 17–40.

Mandava, N., Kozempel, M., Worley, J.F. *et al.* (1978) Isolation of Brassins by extraction of rape (*Brassica napus* L.). *Pollen. Ind. Eng. Chem. Prod. Res. Dev.*, **17**, 351–4.

Marchain, U., Birke, V., Dovtur, A. and Berman, T. (1972) Lucerne saponins as inhibitors of cotton seed germination. *J. Exp. Bot.*, **23**, 302–9.

Martin, P. and Rademacher, B. (1960) Experimentelle Untersuchungen zur Frage der Nachwirkung von Rapswurzelruckstanden. *Z. Acker- und Pflanzenbau*, **111**, 105–15.

Marupov, A. (1986) Influence of intermediate and sideral crops on microsclerocea germination. *Khlopkovodstvo*, **8**, 15–16.

Miller, D. A. (1982) Allelopathic effects of alfalfa. *North Amer. Symp. Allelopathy*, Urbana-Champaign, Illinois (Abstr.).

Miller, D. A. (1983) Allelopathic effects of alfalfa. *J. Chem. Ecol.*, **9**, 1059–72.

Naumova, A. N. (1959) Accumulation of toxic substances in seroziom soils under alfalfa and their influence on the microflora, in *Mikrobiologia na sluzhbie selskogo khoziaistwa*, Selkhozgiz, Moscow, pp. 39–43.

Nielsen, K. F., Cuddy, T. and Woods, W. (1960) Allelopathic potential of ground cover species on *Pinus resinosa* seedlings. *Can. J. Plant Sci.*, **40**, 188–97.

Parishkura, N. C. (1968) Sideral crops and their utilization in Tajik SSR. *Khlopkovodstvo*, **12**, 13–17.

Parishkura, N. C. (1970) Intermediate crops in cotton crop rotations in the Vakhsh valley. *Khlopkovodstvo*, **10**, 22–7.

Pederson, M. W. (1965) Effect of alfalfa saponin on cotton seed germination. *Agronomy J.*, **57**, 516–17.

Persanowski, R. (1970) Mozlivosti rozsherzenia uprawy poplonow w Polsce. *Nowe rol.*, **2**, 1–3.

Pursianova, I. M. and Prutenskaya, N. I. (1987) Allelopathic activity of soils from under cotton and alfalfa, in *Methodological Problems of Allelopathy*, Naukova Dumka, Kiev, pp. 91–100.

Stepanova, L. P. (1971) Effect of inhibitors of alfalfa cultivars AzSKHI-1 and AzNIKhI-262 on the growth and development of successive crops. *Izv. Akad. Nauk Azerbaijan SSR, Biol. Series*, **1**, 43–8.

Stepanova, L. P. and Prutenskaya, N. I. (1975) On allelopathic properties of alfalfa. *Fiziologo-biochemicheskie osnovy vzaimodeistvia pastenii v fitotsenozakh*, Naukova Dumka, **6**, 69–72.

Urupov, I., Mirzabayev, M. and Khasanov, T. (1986) Glauconite and wilt resistance in cotton. *Khlopkovodstvo*, **9**, 23–4.

Uteush, Yu. A. (1979) *Rape and Forage in Forage Production*. Naukova Dumka, Kiev.

Yurchak, L. D., Uteush, Yu. A. and Omelchenko, T. V. (1977) Microflora and specific allelopathic properties of fodder plants from the Cruciferae Family, in *Plant–Microorganism Interactions in Phytocoenoses*, Naukova Dumka, Kiev, pp. 161–8.

CHAPTER 7

Improving yield of corn–soybean rotation: role of allelopathy

E. Sarobol and I. C. Anderson

7.1 INTRODUCTION

The beneficial effects of crop rotations, especially those involving cereals and legumes, have been recognized by agriculturalists and exploited by farmers for centuries. Besides providing reduced nitrogen compounds, leguminous plants may benefit cereals in crop rotations by: improving soil physical properties; reducing soil erosion; and suppressing weeds, insects and diseases. A rotation system, which has been practised world wide, involves corn (*Zea mays* L.) and soybean (*Glycine max* (L.) Merr.). Several reports indicated that continuous cropping of corn and soybean results in yields that are depressed below the level obtained when a crop is rotated with other crops (Slife, 1976; Welch, 1977, 1985; Barber, 1978; Mulvaney and Paul, 1984). The advantage of crop rotation persisted even beyond optimum levels of all management inputs.

The chemical warfare among higher plants has long been implicated and documented mostly in natural ecosystems, and crop plants in agricultural fields are no exception. In fact, there are numerous instances where allelopathic interference has been suggested. Because of the increased interest in cropping system and reduced tillage, either in developed or developing countries, understanding the interactions among agricultural crops would be of importance. The aim of this chapter is to discuss the potential allelopathic effects of corn and soybean in crop rotation with special emphasis on their effects on the subsequent corn crop. Allelopathic activities appear to be greater at higher latitudes and decrease towards the equator. The discussion will be based on information from investigations in the United States as well as in Thailand.

7.2 THE BENEFIT OF CROP ROTATION

Evidence from research data in the corn belt of the United States has shown that yields of corn in rotation have consistently exceeded yields of

Allelopathy: Basic and applied aspects Edited by S. J. H. Rizvi and V. Rizvi
Published in 1992 by Chapman & Hall, London ISBN 0 412 39400 6

TABLE 7.1 *Effect of N-fertilizer on grain yield (t ha^{-1}) of corn in different crop rotation at the Clarion-Webster Research Center in Iowa (mean of 1979–81) (after Voss and Shrader, 1984)*

N (kg ha^{-1})	Cont. C	Crop rotation							
		CCCO$_x$	CCCO$_x$	COCO$_x$	CSCO$_x$	CSCO$_x$	CCOM	CCOM	COMM
0	3.83	7.91	4.90	5.15	7.97	8.22	10.86	7.16	11.05
67	6.65	9.23	7.60	7.03	9.73	10.04	11.11	9.73	10.92
135	9.29	10.48	10.11	9.48	10.74	10.67	11.17	10.55	11.24
200	10.55	10.86	9.86	10.42	10.86	10.86	11.24	10.61	11.24

C = corn, S = soybean, O$_x$ = oat with alfalfa catch crop,
M = alfalfa. The crop in the rotation represented by the yield is shown in italic.

TABLE 7.2 *Corn yields (kg ha^{-1}) when planted following either corn or soybean, on the Morrow Plots at Urbana, Illinois (after Odell* et al.*, 1982)*

Year	Corn yield after		% Increase for corn after soybean
	Corn	Soybean	
1969	8543	9101	6.5
1971	9158	10594	15.7
1973	9321	11259	20.8
1975	10117	12005	18.7
1977	7088	8925	25.9
1979	7270	9496	30.6
1981	10807	11547	6.8
Average	8900	10418	17.0

corn in continuous corn (Shrader and Voss, 1980; Langer and Randall, 1981). From the Iowa rotation studies, Voss and Shrader (1984) reported that yields of corn following legumes were greater than that of continuous corn (Table 7.1). This result occurred regardless of the amount of nitrogen (N) furnished.

Regarding corn–soybean rotation, Odell *et al.* (1982) reported that corn following soybean yielded 17% more than continuous corn (Table 7.2). Higgs *et al.* (1976) also reported that continuous corn yielded 502 kg ha^{-1} less than the yield of corn following soybean. This similar magnitude of yield differences was also obtained by Slife (1976). Data from a four-year corn–soybean rotation at Urbana, Illinois, indicated that the average yield of corn in rotation was 1444 kg ha^{-1} greater than that of continuous corn (Welch, 1977). The benefit of soybean on corn yield in crop rotation was also demonstrated by Barber (1978). Contradictorily, Crookston

(1984) found no benefit of soybean, in crop rotation, on corn yield. Instead, he found that continuous corn yielded better than corn in rotation. This result was also in agreement with data of Suwanarit and Suwanarat (1985), from Thailand, who studied the cumulative effects of crop rotation and reported a yield advantage of the tenth-year corn in the continuous corn system over the tenth-year corn in a corn–soybean rotation system. Additionally, farmers in Pattananikom District, Lop Buri Province, Thailand, observed a yield reduction of corn when planted after sunflower (*Helianthus annuus* L.) as compared with continuous corn.

The rotation effect is not limited to only corn and soybean or cereals and legumes. Several crops such as wheat (*Triticum aestivum* L.), sunflower, cotton (*Gossypium hirsutum* L.), sorghum (*Sorghum bicolor* L.) and barley (*Hordeum vulgare* L.) have been reported to benefit from being preceded by different crops in a previous year (Crookston, 1984). Langer and Randall (1981) stated that preceding corn by wheat without supplemental N was as effective as preceding corn by soybean.

To date, agriculturalists are unable to explain the phenomenon of yield enhancement in crop rotations. The effect is generally called the 'rotation effect' and is sometimes called the 'memory effect' or 'residual effect' (Cruse *et al.*, 1985) of crop rotation. Allelopathic phenomenon may be involved in the residual effect of crop rotation. As suggested by Cruse *et al.* (1985), the most important need, currently, is to determine if the increased yield of corn following soybean is due to (1) the adverse effects of corn on corn, (2) the stimulatory effects of soybean on the subsequent corn, or (3) a combination of both effects.

The authors shall explain the yield enhancement in corn–soybean rotation system based on inhibitory allelopathic chemicals (IACs) from corn and the stimulatory allelopathic chemicals (SACs) from soybean which affect the subsequent corn. One should not overlook the possibilities that corn potentially contains SACs which promote growth and yield of corn or different crops. The opposite is also possible for soybean: that is, soybean potentially contains IACs which suppress growth and the yield of the subsequent soybean or different crops.

7.3 STIMULATORY ALLELOPATHIC POTENTIAL OF SOYBEAN AND OTHER LEGUMES

Few studies on stimulatory effects of soybean and other legumes have been reported. Sukthumrong *et al.* (1985), in Thailand, conducted a long-term experiment to study the effect of green manure–chemical fertilizer combinations for corn cropping system. Corn was planted after fallow (no legume), lab-lab (*Lablab purpureous* L.), thornless mimosa (*Mimosa invisa* L.) and peanut (*Arachis hypogaea* L.) combining with no fertilizer,

TABLE 7.3 *Grain yield of corn (6-year average, 1980–5) as affected by legume mulching and compost. Weight of crop residues (5 year average, 1981–5) is also shown (after Phetchawee* et al., *1985)*

	Grain yield ($kg\ ha^{-1}$)		Crop residue ($kg\ ha^{-1}$)	
Management	+F	−F	+F	−F
No mulch	3420	1928	2880	2680
Rice straw mulch	5192	2529	4880	3600
Crotalaria juncea mulch	3742	2440	4410	3700
Vigna umbellata mulch	5205	3870	8340	7760
Compost incorporate	5205	3600	4880	4180
Mimosa invisa mulch	4841	4050	10640	10220

+F = with fertilizer, 100–100–50 kg $N–P_2O_5–K_2O\ ha^{-1}$.
−F = without fertilizer.

156 kg ha^{-1} of rock phosphate (P), 150 kg ha^{-1} $(NH_4)_2SO_4$ + 150 kg ha^{-1} of rock phosphate (low NP) and 312 kg ha^{-1} of 16–20–0 compound fertilizer (high NP). They reported yield advantage of corn after legumes under high NP treatments over corn after fallow. Additionally, Phetchawee *et al.* (1985) studied the long-term effect of legume mulching on associated corn yields in the system in which legumes were planted simultaneously with, or 1–3 weeks after, corn and were chopped down at a later stage of corn growth. They reported an improvement in the yield of corn with legume mulching over corn with no mulch (Table 7.3). A similar yield improvement of corn with legume mulching and fertilizations over corn with no mulch was also reported, in Thailand, by Watanabe *et al.* (1989). Chopped alfalfa (*Medicago sativa* L.), when applied to soil, was reported to stimulate the growth of tomato (*Lycopersicon esculentum* L.), cucumber (*Cucumis sativum* L.), lettuce (*Lactuca sativa* L.) and several other crops (Einhellig, 1985a). In one test, applications of 117 kg of chopped alfalfa per hectare increased an early tomato yield by 10 metric tonnes per hectare (Einhellig, 1985b). Similarly, Anderson (1984) stated that work done by Ries and his colleagues showed that alfalfa meal placed beside rows of corn increased the yield of corn more than corn fertilized with nitrogen. The extract from alfalfa showed a growth-regulating activity similar to triacontanol (Ries and Houtz, 1983). Thus, several researchers proposed that triacontanol was responsible for the increased corn yield that applied with alfalfa meal. Since, apparently, triacontanol released from the alfalfa was the cause of the residue-induced yield increases, this fits in with the spectrum of allelopathic effects (Einhellig, 1985b). According to Cruse *et al.* (1985), soybean leaves also contain high levels of triacontanol.

Granato *et al.* (1983) identified potential allelochemicals from soybean root extracts of five cultivars at seven stages of growth (emergence to senescence). Fifteen major peaks, identified by using high-pressure liquid chromatography, were found in all cultivars at most growth stages and two peaks out of the fifteen were identified tentatively as daidzein and coumestrol. The concentration of compounds extracted from roots increased dramatically as plants approached nodulation and flowering, and then declined to the initial levels as the plants matured. Nelson (1985) extracted soil samples, either from plots uncovered or covered with plastic, which were used to plant soybean the previous year and bioassayed with duckweed (*Lemna minor* L.) strain 5. The results showed that allelopathic compounds, both inhibitory and stimulatory, could be isolated from the soybean soils. Unexpectedly, more inhibitory than stimulatory activities were observed under the condition of the experiments. The most bioactive soil sample was collected between soybean rows from uncovered plots. However, no attempt was made to identify compounds within active fractions.

Most of the work done by Iowa scientists has emphasized the stimulatory effects of soybean. Assumpcao (1979) conducted a greenhouse experiment by incorporating soybean residues into soil and planted corn. He observed that soybean residues produced stimulatory effects on corn seedling height and weight. Kalantari (1981) also reported that seedling growth of corn was stimulated by soybean residues. He collected soil containing soybean residues from a harvested soybean field and extracted the soil with various solvents. Possible areas of activity from the extracts were separated by column chromatography techniques. The chromatographed fractions were bioassayed by treating germination papers with a fraction and corn seedlings were grown on the papers. Measurements were done on coleoptile and radicle growth. The results revealed that radicle and coleoptile growth were significantly stimulated by compounds in fraction 2, 3 and 12. He also stated that the longer the soybean residues interacted with the soil, the greater was the stimulatory influence. The results from Nelson (1985), Assumpcao (1979) and Kalantari (1981) imply that root exudates, rain-leached substances and residues constitute the sources of stimulatory allelochemicals from soybean.

The previous experiments were done in the greenhouse or laboratory. Field experiments concerning stimulatory effects of soybean, in Iowa, were first done by Sarobol (1986). In the previous year, late, early and mid-season soybean cultivars were planted in a field, and three dates of planting (15 May, 30 June and 30 July) were imposed on the mid-season cultivar. Late and early cultivars were planted on 15 May. At the end of the growing season soybean was harvested and the residues were either removed (− residue) out of the field or shallowly incorporated (+ residue) into the soil with a moldboard plough. Oat (*Avena sativa* L.) was also planted in the same field as the control plots. These sets of

TABLE 7.4 *Soybean cultivar and residue effects on grain yield (%) of subsequent corn (after Sarobol, 1986)*

Treatments	Experimental year		Mean
	1983/1984	1984/1985	
Cultivars			
Early	116.5	102.5	109.5
Mid-season			
15 May	113.4	122.5	117.8
30 June	112.6	105.5	109.1
30 July	111.5	113.4	112.5
Late	115.5	106.0	110.8
$\overline{X}$	114.0	110.0	112.0
SE	4.1	7.8	
LSD (0.05)[a]	12.6[c]	8.9[c]	
LSD (0.05)[b]	6.3	4.5	
Residues			
+ (incorporated)	115.0	107.4	111.2
– (removed)	112.8	112.4	112.6
SE	2.0	2.4	
LSD (0.05)[a]	7.8[c]	7.3[c]	
LSD (0.05)[b]	5.5	5.2	

[a] Least significant difference ($p \leq 0.05$) among treatment means.
[b] Least significant difference ($p \leq 0.05$) between a treatment mean and 100% (the value for the oat control plots).
[c] Not significant according to F-test.

treatments enable one to evaluate cultivar, ontogenic and crop residue effects on the subsequent crop, besides evaluating the stimulatory effect of the previous soybean. In the subsequent year, the entire field (including the oat plots) was planted to a corn hybrid and grain yield, and other characteristics were measured. Grain yields of corn planted on the previous soybean fields were expressed as the percentage of averaged yield of corn (100%) planted on the previous oat plots. The results indicated that all treatments boosted yield of the subsequent corn (Table 7.4). However, the degree of yield stimulation varied between years. Soil samples were taken from the plots planted with soybean in the previous year. The extracts from soil samples, obtained from an extraction method different from that of Nelson (1985), were bioassayed with duckweed. Only in the 1983/1984 experiment the soil extracts from soybean plots stimulated the growth of duckweed. The opposite was true for the 1984/1985 experiment. This inconsistency is unexplained. Although the results from the duckweed bioassay are indefinite, the results from field experiments were directed towards the potential occurrence of SACs from soybean. Evidently, soybean increases the yield of the following corn.

7.4 INHIBITORY ALLELOPATHIC POTENTIAL OF CORN AND OTHER CEREALS

Several cereal crops have been shown to have allelopathic effects among themselves and other crops. Continuous cropping of wheat in the United States, Australia and the Union of Soviet Socialist Republics has sometimes resulted in repression of wheat yield, particularly when the straw was left on the soil surface under a cool, wet season (Einhellig, 1985a). Inhibitory allelochemicals from crop residues have been reported to account for this type of yield repression (Nielson *et al.*, 1960; Lawrence and Kilcher, 1962; Guenzi *et al.*, 1967). A greenhouse study by Assumpcao (1979) also demonstrated that corn residues inhibited corn seedling growth. Earlier, Guenzi *et al.* (1967) reported that corn and sorghum residues had considerable toxic materials at harvest and required about 22–28 weeks of decomposition before the water-soluble portion of the residues was relatively non-toxic. The organic compounds released from the decomposing corn residues in soil, phytotoxic to seed germination, were identified by Chou and Patrick (1976). Under ideal decomposition conditions, *p*-coumaric acid – one of the five major phenolic acids released from corn, wheat, sorghum and oat residues – might be released in sufficient quantity in localized areas of soil to affect plant growth (Guenzi and McCalla, 1966).

Einhellig *et al.* (1985) and Rice (1984) stated that inhibition or stimulation of plant growth by allelochemicals from crop residues depends on the age of the residues, stage of decomposition, concentrations of compounds and cultivars. The effect of corn residue age and placement on early corn growth was studied by Yakle and Cruse (1983). Both fresh and partially decomposed corn residues were collected from the field, oven dried, ground and placed as horizontal bands in sand- or soil-filled acrylic containers. Corn seeds were planted with one residue band placed either above, below or at seed depth. A control treatment, non-residue, was also included. They showed that fresh residues reduced root and shoot weights more than partially decomposed corn residues. The effects were most notable when roots came into contact with a residue layer. This result emphasized the importance of root residue proximity. The reduction of growth could not be explained by nitrogen immobilization by microorganisms during residue decomposition and residue effect on water management (Yakle and Cruse, 1983). The higher salt concentration in the residue extract from corn had little effect on its allelopathic role to corn seedlings (Yakle and Cruse, 1984).

Garcia and Anderson (1984) demonstrated that corn residues varied in its allelopathic effects throughout a growing season. Soil samples were monthly collected, during April to September, from a second-year corn field in Iowa and bioassayed with corn seedling growth. The results showed that the April, August and September soil samples were inhibitory to corn growth, whereas June and July soil samples had stimulatory effects. Patrick *et al.* (1963) have also reported periodic phytotoxic activity from the extracts of wheat and barley residues.

Pollen grains of corn have a high allelopathic potential. Jiménez *et al.* (1983) reported that pollen grains of corn at the concentration that is commonly present in the soil, strongly inhibited the growth of several plants. The least inhibition (28%) was observed in unsterilized soil and the strongest inhibition (54%) was found in sterilized soil. The result stressed the role of microorganisms in dissipating the allelopathic effect of corn pollen.

Phytotoxicity may vary with corn genotypes. Hicks and Peterson (1981) investigated the yield response of a corn–soybean rotation as opposed to continuous corn in southern Minnesota. They planted five corn hybrids with each hybrid following itself, other hybrids and soybean. The objective of their study was to determine if different corn hybrids in rotation would stimulate the yield comparable to that obtained by corn–soybean rotation. They reported (Table 7.5) autotoxicity of hybrids. Corn yields averaged 14% higher for corn following soybean when compared with continuous corn of the same hybrid. The increased yield from rotating corn hybrids was neither large nor consistent. The yield difference was only 2.5% higher when rotating corn hybrids as compared with continuous corn. Variation in the allelopathic effects of different corn hybrids on the subsequent corn has also been reported by Sarobol (1986). With the same experiments as previously mentioned for soybean, late, early and mid-season corn hybrids were planted in a field, the previous year, with three planting dates (15 May, 30 June and 30 July) were imposed on the mid-season corn hybrids. Corn residues were either removed or incorporated into the soil at the end of the season. The following year, the same field was entirely planted with only a corn hybrid. The subsequent corn yields were expressed as the percentage of averaged yield of corn (100%) planted on the previous oat plots. The expectations of the experiments are that the strongest and weakest inhibitions come from late and early corn hybrids, respectively. The moderate inhibition comes from mid-season corn hybrid with May-planted corn > June-planted corn > July-planted corn. The results in Table 7.6 showed that, obviously, the late corn hybrid strongly depressed the yield of the subsequent corn. The expectation was also met by the allelopathic effect of May-planted, mid-season hybrid. It was not surprising that the June- and July-planted, mid-season hybrid appeared to stimulate corn yields, as did the early corn hybrid. The findings of this study are in line with Crookston (1982) in the sense that, depending on hybrid and maturity, the previous corn hybrid

TABLE 7.5 *Grain yield (kg ha^{-1}) of five corn hybrids grown continuously or in rotation with soybean and other corn hybrids (after Hicks and Peterson, 1981)*

Locations and years	Corn hybrids														
	H_1			H_2			H_3			H_4			H_5		
	H_1	S	X	H_2	S	X	H_3	S	X	H_4	S	X	H_5	S	X
Lamberton															
1979	7 716	7 584	7 534	6 605	8 080	7 207	5 901	6 805	5 876	7 596	7 151	6 969	6 178	6 730	6 391
1980	6 253	6 492	5 857	5 908	9 028	6 805	6 718	8 657	6 492	5 977	9 003	5 920	6 027	7 571	6 240
Weseca															
1979	10 246	9 116	10 007	10 547	10 516	10 566	9 875	10 013	10 020	10 509	11 363	10 905	9 260	10 114	10 597
1980	4 545	5 330	4 483	3 315	6 058	5 123	5 776	7 559	6 178	5 443	5 280	4 759	8 306	9 737	8 406
Avg. over locations and years	7 190	7 130	6 970	6 593	8 420	7 425	7 067	8 258	7 141	7 381	8 199	7 138	7 442	8 538	7 908

H_1 = W64A × W117; H_2 = Pioneer brand 3780; H_3 = W64A × OH43; H_4 = A632 × A619; H_5 = Mo17 × B73; S = soybean; X = mean yield following other hybrids.

TABLE 7.6 *Corn hybrid and residue effects on grain yield of the subsequent corn (after Sarobol, 1986)*

Treatments	Grain yield (%)
Hybrids	
Early	102.0ab
Mid-season	
15 May	96.2bc
30 June	101.0ab
30 July	106.1a
Late	90.7c
SE	2.1
LSD (0.05)[a]	6.3
LSD (0.05)[b]	4.5
Residues	
+ (incorporated)	100.8
– (removed)	97.6
SE	1.9
LSD (0.05)[a]	5.5[c]
LSD (0.05)[b]	3.9

[a] Least significant difference ($p \leq 0.05$) among treatment means.
[b] Least significant difference ($p \leq 0.05$) between a treatment mean and 100% (the value for the oat control plots).
[c] Not significant according to F-test.
Grain yields followed by different letters, when compared with treatment means, differ significantly at $p < 0.05$.

may increase the yield of the different, subsequent corn hybrid. Sarobol (1986) concluded that maturity type and hybrid of corn, associated with plant size, were the prime factors affecting the subsequent corn yield. His conclusions support the ideas that the allelopathic effects of corn, either stimulatory or inhibitory, are related to the genotype and growth stage of crops.

Corn residues did not affect the yield of the following corn (Sarobol, 1986). This result is in agreement with Garcia (1983) who concluded that allelochemicals from corn were being released by living corn plants through root exudation or rain-leached substances. On the other hand, Crookston (1982) reported that corn residue did not reduce the grain yield of corn but rather increased it. He did not believe the idea that allelochemicals were involved in the yield enhancement of crop rotation and proposed that 'anti-corn bodies' might play a part in this phenomenon.

Apparently, the allelopathic effects of corn have been mainly inhibitory.

7.5 CROPPING SYSTEM APPROACH

Evidently, the yield difference between corn in continuous corn and corn following soybean is due to both the inhibitory effects of the previous corn and the stimulatory effects of the previous soybean. The relative contribution of IACs from corn and SACs from soybean on this yield difference is equivalent: 10% for late corn (Table 7.5) and, on average, 12% for soybean (Table 7.4). I. C. Anderson is now investigating the variations of inhibitory allelopathic effects of late corn hybrids on a subsequent corn.

Chui and Shibles (1984) intercropped soybean within corn rows under 70 × 30 cm spacing. They found that intercropped corn gave a significantly greater yield than monocropped corn. It is tempting (Chui and Shibles, 1984) to speculate that the presence of soybean actually enhanced the corn yield under this (favourable) spatial arrangement. Their speculation forms the basis of our investigation on a different intercropping experiment. We evaluated whether intrarow corn–soybean association in one year can increase yield of corn in the next year. Three soybean cultivars of maturity groups 00, II and VIII were planted in combination with three soybean plant densities (Sarobol, 1986). The results were inconsistent but were not entirely discouraging. There were cases, in other studies, where overyielding of corn occurred when corn was overseeded with legumes such as soybean and cowpea (Foster, 1986).

7.6 CONCLUSIONS AND PROSPECTS

The evidence led to the conclusion that allelopathy may play a role in agricultural systems. A stimulatory allelopathic role is played mainly by soybean and other legumes, whereas an inhibitory allelopathic potential is exerted by corn and other cereals. This review revealed that the increased yield of corn following soybean, as compared to the yield of continuous corn, was due to the combination of the adverse effects of corn on corn and the stimulatory effects of soybean on the subsequent corn. A problem arises as several attempts to extract or identify the allelochemicals were unsuccessful, despite the high technology available in today's world of science. Nevertheless, 'the rotation effect' still exists and can be utilized world wide.

We suggest that it is now appropriate to investigate intrarow cereal–legume association, living mulch and legume overseeding. The resulting systems of crop production, after being adapted, would be suitable for farmers in both developed and developing countries. Moreover, exploitation of allelopathy through the above-mentioned approaches, and others, is relevant to the 'organic' or 'alternative' agriculture that has received much attention recently. The senior author, in Thailand, is investigating the contribution of grain and forage legumes, as well as native legumes, in association with cereal as a means to boost the grain yield of subsequent crops and to increase the organic matter content of a soil.

The potential applications of allelopathy have been addressed by several reviewers (Rice, 1984; Einhellig, 1985a, b; Mandava, 1985; Putnam, 1985), and include: (1) the use of allelopathic crops to kill weeds; (2) the extraction of natural allelochemicals for use as bioregulators; (3) the coordinated use of herbicide and allelopathic residue; and (4) the use of a cropping sequence to avoid yield reduction or to increase the yield. All the above aspects deserve further exploration and exploitation.

REFERENCES

Anderson, I. C. (1984) From my view. *Soybean News*, **35**, 3.

Assumpcao, L. C. D. (1979) Allelopathic effects of four crop residues on corn seedlings. MS Thesis (unpublished), Iowa State Univ. Ames, Iowa.

Barber, S. A. (1978) Increased corn yields with soybean rotation. *Soybean News*, **29**, 1.

Chou, C. and Patrick, Z. Z. (1976) Identification and phytotoxicity of compounds produced during the decomposition of corn and rye residue in soil. *J. Chem. Ecol.*, **2**, 369–87.

Chui, J. A. N. and Shibles, R. M. (1984) Influence of spatial arrangements of maize on performance of an associated soybean intercrop. *Field Crop Res.*, **8**, 187–98.

Crookston, R. K. (1982) Field studies on the yield effect of corn and soybeans in rotation – are plant growth regulators involved? *Plant Growth Regulatory Soc. Proc.*, **9**, 137–43.

Crookston, R. K. (1984) The rotation effect: What causes it to boost yields? *Crops and Soils*, **2**, 12–14.

Cruse, R. M., Anderson, I. C. and Amos Jr, F. B. (1985) Residual effects of corn and soybean on the subsequent corn crop, in *Proc. World Soybean Research Conf. III* (ed. R. M. Shibles), Westview Press, Inc., Boulder, CO, pp. 1061–7.

Einhellig, F. A. (1985a) Effects of allelopathic chemicals on crop productivity, in *Bioregulators for Pest Control* (ed. P. A. Heldin), ACS Symp. Ser. 276, Amer. Chem. Soc., Washington, DC, pp. 109–30.

Einhellig, F. A. (1985b) Allelopathy – a natural protection, allelochemicals, in *CRC Handbook of Natural Pesticides: Methods. Vol I: Theory, Practice and Detection* (ed. N. B. Mandava), CRC Press, Florida, pp. 161–200.

Einhellig, F. A., Leather, G. R. and Hobbs, L. L. (1985) Use of *Lemna minor* L. as a bioassay in allelopathy. *J. Chem. Ecol.*, **11**, 65–72.

Foster, B. M. (1986) Effects of overseeded legumes on corn growth and grain yield. MS Thesis (unpublished), Iowa State Univ., Ames, Iowa.

Garcia, A. G. (1983) Seasonal variation in allelopathic effects of corn residue on corn and cress seedlings. PhD Dissertation (unpublished), Iowa State Univ., Ames, Iowa.

Garcia, A. G. and Anderson, I. C. (1984) Monthly variation in allelopathic effects of corn residue on corn seedling growth under three tillage practices. *Philipp. J. Crop Sci.*, **9**, 61–4.

Granato, T. C., Banwort, W. L., Porter, P. M. and Hassett, J. J. (1983) Effect of variety and stage of growth on potential allelochemical compounds in soybean roots. *J. Chem. Ecol.*, **9**, 1281–91.

Guenzi, W. D. and McCalla, T. M. (1966) Phenolic acid in oats, wheat, sorghum, and corn residues and their phytotoxicity. *Agron. J.*, **58**, 303–4.

Guenzi, W. D., McCalla, T. M. and Norstadt, F. A. (1967) Presence and persistence of phytotoxicity in wheat, oats, corn, and sorghum residues, *Agron. J.*, **59**, 163–5.

Hicks, D. R. and Peterson, R. H. (1981) Effect of corn variety and soybean rotation on corn yield. *Annual Corn and Sorghum. Res. Conf.*, **36**, 89–93.

Higgs, R. L., Paulson, W. H., Pendleton, J. W. *et al.* (1976) *Crop Rotation and Nitrogen*, Univ. of Wisconsin Res. Bul. R2761.

Jiménez, J. J., Schultz, A., Anaya, A. L. *et al.* (1983) Allelopathic potential of corn pollen. *J. Chem. Ecol.*, **9**, 1001–25.

Kalantari, I. (1981) Stimulation of corn seedling growth by allelochemicals from soybean residue. PhD Dissertation (unpublished), Iowa, State Univ., Ames, Iowa.

Langer, D. K. and Randall, G. W. (1981) Corn production as influenced by previous crop and N rate. *Agron. Abstr.*, 1981, 182.

Lawrence, T. and Kilcher, M. R. (1962) The effect of fourteen root extracts upon germination and seedling length of fifteen plant species. *Can. J. Plant Sci.*, **42**, 308–13.

Mandava, N. B. (1985) Chemistry and biology of allelopathic agents, in *The Chemistry of Allelopathy: Biochemical Interactions Among Plants* (ed. A. C. Thompson), ACS Symp. Ser. 268, Amer. Chem. Soc, Washington, DC, pp. 33–53.

Mulvaney, D. L. and Paul, L. (1984) Rotating crops and tillage. *Crops and Soils*, **36**, 18–19.

Nelson, L. S. (1985) Isolating potential allelochemicals from soybean-soil residues. MS Thesis (unpublished), Iowa State Univ., Ames, Iowa.

Nielson, K. F., Cuddy, T. F. and Woods, W. B. (1960) The influence of the extract of some crops and soil residues on germination and growth. *Can. J. Plant Sci.*, **40**, 188–93.

Odell, R. T., Walker, W. M., Boone, L. V. and Oldham, M. G. (1982) *The Morrow Plots – A Century of Learning*, Illinois Agric. Exp. Stn. Bull. 775.

Patrick, Z. A., Toussoon, T. A. and Snyder, W. C. (1963) Phytotoxic substances in arable soil associated with decomposition of plant residues. *Phytopathology*, **53**, 152–61.

Phetchawee, S., Vibulsukh, N., Theppoolpon, M. and Masarngsan, W. (1985) Long-term effect of mulching with fertilizer under corn–legumes intercropping on crop yield and improvement of soil chemical and physical properties, in *Thailand National Corn and Sorghum Program 1985 Annual Report*, Department of Agriculture, Department of Agricultural Extension, Kasetsart University and International Maize and Wheat Improvement Center, pp. 204–12.

Putnam, A. R. (1985) Allelopathic research in agriculture: past highlights and potential, in *The Chemistry of Allelopathy: Biochemical Interactions Among Plants* (ed. A. C. Thompson), ACS Symp. Ser. 268, Amer. Chem. Soc., Washington, DC, pp. 1–8.

Rice, E. L. (1984) *Allelopathy*, 2nd edn, Academic Press, New York.

Ries, S. and Houtz, R. (1983) Triacontanol as a plant growth regulator, *Hortic. Sci.*, **18**, 654–62.

Sarobol, E. (1986) Allelopathic effects of corn and soybean on a subsequent corn crop. PhD Dissertation (unpublished), Iowa State Univ., Ames, Iowa.

Shrader, W. D. and Voss, R. D. (1980) Crop rotation vs monoculture: soil fertility. *Crops and Soils*, **32**, 8–11.

Slife, F. W. (1976) Economics of herbicide use and cultivars tolerance to herbicides. *Annual Corn and Sorghum Res. Conf.*, **31**, 77–84.

Sukthumrong, A., Chotechaungmanirat, S., Chancharoensuk, J. and Veerasan, V. (1985) Effect of green manure-chemical fertilizer combinations on soil fertility and yield of corn. *Int. Seminar on Yield Maximization of Feed Grains Through Soil and Fertilizer Management*, 12–16 May 1986, Bangkok, Thailand.

Suwanarit, A. and Suwanarat, C. (1985) Cumulative effects of cropping systems involving corn and legumes on the yields of the tenth-year crops, in *Thailand National Corn and Sorghum Program 1985 Annual Report*, Department of Agriculture, Department of Agricultural Extension, Kasetsart Univ. and International Maize and Wheat Improvement Center, pp. 199–203.

Voss, R. D. and Shrader, W. D. (1984) Rotation effects and legume sources of nitrogen for corn, in *Organic Farming: Current Technology and its Role in a Sustainable Agriculture* (eds D. F. Bezdicek *et al.*), ASA Spec. Publ. No. 46. Amer. Soc. of Agronomy, Madison, WI, pp. 61–8.

Watanabe, M., Chairoj, P., Masangsan, W. *et al.* (1989) *Studies on the improvement of soil productivity through incorporation of organic matter into upland soil of Thailand*, Tropical Agriculture Research Center, Ministry of Agriculture, Forestry and Fisheries, Japan and Department of Agriculture, Ministry of Agriculture and Cooperatives, Thailand.

Welch, L. F. (1977) Soybeans good for corn. *Soybean News*, **28**, 1–4.

Welch, L. F. (1985) Rotational benefits to soybeans and following crops, in *Proc. World Soybean Res. Conf. III* (ed. R. M. Shibles), Westview Press, Boulder, CO, pp. 1054–60.

Yakle, G. A. and Cruse, R. M. (1983) Corn plant residue age and placement effects on early corn growth. *Can. J. Plant Sci.*, **63**, 817–77.

Yakle, G. A. and Cruse, R. M. (1984) Effects of fresh and decomposing residue extracts on corn seedling development. *Proc. Soil Sci. Soc. Amer.*, **48**, 1143–6.

CHAPTER 8

Biochemical selection of sorghum crop varieties resistant to sorghum shoot fly (*Atherigona soccata*) and stem borer (*Chilo partellus*): role of allelochemicals

H. Alborn, G. Stenhagen and K. Leuschner

8.1 INTRODUCTION

The Semi Arid Tropics (SAT) is populated by more than 700 million people. Most of them are living on a subsistence level and are, for their food, dependent on the limited production of small farms. To be able to feed the increasing population of the SAT, new high-yielding varieties of traditional crops like sorghum have to be developed. Since most farmers have limited resources, and in order not to run into the same environmental difficulties experienced in the developed countries, insecticide use is mostly not feasible and should be kept at a low level. Host plant resistance is therefore essential for the stability of newly developed varieties and hybrids. At ICRISAT (International Crops Research Institute for the Semi Arid Tropics, Patancheru, Andhra Pradesh, India) a broad germplasm base has been collected. For sorghum alone, 28 000 germplasm accessions are available. The identification of resistance sources in the germplasm material by available methods is both intricate and time consuming.

It is now generally accepted that plant resistance against insects (as well as to other organisms) are, to the greatest extent, chemical in origin (Bernays and Chapman, 1978; Kubo and Nakanishi, 1977; Harborne, 1982; Kubo and Hanke, 1985). The defence system is usually not based on a single chemical compound but on a whole spectrum (Lundgren *et al.*, 1982). Therefore a successful parasite must be able to handle the total chemical defence system of the plant. This means that methods must be designed to enable the plant breeder to distinguish genotypes based on their spectra of chemical compounds.

Allelopathy: Basic and applied aspects Edited by S. J. H. Rizvi and V. Rizvi
Published in 1992 by Chapman & Hall, London ISBN 0 412 39400 6

Breeders have been successful in producing vertically (single-gene) resistant plant material (for definition and further discussion of the terms horizontal and vertical resistance see: van der Plank, 1963; Robinson, 1976; Lundgren *et al.*, 1982). The insect populations in SAT are often very dense and many generations per growing season are produced. Vertically based plant resistance is exposed to fast genetic counteradaptations by the pest organisms and will therefore not be a successful approach to obtain stable resistant cultivars for the SAT. Horizontal resistance has not been used in practical plant breeding. An important reason for this is the lack of tools to characterize the chemical background of the resistance.

To develop new methods which allow single plant selection with a high probability that a specific genotype carries the resistance genes, a collaboration project was initiated between ICRISAT and the Department of Chemical Ecology, University of Göteborg, Sweden. The resistance of sorghum (*Sorghum bicolor*) to two severe pest insects, shoot fly (*Atherigona soccata*) and stem borer (*Chilo partellus*), was selected for the study.

Fourteen sorghum varieties were selected (Table 8.1). Five of them were less susceptible to shoot fly, six were less susceptible to stem borer and three were highly susceptible to both insects. All the 'resistant'

TABLE 8.1 Sorghum *germplasm lines identified as highly and as less susceptible to shoot fly (mean of four seasons) and stem borer (mean of six seasons) at the ICRISAT centre, Patancheru, India*

No.	Pedigree	Origin	Shoot fly incidence (%) Egg laying	Shoot fly incidence (%) Dead hearts	Stem borer incidence (%)
1	IS1082	India	45.3	38.5	45.3
2	IS2123	USA	40.5	35.0	30.6
3	IS4663	India	46.6	38.9	
4	IS5566	India	37.0	36.4	32.9
5	IS18551	Ethiopia	36.8	31.3	36.0
6	CSH-1	India	66.4	67.6	70.3
7	Swarna				
8	IS1044	India			32.9
9	IS2122	USA	45.5	40.7	35.8
10	IS2195	India	43.2	34.5	43.8
11	IS2205	India			40.6
12	IS2291	Sudan	43.5	42.7	31.7
13	IS5469	India	43.9	44.6	28.3
14	IS10795				

cultivars were shown in field tests, including several other cultivars, to be among the least susceptible to both insects, although none of them showed a total resistance. CSH-1 has been used in the field tests as a 'susceptible' reference. Unfortunately, for the cultivars Swarna and IS10795 no similar test data are available; in pilot tests both were considered to be as susceptible as CSH-1 and were therefore not subjected to lengthy field tests.

8.2 CORRELATION BETWEEN ALLELOCHEMICAL CONTENTS AND RESISTANCE

The sorghum shoot fly is specific to sorghum and related species. The females withhold egg laying when presented to other grass species (Ogwaro, 1978). In a number of studies, a non-preference for oviposition has been reported as the primary resistance for shoot fly in sorghum (Sharma *et al.*, 1977; Singh and Jotwani, 1980; Singh *et al.*, 1981; Taneja and Leuschner, 1985). The inference is that sorghum exhibits some specific oviposition-stimulating characteristics that are perceived by the plant at the plant surface. Contact chemoreceptors are present on the tarsi (Ogwaro and Kokwaro, 1981). After hatching the larva moves down the leaf blade and down the culm within the outer sheath, and finally bores into the growing point. Soto (1974), Blum (1972) and Raina *et al.* (1981) suggested the existence of an antibiosis in addition to a non-preference for oviposition.

The stem borer is less specific in its oviposition behaviour than the shoot fly. However, as a preliminary to oviposition, the insect touches the leaf surface with its antennae and tarsi in addition to its ovipositor. The eggs of the stem borer are usually laid on the lower leaves of sorghum, while the first instar larvae feed only on the whorl from the site at which the eggs were laid. During this phase of their life cycle, the larvae are influenced by the surface features of the plant.

The fact that the field tests (Table 8.1) showed a correlation between shoot fly and stem borer resistance makes it likely that the nature of the resistance is largely of the same kind. There are many facts that point to the chemical composition of the plant as the major cause for the resistance of sorghum to oviposition and larvae feeding by shoot fly and stem borer. Sorghum tissue has been shown to contain dhurrin, a non-deterrent cyanogenic glucoside, which, upon being bitten by locusts, gives off hydrogen cyanide and *p*-hydroxybenzaldehyde. Phenolic acid esters with a deterring effect on locusts have also been isolated from sorghum tissue (Woodhead and Bernays, 1978). Woodhead (1983) found *p*-hydroxybenzaldehyde and waxes with short carbon chains in the leaf wax with a contact acting and feeding deterrent effect on locusts.

Our strategy for the screening of the selected cultivars was, first, to select the best way of sampling (plant age, plant parts and sampling technique); second, to obtain chromatographic separation patterns of the samples; third, by the help of analogy analyses of the chromatographic data, to point out the active components; and finally, to identify them.

8.3 ANALYTICAL METHODS

8.3.1 Sampling

Methods were developed for the sampling of spontaneously emitted volatiles, tissue-related compounds and substances located on the leaf surface.

For the collection of leaf volatiles, the plants were placed in glass cylinders flushed with clean, charcoal filtered air. The volatiles emitted from the plants were sampled in adsorption tubes with TENAX TA. The technique is described in Stenhagen *et al.* (1987).

The tissue-related compounds were extracted using a water/methanol mixtures. For each sample a young but fully developed leaf was taken from three different plants. The leaves were cut in five centimetre pieces and extracted with 30 ml 50% water/methanol over night. Prior to analyses by HPLC the extract was concentrated by evaporation under vacuum to 1 ml. The injection volume was 10 µl.

The field samples collected in India were freeze dried before transport to the laboratory. Three different samples were taken from the plants: a young leaf, the growing point and the part just above the growing point. Ten plants from each cultivar were used. Before the extraction the plant material was grinded.

Two slightly different methods were used to collect components on the leaf surface: 'chloroform wash' and 'steam wash'.

To collect leaf waxes the chloroform wash method was used. It was carried out using a micro spray device utilizing the injector principle. Two ejector tubes are mounted at right angles to each other in a bracket. Compressed air or nitrogen is led through the horizontal tube, causing the solvent to be drawn up through the vertical tube to form a fine spray. For each cultivar, ten fully developed young leaves were sprayed on both sides over a beaker where the sample was collected. The beaker was washed by shaking twice with chloroform and the solutions transferred to a screw-capped tube. The final samples were concentrated to 200 µl by a gentle stream of nitrogen.

The steam wash technique was used to collect polar components on the leaf surface. Steam is produced in a pressure vessel and, via a teflon tube, sprayed over the leaf surface. The condensate was collected in a beaker. The solution was then filtered and stored in a screw-capped tube.

8.3.2 Chromatographic methods

Gas chromatographic analyses of the adsorption samples were made using a capillary gas chromatograph (Carlo Erba 2900) equipped with a modified injection system (Stenhagen *et al.*, 1987). The system is based on direct thermal desorption of the sample, via a cold trap, onto the column. The most significant properties of the inlet system are its ability to yield reproducible retention times and the lack of a solvent peak which may hide volatile components.

Liquid chromatography analyses were carried out using a gradient reversed phase system (Hewlett Packard 1084B) and UV detection at 200 nm. To achieve reproducible retention times and to retard polar compounds the mobile phase (water/acetonitrile) was buffered with potassium dihydrogen phosphate. In addition, an accurate equilibrium procedure between every chromatographic run had to be used. The detection at 200 nm was chosen in an effort to detect as much as possible of the content in the sample in a single run.

8.3.3 Analogy analyses

The analogy analysis implies looking for regularities in the observations made. The cultivars with separation patterns that were alike were brought into the same class. In our case there were two classes: 'resistant' and 'susceptible' cultivars. One method that is successfully used for chromatographic data is the SIMCA (Soft Independent Modelling of Class Analogy) method (Wold *et al.*, 1983). We have used this method to determine the existence of relevant chromatographic data and the importance of individual fractions for the class separation.

In spite of all efforts to obtain reproducible retention times, these vary for different chromatographic runs of the same component. Computer programs were developed to solve this problem. The pattern of peaks in the chromatograms is often very much alike. Some peaks exist in all the chromatograms and can be used as internal calibration peaks. For each reference peak, a mean value based on all runs was calculated. New retention time values were calculated for each peak in the data set by the use of straight-line expressions. The coefficients of the expressions were based on the values of the nearest reference peak on both sides of the peak to be calibrated. After calibration, a reference vector was created. This vector consisted of a series of retention time values and was used to match the fractions in a data matrix. For each chromatographic run, the retention time values of the fractions were compared with each value of the reference vector. The peaks that do not match were added to the nearest peak.

The result of the pre-processing was a data matrix with information, for all runs about the amount of each fraction, on one line. This matrix was the base for the multivariate data analysis.

The first question we intended to give an answer to, was: Is there, in the chromatographic separation patterns, any information related to the resistance? The SIMCA method and the principal components (PC) analysis are common methods of obtaining a view of multivariate data. Only a short presentation of the method for multivariate data analysis will be given here.

To obtain an eigenvector projection (also called PC plot or principal vector plot), the expression of the plane that approximates the data set in the best way (in the sense of the least squares) is calculated. In the plane of the projection, the coordinates of each point are calculated and the points are plotted in a diagram. The relevance of the data to the problem can often be seen already in this first plot.

The basic idea of the SIMCA method is that by the use of multivariate data of a group of similar objects, a proper class is well-approximated by a simple PC model. The dimensionality of the model is estimated so as to give the model predictive properties that are as good as possible. The fitting is made using the least squares criterion, i.e. the sum of the squared residuals is minimized for the data set of each class. The appropriate class can be determined when the distance of each object to the class model is compared with the typical distance of the class object to the same model. The SIMCA method provides a measure of relevance for the separated fractions (variables), the modelling power. It is based on how much each variable participates in the modelling of each class.

8.4 RESULTS

8.4.1 Volatiles

Eigenvector projections of the chromatographic data of the volatiles did not show any distinctive groups of stem borer or shoot fly resistant cultivars. This result corresponds well with the field observations because there is no indication of distance attraction of pest insects to sorghum. In other words: 'Arrival at the plant may be the result of a random process' (Chapman and Woodhead, 1985).

8.4.2 Leaf wash with chloroform

No significant difference between the susceptible and the resistant cultivars were shown in the analyses of chloroform leaf wash. A new group of very high boiling leaf surface compounds were detected, but because we could not correlate them to the resistance, they will not be further discussed here. We also included extracts from young seedlings of CSH-1 and IS 2146 (approximately one week old) because the shoot fly starts to attack when the seedlings are very young. These samples clearly differed from the rest mainly due to a less complex pattern. It was also shown that the cultivar CSH-1 contained a higher amount, than the cultivar IS2146, of a component not yet identified.

8.4.3 Leaf extracts

Prior to the field sampling on locations in India, leaf extracts (50% methanol/water) of fresh and dried leaves were compared. The qualitative differences between fresh and freeze-dried plant material were small, but some minor differences were observed. However, these differences were considered mainly to be caused by the much more efficient extraction when dried leaves were used. Due to these results we decided to freeze-dry the field samples.

Figure 8.1 shows the eigenvector projections of LC data obtained from dried leaves. The three susceptible cultivars, objects 6 (CSH-1), 7 (Swarna) and 14 (IS10795), are well separated from the others. Several repeated series were made. In some of them, the stem borer resistant (18–17) and the shoot fly resistant (2–5) cultivars showed a tendency to separate. Due to lack of computer power more than two series could not be compared at the same time. To see if there is a real class separation between these two groups more replicates must probably be included in the calculations. However, the field observations also showed an overlapping of these two groups of cultivars.

The LC data has been further analysed. Figure 8.1A shows a PC plot which includes data from all chromatographic peaks. Most of the variables (peaks) gave no contribution to the class separation and when the number of variables was reduced, according to their modelling power, an even better separation was achieved. The reliability of this reduction is dependent on the number of objects (number of runs) in each class, but the reduction, with the maintenance of the class separations, can in this case be done to less than nine variables, Fig. 8.1B. In the chromatogram in Fig. 8.2 the four best contributing fractions are marked, and their relative amounts (relative to the largest fraction in one of the cultivars) are plotted in Fig. 8.3. Data from two chromatographic series are shown. To eliminate the contribution from system changes within a series, the second series was run starting with cultivar number 14 (see Table 8.1) and ending with number 1. This means (see Fig. 8.3) that nos 1 and 28, nos 2 and 27 . . . nos 14 and 15 are the same cultivar.

The susceptible cultivars, CSH-1, Swarna and IS10795, contained almost twice as much of the dominating fraction 21 as did the others, indicating this to be a behaviour-activating compound. Also, fraction 79 existed in larger amounts in the susceptible than in the non-susceptible cultivars. Fractions 23 and 42 were both less abundant in the susceptible cultivars.

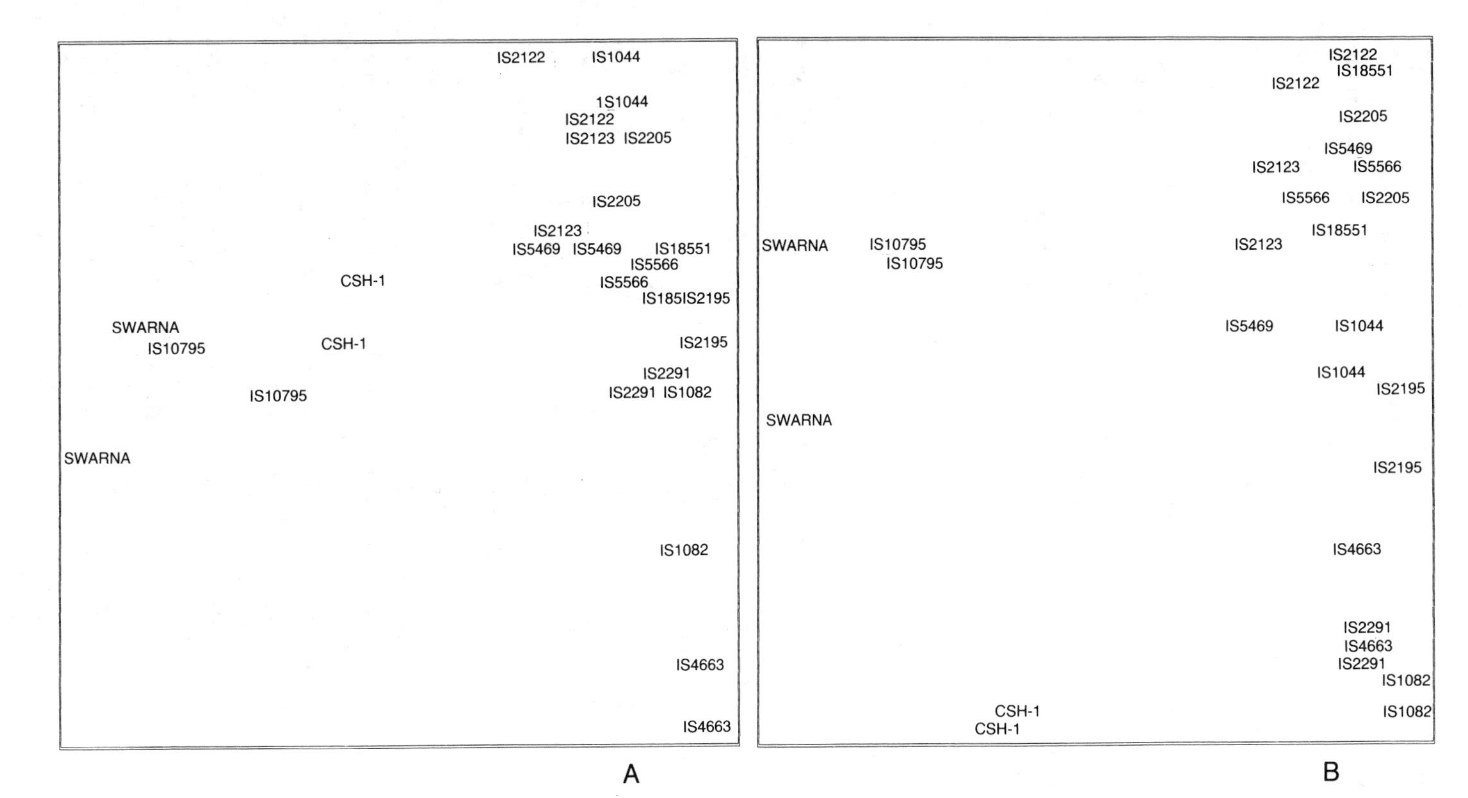

Fig. 8.1 Eigenvector projections (PC plot of two dimensions). Liquid chromatographic analysis (LC) of leaf extracts. A: 175 fractions included. B: Only nine fractions included. UV detection at 200 nm.

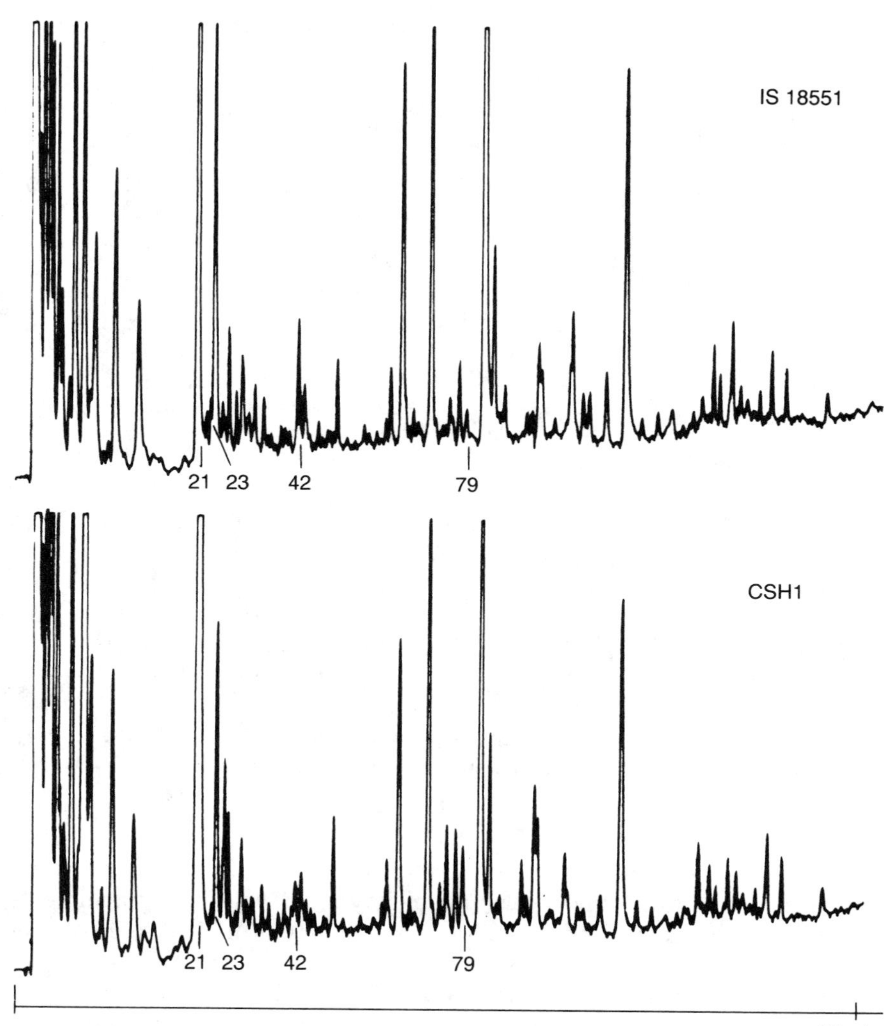

Fig. 8.2 LC separations of leaf samples of the 'resistant' cultivar IS18551 and the susceptible cultivar CSH-1. The four most class-separating fractions are marked in the figure.

8.4.4 Different plant parts

In order to look for differences in the chemistry of the (A) oviposition, (B) larval moving and (C) larval feeding areas of the plants, different parts of the plant were collected and analysed. As seen in Fig. 8.4, the three kinds of samples are gathered in three groups. The reason for these

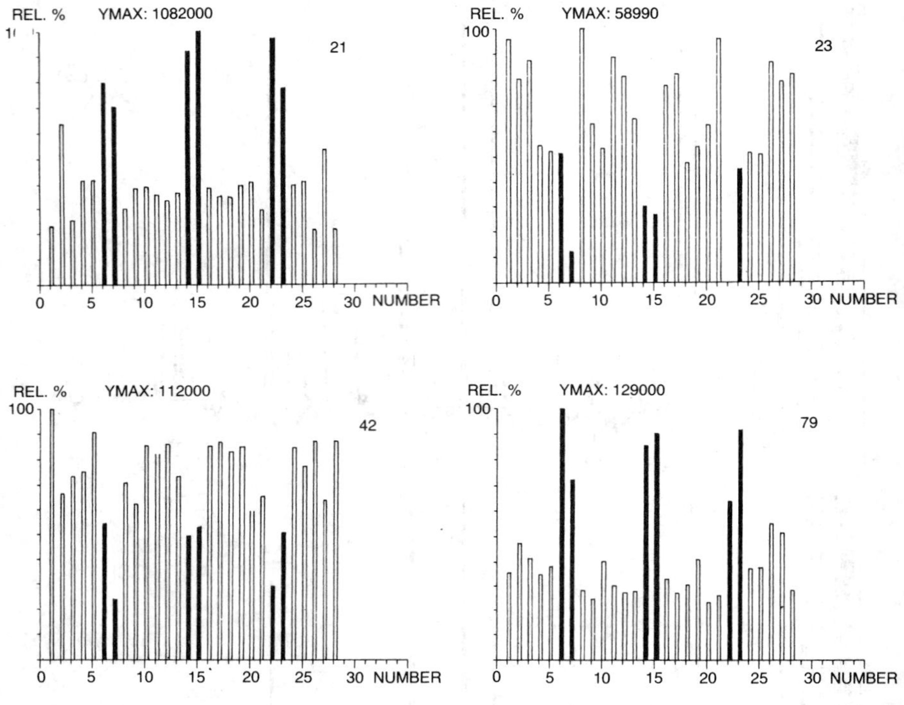

Fig. 8.3 Comparison of the relative amounts of the four most class-separating fractions in the leaf extract. The amount corresponding to the largest peak in each plot is set to 100%.

distinct groups was found in the complexity of the chromatographic pattern, e.g. the number of peaks differ from *c*. 170 in the leaf samples (A) down to two dominating peaks in the samples from the growing area (C). In Fig. 8.5 there is a clear correlation between tissue age and chemical complexity. In all three groups the susceptible cultivars are located at the centre of the figure, which means they are more alike.

8.4.5 Identifications

UV spectra of three of the active fractions and of the two reference substances – *p*-hydroxybenzaldehyde and *p*-hydroxymandelonitrile – were registered (see Fig. 8.6) using the scanning UV detector (stop flow type) integrated in the LC system (Hewlett Packard 1084B). p-Hydroxymandelonitrile is the genin part of dhurrin, a cyanogenic glucoside previously found in sorghum. The spectrum of fraction 21 shows a good resemblance to *p*-hydroxymandelonitrile, indicating that it could be the same or a similar compound, however, with not the same retention time.

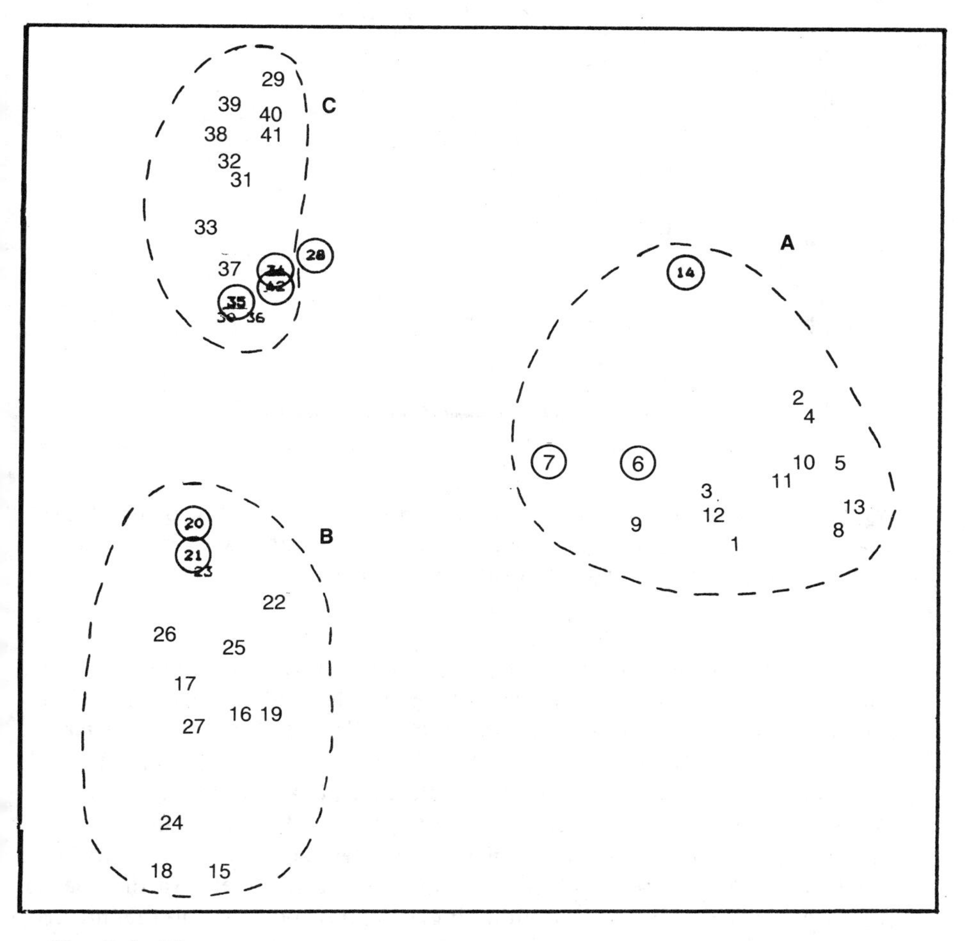

Fig. 8.4 Eigenvector projections (PC plot of two dimensions) of LC data of samples from different plant parts. Susceptible cultivars are encircled. A: The middle part of young fully developed leaves (oviposition area). B: The part of young leaves just above the growing point (moving area). C: The growing point (feeding area).

Also, fraction 79 shows similarities to *p*-hydroxymandelonitrile such as the high adsorption at 200 nm. Fraction 42 shows a spectrum similar to some of the spectra obtained earlier of phenolic acid glycosides. We did not succeed in obtaining a spectrum of fraction 23.

Fraction number 21 was collected by the use of the analytical LC system. The sample was evaporated to dryness and dissolved in methanol. In this way most of the buffer originating from the mobile phase could be removed. The sample was then analysed using a LC-MS system developed and constructed at the laboratory (Alborn and Stenhagen, 1985,

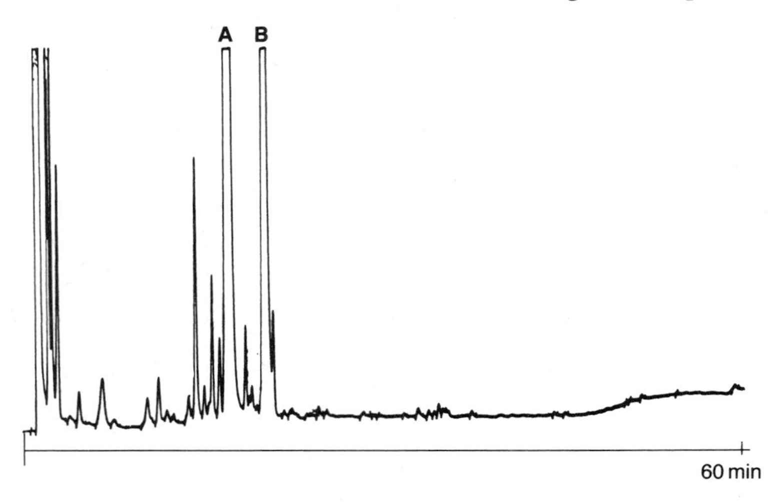

Fig. 8.5 LC separation of extracts obtained from the growing point of the Swarna cultivar. Note the less complex pattern compared with Fig. 8.2. A: Dhurrin (fraction no. 21) and B: *p*-hydroxybenzaldehyde.

1987; Stenhagen and Alborn, 1989). The system was developed for the analysis of plant allelochemicals and gives electron impact ionization spectra with reduced thermal fragmentation. The LC-MS spectra of fraction 21 is shown in Fig. 8.7. We have identified the compound to be dhurrin. The peak at m/z 311 is the molecular peak. The mass spectrum also shows typical glucose peaks at m/z 60, 61, 73, 91, 127, 145 and 163. There are two alternative ways of breakage of the glucosidic bondage, which results in the peaks at m/z 149 and 132, respectively. In the spectrum, the m/z 132 peak is dominating probably because of the nitrile group attached to the same carbon atom as the sugar. Free *p*-hydroxymandelonitrile shows a completely different fragmentation pattern compared with dhurrin. The free molecule is labile and easily rearranges to *p*-hydroxybenzaldehyde by the loss of hydrogen cyanide and giving intense peaks at m/z 121 and 122 which, because of the glucosidic bond, does not occur in the case of dhurrin.

8.4.6 Steam wash

The 'steam wash' extraction was carried out on young seedlings of CSH-1 (1 month), grown in a greenhouse. The following LC analyses gave one dominating component on the leaf surface, dhurrin. Woodhead (1983) reported *p*-hydroxybenzaldehyde to be the dominating compound in the leaf wax of young seedlings. Since we found no *p*-hydroxybenzaldehyde on the surface, we suppose it is not transported to the leaf surface by

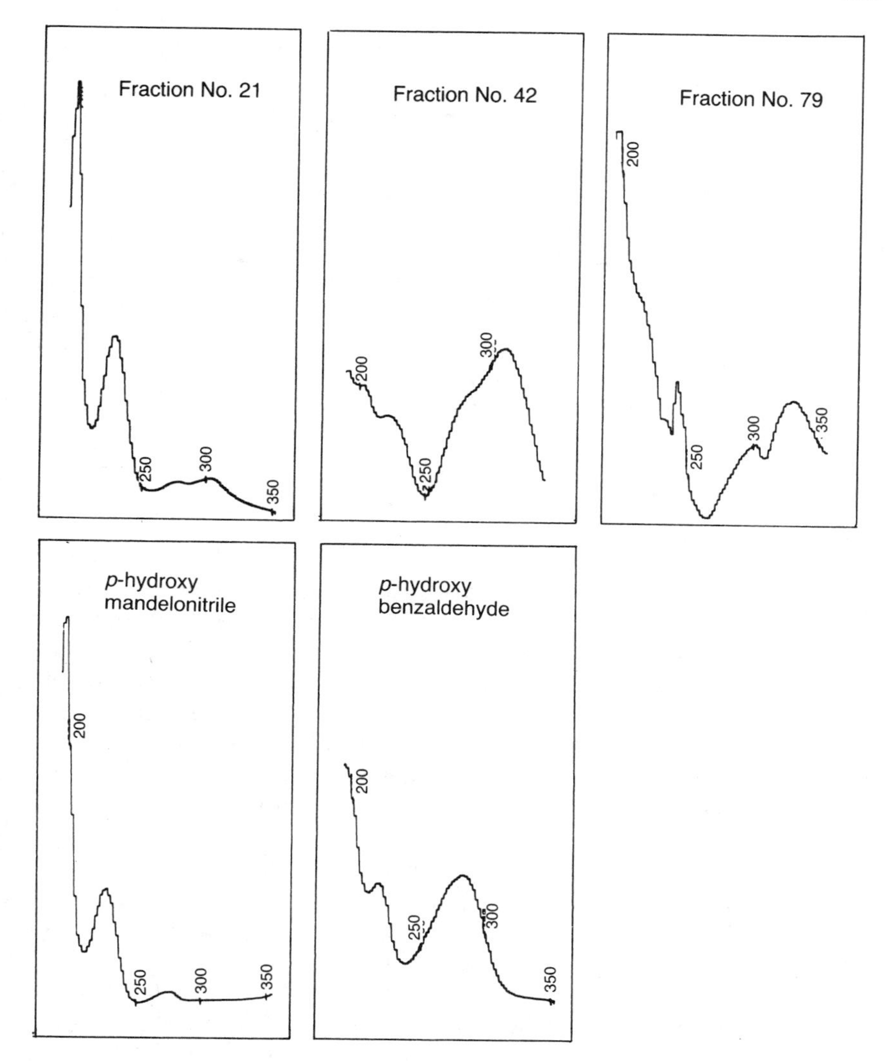

Fig. 8.6 UV spectra of the fraction numbers 21, 42 and 79 in comparison with two reference compounds. Spectra were obtained by use of a scanning LC detector. The reference substances are *p*-hydroxymandelonitrile and *p*-hydroxybenzaldehyde.

the plant, but may be a result of bacterial degradation. At the present moment we cannot say whether the three other compounds found to be active in this investigation also exist on the leaf surface. To answer this question, it is probably necessary to analyse steam wash extraction from older plants.

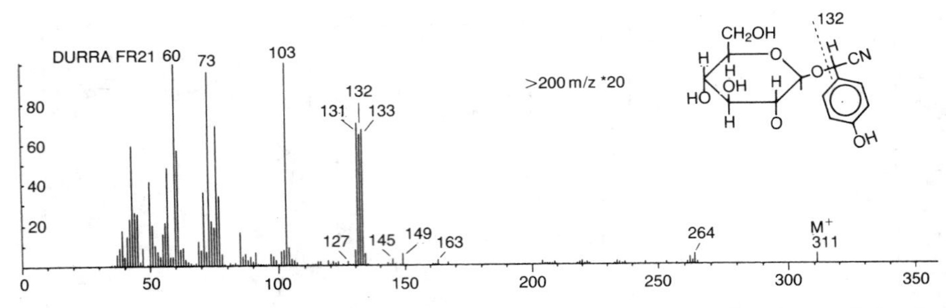

Fig. 8.7 LC mass spectrum of dhurrin (fraction no. 21). Mobile phase: 10% of methanol in water. Ion source temperature: 140°C. Electron energy: 70 eV. (After Stenhagen and Alborn, 1989. Reproduced with permission, Stenhagen, G. and Alborn, H. and Elsevier Science Publishers, Netherlands.)

8.5 DISCUSSION

In this study a sorghum material was used in which no cultivar showed a total resistance. The 'resistant' cultivars have a level of protection considered to be sufficient for cultivation in the SAT without or with a minimum use of pesticides; however, they are all low yielders. The goal of ICRISAT is to obtain high yielders with the same or a better level of resistance. To be able to screen the extensive plant material and to know which characters to incorporate into the high-yielding varieties, it was considered necessary to study the chemical background of resistance and susceptibility.

We found no correlation between volatile compounds and the resistance. This does not mean that volatiles are of no importance for the insects to find their host plants. Volatiles can still be important for the insects to distinguish between sorghum and other plants.

Only the data from the tissue extracts of young leaves, analysed by LC, could be correlated to the field data. This could be done even if the number of variables (components) were reduced to less than nine using their modelling power. The four most important were further studied. One of these was also one of the major components in the extract (fraction 21) and by the help of LC-MS was identified as dhurrin, a cyanogenic glucoside. Because of the detection method, UV at 200 nm, a minor peak is not necessarily a less important allelochemical component. The technique used for the analogy analyses compensates for this. However, the system error in the calculation of the peak area (amount) is larger for a small peak than for a large one and this error can destroy a fine but significant difference between groups of cultivars.

In the samples taken from, and just above, the growing point two compounds, *p*-hydroxybenzaldehyde and dhurrin, were dominant.

p-Hydroxybenzaldehyde is known to be produced as a result of an enzymatic degradation of dhurrin. Woodhead *et al.* (1982) have shown it to exist on the leaf surface of sorghum (in the leaf wax) and to have a deterring effect on locusts. The steam wash of one cultivar (CSH-1) showed that only dhurrin occurred on the leaf surface and in much higher concentrations than any other polar compound. Despite the fact that our examination does not show *p*-hydroxybenzaldehyde to be one of the active components, its high abundance in the larval feeding region still makes it interesting. It is also possible that the high amount of aldehyde is an artefact produced in the freeze-drying procedure. In an intact cell, dhurrin and the degrading enzyme is kept apart. The freeze-drying exposes the plant material to high vacuum and is likely to damage a large amount of cells, and consequently to speed up the production of the aldehyde. The thin leaf samples probably dry out so fast that this does not occur. It could be a problem using soft and thick tissue, like the growing point. On the other hand, larvae feeding on the tissue will also damage the cells, producing both the aldehyde and the hydrogen cyanide.

8.6 CONCLUSIONS AND PROSPECTS

One conclusion of this investigation is that dhurrin probably acts as an oviposition activator for shoot fly and stem borer. It is known that specialized insects use specific plant substances to recognize their hosts; these substances can be effective defence chemicals to other insects (Rothschild *et al.*, 1988). The high concentrations of dhurrin and its degrading products in the growing point of the plant where the larvae feed, proves that the insects are well adapted. Dhurrin probably also acts as a necessary feeding activator for the larvae. Theoretically it is possible to reduce the amount of activator to a level that makes the plant a non-host. However, this will in turn open the gate to a number of new possible attackers that were previously deterred from attacking. We believe that further study of the other compounds that have shown indications of having a deterring effect would be very fruitful.

Another reason to study the plant chemistry – often forgotten in plant breeding for resistance – is to detect substances in the crops that are unsuitable for consumption by humans and animals. This is necessary even if the work is carried out at the genetic level, because resistance to insects does not act at such a level; it is the allelochemicals, the product of the genes, that are the active components.

This investigation has shown the possibility of using chromatographic screening methods in combination with computerized analogy analyses to investigate the chemical based resistance to insects in plants. It is essential to collect the components of importance and study their effect on insect behaviour. We believe that if more were known about the biochemistry which classifies a crop plant as being a host or a non-host, it would be possible to change a crop to a non-host, making it resistant to insects.

ACKNOWLEDGEMENTS

We thank Mr Jonas Levin for the technical assistance. Financial support was given by the Swedish Agency for Research Cooperation with Developing Countries (SAREC).

REFERENCES

Alborn, H. and Stenhagen, G. (1985) Direct coupling of packed fused-silica liquid chromatographic columns to a magnetic sector mass spectrometer and application to polar thermolabile compounds. *J. Chromatogr.* **323**, 47–66.

Alborn, H. and Stenhagen, G. (1987) Micro liquid chromatography – mass spectrometry with low flow gradient elution. Studies of electrostatic nebulization and fused-silica column design. *J. Chromatogr.*, **394**, 35–49.

Bernays, E. A. and Chapman, R. F. (1978) Plant chemistry and acridoid feeding behaviour, in *Biochemical Aspects of Plant and Animal Coevolution* (ed. J. B. Harborne), Academic Press, New York, pp. 99–137.

Blum, A. (1972) Breeding for resistance in crop plants with special reference to sorghum, in *Sorghum in Seventies* (eds N. G. Rao and L. R. House), Oxford and IBH, New Delhi, India, pp. 390–410.

Chapman, R. F. and Woodhead, S. (1985) Insect behavior in sorghum resistance mechanisms, in *Proc. Int. Sorghum Entomology Workshop* (eds K. Leuschner and G. L. Teetes), ICRISAT, Patancheru P. O., Andhra Pradesh, 502 324 India, pp. 137–47.

Harborne, J. B. (1982) *Introduction to Ecological Biochemistry*, Academic Press, New York, pp. 66–7.

Kubo, I. and Hanke, F. J. (1985) Multifaceted chemically based resistance in plants, in *Recent Advances in Phytochemistry, Vol. 19, Chemically Mediated Interactions Between Plants and other Organisms* (eds G. A. Cooper-Driver, T. Swain and E. E. Conn), Plenum, New York, pp. 171–94.

Kubo, I. and Nakanishi, K. (1977) Insect antifeedants and repellents from African plants, in *Host Plant Resistance to Pests* (ed. P. A. Hedin), ACS Symp. Ser. 62 Amer. Chem. Soc., Washington, DC, pp. 165–78.

Lundgren, L., Norelius G. and Stenhagen, G. (1982) Prospects of a holistic approach to the biochemistry of pest and disease resistance in crop plants. *Hereditas*, **97**, 115–22.

Ogwaro, K. (1978) Ovipositional behavior and host plant preference of the sorghum shoot fly, *Atherigona soccata. Entomol. Exp. Appl.*, **23**, 189–99.

Ogwaro, K. and Kokwaro, E. D. (1981) Morphological observations on sensory structures on the ovipositor and tarsi of its female and on the head capsule of the larva of the sorghum shoot fly *Atherigona soccata. Insect Sci. Appl.*, **2**, 25–32.

Raina, A. K., Thindwa, H. Z., Othieno, S. M. and Corkhill, R. T. (1981)

Resistance in sorghum to the sorghum shoot fly: Larval development and adult longevity and fecundity on selected cultivars. *Insect Sci. Appl.*, **2**, 99–103.

Robinson, R. A. (1976) *The Patho-system Concept. Plant Patho-system*, Springer-Verlag, Berlin.

Rothschild, M., Alborn, H., Stenhagen, G. and Schoonhoven, L. M. (1988) A strophantidin glycoside in Siberian wallflower: A contact deterrent for the large white butterfly. *Phytochemistry*, **27**, 101–8.

Sharma, G. C., Jotwani, M. G., Rana, B. S. and Rao, N. G. P. (1977) Resistance in sorghum shoot fly *Atherigona soccata* and its genetic analysis. *J. Entomol. Res.*, **1**, 1–12.

Singh, B. U., Rana, B. S. and Rao, N. G. P. (1981) Host plant resistance to mite *Oligonychus indicus* and its relationship with shoot fly *Atherigona soccata* resistance in sorghum. *J. Entomol. Res.*, **5**, 230.

Singh, S. P. and Jotwani, M. G. (1980) Mechanism of resistance in sorghum to shoot fly. 1. Ovipositional non-preference. *Indian J. Entomol.*, **42**, 240–7.

Soto, P. E. (1974) Ovipositional preference and antibiosis in relation to resistance to sorghum shoot fly. *J. Econ. Entomol*, **67**, 165–7.

Stenhagen, G. and Alborn, H. (1989) Developments of micro liquid chromatography – mass spectrometry with gradient elution. Improvements to obtain less thermal decomposition of labile compounds. *J. Chromatogr.*, **474**, 285–300.

Stenhagen, G., Alborn, H. and Lundborg, T. (1987) Variation of root and microflora rhizosphere exudates in genotypes of barley, in *Allelochemicals: Role in Agriculture and Forestry*, (ed. G. R. Waller) Amer. Chem. Soc. Symp. Series 330, Washington, DC, pp. 76–88.

Taneja, S. L. and Leuschner, K. (1985) Resistance screening and mechanisms of resistance in sorghum to shoot fly, in *Proc. Int. Sorghum Entomology Workshop* (eds K. Leuschner and G. L. Teetes), ICRISAT, Patancheru P. O., Andhra Pradesh 502 324, India, pp. 115–36.

van der Plank, J. E. (1963) *Plant Diseases: Epidemics and Control*, Academic Press, New York.

Wold, S., Albano, C., Dunn, W. J. *et al.* (1983) Pattern recognition: finding and using regularities in multivariate data, in *Food Research and Data Analysis* (eds H. Martens and H. Russwurm), Elsevier, Amsterdam.

Woodhead, S. (1983) Surface chemistry of *Sorghum bicolor* and its importance in feeding by *Locusta migratoria. Phys. Entomol.*, **8**, 345–52.

Woodhead, S. and Bernays, E. A. (1978) The chemical basis of resistance of *Sorghum bicolor* to attack by *Locusta migratoria. Entomol. Expl. Appl.* **24**, 123–44.

Woodhead, S., Galeffi, C. and Marini Bettolo, G. B. (1982) *p*-Hydroxybenzaldehyde as a major constituent of the epicuticular wax of seedling *Sorghum bicolor. Phytochemistry*, **21**, 455–6.

CHAPTER 9

Allelochemicals in wheat and barley: role in plant–insect interactions

L. J. Corcuera, V. H. Argandoña and G. E. Zúñiga

9.1 INTRODUCTION

Allelochemicals in plants may be important in allelopathy, plant–insect interactions, plant–herbivore interactions, etc. (Waller, 1987). Wheat is known to accumulate several hydroxamic acid glycosides during its early stages of development (Fig. 9.1) (Argandoña *et al.*, 1981; Wahlroos and Virtanen, 1959; Willard and Penner, 1976). Barley seedlings accumulate gramine, an indole base (Hanson *et al.*, 1983; Smith, 1977). Although the concentrations of hydroxamic acids in wheat and of gramine in barley reach up to about $5\,\mathrm{mmol\,kg^{-1}}$ of fresh mass, little is known about the importance of these compounds for the plant. Hydroxamic acids appear to determine the resistance of maize to the European corn borer *Ostrinia nubilalis* (Klun *et al.*, 1967) and of several members of Gramineae to aphids (Argandoña *et al.*, 1980; Corcuera *et al.*, 1982; Long *et al.*, 1977). Gramine may be responsible for possible allelopathic effects attributed to barley (Overland, 1966) and may cause toxicity in ruminants feeding on other members of Gramineae and Leguminoseae (Gallagher *et al.*, 1964; Marten *et al.*, 1981). Gramine also increases the resistance of barley to aphids (Zúñiga and Corcuera, 1986; Zúñiga *et al.*, 1985). The existing evidence for the participation of hydroxamic acids and gramine in the resistance of wheat and barley to aphids will be discussed in this chapter.

9.2 RESISTANCE OF WHEAT AND BARLEY TO APHIDS

Wheat and barley seedlings have been infested with various species of aphids. Inverse correlations have been found between concentrations of hydroxamic acids or gramine and population growth rate of aphids on the plant (Fig. 9.2). Similar correlations have been observed with rye, triticale and maize, which also contain hydroxamic acids (Argandoña *et al.*, 1980). When hydroxamic acids or gramine were added to detached leaves of

Allelopathy: Basic and applied aspects Edited by S. J. H. Rizvi and V. Rizvi
Published in 1992 by Chapman & Hall, London ISBN 0 412 39400 6

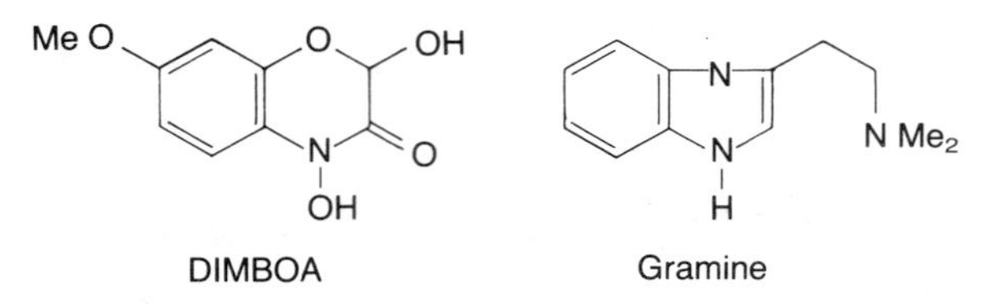

Fig. 9.1 Main secondary metabolites involved in resistance of wheat and barley to aphids. DIMBOA is the main hydroxamic acid present in wheat and gramine is found in barley.

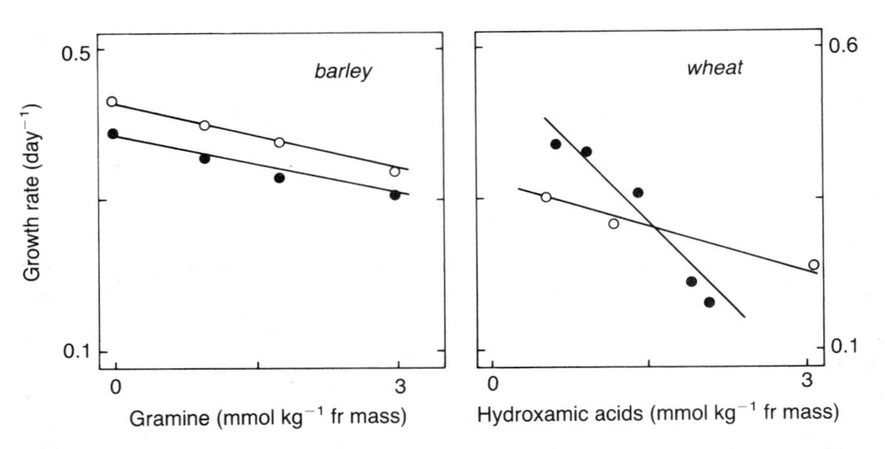

Fig. 9.2 Performance of aphids on wheat and barley seedlings. Wheat seedlings contained hydroxamic acids and barley seedlings contained gramine. Ten-day-old seedlings were infected with non-alate nymphs and six days later population growth rate (r) was calculated by $r = \ln (N_f/N_i)/t$, where N_f and N_i are final and initial number of aphids and $t = 6$ days. (◯) *Metopolophium dirhodum*; (●) *Schizaphis graminum*.

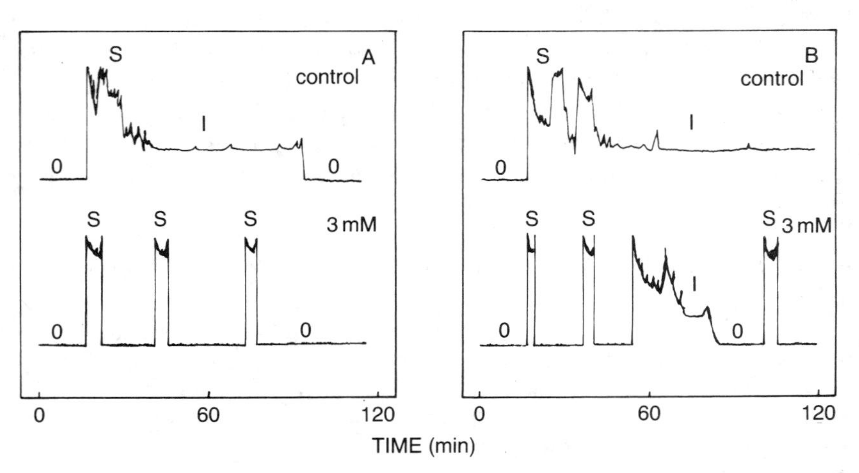

Fig. 9.3 Effect of DIMBOA and gramine on the feeding behaviour of *Schizaphis graminum* in artificial diets. Waveforms are O = base line; S = salivation; I = ingestion. (A) gramine; (B) DIMBOA.

wheat or barley, respectively, the population growth rate of aphids on these leaves also decreased. It appears that these compounds produced effects on aphids. To conclude, however, that the compounds are actually involved in aphid resistance, it should be shown that the compounds have toxic properties or produce feeding deterrency or any other effect that would reduce aphid performance on the plant. In addition, it should be shown that the compounds are present in the tissues where the aphids normally feed and that the concentration of the compounds in these tissues is sufficient to cause the effects on aphids.

9.3 EFFECTS OF HYDROXAMIC ACIDS AND GRAMINE ON FEEDING OF APHIDS

The biological activity of compounds on aphids can be measured by offering an artificial diet with the compounds to the insects. Using such a system, DIMBOA and gramine decreased survival, feeding and reproduction of the aphids *Schizaphis graminum*, *Metopolophium dirhodum*, *Rhopalosiphum padi* and *R. maidis* at concentrations lower than those normally found in plant leaves (Argandoña *et al.*, 1983; Corcuera *et al.*, 1982, 1987). Both compounds appeared to be toxic and feeding deterrents. The feeding behaviour of *S. graminum* on diets containing DIMBOA or gramine or neither of these two compounds was, measured by using an electronic feeding monitoring apparatus. Aphids feeding on diets without a test compound showed S waveforms (salivation) followed by long periods of diet ingestion (Fig. 9.3). The presence of DIMBOA or gramine decreased ingestion time and increased the frequency of salivation (Fig. 9.3 and Table 9.1).

TABLE 9.1 *Feeding of* Schizaphis graminum *in diets with DIMBOA or gramine*[a]

Compound	Concentration (mM)	Ingestion time (min)
Control (without gramine or DIMBOA)	0	46 ± 6
Gramine	0.5	20 ± 2
	1.0	6 ± 1
	3.0	0
DIMBOA	0.5	25 ± 3
	1.0	15 ± 3
	3.0	8 ± 3

[a] Feeding behaviour was electronically monitored over a period of 2 h. Values are means of ten individuals ± standard error.

TABLE 9.2 *Behaviour of aphids feeding on wheat and barley seedlings*

	Compound (mmol kg^{-1} fresh mass)		Ingestion time (min)				
Species	Gramine	Hx	Ip	Inp	S	NI	Tx
S. graminum							
Wheat							
cv. Aurifen	nd	0.5	45	15	37	83	51
cv. SNA-3	nd	2.1	24	26	73	66	81
Barley							
cv. F. Union	nd	nd	42	16	48	73	41
cv. Datil 'S'	3.2	nd	36	0	80	70	105
R. padi							
Barley							
cv. F. Union	nd	nd	31	71	35	44	57
cv. Datil 'S'	3.2	nd	30	0	46	104	104

Hx = hydroxamic acids; ND = not detected; Ip = ingestion from phloem; Inp = ingestion from non-phloem; S = salivation; NI = non-ingestion; Tx = time to produce the first X wave I (initiation of phloem ingestion). Adult aphids (n = 12) fed on 10-day-old seedlings over a period of 3 h.

TABLE 9.3 *Distribution of gramine and hydroxamic acids in barley and wheat leaves (after Argandoña* et al., 1987)

	Compound concentration	
Plant (compound)	(mmol kg^{-1} fresh mass)	(mmol kg^{-1} dry mass)
Barley (gramine)		
Complete leaf	1.3 ± 0.1	14.1 ± 0.1
Mesophyll protoplasts	3.1 ± 0.2	–
Lower epidermis	2.2 ± 0.1	44.6 ± 0.8
Upper epidermis	2.0 ± 0.1	39.2 ± 0.8
Vascular bundles	nd	nd
Wheat (hydroxamic acids)		
Complete leaf	3.0 ± 0.5	37.6 ± 0.1
Mesophyll protoplasts	4.1 ± 0.5	–
Lower epidermis	nd	nd
Vascular bundles	6.8 ± 0.4	40.0 ± 0.8
Xylem exudates	nd	nd
Guttation drops	nd	nd

Values are means of three replicates ±standard error. nd = not detected.

It is also possible, using the same electronic feeding monitor, to study the feeding behaviour of aphids in leaves of wheat and barley (Zúñiga *et al.*, 1988). Waveforms for non-ingestion, salivation and ingestion from phloem were distinguished and measured. *S. graminum* ingested preferentially from phloem in a wheat variety with low content of hydroxamic acids (Table 9.2). In a variety with a higher content of hydroxamic acids, ingestion from phloem was reduced and ingestion from non-phloem increased. The time of salivation was nearly doubled in the second variety. This may be due to the feeding deterrent effect of hydroxamic acids. *S. graminum* feeding behaviour was also studied in barley varieties with (cv. Datil 'S') or without gramine (cv. F. Union). In the variety with natural gramine in the leaves only ingestion from phloem was observed. A second aphid, *Rhopalosiphum padi*, ingested preferentially from non-phloem in a variety without gramine in the leaves. In barley cv. Datil 'S', which contains gramine, ingestion from non-phloem was not observed. These different feeding behaviours can be explained by the location and concentration of hydroxamic acids and gramine in the leaves. This is discussed below.

9.4 DISTRIBUTION OF COMPOUNDS IN LEAVES

It is possible to isolate different tissues of barley and wheat leaves and, therefore, to determine the distribution of different compounds in the leaves (Argandoña *et al.*, 1987). Gramine in barley was found preferentially in mesophyll cells and in the epidermis of barley and was absent in vascular bundles (Table 9.3). Hydroxamic acids were present in vascular bundles and in mesophyll cells of wheat leaves. No hydroxamic acids were detected in epidermis, guttation drops or xylem exudates. Concentrations of gramine and hydroxamic acids in tissues where they were present were similar to those necessary to affect the feeding behaviour and reproduction of the aphids or even to cause toxic effects in artificial diets.

Aphids on barley cv. Datil 'S' (with gramine) fed exclusively from phloem. Gramine was absent in phloem and was present in the other tissues. Aphids on wheat cv. Aurifen fed from phloem and from non-phloem. Hydroxamic acids were found in vascular bundles and in mesophyll cells. When wheat cv. SNA-3 (with higher content of hydroxamic acids) was used, ingestion from phloem was decreased. Hydroxamic acids are more concentrated in vascular bundles than in mesophyll cells. Therefore, the changes observed in the feeding behaviour of aphids in different cultivars of wheat and barley could be caused by hydroxamic acids and gramine.

TABLE 9.4 *Effects of* Schizaphis graminum *on wheat cultivars with different concentrations of hydroxamic acids*

Cultivar	Hydroxamic acids (mmol kg^{-1} fr mass)	Reproduction (Nymphs/Adults)	ADS[a] (%)	Weight index[b]	Chlorophyll content (μg g^{-1} fr mass)
SNA-3	2.1	2.2	65	0.008	54
Quilafen	1.5	1.4	55	0.006	43
Millaleu	0.6	5.0	33	0.005	76
Aurifen	0.5	6.1	28	0.004	82

[a] Average damage score (ADS) = (Ah/Ad) × 100. Ah is the area of the leaf estimated to be healthy. Ad is the area of the leaf that was estimated to be damaged.

[b] Weight index = (Wni − Wi)/Na × 100. Wni is dry weight (g) of the non-infested leaves; Wi is the dry weight of the infested leaves; Na is the number of aphids on the leaves.

9.5 PERFORMANCE OF PLANTS WITH PROTECTIVE COMPOUNDS

Wheat varieties with different contents of hydroxamic acids were infested with *S. graminum*. After 10 days, the number of aphids on the leaves, chlorophyll content, damage and weight index of the leaves were measured. As expected, those cultivars with higher content of hydroxamic acids had the lowest number of aphids (Table 9.4). Thus, aphids perform poorly on cultivars which have more hydroxamic acids. However, after 10 days of aphid infestation, leaves of cultivars with a higher content of hydroxamic acids had a lower chlorophyll content, higher leaf damage and higher weight index. A higher weight index indicates less tolerance of the plant to the aphids. Thus, seedlings with a higher content of hydroxamic acids suffer more damage than those seedlings with lower content of these compounds. These apparently contradictory results could be explained by the feeding behaviour of the aphids on the seedlings. Aphids on leaves with a higher content of hydroxamic acids, salivated twice as much time as those aphids on leaves of seedlings with a lower content of those compounds (Fig. 9.4). This means that aphids probed for a longer time, and therefore made more direct damage to the leaves. A longer salivation time may also have devastating consequences for plants because these aphids are vectors of the barley yellow dwarf virus (Eastop, 1977). This virus also affects wheat and oats and it is transmitted during phloem probing and salivation (Gildow, 1982; Scheller and Shukle, 1986). This

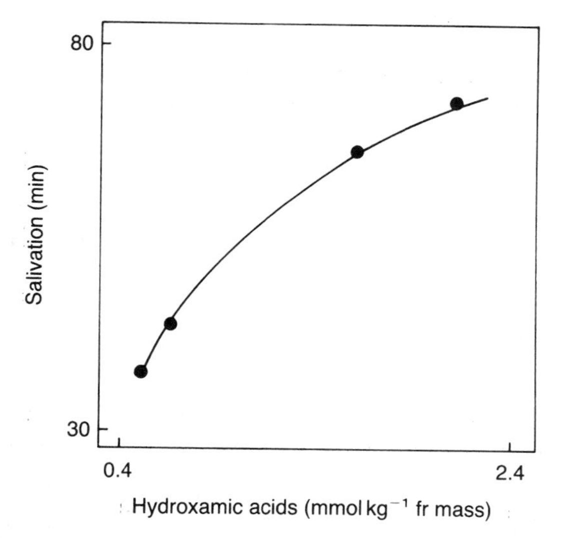

Fig. 9.4 Effect of hydroxamic acids in wheat leaves on salivation of *Schizaphis graminum*. Salivation time was measured with an electronic feeding monitor over a period of 3 h.

suggested that feeding deterrents present in leaves (such as gramine) may increase the chances of virus transmission to the plant.

9.6 CONCLUSIONS AND PROSPECTS

It is clear that hydroxamic acids affect aphids' performance on the plant negatively. Since seedlings with hydroxamic acids are more damaged by aphids, the agricultural value of using hydroxamic acids as a resistance factor towards aphids is questionable. This may also apply to gramine and to other plant metabolites which are feeding deterrents. The experiments presented here have been performed with non-alate nymphs which cannot move easily to another plant. Experiments are needed to test the performance of barley and wheat cultivars with various amounts of gramine or hydroxamic acids when they become infested with aphids (with and without virus).

ACKNOWLEDGEMENT

This work was partially funded by Universidad de Chile (B-2806-8813) and FONDECYT (1126-89).

REFERENCES

Argandoña, V. H., Corcuera, L. J., Niemeyer, H. M. and Campbell, B. C. (1983) Toxicity and feeding deterrency of hydroxamic acids from Gramineae in synthetic diets against the greenbug, *Schizaphis graminum. Entomol. Exp. Appl.*, **34**, 134–8.

Argandoña, V. H., Luza, J. G., Niemeyer, H. M. and Corcuera, L. J. (1980) Role of hydroxamic acids in the resistance of cereals to aphids. *Phytochemistry*, **19**, 1665–8.

Argandoña, V. H., Niemeyer, H. M. and Corcuera, L. J. (1981) Effect of content and distribution of hydroxamic acids in wheat on infestation by the aphid *Schizaphis graminum. Phytochemistry*, **30**, 673–6.

Argandoña, V. H., Zúñiga, G. E. and Corcuera, L. J. (1987) Distribution of gramine and hydroxamic acids in barley and wheat leaves. *Phytochemistry*, **26**, 1917–18.

Corcuera, L. J., Argandoña, V. H., Peña, G. F. *et al.* (1982) Effect of benzoxazinone from wheat on aphids. *Proc. 5th Int. Symp. Insect–Plant Relationships*, Wageningen, Pudoc, Wageningen, pp. 33–9.

Corcuera, L. J., Argandoña, V. H. and Zúñiga, G. E. (1987) Resistance of cereal crops: role of allelochemicals, in *Allelochemicals: Role in Agriculture and Forestry* (ed. G. R. Waller), ACS Symp. Ser. 330, Amer. Chem. Soc., Washington, DC, pp. 129–35.

Eastop, V. F. (1977) Aphid vectors, in *Aphids as Virus Vectors* (eds K. F. Harris and K. Maramorosch), Academic Press, New York, pp. 3–47.

Gallagher, G. H., Koch, J. H., Moore, R. M. and Steel, J. D. (1964) Toxicity of *Phalaris tuberosa* for sheep. *Nature*, **204**, 542–5.

Gildow, F. E. (1982) Coated-vesicle transport of luteo viruses through salivary glands of *Myzus persicae. Phytopathology*, **72**, 1289–96.

Hanson, A. D., Ditz, K. M., Singletary, G. W. and Leland, T. J. (1983) Gramine accumulation in leaves of barley grown under high temperature stress. *Plant Physiol.*, **71**, 849–904.

Klun, J. A., Tipton, C. L. and Brindley, T. A. (1967) 2,4-Dihydroxy-7-methoxy-1,4-benzoxazin-3-one (DIMBOA), an active agent in the resistance of maize to the European corn borer. *J. Econ. Entomol.*, **60**, 1529–33.

Long, B. J., Dunn, G. M., Bowman, J. S. and Routley, D. G. (1977) Relationship of hydroxamic acid content in corn and resistance to the corn leaf aphid. *Crop Sci.*, **17**, 55–8.

Marten, G. C., Jordan, R. M. and Hovin, A. W. (1981) Improved lamb performance associated with breeding for alkaloid reduction in reed canary grass. *Crop Sci.*, **21**, 295–8.

Overland, L. (1966) The role of allelopathic substances in the 'smother crop' barley. *Amer. J. Bot.*, **53**, 423–32.

Scheller, V. H. and Shukle, R. H. (1986) Feeding behavior and transmission of barley yellow dwarf virus by *Sitobion avenae* on oats. *Entomol. Exp. Appl.*, **40**, 189–95.

Smith, T. A. (1977) Tryptamine and related compounds in plants. *Phytochemistry*, **16**, 171–5.

Wahlroos O. and Virtanen, A. I. (1959) The precursors of 6-methoxybenzoxazolinone in maize and wheat plants, their isolation and some of their properties. *Acta Chem. Scand.*, **13**, 1906–8.

Waller, G. R. (ed.) (1987) *Allelochemicals: Role in Agriculture and Forestry*, ACS Symp. Ser. 330, Amer. Chem. Soc., Washington, DC.

Willard, J. I. and Penner, D. (1976) Benzoxazinones: cyclic hydroxamic acids found in plants. *Residue Rev.*, **64**, 67–76.

Zúñiga, G. E. and Corcuera, L. J. (1986) Effect of gramine in the resistance of barley seedlings to the aphids *Rhopalosiphum padi. Entomol. Exp. Appl.*, **40**, 259–62.

Zúñiga, G. E., Salgado, M. S. and Corcuera, L. J. (1985) Role of an indole alkaloid in the resistance of barley seedlings to aphids. *Phytochemistry*, **24**, 945–7.

Zúñiga, G. E., Varanda, E. M. and Corcuera, L. J. (1988) Effect of gramine on the feeding behavior of the aphids *Schizaphis graminum* and *Rhopalosiphum padi. Entomol. Exp. Appl.*, **47**, 161–5.

CHAPTER 10

Allelochemical properties of alkaloids. Effects on plants, bacteria and protein biosynthesis

M. Wink and T. Twardowski

10.1 INTRODUCTION

A characteristic feature of plants is their ability to synthesize a wide range of natural products, the so-called secondary metabolites. Up till now more than 4500 terpenoids, 700 polyketides, 750 polyacetylenes, 500 phenylpropanoids, 1200 flavonoids, 400 non-protein amino acids, 100 glucosinolates, 50 cyanogenic glycosides, 100 amines and over 400 alkaloids have been described. Many of these compounds are used by man as pharmaceuticals, flavours, fragrances, colours, stimulants, hallucinogens, poisons, pesticides or as lead structure for the organic chemists to other more powerful substances and, therefore, plant allelochemicals are often economically important.

For more than 100 years biologists and chemists have tried to answer the question as to why plants invest so much energy and care in the formation of these secondary products. For some time it was believed that these compounds are waste products or otherwise useless substances (Paech, 1950; Mothes, 1955). But for 2–3 decades it has generally been accepted that the secondary metabolites play an important role in the ecology of plants and are therefore crucial for their survival and fitness (Stahl, 1888; Harborne, 1982, 1986; Swain, 1977; Levin, 1976; Rosenthal and Janzen, 1979; Wink, 1988). The main function seems to be chemical defence against microorganisms (viruses, bacteria, fungi), phytophagous animals (nematodes, molluscs, insects, vertebrates) and against other competing plant species ('allelopathy'). In addition, some of these compounds attract pollinating or seed-dispersing animals or display some metabolic functions. These functions are not exclusive, e.g. a flavonoid may function as a coloured attractant for insects, but is concomitantly antibiotic, insect-repellent and a UV-protectant.

Allelopathy: Basic and applied aspects Edited by S. J. H. Rizvi and V. Rizvi
Published in 1992 by Chapman & Hall, London ISBN 0 412 39400 6

Chemical defence is not absolute. And most plant species which are protected against the majority of enemies due to their chemistry, have a few 'parasites' which have overcome the defence system and specialized on a particular plant species. These specialists often use the plants' chemistry for their own defence or they may develop these defence compounds into signal compounds, e.g. pheromones (Boppré, 1986; Harborne, 1982, 1986; Schneider, 1987).

Owing to the enormous number of secondary metabolites, we do not understand the underlying 'chemical ecology' in most cases, which has to take into account:

(1) the chemistry, biochemistry, physiology, molecular biology of the production of these metabolites by the plant;
(2) the biochemical effects of these metabolites on non-specialized and adapted microorganisms, other plants or herbivores;
(3) the metabolism of these compounds by specialized/unspecialized organisms;
(4) the ecological interactions between a given plant species and its herbivores, microorganisms and competing plants.

We need to understand the principles underlying all these interactions if we want to exploit secondary metabolites as natural pesticides (herbicides, insecticides, fungicides, etc.) (Waller, 1987) or if we wish to improve the resistance of plants, especially of our food plants, against pathogens and herbivores, instead of using more and more man-made agrochemicals (Wink, 1987, 1988).

In this chapter we have tested the allelochemical properties of some alkaloids in testing the inhibition of bacterial growth, the germination and growth of *Lactuca sativa* and *Lepidium sativum* seeds or seedlings. In addition, the toxicity of alkaloids for *Lemna* was determined. In order to find a potential target of these metabolites, we have studied in preliminary experiments their action on the essential process of protein biosynthesis.

For comparison we have additionally tested a series of other non-alkaloidal secondary metabolites and metabolic inhibitors under identical conditions.

10.2 INHIBITION OF SEED GERMINATION AND SEEDLINGS' GROWTH

In order to determine the allelopathic potential of alkaloids and other natural products, we studied the inhibition of seed germination in *Lactuca sativa* and in *Lepidium sativum* by alkaloids, and in comparison by phenolic compounds, metabolic inhibitors and 14 crude essential oils (Fig. 10.1, Table 10.1). In addition, the inhibition or stimulation of root or hypocotyl growth was determined (Table 10.2) which is often a more sensitive parameter than the germination rate alone. At a concentration of 0.1% many terpenes and some alkaloids (8-oxychinoline, aconitine, nicotine, cycloheximide) showed substantial inhibitions of germination. Growth of roots and hypocotyls was affected by many compounds, often even at 0.01% concentration. Whereas most compounds were inhibitory, a few substances resulted in an increased growth of roots or hypocotyls, which could be also detrimental for the seedling.

10.3 TOXICITY OF ALKALOIDS AND OTHER INHIBITORS ON *LEMNA GIBBA*

Lemna gibba is a convenient test organism to determine the toxicity of chemicals (Einhellig *et al.*, 1985; Leather and Einhellig, 1988). As shown in Table 10.3, a number of alkaloids lead to death and bleaching of *Lemna* at concentrations of 0.4%. Especially active compounds were quinine, cinchonidine, nicotine, boldine, lobeline, coniine and harmaline.

10.4 INHIBITION OF BACTERIAL GROWTH BY SECONDARY METABOLITES

Parallel to the experiments with plants, we determined whether these compounds inhibited bacterial growth and took as test organisms a Gram-positive and a Gram-negative bacterium, *Bacillus subtilis* and *Escherichia coli* (Table 10.4). Antibiotic activity was clearly exhibited at 0.1% concentration by quinidine, quinine, cinchonine, strychnine, berberine, tubocurarine, sanguinarine, 8-oxychinoline and ephedrine.

10.5 INSECT-REPELLENT PROPERTIES

As examples for biological activities of these compounds against herbivores we have tested larvae of the polyphagous moth, *Syntomis mogadorensis*. Larvae had the choice to feed on untreated controls or leaves onto which solutions of allelochemicals (0.01%, 0.1 and 1% concentrations) were applied. At the 1% level most compounds were both repellent and toxic; at lower concentrations (i.e. 0.01 and 0.1%) the essential oils showed the highest repellency.

10.6 MODULATION OF PROTEIN BIOSYNTHESIS BY SECONDARY METABOLITES

Secondary metabolites are obviously directed against a wide variety of organisms ranging from viruses, bacteria to insects and mammals (Wink,

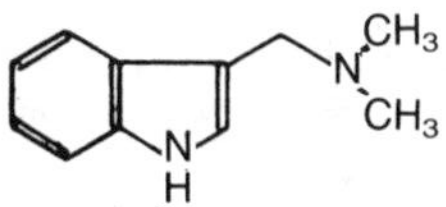

gramine
ajmalicine 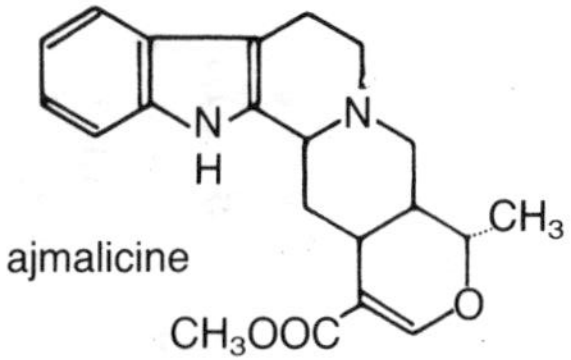
strychnine 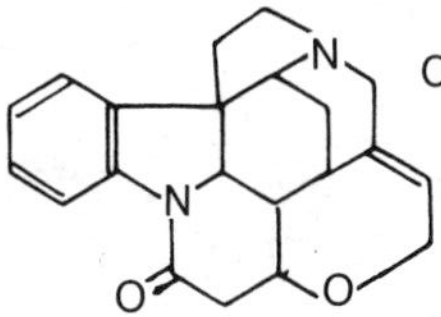
brucine 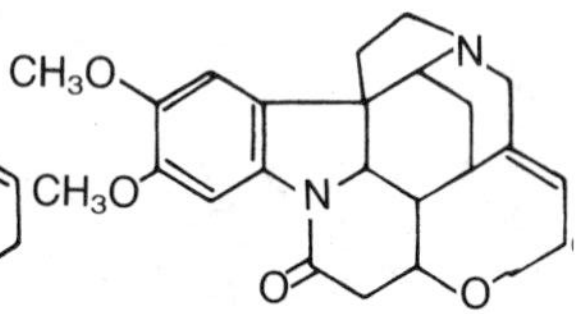
harmaline 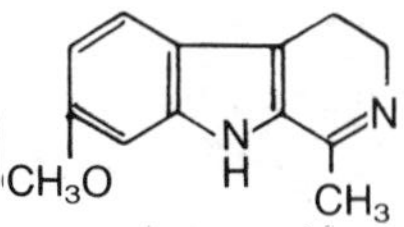

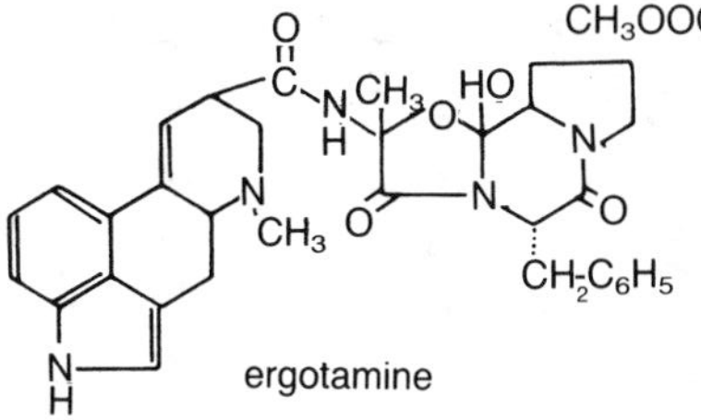

ergotamine
yohimbine 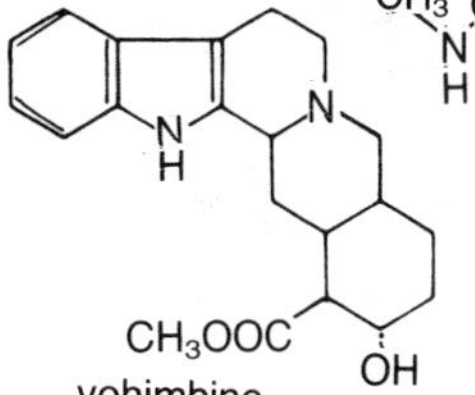
eserine 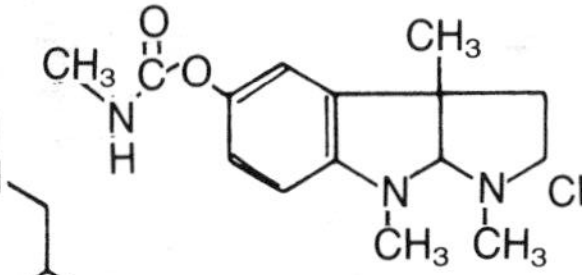
reserpine 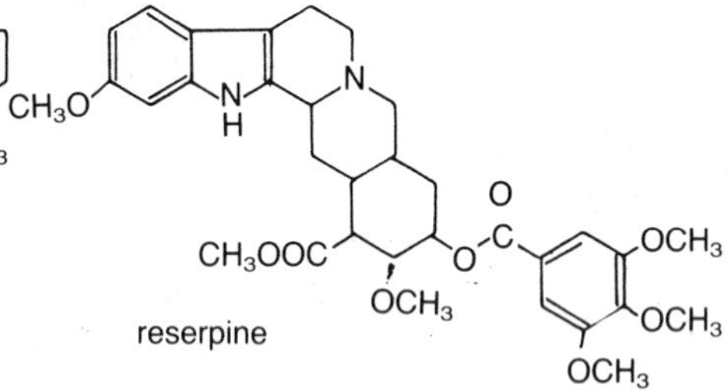

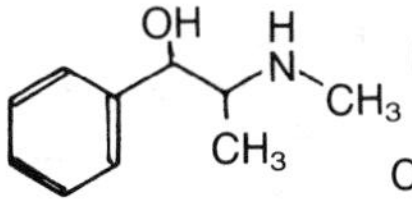

ephedrine
boldine 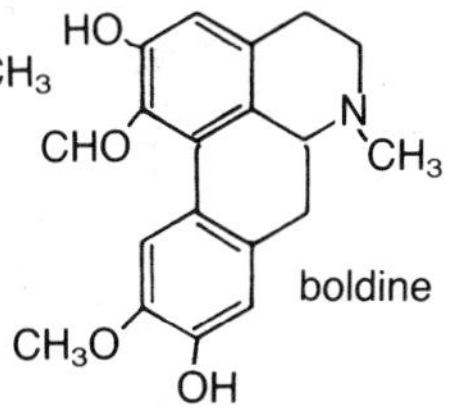
berberine 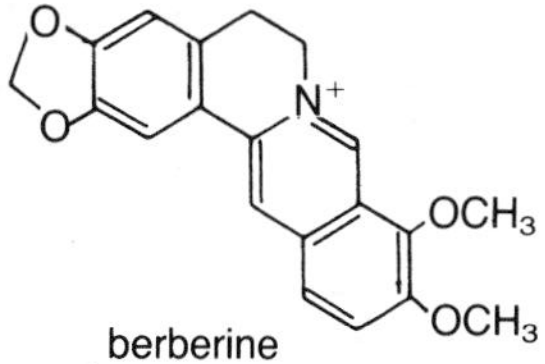
chelidonine 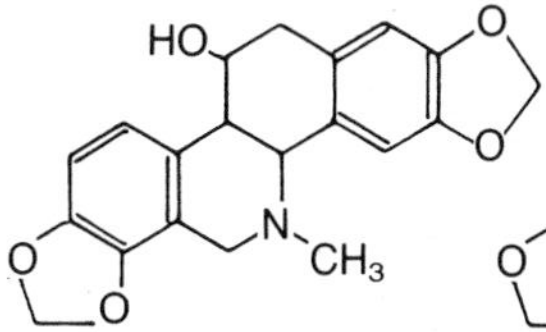
sanguinarine 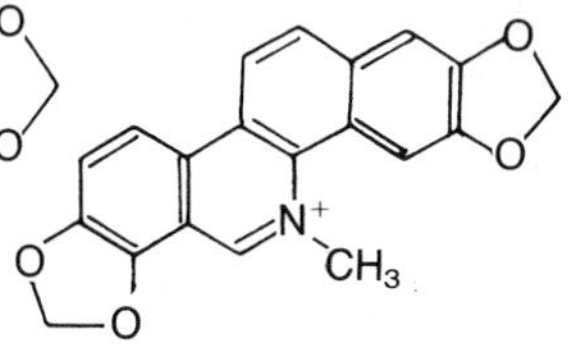

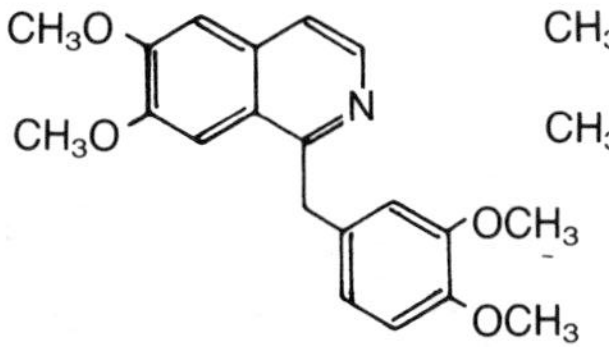

papaverine
emetine 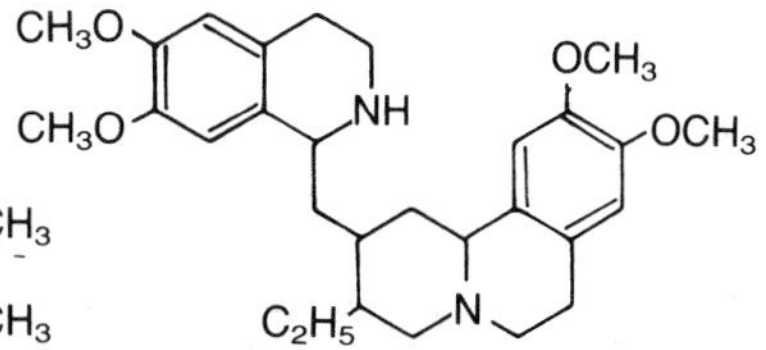
8-hydroxy-quinoline 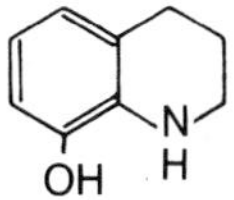
salsoline 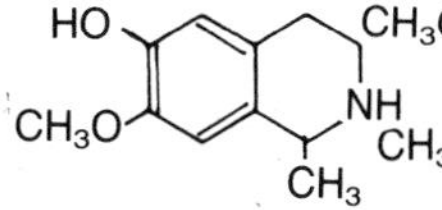
colchicine

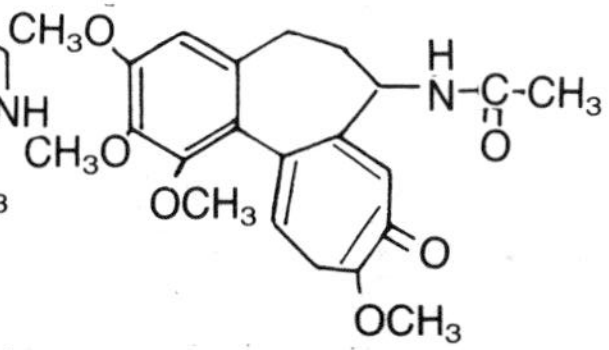

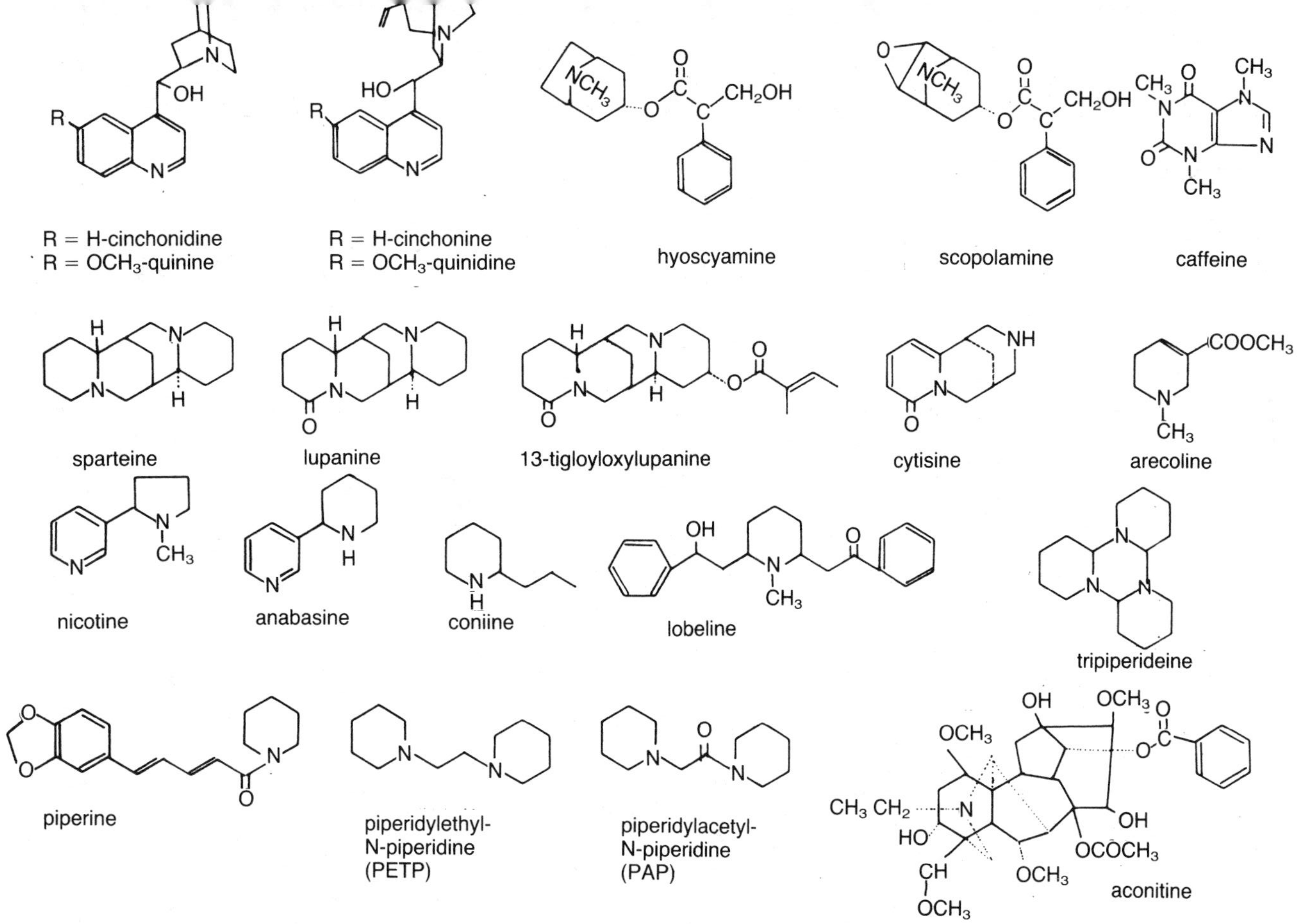

R = H-cinchonidine
R = OCH_3-quinine

R = H-cinchonine
R = OCH_3-quinidine

hyoscyamine

scopolamine

caffeine

sparteine

lupanine

13-tigloyloxylupanine

cytisine

arecoline

nicotine

anabasine

coniine

lobeline

tripiperideine

piperine

piperidylethyl-
N-piperidine
(PETP)

piperidylacetyl-
N-piperidine
(PAP)

aconitine

Fig. 10.1 Structures of the secondary metabolites tested.

TABLE 10.1 *Inhibition of seed germination of* Lactuca sativa *and of* Lepidium sativum *by secondary metabolites*

	Germination rate (% of control)			
	Lactuca		*Lepidium*	
Compounds	0.1% conc.	0.01% conc.	0.1% conc.	0.01% conc.
Untreated control	100	100	100	100
1. *Alkaloids*				
α-Tripiperideine	–	–	98	98
8-Oxychinolin	88	88	0	65
Aconitine	–	–	20	98
Ajmalicine	–	–	94	97
Anabasine	–	–	88	98
Berberine	57	89	87	87
Cadaverine	72	96	94	100
Caffeine	–	–	92	99
Canadine	–	–	98	100
Chelidonine	–	–	99	98
Cinchonidine	0	75	87	100
Colchicine	–	–	97	100
Cytisine	–	–	91	97
D-Ephedrine	–	–	100	92
Ergometrine	–	–	98	100
Ergotamine	–	–	98	100
Gramine	81	93	80	79
Harmaline	–	–	100	95
Lobeline	–	–	98	98
Nicotine	60	75	65	97
Piperine	–	–	18	99
Putrescine	84	81	85	94
Quinidine	42	100	74	88
Salsoline	–	–	95	94
Sanguinarine	53	100	100	100
Scopolamine	82	100	100	80
Sparteine	92	77	96	93
Strychnine	82	89	83	87
Tomatine	–	–	98	99
2. *Phenolics*				
Coumarin	0	0	0	0
Naringenin	–	–	100	96
Pyrogallol	0	88	0	62
Quercetin	80	69	89	100
Tannin	100	100	89	100

TABLE 10.1 Cont'd

	Germination rate (% of control)			
	Lactuca		*Lepidium*	
Compounds	0.1% conc.	0.01% conc.	0.1% conc.	0.01% conc.
3. *Essential oils*				
Balm mint	78	88	47	100
Camphor	0	61	0	81
Cassia	1	0	0	0
Chamomilla	22	100	0	100
Citrus	100	100	47	100
Eucalyptus	100	100	52	100
Foeniculum	100	100	71	100
Hypericum	0	100	0	100
Mint	0	88	0	100
Orange	89	100	81	100
Picea	84	96	100	48
Rosmarin	48	100	42	93
Sage	73	96	67	90
Thymol	0	27	0	0
Thymus	0	100	0	77
Trans-2-hexenal	0	91	0	80
Valeriana	5	100	0	100
4. *Metabolic inhibitors*				
5-Azacytidine	–	–	85	98
Ampillicilin	–	–	100	100
Chloramphenicol	63	80	92	100
Cycloheximide	0	0	0	0
Ethidiumbromid	–	–	0	100
Indomethacine	21	100	0	2
Streptomycin	–	–	100	100

Experimental: About 40 seeds were placed on a filter paper disc in a petri dish which contained 5 ml tap water and natural products in a concentration of 0.1% and 0.01%. To avoid evaporation the petri dishes were sealed with parafilm. All compounds were adjusted to neutral pH, if necessary. All experiments were performed in duplicate, untreated controls in triplicate. When the compounds were dissolved in ethanol or methanol, the controls were treated accordingly. Experiments were evaluated after 3–6 days, when all the control seeds had germinated. Germination rate of controls was set at 100%, (– = not determined).

1988). How can these molecules affect all these different organisms? Alkaloids can interact with DNA by intercalation or generally with nucleic acids by ionic bonding of the positively charged heterocyclic nitrogen of most alkaloids with the negatively charged phosphate groups of nucleic acids. Theoretically, replication or translation (protein biosynthesis) could

TABLE 10.2 *Effect of natural products on the growth of roots and hypocotyls of* Lactuca sativa *and* Lepidium sativum *seedings*

	Effect on root or hypocotyl length (in % of control)							
	Lactuca				*Lepidium*			
	Root		Hypocotyl		Root		Hypocotyl	
Compounds	0.1% conc.	0.01% conc.	0.1% conc.	0.01% conc.	0.1% conc.	0.01% conc.	0.1% conc.	0.01% conc.
Controls	100	100	100	100	100	100	100	100
1. *Alkaloids*								
α-Tripiperideine	–	–	–	–	62	46	90	80
8-Oxychinolin	10	110	23	127	0	27	0	40
Aconitine	–	–	–	–	10	75	0	125
Ajmalicine	–	–	–	–	110	104	108	116
Anabasine	–	–	–	–	89	108	70	120
Berberine	12	43	14	73	16	79	38	101
Cadaverine	44	91	92	171	105	111	119	117
Caffeine	–	–	–	–	25	41	91	98
Canadine	–	–	–	–	93	111	111	118
Chelidonine	–	–	–	–	71	114	117	119
Cinchonidine	0	68	0	74	6	121	16	110
Colchicine	–	–	–	–	8	10	31	35
Cytisine	–	–	–	–	46	79	77	100
D-Ephedrine	–	–	–	–	64	87	105	115
Ergometrine	–	–	–	–	68	29	115	65
Ergotamine	–	–	–	–	75	83	130	125

Gramine	43	99	101	146	44	70	70	94
Harmaline	–	–	–	–	8	19	1	70
Lobeline	–	–	–	–	10	83	9	95
Nicotine	67	85	126	139	60	80	93	104
Piperine	–	–	–	–	16	53	39	103
Putrescine	74	82	94	156	84	121	114	110
Quinidine	10	14	104	128	11	76	8	111
Salsoline	–	–	–	–	13	50	64	89
Sanguinarine	19	146	44	100	15	82	57	86
Scopolamine	47	110	46	93	131	132	103	105
Sparteine	70	123	93	129	35	40	60	49
Strychnine	16	88	17	56	73	146	76	114
Tomatine	–	–	–	–	86	115	96	105
2. *Phenolics*								
Coumarin	0	0	0	0	0	0	0	0
Naringenin	–	–	–	–	86	119	90	110
Pyrogallol	0	42	0	90	0	17	0	30
Quercetin	95	97	136	131	95	115	104	106
Tannin	38	69	89	101	24	100	63	100
3. *Essential oils*								
Balm mint	68	142	58	145	71	70	73	87
Camphor	0	0	63	71	0	0	0	0
Cassia	0	0	0	0	0	0	0	0
Chamomilla	47	119	23	129	0	106	0	89
Citrus	107	125	87	110	71	93	71	95
Eucalyptus	104	121	101	125	108	98	115	87
Foeniculum	98	116	80	132	85	89	89	104
Hypericum	0	83	0	105	0	96	0	85

TABLE 10.2 Cont'd

	Effect on root or hypocotyl length (in % of control)							
	Lactuca				*Lepidium*			
	Root		Hypocotyl		Root		Hypocotyl	
Compounds	0.1% conc.	0.01% conc.	0.1% conc.	0.01% conc.	0.1% conc.	0.01% conc.	0.1% conc.	0.01% conc.
Mint	0	110	0	94	0	87	0	86
Orange	127	107	114	131	124	61	125	84
Picea	103	78	90	129	93	77	112	90
Rosmarin	52	104	50	118	92	99	82	109
Sage	81	114	96	115	67	88	73	92
Thymol	0	0	35	21	0	0	0	0
Thymus	0	97	0	95	0	81	0	83
Trans-2-hexenal	0	134	0	114	0	29	0	45
Valeriana	0	79	0	105	0	93	0	98
4. *Metabolic inhibitors*								
5-Azacytidine	–	–	–	–	15	23	13	85
Ampillicilin	–	–	–	–	100	113	108	112
Chloramphenicol	37	35	46	101	15	24	37	72
Cycloheximide	0	0	0	0	0	0	0	0
Ethidiumbromid	–	–	–	–	0	11	0	42
Indomethacine	69	161	84	97	0	97	0	40
Streptomycin	–	–	–	–	32	32	117	125

Experimental: As in Table 10.1, length of roots and hypocotyls was measured after 4–8 days; the mean value of all individuals that germinated was determined. The length of controls was set at 100%.

TABLE 10.3 *Toxicity of natural products and inhibitors for* Lemna gibba

Compounds	Effect[a]	
	0.4% conc.	0.04% conc.
Untreated control	–	–
Alkaloids		
α-Tripiperideine	⊕	–
Ajmaline	–	–
Anabasine	*	–
Arecoline	–	–
Berberine	–	–
Boldine	⊕	*
Brucine	–	–
Cinchonidine	⊕	*
Cinchonine	–	–
Colchicine	–	–
Coniine	*	*
D-Ephedrine	–	–
Emetine	*	–
Ergometrine	–	–
Ergotamine	–	–
Eserine	*	–
Harmaline	⊕	⊕
Hyoscyamine	*	–
L-Ephedrine	–	–
Lobeline	*	*
Lupanine	–	–
Narcotine	–	–
Nicotine	*	–
Papaverine	–	–
Protoveratrine B	–	–
Quinidine	*	–
Quinine	⊕	*
Reserpine	–	–
Sanguinarine	*	–
Scopolamine	–	–
Sparteine	–	–
Strychnine	*	–
Yohimbine	⊕	–
Other inhibitors		
5-Azacytidine	–	–
Allopurinol	⊕	–
Cycloheximide	–	–
Quercetin	–	–
Tannin	*	–

[a] About 5–10 *Lemna* plants were placed in the well of multi-titre plates containing 1 ml tap water and 0.4 or 0.04% of compounds. After 2–6 days it was determined whether the plants died (⊕†), lost their chlorophyll (*) or were at least superficially intact (–).

TABLE 10.4 *Inhibition of bacterial growth (Bacillus* subtilis, Escherichia coli*) by natural products and by some inhibitors*

Compounds	Inhibition of bacterial growth[a]					
	B. subtilis			*E. coli*		
	0.1% conc.	0.01% conc.	0.001% conc.	0.1% conc.	0.01% conc.	0.025% conc.
Untreated control	100	100	100	100	100	100
1. *Alkaloids*						
α-Tripiperideine	–	–	–	47	109	85
8-Oxychinolin	23	23	55	–	–	–
Aconitine	–	–	–	133	94	123
Ajmaline	–	–	–	–	–	90
Anabasine	–	–	–	83	110	–
Arecoline	79	96	101	–	–	111
Berberine	20	18	53	–	–	–
Boldine	–	–	–	–	–	96
Brucine	56	97	110	–	–	93
Caffeine	83	96	139	–	–	83
Cinchonidine	75	95	100	–	–	90
Cinchonine	52	87	92	–	–	–
Colchicine	–	–	–	128	97	98
Coniine	–	–	–	–	–	72
D-Ephedrine	42	89	104	103	104	95
Emetine	76	94	98	–	–	119
Ergometrine	–	–	–	111	114	116
Ergotamine	–	–	–	114	107	148
Eserine	85	94	110	–	–	102

Eukatropine	92	100	103	–	–	–
Harmaline	–	–	–	5	77	5
Hyoscyamine	–	–	–	–	–	82
L-Ephedrine	–	–	–	–	–	78
Lobeline	–	–	–	41	118	87
Lupanine	–	–	–	–	–	101
Narcotine	–	–	–	–	–	103
Nicotine	75	100	103	–	–	87
Papaverine	–	53	93	–	–	92
Protoveratrine B	–	–	–	–	–	112
Quinidine	5	92	98	–	–	89
Quinine	22	28	87	–	–	91
Reserpine	–	–	–	–	–	126
Sanguinarine	0	7	3	–	–	–
Scopolamine	70	99	109	–	–	87
Sparteine	80	99	99	–	–	96
Strychnine	23	89	95	–	–	71
Tubocurarine	15	72	92	–	–	–
Yohimbine	–	–	–	–	–	119
2. *Other compounds*						
5-Azacytidine	–	–	–	37	82	50
Allopurinol	96	98	99	–	–	107
Cadaverine	101	103	97	–	–	–
Cycloheximide	–	–	–	–	–	97
Naringenin	–	–	–	109	109	–

[a] Compounds (pH-adjusted and filter-sterilized) were added to 5 ml bacterial growth medium. Experiments were started by adding 50 μl of a bacterial suspension. Test tubes were incubated at 37°C (*E. coli*) or 30°C (*B. subtilis*). The experiments were evaluated when the control tubes had reached an optical density of about 0.5–1.2 (600 nm). All experiments were performed in duplicate or triplicate; the density of control cultures was set at 100%.

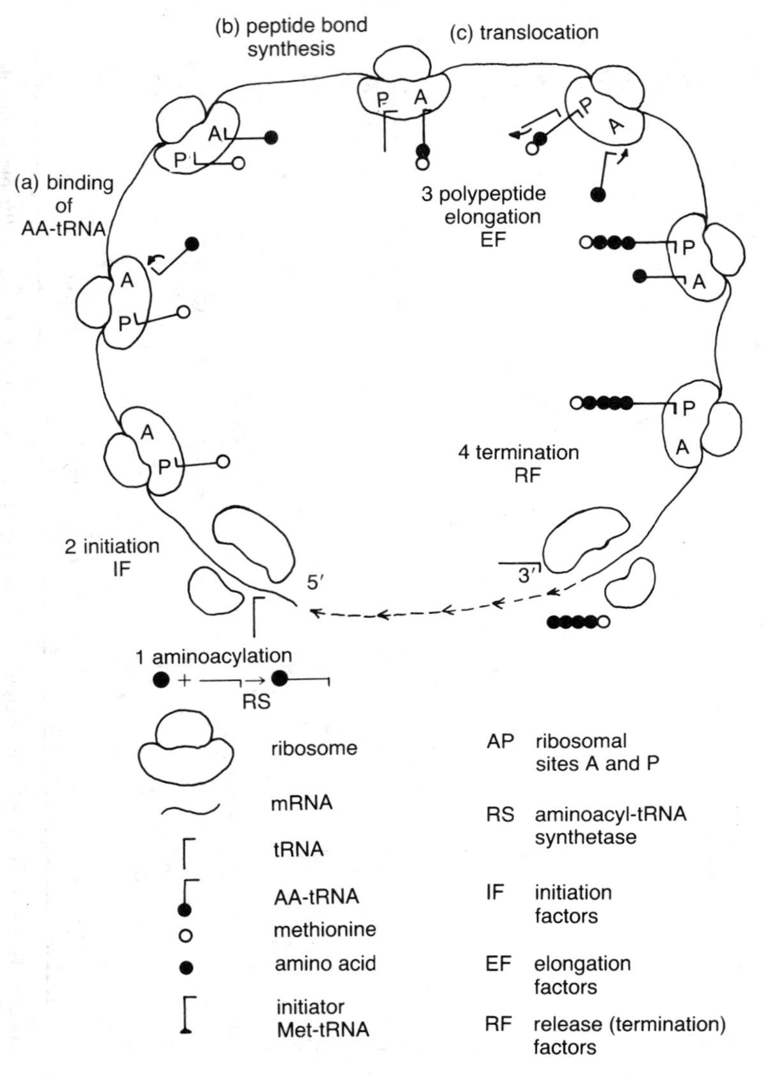

Fig. 10.2 General scheme of protein biosynthesis. The multi-step process of protein biosynthesis is presented as a chain of events with special emphasis on the steps investigated *in vitro*.

1. Aminoacylation of tRNA to form AA-tRNA in the presence of aminoacyl-tRNA-synthetase.
2. Initiation of protein biosynthesis.
3. Elongation of the polypeptide chain:
   (a) binding of aminoacyl-tRNA to ribosomal A-site which is catalysed by elongation factor EF1;
   (b) peptide bond synthesis;
   (c) translocation of newly synthesized peptidyl-tRNA catalysed by elongation factor EF2.
4. Termination of protein biosynthesis.

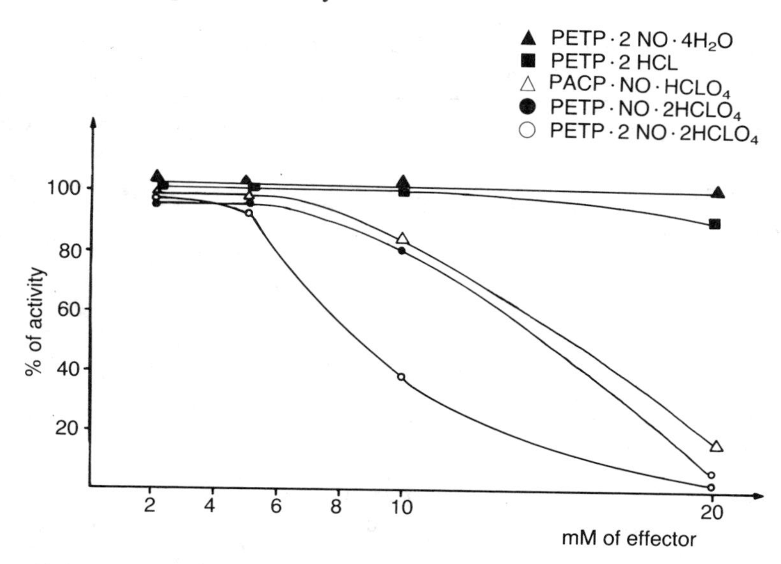

Fig. 10.3 Inhibition of aminoacylation by alkaloids. In a volume of 50 μl: 100 mM Hepes (pH 7.8); 4 mM ATP, 6 mM $MgCl_2$, 2 mM β-mercaptoethanol, 120 nM phenylalanine (1 μCi); 70 mM KCl, 50 μg tRNA phe, 10 μg synthetase. The mixture was incubated at 37°C for 15 min and terminated by phenol extraction. Phe-tRNA was precipitated from the supernatant (K-acetate, ethanol).

therefore be affected by these molecules since they are both a potential target and basic to all living systems. In this communication we report on preliminary experiments concerning the interaction of alkaloids with defined steps of protein biosynthesis.

Protein biosynthesis is a complex multi-step process (Fig. 10.2). We have established *in vitro* assays for various steps (Twardowski *et al.*, 1981) and can determine:

(1) the charging of transfer RNA by cognate amino acids;
(2) the binding of phe-tRNA to poly-U programmed ribosomes;
(3) total protein biosynthesis.

1. Transfer RNA specific for phenylalanine and phe-tRNA-synthetase were purified from seeds of *Lupinus luteus* (Rafalski *et al.*, 1977; Joachimiak *et al.*, 1978). The effect of five model alkaloids on the aminoacylation of tRNA is illustrated in Fig. 10.3, indicating effective inhibition at concentrations between 10 and 20 mmol/l.

2. For binding of phe-tRNA to poly-U programmed eucaryotic ribosomes the protocol given in Pulikowska *et al.* (1979) and Pulikowska and Twardowski (1982) was followed. The elongation factor EF1 was purified from wheatgerm. The influence of time, temperature and the EF1 concentration on phe-tRNA binding is illustrated in Fig. 10.4. Using this

TABLE 10.5 *Conditions of phe-tRNA binding and elongation in poly-U programmed eucaryotic ribosomes*

Conditions[a]		Binding of phe-tRNA (%)
Complete assay		100
without ribosomes		0
poly-U		0
EF1		0
GTP		30
Mg^{2+}		10
K^+		10
plus effectors		
+ cytisine	2.6 mM	11
	0.3 mM	83
+ lupanine	2.2 mM	55
	0.2 mM	100
+ sparteine	2.7 mM	59
	0.3 mM	100
+ 13-hydroxylupanine	2.7 mM	52
	0.3 mM	98
+ 13-tigloyloxylupanine	1.8 mM	32
	0.2 mM	94
+ 17-oxosparteine	3.1 mM	47
	0.3 mM	87
+ quinine	1.8 mM	62
	0.2 mM	84
+ emetine	0.9 mM	46
	0.09 mM	77
+ sanguinarine	1.3 mM	22
	0.13 mM	96
+ papaverine	1.3 mM	87
	0.13 mM	100
+ scopolamine	2.9 mM	89
	0.3 mM	96
+ berberine	1.7 mM	41
	0.17 mM	82
+ strychnine	1.6 mM	84
	0.2 mM	99
+ cycloheximide	1.7 mM	7
	0.17 mM	11
+ tannine	0.05%	0
	0.005%	0
+ quercetine	1.6 mM	25
	0.17 mM	79
+ thymol	4.4 mM	27
	0.4 mM	100
+ camphor	3.3 mM	85
	0.3 mM	100

TABLE 10.5 Cont'd

Conditions		Binding of phe-tRNA (%)
+ essential oils		
Valeriana	0.05%	85
	0.005%	100
Chamomilla	0.05%	85
	0.005%	100
Thymian	0.05%	73
	0.005%	88
Cassia	0.05%	76
	0.005%	100
Picea	0.05%	89
	0.005%	100
Citrus	0.05%	82
	0.005%	100
Mentha	0.05%	76
	0.005%	100
Eucalyptus	0.05%	65
	0.005%	82

[a] According to Pulikowska *et al.* (1979) and Pulikowska and Twardowski (1982) the assay system consisted of:

In a total volume 50 µl: 50 mM Tris-HCl (pH 7.5); 60 mM KCl, 5 mM $MgCl_2$; 3 mM DTT; 0.1 mM GTP; 10 ng poly-U; 30 pmol ^{14}C-phe-tRNA (300 cpm $pmol^{-1}$); 0.3 A_{260} units of ribosomes from wheatgerm and 1–10 µg elongation factor EF1.

Incubation was 5 min at 37°C. The reaction was terminated by adding 1 ml ice-cold 10 mM Tris (pH 7.6) buffer containing 80 mM KCl and 10 mM $MgCl_2$. The mixture was filtered under vacuum through a Millipore nitrocellulose filter (0.45 µm) and washed three times with the buffer. The radioactivity bound on the filter was determined by scintillation counting.

Essential oils were purchased from Serva (Heidelberg); the other compounds from Sigma or Roth (Karlsruhe). Quinolizidine alkaloids were isolated from plants (after Wink, 1983).

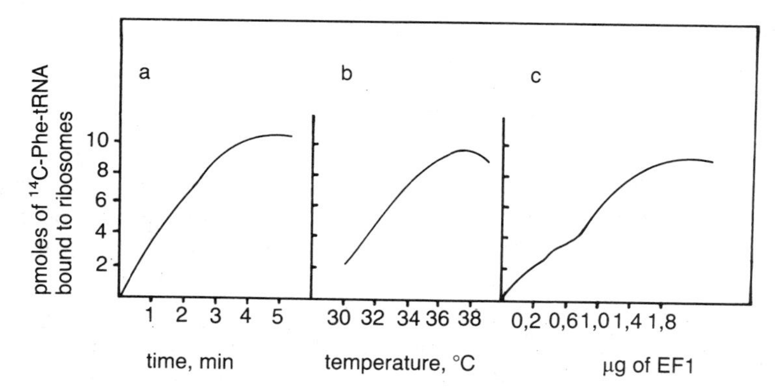

Fig. 10.4 Dependence of phe-tRNA binding on (a) time, (b) temperature and (c) the concentration of EF1. Assay conditions as in Table 10.5.

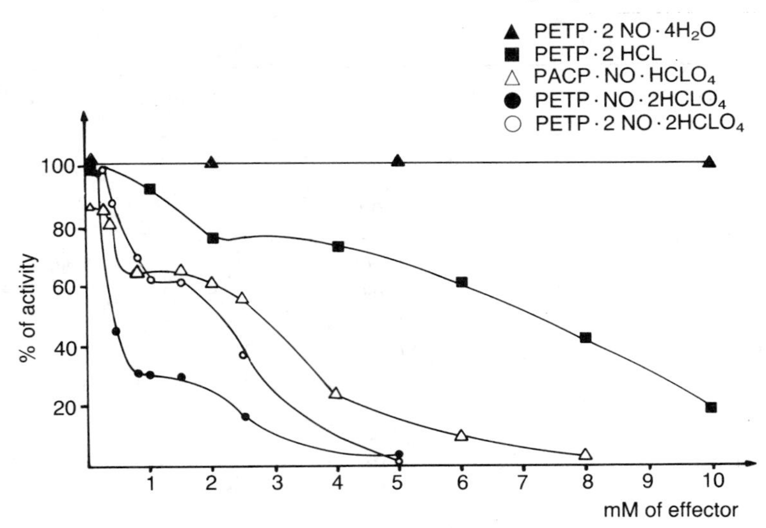

Fig. 10.5 Inhibition of phe-tRNA binding by model alkaloids. Assay conditions as in Table 10.5.

system we have tested a number of alkaloids and crude essential oils (Table 10.5, Figs 10.5, 10.6). The highest inhibitory activity had tannin, a strong inhibitor of enzymes, such as reverse transcriptase (Kakiuchi *et al.*, 1985), followed by cycloheximide, a known translation inhibitor. Most alkaloids affected the phe-tRNA binding at a concentration of 0.05%, which seems to be significant since other secondary products, e.g. essential oils, were rather inactive at these concentrations.

3. In order to measure the complete protein biosynthesis, a commercial wheat germ lysate (Amersham-Buchler) was employed. As shown in Table 10.6, this assay system is also suitable for studying the interaction of natural products with the translation process.

10.7 DISCUSSION AND PROSPECTS

Alkaloids are usually considered as toxic molecules directed against mammals and other herbivore vertebrates and their interactions with the nervous system are comparatively well studied. In our investigation we tried to show that alkaloids can also interact with other plants, microorganisms and insects. As shown in Tables 10.1–10.4, a number of alkaloids are obviously allelopathic, although other metabolites, such as essential oils, seem to be more active in this context. It remains to be studied whether these molecules virtually play an allelopathic role in nature. Are these alkaloids excreted by roots or are they leached from

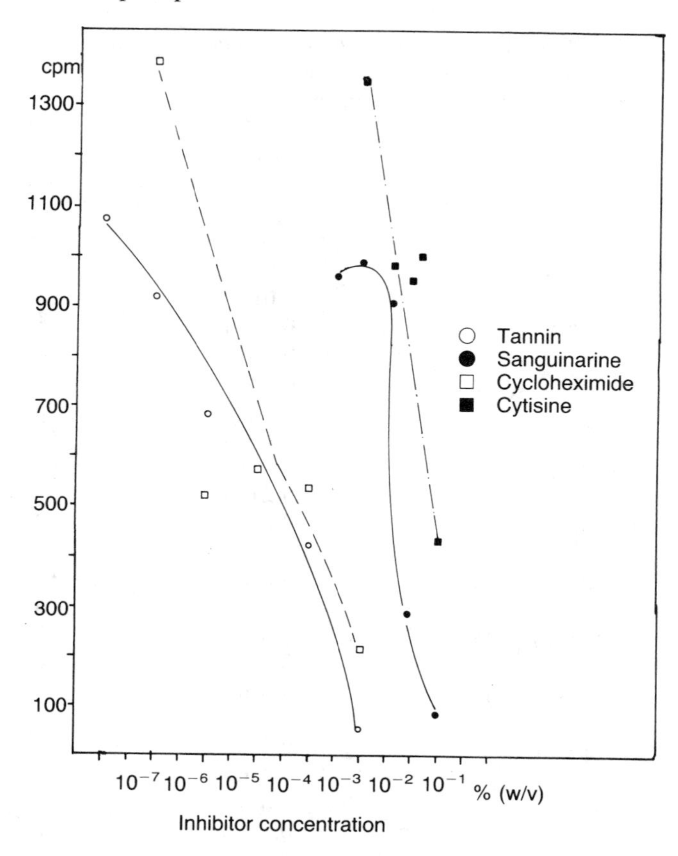

Fig. 10.6 Inhibition of phe-tRNA binding by secondary metabolites. Assay conditions as in Table 10.5.

aerial plant parts? Are the concentrations *in vivo* and in the soil high enough to promote their inhibitory action?

For quinolizidine alkaloids we have previously shown that they can inhibit the germination of *Lactuca* and grass seeds (Wink, 1983) in a dose-dependent fashion. During germination of *Lupinus albus*, these alkaloids are excreted in the rhizosphere in considerable amounts (Wink, 1983). *In vivo* the growth of *Lepidium* seedlings was inhibited when they were in the direct neighbourhood of lupin seedlings (Wink, 1985). Evidence for other allelopathic incidences of alkaloids has been reviewed in Rice (1984) and Levitt and Lovett (1985).

Considering these activities the question is how the alkaloids manage to affect the inhibition of seed germination, of seedlings, and of bacterial growth. Our *in vitro* experiments on protein biosynthesis show that this elementary process can be influenced by alkaloids. This would also ex-

TABLE 10.6 *Effect of secondary metabolites on protein biosynthesis (*in vitro *translation in wheatgerm system)*

Compound		Activity (%)
Control		100
+Tannine	0.03%	9
	0.01%	24
+Quercetine	0.6 mM	5
	0.2 mM	18
+Thymol	0.6 mM	10
	0.2 mM	22
+Camphor	0.6 mM	64
+Cytisine	0.6 mM	68
	0.2 mM	65
+Lupanine	0.6 mM	62
	0.2 mM	98
+Sparteine	0.6 mM	62
	0.2 mM	64
+13-Hydroxylupanine	0.6 mM	57
	0.2 mM	86
+13-Tigloyloxylupanine	0.6 mM	63
	0.2 mM	71
+17-Oxosparteine	0.6 mM	65
	0.2 mM	66
+Quinine	0.6 mM	57
	0.2 mM	60
+Emetine	0.6 mM	70
	0.2 mM	70
+Sanguinarine	0.6 mM	60
	0.2 mM	69
+Papaverine	0.6 mM	55
	0.2 mM	63
+Berberine	0.6 mM	64
	0.2 mM	64
+Scopolamine	0.6 mM	44
	0.2 mM	57

Assay condition: 30 µl total volume, containing wheatgerm lysate (including mRNA, KCl, 19 amino acids; ^{35}S-methionine) and effectors. Incubation 60 min at 25°C. The reaction was terminated by adding 200 µl 1 M NaOH and kept at 37°C for 10 min. Then 1 ml 25% trichloroacetic acid was added and kept on ice for 10 min. The radioactive protein was collected on nitrocellulose filters, which were washed four times with 10% TCA. Radioactivity was determined by liquid scintillation counting.

plain the wide range of activities found for alkaloids since protein biosynthesis is a process common to all organisms. Alkaloids and other secondary metabolites are often present in plants in concentrations between 0.5 and 200 mM or 0.1–5%, which would mean that these concentrations should be high enough for the effects observed *in vitro* (Wink, 1985, 1987, 1988). Certainly we need much more data to understand whether protein biosynthesis is indeed affected *in vivo* and how many other processes are additionally inhibited.

ACKNOWLEDGEMENTS

The authors thank Prof. Dr M. Wiewiorowski for encouragement and interest in this work. We also thank K. Ludwiczak, S. Schmidt, U. Schade, M. Weyerer and J. Wink for technical assistance. This work was supported by the Deutsche Forschungsgemeinschaft and the Fonds der Chemischen Industrie (to M.W.) and the Polish Academy of Sciences (research grant 09.7.1, to T.T.).

REFERENCES

Boppré, M. (1986) Insects pharmacophagously utilizing defensive plant chemicals. *Naturwiss.*, **73**, 17–26.

Einhellig, F. A., Leather, G. R. and Hobbs, L. L. (1985) The use of *Lemna minor* as a bioassay in allelopathy. *J. Chem. Ecol.*, **11**, 65–72.

Harborne, J. B. (1982) *Introduction to Ecological Biochemistry*, Academic Press, London, New York.

Harborne, J. B. (1986) Recent advances in chemical ecology. *Nat. Prod. Rep.*, **3**, 323–44.

Joachimiak, A., Barciszewski, J., Barciszewska, M. *et al.* (1978) Purification and properties of methionyl-tRNA-synthetase from yellow lupine seeds. *FEBS Lett.*, **93**, 51–4.

Kakiuchi, N., Hattori, M. and Namba, T. (1985) Inhibitory effect of tannins on reverse transcriptase from RNA tumor virus. *J. Nat. Prod.*, **48**, 614–21.

Leather, G. R. and Einhellig, F. A. (1988) Bioassay of naturally occurring allelochemicals for phytotoxicity. *J. Chem. Ecol.*, **14**, 1821–8.

Levin, D. A. (1976) The chemical defences of plants to pathogens and herbivores. *Ann. Rev. Ecol. Syst.*, **7**, 121–59.

Levitt, J. and Lovett, J. V. (1985) Alkaloids, antagonisms and allelopathy, *Biol. Agr. Hort.*, **2**, 289–301.

Mothes, K. (1955) Physiology of alkaloids. *Ann. Rev. Plant Physiol.*, **6**, 393–432.

Paech, K. (1950) *Biochemie und Physiologie sekundärer Pflanzenstoffe*, Springer, Berlin, Heidelberg.

Pulikowska, J. and Twardowski, T. (1982) The elongation factor 1 from wheat germ: structural and functional properties. *Acta Biochem. Pol.*, **91**, 245–57.

Pulikowska, J., Barciszewski, J., Barciszewska, M. *et al.* (1979) Effect of elastase on elongation factor 1 from wheat germ. *Biochem. Biophys. Res. Comm.*, **91**, 1011–17.

Rafalski, A., Barciszewski, J., Gulewicz, K. *et al.* (1977), Nucleotide sequence of $tRNA^{PHE}$ from seeds of lupin (*Lupinus luteus*). Comparison of the major species with wheat germ tRNA. *Acta Biochem. Pol.*, **24**, 301–18.

Rice, E. L. (1984) *Allelopathy*, 2nd edn, Academic Press, Orlando, Florida.

Rosenthal, G. A. and Janzen, D. H. (1979) *Herbivores, Their Interaction with Secondary Plant Metabolites*, Academic Press, London, New York.

Schneider, D. (1987) In *Perspectives in Chemoreception and Behavior* (eds R. F. Chapman, E. A. Bernays, and J. G. Stoffolano), Springer, New York, pp. 123–42.

Stahl, E. (1888) Pflanzen und Schnecken. *Jenaer Z. Naturwiss.*, **22**, 557.

Swain, T. (1977) Secondary compounds as protective agents. *Ann. Rev. Plant Physiol.*, **28**, 479–501.

Twardowski, T., Pulikowska, J. and Wiewiorowski, M. (1981) Inhibitory effect of selected quinolizidine alkaloids and their derivatives and analogues on the phe-tRNA binding to ribosomes. *Bull. Pol. Acad. Sci.*, **XXIX**, 129–40.

Waller G. R. (ed.) (1987) *Allelochemicals: Role in Agriculture and Forestry*, ACS Symp. Ser. 330, Amer. Chem. Soc., Washington, DC.

Wink, M. (1983) Inhibition of seed germination by quinolizidine alkaloids. Aspects of allelopathy in *Lupinus albus* L. *Planta*, **158**, 365–8.

Wink, M. (1985) Chemische Verteidigung der Lupinen: Zur biologischen Bedeutung der Chinolizidinalkaloide. *Pl. Syst. Evol.*, **150**, 65–81.

Wink, M. (1987) Chemical ecology of quinolizidine alkaloids, in *Allelochemicals: Role in Agriculture, Forestry and Ecology* (ed. G. R. Waller), ACS Symp. Ser. 330, Amer. Chem Soc., Washington, DC, pp. 524–33.

Wink, M. (1988) Plant breeding: importance of plant secondary metabolites for protection against pathogens and herbivores. *Theor. Appl. Gen.*, **75**, 225–33.

CHAPTER 11

Alfalfa saponins – the allelopathic agents

W. Oleszek, M. Jurzysta and P. M. Górski

11.1 INTRODUCTION

The legume–grain crop sequences were commonly used throughout the beginning of this century. The practice of including forage legumes in crop rotations declined due to the increased availability and relatively low cost of nitrogen fertilizers (CAST, 1980; Hesterman *et al.*, 1986). There is renewed interest among producers in using forage legumes as inexpensive sources of N in crop production due to the increasing price of N fertilizer (Heichel and Barnes, 1984) and to the increasing interest in organic farming techniques.

Alfalfa has been recognized as a valuable legume crop, yielding high protein pasture. Nitrogen fixed by its symbiotic *Rhizobium* bacteria is contributed to the succeeding non-fixing plant upon the decomposition of legume top or root material. The deep root system of alfalfa plants enables the transportation of calcium from subsoil to the surface 0–20 cm layer, thus improving its physical properties. Alfalfa ploughdown also provides some benefits in the breaking of disease cycles and accumulation of soil organic matter. The incorporation of organic matter from alfalfa stands ranges from 10 up to 20 tonnes per hectare and is equivalent to that of 90–125 $kg\,N\,ha^{-1}$ (Bruulsema and Christie, 1987). This value can be much higher (300 $kg\,N\,ha^{-1}$) when herbage is harvested several times during the season and retained on the soil surface for incorporation in the fall (Hesterman *et al.*, 1987). Thus, in most cases legume N availability is greater from alfalfa than from any other legume crop. However, the larger N supply does not always promote a higher grain yield from the succeeding crop, suggesting that some factors other than N availability – e.g. soil, climatic and stand age – should be considered (Bruulsema and Christie, 1987; Hesterman *et al.*, 1986).

There is some evidence that some phytotoxic substances are involved in this phenomenon and saponins were identified to be the growth inhibitor.

Allelopathy: Basic and applied aspects Edited by S. J. H. Rizvi and V. Rizvi
Published in 1992 by Chapman & Hall, London ISBN 0 412 39400 6

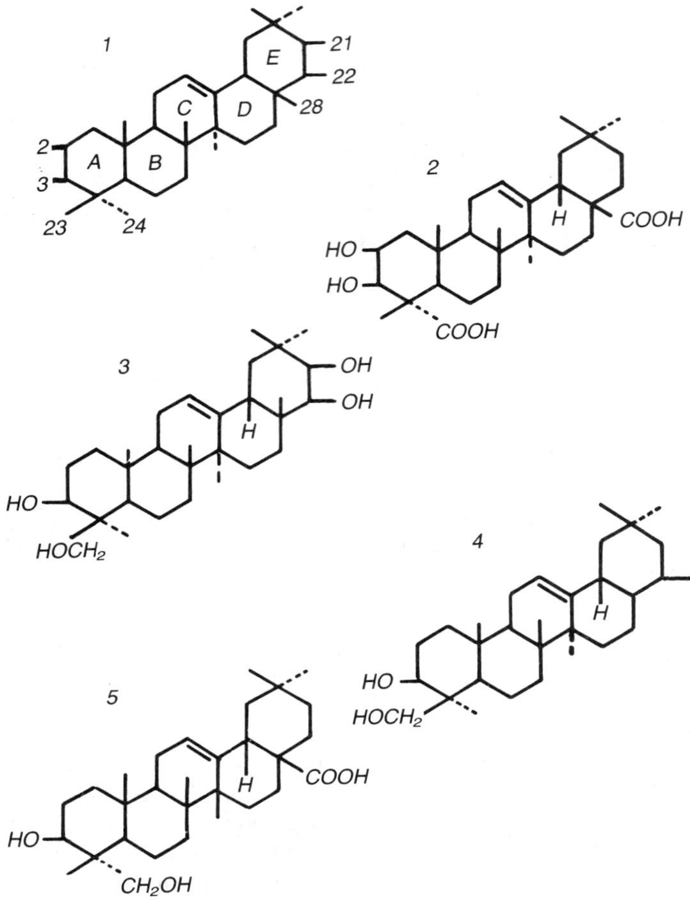

Fig. 11.1 Structures of some alfalfa aglycones: 1, general structure; 2, medicagenic acid; 3, soyasapogenol A; 4, soyasapogenol B; 5, hederagenin.

This chapter presents an overview of research on the germination and growth regulation of plants by alfalfa saponins.

11.2 CHEMICAL CHARACTERISTICS OF ALFALFA SAPONINS

Alfalfa saponins are a mixture of many glycosides which on acid hydrolysis yield pentoses, hexoses, uronic acids and aglycones, the non-sugar parts of saponin moiety. The aglycones of alfalfa saponins are pentacyclic triterpenoids that originate from oleane skeleton (Fig. 11.1). Depending on the substitution at skeleton, they can be recognized as soyasapogenol

A, B, hederagenin and medicagenic acid. In hydrolysates some other soyasapogenols (C, D, E and F) have also been found but these compounds were proved to be artefacts arising in acidic condition from soyasapogenol B (Gestetner *et al.*, 1970; Jurzysta, 1982, 1984; Price *et al.*, 1986).

The make-ups of particular aglycones strongly differ depending on the plant part and plant species (Birk, 1969; Jurzysta, 1982; Pedersen, 1965; Price *et al.*, 1987b). It has been documented that alfalfa roots are much more abundant in medicagenic acid glycosides than any other plant parts (Gestetner *et al.*, 1970; Oleszek and Jurzysta, 1986a; Oleszek *et al.*, 1987a). On the other hand, the seeds of some alfalfa species, e.g. *Medicago sativa*, are completely free of these compounds whereas some other species, like *M. lupulina*, possess them in quite high quantities (Jurzysta, 1973a, b).

Medicagenic acid glycosides rarely occur in the plant kingdom, but to date they have been found in numerous species of the genus *Medicago* (Papilionaceae) (Jurzysta *et al.*, 1988a, b), *Medicago* (former *Trigonella*, Small *et al.*, 1987) (Papilionaceae) (Oleszek *et al.*, 1987b; Jurzysta *et al.*, 1988a, b), *Herniaria* (Caryophyllaceae) (Bukharov and Shcherbak, 1970; Klein *et al.*, 1982) and in *Castanospermum australe* (Papilionaceae) (Hiller and Voigt, 1977). In the family Papilionaceae the medicagenic acid glycosides proved to be the chemotaxonomic marker distinguishing the genus *Medicago* from other closely related genera, e.g. *Trigonella* and *Melilotus* (Jurzysta *et al.*, 1988a). The full chemical structures of most of the particular saponin glycosides remain unknown. Up to date only a few structures have been fully elucidated. Morris *et al.* (1961) reported the presence of medicagenic acid 3-*O*-glucopyranoside in alfalfa roots. This compound, as well as bisdesmoside 3,28-di(*O*-glucopyranoside) of medicagenic acid, was identified by Timbekova and Abubakirov (1984) in *M. sativa* and confirmed by Levy *et al.* (1986) and Oleszek (1988) in *M. media* roots.

Moreover, their presence was also confirmed in *M. lupulina* roots (Oleszek *et al.*, 1988a). It has been shown that while *M. media* contained in its roots at least 11 medicagenic acid glycosides (Oleszek and Jurzysta, 1986a), the 3-*O*-mono- and 3,28-di-(*O*-glucopyranoside) of medicagenic acid were the only compounds of this kind in *M. lupulina* roots.

The structures of two additional alfalfa medicagenic acid glycosides have been established. Morris and Hussey (1965) isolated from alfalfa flowers a medicagenic acid glycoside with trisaccharide rhamnose-glucuronic acid-glucose linked at the 3-*O* position. Gestetner (1971) reported the isolation of medicagenic acid 3-*O*-glucopyranoside-(1→6)-glucopyranoside-(1→3)-glucopyranoside from alfalfa tops. This compound exerted strong physiological activity. Two hederagenin glycosides: 3-*O*-glucopyranoside-(1→2)-arabinopyranoside and 3-*O*-arabinopyranoside-(1→2)-glucopyranoside-(1→2)-arabinopyranoside – were identified by

Timbekova and Abubakirov (1985) in *M. sativa*, and 3-*O*-glucopyranoside of hederagenin and soyasaponin I (tetrasaccharide of soyasapogenol B) was found in *M. lupulina* (Oleszek *et al.*, 1988a). The isolation and identification of individual glycosides from alfalfa plant material has been complicated because of the multiplicity of the forms present. Some plant parts, e.g. leaves, contain as many as 30 individual compounds (Jurzysta, 1982), polarity of which is very similar, which makes separation extremely difficult. Recent developments in separation techniques like flash chromatography (Price *et al.*, 1987a) and the improvement in fractionation on the normal and reversed phase supports (Oleszek, 1988), give some new hopes of progress in this field.

11.3 GERMINATION AND PLANT GROWTH REGULATION BY SAPONINS

It has been documented that the biological activity of alfalfa saponins, e.g. haemolytic and fungitoxic, can be attributed mostly to the presence of medicagenic acid glycosides. From a saponin mixture they can be separated by precipitation with cholesterol, due to their high affinity to this compound. This method allowed separation of *M. media* root saponins into cholesterol precipitable (medicagenic acid glycosides) and non-precipitable (hederagenin and soyasapogenol glycosides) fractions. These fractions were used in petri dish tests to determine their allelopathic potential against wheat (Oleszek and Jurzysta, 1986a). Medicagenic acid glycosides inhibited wheat root growth by 50% when present in the growth medium at the concentration of 100 ppm (Table 11.1). Wheat shoots at this concentration also suffered some retardation, however to a much lower degree than roots. At higher saponin concentration levels wheat germination also fell severely if compared with the distilled water

TABLE 11.1 *Inhibition of wheat seedling growth by alfalfa root saponins*

	Length of wheat seedlings (mm)			
	Cholesterol (precipitable)		Cholesterol (non-precipitable)	
Saponin concentration (ppm)	Tops	Roots	Tops	Roots
100	72.0	49.3	77.7	68.3
500	52.0	21.3	62.0	41.3
1000	31.1	9.3	60.3	31.0
Control (H_2O)	87.7	88.3		

as a control treatment. Non-precipitable saponins also retarded wheat seedling growth, but to a degree a few times lower than precipitable ones. This fraction consisted of soyasapogenol B and hederagenin glycosides and it is difficult to judge unequivocally which glycosides bear responsibility for this slight growth retardation. It seems, however, very probable that the harmful effect can be attributed rather to hederagenin glycosides. This conclusion comes from the fact that two soyasapogenol B glycosides isolated from red clover roots (*Trifolium pratense*) did not show any harmful effect on wheat seedlings (Oleszek, 1985; Oleszek and Jurzysta, 1986b). It becomes unquestionably clear that, of alfalfa saponins, medicagenic acid glycosides possess an inhibiting effect on the growth of other plants.

The cholesterol precipitable fraction of saponins isolated from the cimmaron variety of alfalfa depressed the germination and the growth of the following weeds: cheat (*Bromus secalimus*), barnyardgrass (*Echinochloa crus-galli*), pigweed (*Amaranthus retroflexus*), dandelion (*Taraxacum vulgare*) and coffeeweed (*Sesbania exaltate*) (Jurzysta and Waller, unpublished). The most sensitive species were cheat and barnyardgrass: saponins at 0.001% concentration depressed the growth of roots and shoots to 19% and 28% for cheat, and 11% and 17% for barnyardgrass, respectively. The growth of the seedlings of coffeeweed was inhibited at a higher level of saponins (0.1%).

The role of alfalfa saponin as an allelopathic agent was first reported by Mishustin and Naumova (1955), who showed that the yield of cotton grown as a succeeding crop after alfalfa was lower than in other stands. This decrease was higher after older stands, which suggested that some allelochemicals accumulated in the soil were involved. Mishustin and Naumova (1955) further suggested that saponins leached into the soil from alfalfa roots were responsible for the yield reduction.

These suggestions were futher supported by laboratory tests showing that alfalfa saponins were detrimental to the germination of cotton seeds (Marchaim *et al.*, 1975). A decrease in the percentage of germinated seeds was also observed when germination was performed on the soils taken from 2-, 3- and 4-year-old alfalfa stands. In this case the percentage of cotton germination was 63, 61 and 51%, respectively, and was correlated to the *Trichoderma viride* growth retardation by water extracts of particular soils.

These observations provoked several other researchers to look closer at the allelopathic potential of alfalfa plants, though Lawrance and Kichler (1962) showed that alfalfa roots contained water-soluble substances toxic to creasted wheatgrass, Russian wild ryegrass, intermediate wheatgrass, sorghum, alfalfa, sweetclover, wild barley, dandelion, wheat, oats and barley seedlings. Guenzi *et al.* (1964) studied the influence of water extracts of samples from Buffalo and Ranger alfalfa varieties at three cuttings and at six stages of growth, on the germination and seedling growth of corn. They showed that all the extracts contained water-soluble substances that reduced shoot and root growth of corn seedlings. While they were not able to find any differences with respect to phytotoxicity of varieties, they found significant differences between extracts from plants at various stages of growth and cuttings.

Variety-dependent differences in the allelopathic potential of alfalfa were, however, reported by Pedersen *et al.* (1966). They have additionally shown that leaf saponins were more toxic than those isolated from the stems.

Allelopathic potential was also a characteristic feature of saponin isolated from alfalfa seeds. Jurzysta (1970) isolated saponin fractions from the seeds of two lucerne species: *M. media* and *M. lupulina*. Saponins from *M. lupulina* seeds reduced the germination and seedling growth of barley, oat, wheat and rye when applied at the concentration of 0.1 and 0.5% in petri dish tests. The same concentrations of *M. media* saponins retarded the growth of barley and oat seedlings, but had no influence on wheat and rye growth. Only the oat seedlings responded in a similar way to the presence of saponins from the seeds of both species. These results suggested that some qualitative differences exist among saponins from these two species and thus their allelopathic potentials are different.

Differences in the suppressive activity of alfalfa species were also noted when *M. lupulina* and *M. media* finely powdered roots were incorporated into the sand at the rate of 0.25 and 0.5% and then wheat seeds were planted (Oleszek *et al.*, 1987a; Table 11.2). Roots and shoots of wheat seedlings were more strongly retarded in the sand amended with *M. media* than with *M. lupulina* roots. This suppressive activity against wheat seedlings was not correlated with an inhibiting effect on *T. viride*. This finding supports the view that there are some structural differences among root saponins of these species and different biological activities are structurally dependent. As was noted above, most of those structures remain unknown. Thus most of the allelopathic tests which have been performed so far applied purified saponin mixtures or fractions. In some cases finely powdered alfalfa roots can be used and in many cases they can provide some satisfactory results. This technique was successfully used by Oleszek and Jurzysta (1987) to study the behaviour and allelopathic potentials of alfalfa root saponins in different soil environments. The incorporation of alfalfa roots at different rates into sand resulted in a severe decrease of wheat germination and seedling growth (Table 11.3). Seedling roots suffered more severe damage than shoots in all treatments. Wheat roots were completely retarded at the concentration of 0.25%. This is approximately the concentration that would occur in field conditions if root dry matter of an alfalfa stand were uniformly distributed in the 0–20 cm soil layer. This high allelopathic potential is, however, reduced in heavier soils (Table 11.4) because of their higher absorptive capacity and higher activity of soil microorganisms.

TABLE 11.2 *Wheat growth retarding, antifungal and haemolytic activity of* M. lupulina *and* M. media *roots*

Biotest		*Medicago lupulina*	*Medicago media*
Seedling root	a*	35	69
Inhibition, % control	b	52	75
Seedling shoot	a	5	25
Inhibition, % control	b	24	42
Trichoderma viride	c	29	1
Growth inhibition	d	54	45
% control	e	62	52
Haemolytic activity		288	432

* a and b – rate of root incorporation 0.25 and 0.5% respectively; c, d and e – amount of roots in fungus growth medium 30, 50 and 100 mg per 100 ml, respectively.

TABLE 11.3 *The effect of the rate of incorporation of alfalfa roots into silica sand on wheat seedling growth inhibition (%)*

	Residue rates (%)		
Wheat growth % of control	0.25	0.5	1.0
Wheat root growth inhibition	71	78	87
Wheat shoot growth inhibition	28	42	50

TABLE 11.4 *The effect of alfalfa root incorporated at the rate of 1% into soils with different textures on wheat seedling growth*

	Wheat seedling growth inhibition (%)	
Soil type	Root	Shoot
Loose sand	86	52
Coarse sand	53	41
Loamy sand	47	41
Clay loam	17	27

The amount of pure medicagenic acid glycoside fraction adsorbed in loose sand, coarse sand, loamy sand and clay loam was, respectively, 0, 6, 24 and 78 mg per 100 g of the soil. The process of absorption is one of the natural means of saponin detoxification. Absorbed saponins, removed from the soil solution, are probably not available for the plant receiver and thus present no hazard for plant seedlings. It seems unlikely that saponins once absorbed in the soil colloids can be released into the soil solution again. It is probable that, while absorbed, they are decomposed by soil microorganisms and utilized as carbon resources.

In this light the claims of some authors, that in alfalfa stands saponins released into the soil can be accumulated there, making older alfalfa stands more suppressive of the following crop, seem very doubtful. Such a phenomenon, however, was reported by Mishustin and Naumova (1955) and by Shany *et al.* (1970) who found that, in the soil in which alfalfa was grown for three consecutive years, 25% of cotton seeds did not germinate. It seems reasonable to assume that older stands introduce more root dry matter, decomposition of which can release more allelochemicals, creating in this way more unfavourable conditions for the succeeding crop (Oleszek and Jurzysta, 1987). This view was further supported recently by Read and Jensen (1989) who showed that water extracts of crop residues caused a greater inhibition of seedling development than either whole or screened soil extracts. Moreover, they reported that residues from soil from old alfalfa stands were more inhibitory than residues from a soil cropped to alfalfa for one year.

In model experiments, where roots are incorporated into the soil as a fine powder, release of saponins is immediate. In the soil condition this release can be long-lasting depending on the environmental conditions, water availability, temperature, aeration and microorganism activity. In such circumstances the concentration of saponins may not be high enough to harm plant growth. In the soil, however, distribution of alfalfa roots in a ploughed alfalfa stand is not uniform and local concentrations of an inhibitor may be extremely high. This may lead to variable succeeding plant performance all over the field, as severity of injury to seedling roots is related to the amount of plant residue in the immediate vicinity of the root (Patrick, 1971).

The inhibiting effect is not long-lasting. Released saponins are immediately attacked by soil microorganisms and the soil solution becomes free of inhibitors quite quickly. The rate of the saponin decomposition is strongly dependent on the soil structure, and this can be attributed to the number of microorganisms present (Fig. 11.2).

The decomposition of saponins proceeds through the cleavage of sugar components from the glucosidic chain to the aglycone. In the next step, aglycone is probably reduced to simple compounds that can be utilized by microorganisms.

Several reports related to alfalfa stand establishment and production

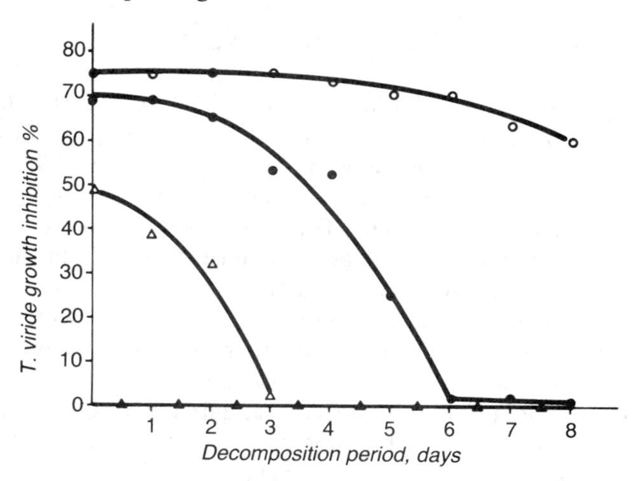

Fig. 11.2 *T. viride* growth inhibition as affected by water extracts of alfalfa roots decomposed in differently textured soils. Symbols: ○—○ loose sand, ●—● coarse sand, △—△ loamy sand, ▲—▲ heavy loam (After Oleszek and Jurzysta, 1987. Reproduced with permission, Oleszek, W. and Jurzysta, M. and Kluwer Academic Publishers, The Netherlands.)

referred to allelopathy. Thickening an established alfalfa stand by interseeding or overseeding has not in general been successful. This alfalfa 'soil sickness' may be related to nutrient depletion or overshading, but accidental incorporation into soil of young alfalfa forage from ageing plants while interseeding alfalfa to thicken old stands may introduce water-soluble substances toxic to the new alfalfa seedling (Webster *et al.*, 1967; Kehr *et al.*, 1983). Toxic substances from old stems, crowns and roots of alfalfa may be detrimental to the establishment of new seedlings. Alfalfa autotoxicity for seedling growth was reported in laboratory and greenhouse studies (Nielsen *et al.*, 1960; McElgunn and Heinricks, 1970; Ream *et al.*, 1977; Walgenbach *et al.*, 1977; Klein and Miller, 1980; Jensen, 1984). An autotoxic agent was reported to be present in fresh alfalfa leaves, stems, crowns, and in dry hay, old roots, soil and soil residues. Exudates of different saponin level cultivars indicated that saponin content is not the suspected phytotoxic factor (Miller, 1983a, b). This view was further supported by experiments performed by Jurzysta and Waller (unpublished data) showing that the allelopathic potential of alfalfa root saponins was several times lower against alfalfa than against wheat seedling growth. They also found that young seedlings of alfalfa contained high levels of asparagin, the amino acid strongly retarding alfalfa seedling growth (Jurzysta and Waller, unpublished data). This finding corresponds to the data of Miller (1983a, b), who showed by laboratory experiments that three-day old alfalfa seedlings release a toxic compound that inhibits germination and retards seedling elongation.

11.4 FUNGI GROWTH RETARDING ACTIVITY

An antifungal activity of alfalfa was recognized very early. The allelopathic action of alfalfa meal in the control of avocado root rot caused by *Phytophthora cinnamoni* was found by Zentmeyer (1963). Then the same author found that saponins are the main factors responsible for the fungitoxic effect (Zentmeyer and Tompson, 1967). Among other fungi, *Trichoderma viride* was found to be very sensitive to alfalfa saponins and Zimmer *et al.* (1967) have developed the bioassay for saponin quantification. This bioassay in several modifications is still widely used for saponin evaluation (Jurzysta, 1979). Further investigations into the fungistatic activity of different alfalfa extracts and saponin fractions revealed that, similar to haemolytic activity, antifungal activity is characteristic of medicagenic acid glycosides but not of soyasapogenol ones (Gestetner *et al.*, 1970; Assa *et al.*, 1972). However, Scardavi and Elliot (1967) claimed that the sugar chain of the saponin moiety is an important factor in determining its water solubility, an essential condition for biological activity.

The question whether antifungal activity of alfalfa saponins depends on aglycone structure or sugar chain composition was open as long as investigation was carried out on alfalfa extracts, and not on pure isolated saponins. We found that solubility is very important in the process of *in vitro* haemolyses. Insoluble medicagenic acid caused practically no haemolyses while soluble sodium salt of medicagenic acid caused at the same concentration massive haemolyses. This is not so with fungus *T. viride* growth retardation (Oleszek *et al.*, 1988b). However, Nonaka (1986) has recently shown that medicagenic acid exhibits low antifungal activity towards *T. viride* as compared with its glycosides.

New data concerning saponin antifungal activity and its chemical and structural composition were obtained with a number of isolated and synthetic saponins. The isolated saponins (chemical formula presented in Fig. 11.3) were subjected to *T. viride* biotests (Oleszek *et al.*, 1988b). It was found that compounds 1, 2 and 6 – namely medicagenic acid, sodium salt of medicagenic acid and medicagenic acid 3-*O*-glucopyranoside – all possess the highest and similar antifungal activity (Fig. 11.4). They completely retarded *T. viride* growth at the concentration of 0.2 mg $100\,ml^{-1}$. In contrast, the bisdesmoside, medicagenic acid-3,28-di(*O*-glucopyranoside) retarded growth to a much lower degree than the above compounds. Complete inhibition was observed at a concentration of 5 mg $100\,ml^{-1}$. The sharp drop in biological activity following the binding of glucose moiety to the C-28 carboxylic acid group may be due to blockage of an important active site. This possibility was examined in more detail by the bioassay performed with a number of derivatives of the compounds mentioned above. In all cases blockage of the carboxylic acid groups (compounds 4, 5, 7, 11) led to impaired growth, so that 55–60%

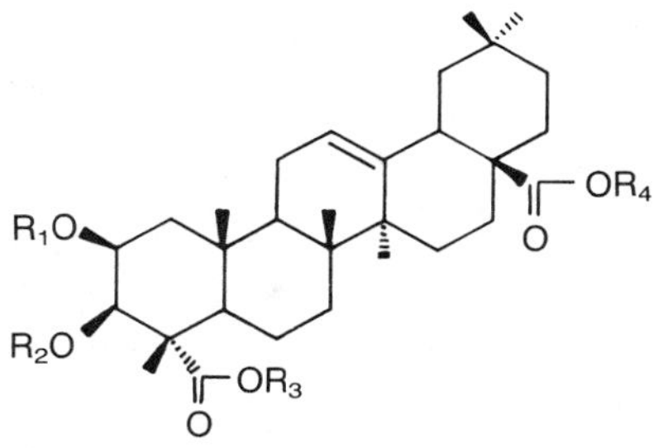

1: $R_1 = R_2 = R_3 = R_4 = H$
2: $R_1 = R_2 = H$, $R_3 = R_4 = Na$
3: $R_1 = R_2 = Ac$, $R_3 = R_4 = H$
4: $R_1 = R_2 = H$, $R_3 = R_4 = Me$
5: $R_2 = H$, $R_1 = R_3 = R_4 = Me$
6: $R_1 = H$, $R_2 = \beta$-D-Glc-p, $R_3 = R_4 = H$
7: $R_1 = H$, $R_2 = \beta$-D-Glc-p, $R_3 = R_4 = CH_3$
8: $R_1 = H$, $R_2 = \beta$-D-Glc-p, $R_3 = H$, $R_4 = CH_3$
9: $R_1 = Ac$, $R_2 = 4\ Ac\ \beta$-D-Glc-p, $R_3 = R_4 = H$
10: $R_1 = H$, $R_2 = \beta$-D-Glc-p, $R_3 = H$, $R_4 = \beta$-D-Glc-p
11: $R_1 = H$, $R_2 = \beta$-D-Glc-p, $R_3 = CH_3$, $R_4 = \beta$-D-Glc-p
12: $R_1 = Ac$, $R_2 = 4\ Ac\ \beta$-D-Glc-p, $R_3 = H$, $R_4 = 4\ Ac\ \beta$-D-Glc-p

Fig. 11.3 Schematic diagrams of compounds tested for activity against *T. viride* growth.

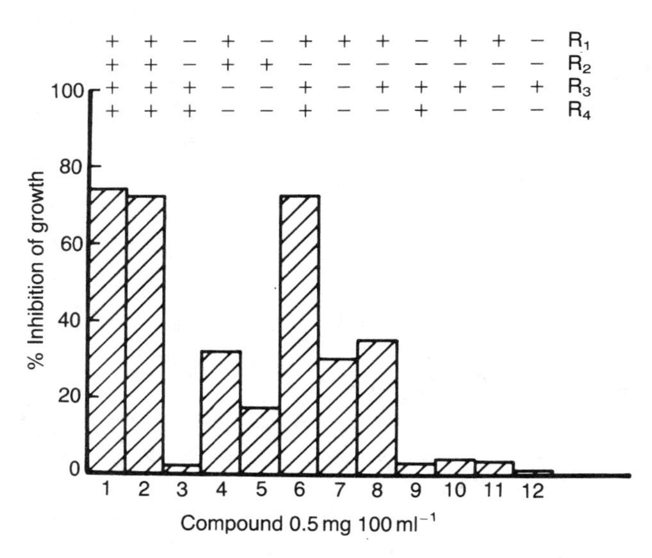

Fig. 11.4 Inhibition of *T. viride* growth by medicagenic acid, medicagenic acid-3-(*O*-glucopyranoside), medicagenic acid-3,28-di-(*O*-glucopyranoside) and their derivatives at a concentration 0.5 mg 100 ml^{-1} of growth medium. (For chemical formula see Fig. 11.3.) Symbols: + active group blocked; − active group free.

of the growth was maintained even at concentrations of 10 mg 100 ml^{-1}. Blocking of the –OH groups in the aglycone (compounds 3, 9, 12) or both aglycone and sugar moieties (compounds 9, 12) produced a very significant reduction in antifungal activity. These results indicated that the antifungal activity of medicagenic acid and its derivatives against *T. viride* was markedly dependent upon free $-CO_2H$ and –OH groups. Free –OH groups, however, seem to be more important than $-CO_2H$ ones in this respect (Oleszek *et al.*, 1988b).

Medicagenic acid 3-(*O*-glucopyranoside) isolated from alfalfa roots demonstrates high activity against *T. viride* as well as against some important plant pathogens: *Sclerotium rolfsii*, *Rhizopus mucco*, *Aspergillus niger*, *Phytophthora cinnamoni*, *Fusarium oxysporum* f. sp. *lycopersici* (Levi *et al.*, 1986). The same compound was toxic to ten medically important yeasts (*Candida* sp., *Toruloppsis* sp. and *Geotrichum candidum*) (Polachek *et al.*, 1986). The considerable stability of that compound, absence from it of haemolytic activity (Polachek *et al.*, 1986) and strong antifungal activity against a broad range of yeasts might be useful in the treatment of mycotic infections.

Comparing the chemical structure and antifungal activity of medicagenic acid and its derivatives, it was suggested that this activity is markedly dependent upon free OH and CO_2H groups. It seems that the maximum activity is associated with free R_1, R_2 (OH) and R_3, R_4 (CO_2H) groups (Oleszek *et al.*, 1988b).

It should be stressed again that the antifungal activity of alfalfa saponins is connected with medicagenic acid and its glycosides, while the activity of hederagenin is low and that of soyasapogenol is negligible (Shany *et al.*, 1970; Gestetner *et al.*, 1971). Due to this activity, some alfalfa saponins might be useful in agriculture for plant protection, as well as in human and animal treatments.

11.5 THE MODE OF ACTION

The germination inhibiting effect of alfalfa saponins does not seem to be biochemically specific. Pre-immersion of cotton seeds in saponin solutions resulted in the decrease of oxygen diffusion through the seed coat and in distinct decrease of the rate of respiration if compared with the seeds immersed in distilled water (Marchaim *et al.*, 1970, 1972, 1974; Mayevsky and Marchaim, 1972). At the same time embryos separated from the seeds pre-immersed either in water or in saponin solution had identical respiration rates. This undoubtedly suggests that the primary site of action of saponins has been located not in the embryo but in the seed coats. The effect was similar to those expressed by aqueous solutions of anionic, non-ionic or cationic commercial surfactants.

Observations with a light and scanning microscope showed that pre-immersion of cotton seeds in lucerne saponins, as well as in other surfactant solutions, caused structural changes in the membranes, showing increased swelling at the fringe and living cell walls, which might have affected their permeability by oxygen (Marchaim *et al.*, 1974). The germination inhibiting effect was irreversible after pre-immersion in 0.5% saponin solution for 6 h or more. Exogenous gibberellic acid (GA_3) did not overcome the inhibition of seed germination due to pre-immersion in saponins.

The seedling growth retardation mechanism was not precisely researched, but presumably it undergoes similar processess. Wheat seedlings exposed to the presence of alfalfa root saponins suffered severe damage, especially in the root system. The first symptom of saponin action was the browning of the meristemic areas, followed by their slow decay. This progressed through the whole root system (Oleszek and Jurzysta, 1987).

The fungistatic activity has been initially attributed to the interaction of saponins with membrane cholesterol (Assa *et al.*, 1972). Later Assa *et al.* (1975) found in experiments with myxomycetes *Physarum polycephalum* that alfalfa saponins interact with membrane sterols, proteins and phospholipids. The interaction with proteins also affected certain membrane enzymatic activities such as NADH oxidase and malate dehydrogenase. The attack of saponins on various membrane constituents caused changes in membrane permeability and lysis.

11.6 CONCLUSIONS AND PROSPECTS

The presented data clearly show that alfalfa saponins possess high allelopathic potential against plants and soil microorganisms. Their excretion from alfalfa roots into the soil has not been fully proved and thus the degree of their accumulation in soil sorption complex cannot be clearly estimated. However, it seems to be evident that saponins are potentially active against other plants only when they are present in soil solution. They lose such an activity when bound in the soil. The fate of bound saponins remains unclear and this could not be unequivocally stated that in some favourable environmental conditions saponins are not released and reactivated. This needs more basic research which would clarify the phenomenon of the expression of higher phytotoxicity by older alfalfa stands in comparison to younger stands, as observed in field practice.

The available data do not provide information as to the mechanism and dynamics of the decomposition of saponins when introduced into the soil with live alfalfa roots. In the model experiments roots were usually incorporated into the soil as a fine powder, from which saponins, water-soluble compounds, can be readily released into the environment. In this case, the process of decomposition is probably much slower and the availability of phytotoxins strongly delayed. To avoid allelopathic consequences of saponin presence, and to optimize the utilization of alfalfa residues, there is a need to propose precisely proper agricultural practices for the preparation of alfalfa stands for following crops. Thus more complex interdisciplinary studies are still highly desirable.

REFERENCES

Assa, Y., Gestetner, B., Chet, J. and Henis, Y. (1972) Fungistatic activity of lucerne saponins and digitonin as related to sterols. *Life Sci.*, **11**, 637–47.

Assa, Y., Chet, J., Gestetner, B. *et al.* (1975) The effect of alfalfa saponins on growth and lysis of *Physarum polycephalum. Arch. Microbiol.*, **103**, 77–80.

Birk, Y. (1969) Saponins, in *Toxic Constituents of Plant Foodstuffs* (ed. I. E. Lienear), Academic Press, New York, pp. 169–210.

Bruulsema, T. W. and Christie, B. R. (1987) Nitrogen contribution to succeeding corn from alfalfa and red clover. *Agron. J.*, **79**, 96–100.

Bukharov, B. G. and Shcherbak, C. P. (1970) Triterpenovye glikozydy *Herniaria glabra. Izv. Prir. Soed.*, **3**, 307–11.

Council for Agricultural Science and Technology (1980) Organic and conventional farming compared. CAST Report 84. Council for Agriculture Science and Technology, Ames, IA.

Gestetner, B. (1971) Structure of a saponin from lucerne (*Medicago sativa*). *Phytochemistry*, **110**, 2221–3.

Gestetner, B., Assa, Y., Henis, Y. *et al.* (1971) Lucerne saponins. IV. Relationship between their chemical constitution and haemolytic and antifungal activity. *J. Sci. Food Agric.*, **22**, 168–72.

Gestetner, B., Shany, S., Tencer, Y. *et al.* (1970) Lucerne saponins. II. Purification and fractionation of saponins from lucerne tops and roots and characterization of the isolated fractions. *J. Sci. Food Agric.*, **21**, 501–10.

Guenzi, W. D., Kehr, W. R. and McCalla, T. M. (1964) Water soluble phytotoxic substances in alfalfa forage; variation with variety, cutting, year and stage of growth. *Agron. J.*, **56**, 499–500.

Heichel, G. H. and Barnes, D. K. (1984) Opportunities for meeting crop nitrogen needs from symbiotic nitrogen fixation, in *Organic Farming: Current technology and its role in a sustainable agriculture* (eds D. A. Bezdicek *et al.*), Spec. Pub. 46, Amer. Soc. of Agronomy, Madison, WI, pp. 49–59.

Hesterman, O. B., Russelle, M. P., Sheaffer, C. C. and Heichel, G. H. (1987) Nitrogen utilization from fertilizer and legume residues in legume–corn rotations. *Agron. J.*, **79**, 726–31.

Hesterman, O. B., Sheaffer, C. C., Barnes, D. K. *et al.* (1986) Alfalfa dry matter and nitrogen production, and fertilizer nitrogen response in legume–corn rotations. *Agron. J.*, **78**, 19–23.

Hiller, K. and Voigt, G. (1977) Neue Ergebnisse in der Erforschung der Triterpenesaponine. *Die Pharmazie*, **32**, 365–93.

Jensen, E. H. (1984) Problems of continuous alfalfa, in *Proc. 3rd Western Alfalfa Symp.*, University of Nevada-Reno, College of Agriculture Cooperative Extension Service, MS 165, pp. 1–8.

Jurzysta, M. (1970) Effect of saponins isolated from seeds of lucerne on germination and growth of cereal seedlings. *Zesz. Nauk. UMK Torun*, **13**, 253–6.

Jurzysta, M. (1973a) Chemical characteristics of saponins from *Medicago lupulina* seeds. *Proc. 11th Meeting Polish Biochem. Soc., Bialysto*, p. 81.

Jurzysta, M. (1973b) Isolation and chemical characterization of saponins from lucerne seeds (*Medicago media* Pers.). *Acta Soc. Bot. Pol.*, **42**, 201–7.

Jurzysta, M. (1979) A simple method of quantification of biologically active alfalfa saponins by *Trichoderma viride* growth. *Biul. Branz. Hod. Roslin*, **1**, 16–18.

Jurzysta, M. (1982) Investigation of saponins of native lucerne populations (*Medicago sativa* Pers.). *R(170)*, IUNG, Pulawy, Poland, p. 64.

Jurzysta, M. (1984) Transformation of soyasapogenol B into soyasapogenols C, D and F under acidic conditions. *Proc. 14th Int. Symp. Nat. Prod.*, p. 127.

Jurzysta, M., Burda, S., Oleszek, W. and Ploszynski, M. (1988a) The chemotaxonomic significance of larycitrin and medicagenic acid in the tribe Trigonellae. *Can. J. Bot.*, **66**, 363–7.

Jurzysta, M., Small, E. and Nozzolillo, C. (1988b) Hemolysis, a synapomorphic discriminator of an expanded genus *Medicago* (Leguminosae). *Taxon*, **37**, 354–63.

Kehr, W. R., Watkins, J. E. and Ogden, R. L. (1983) Alfalfa establishment and production with continuous alfalfa and following soybeans. *Agron. J.*, **75**, 435–8.

Klein, R. R. and Miller, D. A. (1980) Allelopathy and its role in agriculture. *Commun. Soil Sci. Plant Anal.*, **11**, 43–6.

Klein, G., Jurenitsch, J. and Kubelka, W. (1982) Structur der Sapogenine von Herba *Herniariae* (*Herniaria glabra* L. und *Herniaria hirsuta* L.). *Sci. Pharm.*, **50**, 216–33.

Lawrance, T. and Kichler, M. R. (1962) The effect of fourteen root extracts upon germination and seedling length of fifteen plant species. *Can J. Plant Sci.*, **42**, 308–13.

Levy, M., Zehavi, U., Naim, M. and Polachek, J. (1986) An improved procedure for the isolation of medicagenic acid 3-O-D-glucopyranoside from alfalfa roots and its antifungal activity on plant pathogen. *J. Agric. Food Chem.*, **34**, 960–3.

Marchaim, U., Birk, Y., Dovrat, A. and Berman, T. (1970) Alfalfa saponins as inhibitors of cotton seed germination. *Plant Cell Physiol.*, **11**, 511–14.

Marchaim, U., Birk, Y., Dovrat, A. and Berman, T. (1972) Lucerne saponins as inhibitors of cotton seed germination; their effect on diffusion of oxygen through the seed coat. *J. Exp. Bot.*, **23**, 302–9.

Marchaim, U., Weker, E. and Thomas, W. D. E. (1974) Changes in the anatomy of cotton seed coats by lucerne saponins. *Bot. Gaz.*, **135**, 139–46.

Marchaim, U., Birk, Y., Dovrat, A. and Berman, T. (1975) Kinetics of the inhibition of cotton seeds germination by lucerne saponins. *Plant Cell Physiol.*, **16**, 857–64.

Mayevsky, A. and Marchaim, U. (1972) Studies on the effect of lucerne saponins on the respiration of cotton seeds using $^{18}O_2$. *Plant Cell Physiol.*, **13**, 927–30.

McElgunn, J. D. and Heinricks, D. H. (1970) Effects of root temperature and a suspected phytotoxic substance on the growth of alfalfa. *Can. J. Plant Sci.*, **50**, 307–11.

Miller, D. A. (1983a) Allelopathic effects of alfalfa. *USDA Report Alfalfa Impr. Conf.*, p. 26.

Miller, D. A. (1983b) Allelopathic effects of alfalfa. *J. Chem. Ecol.*, **9**, 1059–72.

Mishustin, B. N. and Naumova, A. N. (1955) Secretion of toxic substances by alfalfa and their effect on cotton and soil microflora. *Akad. Nauk USSR Izvestija, Ser. Biol.*, **6**, 3–9.

Morris, R. J., Dye, W. B. and Gisler, P. S. (1961) Isolation, purification, and structural identity of an alfalfa root saponin. *J. Org. Chem.*, **26**, 1241–3.

Morris, R. J. and Hussey, E. W. (1965) A natural glycoside of medicagenic acid; an alfalfa blossom saponin. *J. Org. Chem.*, **30**, 166–8.

Nielsen, K. F., Cuddy, T. and Woods, W. (1960) The influence of the extract of some crops and soil residues on germination and growth. *Can. J. Plant Sci.*, **40**, 188–97.

Nonaka, M. (1986) Variable sensitivity of *Trichoderma viride* to *Medicago sativa* saponins. *Phytochemistry*, **25**, 73–5.

Oleszek, W. (1985) Saponins from alfalfa (*Medicago media.* Pers.) and red clover (*Trifolium pratense* L.) roots and their allelopathic effect on winter wheat seedling (*Triticum aestivum*) growth. PhD Thesis, IUNG Pulawy, Poland.

Oleszek, W. (1988) Solid-phase extraction-fractionation of alfalfa saponins. *J. Sci. Food Agric.*, **44**, 43–9.

Oleszek, W. and Jurzysta, M. (1986a) Isolation, chemical characterization and biological activity of alfalfa (*Medicago media* Pers.) root saponins. *Acta Soc. Bot. Pol.*, **55**, 23–33.

Oleszek, W. and Jurzysta, M. (1986b) Isolation, chemical characterization and biological activity of red clover (*Trifolium pratense* L.) root saponins. *Acta Soc. Bot. Pol.*, **55**, 247–52.

Oleszek, W., Jurzysta, M., Górski, P. *et al.* (1987a) Studies on *Medicago lupulina* saponins. 6. Some chemical characteristics and biological activity of root saponins. *Acta Soc. Bot. Pol.*, **56**, 119–26.

Oleszek, W., Jurzysta, M., Burda, S. and Ploszynski, M. (1987b) Triterpene saponins of *Trigonella monspeliaca* L. *Acta Soc. Bot. Pol.*, **56**, 281–5.

Oleszek, W. and Jurzysta, M. (1987) The allelopathic potential of alfalfa root medicagenic acid glycosides and their fate in soil environments. *Plant and Soil*, **98**, 67–80.

Oleszek, W., Price, K. R. and Fenwick, G. R. (1988a) Triterpene saponins from the roots of *Medicago lupulina* L. (Black medick trefoil). *J. Sci. Food Agric.*, **43**, 289–97.

Oleszek, W., Price, K. R. and Fenwick, G. R. (1988b) The sensitivity of *Trichoderma viride* to medicagenic acid, its natural glucosides (saponins) and derivatives. *Acta Soc. Bot. Pol.*, **57**, 361–70.

Patrick, Z. A. (1971) Phytotoxic substances associated with the decomposition in soil of plant residues. *Soil Sci.*, **111**, 13–18.

Pedersen, M. W. (1965) Effect of alfalfa saponin on cottonseed germination. *Agron. J.*, **57**, 516–17.

Pedersen, M. W., Zimmer, D. E., Anderson, J. O. and McGuire, C. F. (1966) A comparison of saponins from Du Puits, Lahontan, Ranger and Uinta alfalfas. *Proc. 10th Int. Grassland Cong. Helsinki*, pp. 266–9.

Polachek, J., Zehavi, U., Naim, M. *et al.* (1986) Activity of compound G2 isolated from alfalfa roots against medically important yeasts. *Antimic.*

Agents Chemoth., **30**, 290–4.

Price, K. R., Fenwick, G. R. and Jurzysta, M. (1986) Soyasapogenols – separation, analysis and interconversions. *J. Sci. Food Agric.*, **37**, 1027–34.

Price, K. R., Curl, C. L. and Fenwick, G. R. (1987a) Flash chromatography – a simple technique of potential value to food chemists. *Food Chem.*, **25**, 145–53.

Price, K. R., Johnson, I. T. and Fenwick, G. R. (1987b) The chemistry and biological significance of saponins in food and feeding stuffs. *CRC Crit. Rev. Fd. Sci. Nutr.*, **26**, 27–135.

Read, J. J. and Jensen, E. H. (1989) Phytotoxicity of water-soluble substances from alfalfa and barley soii extracts on four crop species. *J. Chem. Ecol.*, **15**, 619–28.

Ream, H. W., Smith, D. and Walgenbach, R. P. (1977) Effects of deproteinized alfalfa juice applied to alfalfa-bromegrass, bromegrass, and corn. *Agron. J.*, **69**, 685–9.

Shany, S., Birk, Y., Gestetner, B. and Bondi, A. (1970) Preparation, characterization and some properties of saponins from lucerne tops and roots. *J. Sci. Food Agric.*, **21**, 131–4.

Scardavi, A. and Elliot, F. C. (1967) A review of saponins in alfalfa and their bioassay utilizing *Trichoderma* ssp. *Bull. Michigan Agric. Exp. St., East Lansing*, **50**, 163–77.

Small, E., Lassen, P. and Brookes, B. C. (1987) An expanded circumscription of *Medicago* (Leguminosae, Trifolieae) based on explosive flower tripping. *Willdenowia*, **16**, 415–39.

Timbekova, A. E. and Abubakirov, N. K. (1984) Triterpene glycosides of alfalfa. I. Medicoside G – a novel bisdesmoside from *Medicago sativa. Khim. Prir. Soedin.*, **4**, 451–8.

Timbekova, A. E. and Abubakirov, N. K. (1985) Triterpene glycosides of alfalfa. II. Medicoside C. *Khim. Prir. Soedin.*, **6**, 805–8.

Walgenbach, R. P., Smith, D. and Ream, H. W. (1977) Growth and chemical composition of alfalfa fertilized in greenhouse trials with deproteinized alfalfa juice. *Agron. J.*, **69**, 690–4.

Webster, G. R., Khan, S. V. and Moore, A. W. (1967) Poor growth of alfalfa (*Medicago sativa*) on some Alberta soils. *Agron. J.*, **59**, 37–41.

Zentmeyer, G. A. (1963) Biological control of *Phytophthora* root rot of avocado with alfalfa meal. *Phytopathology*, **53**, 1383–6.

Zentmeyer, G. A. and Tompson, G. R. (1967) The effect of saponins from alfalfa on *Phytophthora cinnamoni* in relation to control of root rot of avocado. *Phytopathology*, **57**, 1278–90.

Zimmer, D. E., Pedersen, M. W. and McGuire, D. F. (1967) A bioassay for alfalfa saponins using the fungus *Trichoderma viride* pers. ex. fr. *Crop Sci.*, **7**, 223–4.

CHAPTER 12

Allelopathy in alfalfa and other forage crops in the United States

D. A. Miller

12.1 INTRODUCTION

Inhibitory effects of crop plants on other crop plants were observed over 2000 years ago, but no scientific studies were done on the subject until early in the twentieth century. Shreiner and his associates published a series of papers starting in 1909 in which they presented evidence that some soils will cease to provide the life support of a continuous single cropping system due to an addition of growth inhibitors released into the soil by certain crop plants (Rice, 1984). This was the early beginnings of allelopathy and autotoxicity.

It has been observed for many years that alfalfa cannot be successfully re-established into an existing alfalfa stand that may be beginning to die out (Boknenblust, 1983; Guenzi *et al.*, 1964; Jensen *et al.*, 1981; Kehr *et al.*, 1983; Klein and Miller, 1980; Rice, 1984). Some forage grass species have shown similar responses when re-establishment has been attempted. Poor establishment of many forage crops have been observed when the former crop was of the same species even following proper tillage and seedbed preparation. In many areas of the USA and other countries, a continuous sod or forage may be needed for both an efficient enterprise and for conservation purposes.

12.2 ALFALFA AUTOTOXICITY

Re-establishing alfalfa into an old stand or following the tillage of an old stand has resulted in poor stand establishment. We observed this problem in the late 1960s when we began our annual alfalfa production trial. After three years of continuously re-establishing alfalfa in the same field we observed a reduction in germination and reduced growth during and after

Allelopathy: Basic and applied aspects Edited by S. J. H. Rizvi and V. Rizvi
Published in 1992 by Chapman & Hall, London ISBN 0 412 39400 6

TABLE 12.1 *Survey of various alfalfa regions in the United States for autotoxicity*

State	Occurrence of alfalfa autotoxicity
California	Possible
Illinois	Yes
Iowa	Yes – spring seeding
	Questionable fall seeding
Michigan	Yes – immediately
	No – 2 weeks or later
Nebraska	Yes
Nevada	Yes
New Mexico	Probably
New York	Yes
Ohio	Yes
Oregon	Yes
South Dakota	Yes
Utah	Yes
Virginia	Yes
Wisconsin	Yes
Wyoming	Yes

germination. Various alfalfa regions in the USA were surveyed to see if alfalfa-autotoxicity does occur and what the researchers have observed or found (Table 12.1).

Numerous researchers have observed phytotoxicity of alfalfa (Table 12.1). Alfalfa contains water-soluble substances which are toxic to itself and to other species. It has been found that the water extracts from immature alfalfa forage produced the highest phytotoxic effects on corn seedlings in a laboratory (Guenzi *et al.*, 1964). Soil incorporation of young alfalfa forage from old plants when reseeding or interseeding alfalfa to thicken the stand may introduce water-soluble substances toxic to the new alfalfa seedlings (Guenzi *et al.*, 1964; Miller, 1986, 1987).

Early research indicated that there was a 'soil-sickness' related to soil continuously seeded to alfalfa (Schreiner and Lathrop, 1911). Phytotoxic substances have been found in the soil from continuous seeded alfalfa (Goplen and Webster, 1969; Jensen *et al.*, 1984; McElgunn *et al.*, 1970; Webster *et al.*, 1967).

Some of the earlier work on autotoxicity was done in Nebraska (Grandfield and Metzger, 1936; Kiesselbach and Anderson, 1926; Kiesselbach *et al.*, 1929, 1934). More recently, research was first reported in Illinois (Klein and Miller, 1980) followed by research done in many other states – Michigan, Nebraska, Nevada, Ohio, Wisconsin and Wyoming.

It was found that three-day old alfalfa seedlings release toxic compound(s) that inhibit alfalfa (1) germination and (2) seedling elongation (Guenzi *et al.*, 1964; Jensen *et al.*, 1984; Miller, 1983). It was also found that a relationship exists between (1) radicle elongation and (2) autumn growth. A phytotoxic exudate release may play a role in fall dormancy (Miller, 1983). In addition, there appears to be no major genetic differences among alfalfa cultivars and autotoxicity resistance.

Autotoxicity of alfalfa is where alfalfa produces compounds that are toxic to itself. Alfalfa autotoxicity on germination and seedling growth has been shown in the greenhouse and laboratory (Boknenblust, 1983; Hegde and Miller, 1989; Jensen *et al.*, 1981, 1984; Liebl and Worsham, 1983; McElgunn *et al.*, 1970; Miller, 1983, 1987; Nielsen *et al.*, 1960). These compounds are found in the seed coat, fresh alfalfa leaves, stems and crowns, plus in dry hay, old roots and soil residues. Fungicide seed treatments will effectively control damping-off, but a preplant incorporation of a fungicide did not improve alfalfa stands in a prepared seedbed in Illinois (Faix *et al.*, 1979).

In Nebraska, they compared relative five year total alfalfa yields grown on land not previously planted to alfalfa, land continued in an old alfalfa stand, and land reseeded to alfalfa, and found them to be 100, 69 and 68%, respectively. We found in Illinois the relative stands of alfalfa after six years of corn–alfalfa, corn–soybean–alfalfa and alfalfa–alfalfa cropping sequences were 100, 83 and 43%, respectively, while the relative yields were 100, 92 and 50%, respectively (Table 12.2) (Klein and Miller, 1980). In Virginia the average yield of eight cultivars grown on soil not previously seeded to alfalfa was 27% higher than the yield of the same cultivars grown on soil previously seeded to alfalfa (Bryant and Hammes, 1981).

In a later study in Nebraska, a five-year-old stand of alfalfa was ploughed under and an alfalfa seeding was made the following year on part of it, plus soybeans established on another part of it for one and two years and then seeded to alfalfa following the soybeans. Average yield of the first year seeding was 26% higher than the re-established seeding one year following alfalfa, and 40 and 35% higher in the first and second years after seeding, respectively (Kehr *et al.*, 1983). It was suggested that autotoxicity or allelopathy may be involved in the lower yields.

TABLE 12.2 *The effects of various crop sequences on alfalfa yields and stand count after 6 years*

Cropping sequence	Tons DM/A	Plants ft^{-2}
Corn–alfalfa	3.79	4.6
Corn–soybean–alfalfa	3.48	3.8
Alfalfa–alfalfa	1.92	2.0

TABLE 12.3 *Percent germination of alfalfa when grown in various stand densities of different plant parts mixed in the soil*

	Plant density		
Plant parts returned	3 ft^{-2}	6 ft^{-2}	9 ft^{-2}
Roots and crowns	80	71	62
Tops only	47	53	22
Whole plant	57	32	20

It was found in Illinois that there was much more phytotoxic activity in the forage portion than the crown and root portion. This alone suggests that one should remove as much of the forage before ploughing it under if alfalfa is to be re-established. We have tried to simulate in the greenhouse the ploughing down of various alfalfa stand densities composed of whole plant, crown and roots, and tops alone from varying densities of 3, 6 or 9 plants per square foot into the soil and reseeding alfalfa (Table 12.3). We found greater autotoxicity activity in the tops than the roots and crown, under any stand density. When the entire alfalfa plant from a poor stand of 3 plants per square foot was ploughed down, 57% germination occurred, but decreased to 32 and 20% germination, with 6 and 9 plants per square foot, respectively. We would estimate 9 mature plants per square foot to be a fairly good stand of over 3 years old, while 6 plants per square foot to be about a 60% stand and 3 plants per square foot approximately a 30% stand. If one were to re-establish alfalfa in such densities, it is obvious that germination and establishment may be greatly reduced. Therefore, the greatest amount of allelochemicals are present in the top growth. With recent work at Illinois and Michigan, it appears that one might be able to harvest the forage from a depleting stand then apply Roundup, followed with a no-till seeder to re-establish alfalfa into a poor stand. By not ploughing down the crown and roots or mixing them into the soil, it appears that any phytotoxic substance(s) that might be released from the intact crown and roots is much less than when the plant tissue is mixed into the soil.

Recently laboratory and greenhouse studies were conducted to determine if autotoxicity in alfalfa due to water-soluble compounds is concentration dependent. It was found that alfalfa radicle elongation was inhibited beyond the 2% (w/v, on fresh weight basis) concentration. Aqueous shoot extracts from bud stage alfalfa was more inhibitory than from the 3-week old stage. Incorporation of fresh 3-week old herbage (aboveground portion) at 9 shoots per square foot, 4 inch tall severely reduced alfalfa emergence, while 14 shoots per square foot completely

TABLE 12.4 *Dry matter yields and seedling populations under two rotation sequences*

Rotation sequence	Seedling population[a] (plants ft^{-2})	Dry matter yields[c] (t acre^{-1})		
		9/78	5/79	8/79
Alfalfa–alfalfa	16	0.89	2.23	2.05
Forage grasses–alfalfa	19	0.98	2.45	2.36
Difference	−2.3*[b]	−0.09*	−0.22*	−0.31*
Average	17.5	0.94	2.34	2.21

[a] Counts taken 2 weeks following seeding.
*[b] Significant, based on one-tailed *t* test, at the 0.05 level.
[c] Seeding 15 January 1978.

prevented pre-germinated seeds from emerging at all (Hegde and Miller, 1989).

Tesar (1984) at Michigan has conducted considerable research in re-establishing alfalfa into old stands. His data do not support the theory that there are considerable phytotoxins present inhibiting alfalfa germination and establishment. Tesar summarizes his data by saying that autotoxicity does occur when immediately reseeding alfalfa, but by delaying the seeding for two weeks or more, one can safely seed alfalfa after ploughing or spraying the old stand with Roundup. Tesar prefers autumn ploughing and spring seeding with a second option of spring or summer ploughing followed by seeding in two or more weeks. A third choice would be a sod seeding on land too steep to plough, by applying Roundup in the autumn and spring sod seeding or a spring or summer Roundup treatment followed by sod seeding in two weeks or more.

Studies at Illinois still support the observation that alfalfa yields and stand densities are greater when alfalfa is rotated with soybeans, corn, or forage grasses, compared with continuous cropping of alfalfa (Miller, 1983) Tables 12.2 and 12.4). Illinois data plus others are too strong an evidence completely to eliminate autotoxicity in alfalfa.

12.3 OTHER FORAGE CROPS

No-till planting of alfalfa into grass sods are important today to reduce soil erosion and maintain soil productivity. Tall fescue (*Festuca arundinacea* Shreb.), orchardgrass (*Dactylis glomerata* L.), smooth bromegrass (*Bromus inermis* Leyss.) and Timothy (*Phleum pratense* L.) were studied because of widespread use in agriculture. Germination studies were conducted in the laboratory using cold water extracts. Extracts were made by soaking 1.25, 2.5, 5.0, 10.0 g of dried shoot and root tissue in 100 ml distilled water for 48 h at 25°C. All shoot extracts radically reduced germination with all grasses; at 2.5% level and above 0% germination resulted with smooth bromegrass and tall fescue, root extracts reduced germination 50–70%. Greenhouse studies conducted using the same four levels of extracts added to pots at the rate of 30 ml per pot three times, 7 days apart, showed no germination difference at 10 days. Dry matter yields at 8 weeks increased as level of extract increased. Testing of extracts showed a significant level of nitrogen which may have caused the yield increase. Field studies at three locations and three soil types were conducted using grasses established for two years. Treatments were bare soil (grass killed, burned off six months previously), light mulch (top growth removed two weeks prior to planting and allowed to regrow), and heavy mulch (grass allowed to grow until planting). Tall fescue and smooth bromegrass treatments resulted in lower populations at all three locations and orchardgrass reduced population at one location. Heavy mulch smooth bromegrass and orchardgrass reduced yields at one location.

12.4 IDENTIFIABLE CHEMICALS

There is no evidence supporting saponins as being phytotoxic to alfalfa (Miller, 1983) (Table 12.5). There is some evidence that saponins may have an allelopathic effect on wheat (*Triticum aestivum*). Additionally,

TABLE 12.5 *Dry matter yields (tons acre^{-1}) of 12 alfalfa cultivars for two rotation sequences*

	Rotation sequence	
	Alfalfa–alfalfa	Soybean–corn–alfalfa
Ladak (low saponin)	0.80	1.12
Ladak (high saponin)	1.16	1.29
Uinta (low saponin)	1.16	1.43
Uinta (high saponin)	1.07	1.38
DuPuits (low saponin)	0.71	1.16
DuPuits (high saponin)	0.94	1.20
Ranger (low saponin)	1.12	1.38
Ranger (high saponin)	1.29	1.38
BWR syn. (low saponin)	1.02	1.61
BWR syn. (high saponin)	1.25	1.47
Saranac AR	1.38	1.61
Vernal	1.34	1.34

there is no evidence of glyphosate (Roundup) phytotoxicity when sprayed on sod for germinating alfalfa (Eltun *et al.*, 1985).

Ferulic acid is one of the allelopathic compounds that has been identified (Liebl and Worsham, 1983). It is found in plants and has been isolated from soil. It is a product of lignin degradation, which is associated with alfalfa degradation.

Trials conducted at Ohio State University have found autotoxic compounds(s) in alfalfa (Hall and Henderlong, 1989). Using paper chromatographic separation the autotoxic compound has an Rf characterization similar to phenolic acid. However, using phenolic-absorbent polyvinylpyrrolidone did not affect the autotoxic response. Their results indicate that alfalfa contains an autotoxic compound that characterizes a phenolic compound. This compound(s) is found predominantly in the water soluble fraction of alfalfa tissue and appears not to be a direct result of microbial activity.

Recently other chemical compounds have been isolated using the GC-MS as those that contribute to alfalfa autotoxicity. Medicarpin (3-hydroxy-9-methoxypterocarpan) inhibits alfalfa germination. Medicarpin, 4-methoxymedicarpin, sativan, and 5-methoxysativan have been isolated from alfalfa foliage. Medicarpin may reduce alfalfa and velvetleaf (*Abutilon theophrastii* Medic) seedling length. Sativan and both methoxy derivatives showed no effect on germination or seedling growth. There is some thought that medicarpin released by alfalfa will help control the ecology immediately around an alfalfa plant and inhibit alfalfa seedling establishment plus control some weeds.

12.5 CONCLUSIONS AND PROSPECTS

In summary, there is a problem when one seeds alfalfa directly back into an old alfalfa stand, especially if the old stand is around 50% or more alfalfa. If one wishes to reseed an old field of alfalfa back into alfalfa there are several recommendations. The soil fertility should be sufficiently high for excellent alfalfa growth – i.e. a pH of 6.8–7.0, P_1 test of 55–75, and K for a soil test of 400. Remove all of the alfalfa top growth from a 50% stand or less as forage and plough under the crown and roots. It would be best if this could be done at least the preceding autumn if a spring seeding of alfalfa is to occur. We would suggest a one-year grass rotation, such as corn or some other crop to be seeded in the field before reseeding it to alfalfa. If corn is to be seeded in the old alfalfa field, one might plough under the last growth of alfalfa as a N source for the corn. The one year out of alfalfa production is long enough for an excellent alfalfa re-establishment.

If the length of time between harvesting and/or reseeding alfalfa is too short due to the topography or the need for alfalfa forage regrowth, then one should definitely remove the last top growth, apply Roundup and wait at least two weeks or more before re-establishing alfalfa with a no-till seeder. We believe a month between spraying and reseeding is better.

REFERENCES

Boknenblust, K. E. (1983) Effect of allelopathy or autotoxicity on alfalfa seedling establishment. *Report 28th Alfalfa Improv. Conf.*, pp. 25.

Bryant, H. T. and Hammes, R. C., Jr (1981) Performance of alfalfa grown on land previously and not previously in alfalfa. *Agron. Abstr.*, p. 101.

Eltun, R., Wakefield, R. C. and Sullivan, W. M. (1985) Effect of spray/planting intervals and various grass sods on no-till establishment of alfalfa. *Agron. J.*, **77**, 5–8.

Faix, J. J., Graffis, D. W. and Kaiser, C. J. (1979) Conventional and zero-till planted alfalfa with various pesticides. *Ill. Agric. Exp. Sta. UPDATE Dixon Springs Agric. Center*, pp. 117–23.

Goplen, B. P. and Webster, G. R. (1969) Selection in *Medicago sativa* L. for tolerance to alfalfa-sick soils in central Alberta. *Agron. J.*, **61**, 589–90.

Grandfield, C. O. and Metzger, W. H. (1936) Relation of fallow to restoration of subsoil moisture in an old alfalfa field and subsequent depletion after reseeding. *J. Amer. Soc. Agron.*, **28**, 115–23.

Guenzi, W.D., Kehr, W. R. and McCalla, T. M. (1964) Water-soluble phytotoxic substances in alfalfa forage: variation with variety, cutting, year, and stage of growth. *Agron. J.*, **56**, 499–500.

Hall, M. H. and Henderlong, P. R. (1989) Alfalfa autotoxic fraction characterization and initial separation. *Crop Sci.*, **29**, 425–8.

Hegde, R. S. and Miller, D. A. (1989) Concentration dependency of water-soluble compounds in alfalfa (*Medicago sativa* L.) autotoxicity. *Agron. Abstr.*, p. 133.

Jensen, E. H., Hartman, B. J., Lundin, F. *et al.* (1981) Autotoxicity of alfalfa. *Nevada Agric. Exp. Sta.*, R144.

Jensen, E. H., Meyers, K. D., Jones, C. L. and Leedy, C. D. (1984) Effect of alfalfa foliage and alfalfa soil extracts on alfalfa seedling vigor. *Report 29th Alfalfa Improv. Conf.*, p. 38.

Kehr, W. R., Watkins, J. E. and Ogden, R. L. (1983) Alfalfa establishment and production with continuous alfalfa and following soybeans. *Agron. J.*, **75**, 435–8.

Kiesselbach, T. A. and Anderson, A. (1926) Alfalfa investigation. *Nebr. Agric. Exp. Stn. Res. Bull.*, 36.

Kiesselbach, T. A., Anderson, A. and Russell, J. C. (1934) Soil moisture and crop sequence in relation to alfalfa production. *J. Amer. Soc. Agron.*, **26**, 422–42.

Kiesselbach, T. A., Russell, J. C. and Anderson, A. (1929) The significance of subsoil moisture in alfalfa production. *J. Amer. Soc. Agron.*, **21**, 241–68.

Klein, R. R. and Miller, D. A. (1980) Allelopathy and its role in agriculture. *Commun. Soil Plant Anal.*, **11**, 43–56.

Liebl, R. A. and Worsham, A. D. (1983) Inhibition of pitted morning glory (*Ipomoea lacumosa* L.) and certain weed species by phytotoxic components of wheat straw. *J. Chem. Ecol.*, **9**, 1027–43.
McElgunn, J. D. and Heinricks, D. H. (1970) Effects of root temperature and a suspected phytotoxic substance on the growth of alfalfa. *Can. J. Plant Sci.*, **50**, 307–11.
Miller, D. A. (1983) Allelopathic effects of alfalfa. *J. Chem. Ecol.*, **9**, 1059–71.
Miller, D. A. (1986) Alfalfa autotoxicity. *National Alfalfa Symp.*, pp. 134–42.
Miller, D. A. (1987) Allelopathy in alfalfa and other forage crops. *North Central Alfalfa Improv. Conf.*, Urbana, Ill.
Nielsen, K. F., Cuddy, T. and Woods, W. (1960) The influence of the extract of some crops and soil residues on germination and growth. *Can. J. Plant Sci.*, **40**, 188–97.
Rice, E. L. (1984) *Allelopathy*. Academic Press, Orlando, Florida.
Schreiner, O. and Lathrop, E. C. (1911) Examination of soils for organic constituents, expecially dihydroxystearic acid. *USDA Bur. Soils Bull.*, 80.
Tesar, M. B. (1984) Establishing alfalfa after alfalfa without autotoxicity. *Report 29th Alfalfa Improv. Conf.*, pp. 39.
Webster, G. R., Khan, S. V. and Moore, A. W. (1967) Poor growth of alfalfa (*Medicago sativa* L.) on some Alberta soils. *Agron. J.*, **59**, 37–41.

CHAPTER 13

Allelopathy in relation to agricultural productivity in Taiwan: problems and prospects

Chang-Hung Chou

13.1 INTRODUCTION

Allelopathy has been increasingly recognized as an important ecological mechanism which interprets plant dominance, succession, formation of plant community and climax vegetation, and crop productivity (Muller, 1966, 1971; Whittaker and Feeney, 1971; Rice, 1984; Chou, 1987; Putnam and Tang, 1986; Waller, 1987; Chou and Waller, 1923, 1989). The allelopathic phenomenon is always related to environmental parameters and cannot be singled out from the environmental complex (Muller, 1974). Koeppe *et al.* (1971, 1976) demonstrated several cases of allelopathic effects in relation to environmental stresses. Putnam and Duke (1974) firstly introduced the concept into agricultural practice to select a crop variety with high allelopathic potential in order to lessen the use of herbicide. Chou *et al.* (1987, 1989) have elaborated a practical system of pasture–forest intercropping in fields to avoid herbicide application and to lessen the labour cost based on an allelopathic approach. Substantial information concerning the allelopathic study in a subtropical and tropical region such as Taiwan has been accumulated, based primarily on the author's findings since 1972. In particular, the investigation of the allelopathic mechanism in relation to agriculture productivity in Taiwan has been emphasized due to the unique agricultural situation there. This chapter thus describes the research carried out in Taiwan since 1972.

13.2 ALLELOPATHIC INFLUENCES ON CROP PRODUCTIVITY IN TAIWAN

13.2.1 The adaptive autointoxication of rice plants

According to data of rice production in Taiwan from 1910 to 1975, rice yield of the second crop (from August to December) has been generally

Allelopathy: Basic and applied aspects Edited by S. J. H. Rizvi and V. Rizvi
Published in 1992 by Chapman & Hall, London ISBN 0 412 39400 6

lower by 25% than that of the first crop (from March to July), and the reduction of rice productivity (about 1000 kg ha^{-1}) has been particularly pronounced in areas of poor water drainage (Wu and Antonovics, 1979). The cropping system of rice in Taiwan is different from that of other countries. For example, a 3-week fallowing period between the first crop and the second crop is different in comparison to the 10-week period between the second crop and the first crop of the following year. Between two crops, the farmers always left rice stubbles in the field after harvesting, and submerged these residues into the soil for decomposition during the fallowing time. During the second crop season, the typhoon (or monsoon) brings a great amount of rainfall, leading to a high water table in some areas where the water drainage is rather poor. This is one of the limiting factors that causes a reduction in rice production in the second crop season.

In a series of experiments a soil–rice straw and root mixture (3 kg : 200 g) was saturated with distilled water and allowed to decompose for the time intervals of 1, 2 and 4 weeks under greenhouse conditions. Soil alone was treated in the same manner as a control. The extract from soil alone showed no phytotoxicity, however, extracts from the decaying rice residues revealed more than 70% inhibition. Extracts made from the same pot 1 week later also showed significant inhibition and subsequent extracts in the following 4 weeks continued to show inhibition. In another experiment, similar to the former, the toxicity was obviously high in the straw–soil extract, but gradually decreased with increased decomposition time and subsequent extractions. Nevertheless, the phytotoxicity was still above 40% after 16 weeks of decomposition (Fig. 13.1) (Chou and Lin, 1976). The third experiment was to evaluate the effect of temperature on phytotoxin production in the soil–straw mixture; the experiment was described in detail by Chou *et al.* (1981). The bioassay of aqueous extracts from the soil with seeds of lettuce and rice indicated that soil–straw extracts exhibited a significantly higher inhibition than the control soil extract. Phytotoxicity was highest at 20–25°C and decreased at both lower and higher temperatures. A decrease in the soil phytotoxicity with time was more noticeable at temperature above 30°C (Fig. 13.2). This shows that at such temperatures the decomposition of rice straw in soil would be enhanced but the resulting phytotoxicity to soil was of shorter duration (Chou *et al.*, 1981). The yield components of rice plants were also affected by phytotoxins produced during the period rice residues were decaying in the soil. The number of tillers and panicle and weight of grain and yield were much lower in the soil–rice residue mixture than in the control soil (Table 13.1) (Chou and Chiou, 1979).

The decomposing rice straw was extracted with ethanol and re-extracted with ethyl ether, then the phytotoxins present in the ether fraction of the extract were identified by chromatography. The identified compounds were *p*-hydroxybenzoic, *p*-coumaric, syringic, vanillic,

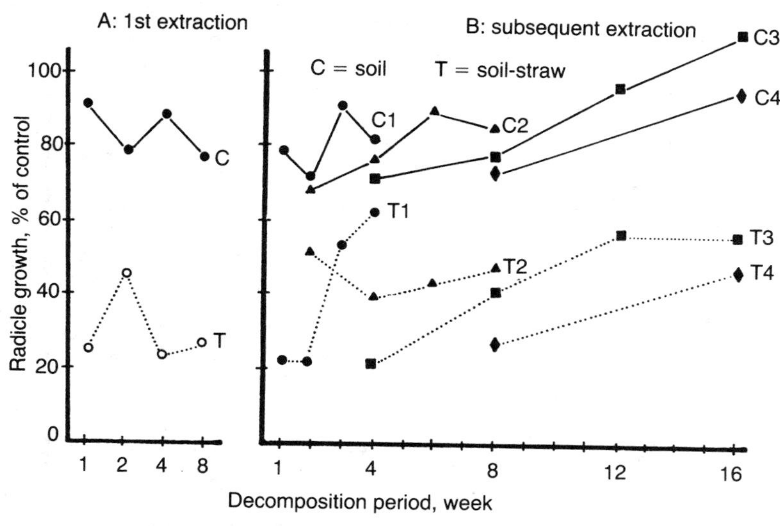

Fig. 13.1 Inhibitory effect of extracts from decomposition of rice residues in soil on radicle growth of lettuce. (A) Extracts were obtained from the soil alone, C, and from soil–straw, T, after the indicated period of decomposition. (B) Subsequent extractions were made from the soil and soil–straw samples. On the C1, T1 set of pots, extractions began at the end of the first week; on the C2, T2 at the end of second week; on the C3, T3 at the end of the fourth week; and on the C4, T4 at the end of the eighth week. After each extraction, the residues were returned to the same pot for a period of further decomposition. (After Chou and Lin, 1976. Reproduced with permission, Chou, Cott. and Plenum Publishing Corpn, New York.)

o-hydroxyphenylacetic, and ferulic acids (Table 13.2) (Chou and Lin, 1976), and propionic, acetic and butyric acids (Wu *et al.*, 1976b). In particular, *o*-hydroxyphenylacetic acid, our first reported phytotoxin, was toxic to rice growth at a concentration of 25 ppm. We found that the concentration of *o*-hydroxyphenylacetic acid reached about 10^{-2} M where the rice residues were decomposing in the soil.

The phytotoxins produced during rice residue decomposing can only suppress the growth of rice plants to some extent, and would not completely retard the growth of rice plants. The self-thinning of rice plants during the period of rice growth was also found in fields (Oka, 1976; Chou *et al.*, 1984). The above phenomenon of the suppression pattern of rice plants is a sort of adaptive auto-inhibition. Therefore, the 25% reduction in yield in the second crop is meaningful and can be interpreted as a self-regulation mechanism to reduce energy consumption when rice plants are under stress conditions (Chou, 1987). Further evidence of adaptive auto-intoxication can also be found in many perennial weeds or in climax vegetation. The stable and maximum productivity of vegetation

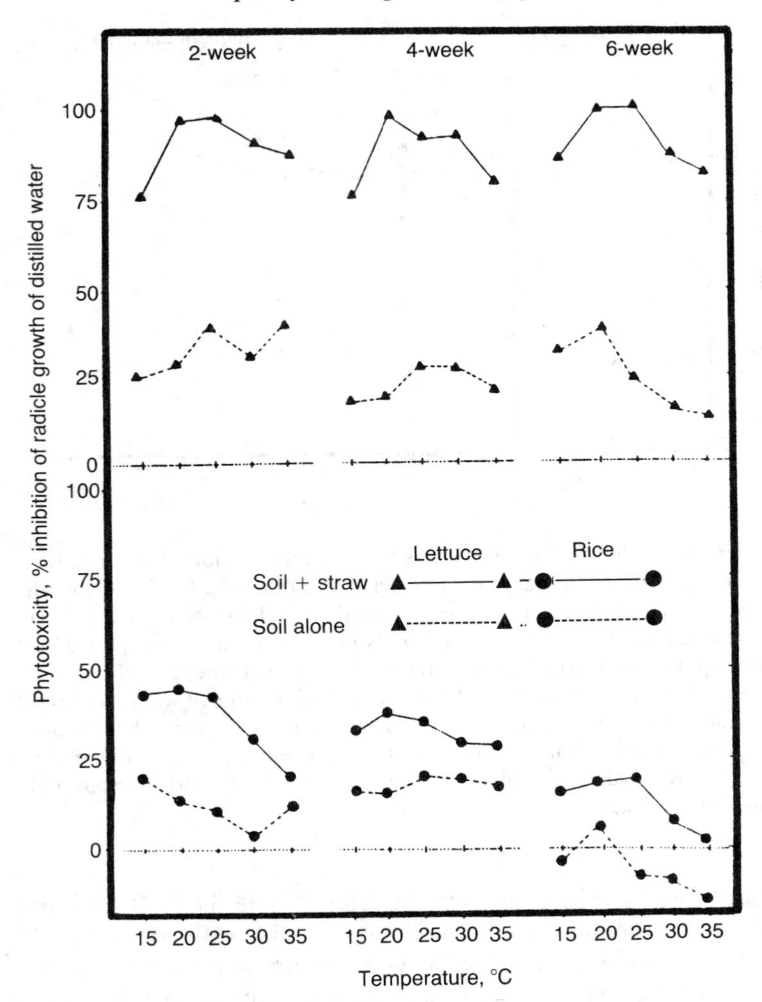

Fig. 13.2 Effect of temperature on phytotoxicity produced during rice straw decomposition in soil. The phytotoxicity is expressed as a percentage of inhibition of growth of lettuce or rice radicle in various soil extracts as compared with distilled water control. (After Chou *et al.*, 1981. Reproduced with permission, Chou, C. H. and Plenum Publishing Corpn, New York.)

is actually self-regulated by means of auto-inhibitors that are also important in the natural selection and evolution of plant species.

13.2.2 Phytotoxic effects of decomposing sugar cane residues in soil

Another case of adaptive auto-inhibition is described by Wang and associates (1984). The reduced productivity of ratoon sugar cane, as a result of

TABLE 13.1 *Comparison of yield components of rice plants growing in pots under different soil treatments (after Chou and Chiou, 1979)*

	A[a]			B			Ratio of A/B	
Yield component (per hill)	S[b]	SR	% decrease	S	SR	% decrease	S	SR
Panicle number	38.75	32.25	17	26.25	19.50	26[c]	1.53[d]	1.65[d]
Total panicle weight (g)	64.33	58.24	10	52.25	42.26	20	1.23	1.37[c]
Ripening rate (%)	90.90	89.13	1	97.65	92.70	5	0.93	0.96
Testing weight (g/1000 seeds)	27.27	27.16	0.5	28.01	26.01	7	0.97	1.01
Grain weight (g)	61.85	55.24	11	50.26	40.14	21[c]	1.23[c]	1.37[c]
Yield (g m^{-2})	1546.30	1380.90	11	1256.40	1003.57	20[c]	1.23[c]	1.37[c]

[a] A, B: Top dressing of nitrogen fertilizer (NH_4^+–N) and (NO_3^-–N), respectively.
[b] S, SR: The pot was filled with 10 kg soil alone (S), and the soil was mixed with 100 g chopped rice straw (SR).
[c,d] Statistical significance at 5% and 1% level, respectively.

TABLE 13.2 *Allelopathic substances found in different plants growing in various environments at the present studies*

Compound	Rice (1, 2)	Sugar cane (2, 3)	Asparagus (4)	Pangola (5)	Leucaena (6)
Formic acid		+			
Acetic acid	+	+			
Propionic acid	+				
Butyric acid	+				
Malic acid	+				
Oxalic acid	+				
Tartaric acid	+				
Lactic acid	+				
p-Hydroxybenzoic acid	+	+		+	+
p-Coumaric acid	+	+		+	+
Syringic acid	+	+		+	+
Vanillic acid	+	+		+	+
Ferulic acid	+	+		+	+
p-Hydroxyphenylacetic acid					+
o-Hydroxyphenylacetic acid	+			+	
3,4,5-Trihydroxybenzoic acid				+	+
3,4-Dihydroxyphenylacetic acid			+	+	+
3,4-Dihydroxybenzoic acid			+		
2,5-Dihydroxybenzoic acid			+		
β-*m*-Hydroxphenylpropionic acid			+		
3,4-Dimethoxyacetophenone			+		
3,4-Dimethoxybenzoic acid			+		

Sources: (1) Chou and Lin (1976), Chou and Chiou (1979); (2) Wu *et al.* (1976a); (3) Wang *et al.* (1967, 1984); (4) Young (1986); (5) Chou and Young (1975), Chou (1987); (6) Chou and Kuo (1987).

its poor germination, has occurred in many farms of the Taiwan Sugarcane Corporation (TSC) for several decades. Many studies have been conducted to elucidate the mechanism of yield reduction, yet no single factor causing the reduction can be found. Wang *et al.* (1984) have demonstrated field and laboratory experiments and concluded that the phytotoxic effect is one of the important factors involved. Five phenolic acids – *p*-hydroxybenzoic, *p*-coumaric, syringic, ferulic and vanillic – were found in the decomposing sugar cane root in soil (Chou, 1968). In addition to these phenolics, formic, acetic, oxalic, malonic, tartaric and malic acids were identified in the decomposing sugar cane leaves in soil under waterlogged condition (Table 13.2). At 50 ppm of these phenolic acids in water culture, the growth of young sugar cane root was significantly inhibited. On the other hand, the aforementioned aliphatic acids can inhibit the growth of ratoon sugar cane at 10^{-3} M. Furthermore, Wu *et al.* (1976a) found that the population of *Fusarium oxysporum* in the rhizosphere soil of poor ratoon cane roots was much greater than that of good growing ratoon or of newly planted sugar cane roots. They found that fusaric acid, the secondary metabolite of the organism, was toxic to the growth of young sugar cane plant *in vitro* (Wu *et al.*, 1976a).

13.2.3 The mechanism of yield reduction in asparagus plantation

A serious reduction in yield and quality of *Asparagus officinalis* often occurs in old plantation soil (Chen, 1978; Young, 1986). The wilting of asparagus plants has been found to be due to the monoculture of the crop. It has been suggested that asparagus in Taiwan needs replanting 5 years after planting in some areas. Asparagus plants thus grow relatively well in sandy and well-drained sandy loam soil, or in seacoast or riverside areas from the south of Europe to the south of Russia (Hung, 1976). Young (1986) thus conducted experiments to find out the cause of the weak plantations found in Taiwan. He used a simplified vermiculite-nutrient culture to determine whether root exudates of asparagus affected other asparagus growth. The results showed that the growth of asparagus seedlings receiving root exudates from three asparagus varieties, namely, Mary Washington, California 309, and California 711 was significantly lower than that of controls during the early growth stages. However, there were no significant differences in the toxicity of root exudates from these three varieties (Fig. 13.3). The growth of both tops and roots was significantly inhibited by asparagus root exudates. The inhibition was not due to the deficiency of mineral nutrients or to the inhibition of uptake and transport of N, P, K (Young, 1986). Exduate collected by use of the circular trapping with a XAD-4 resin significantly retarded radicle and shoot growth of asparagus seedlings. Six phytotoxic phenolics – namely, 3,4-dihydroxybenzoic, 3,4-dimethoxybenzoic, 2,5-dihydroxybenzoic, 3,4-dihydroxyphenylacetic, β-*m*-hydroxyphenylpropionic acid and 3,4-dimethoxyacetophenone – were found in the extracts and exudates of asparagus plant parts (Table 13.2). The amount of phytotoxins identified was significantly higher in the stem than in the root, and was well correlated to phytotoxicity (Young, 1986).

13.3 APPLICATION OF ALLELOPATHY IN AGRICULTURAL PRACTICES IN TAIWAN

13.3.1 Selecting allelopathic cultivars for pasture

Pangola grass, *Digitaria decumbens*, is one of the most dominant forage grasses in Taiwan. The grass grows in many pasture areas and possesses

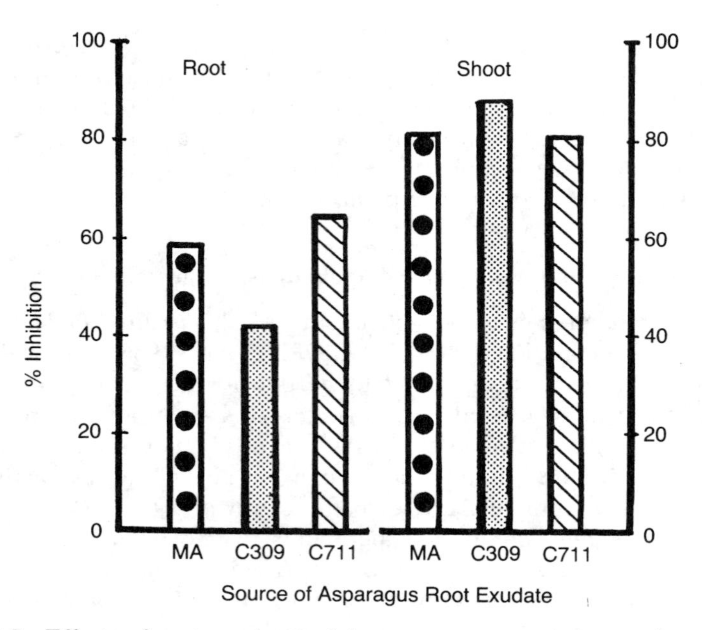

Fig. 13.3 Effects of root exudates of three asparagus varieties on the root and shoot growth of asparagus seedlings. (After Young, 1986).

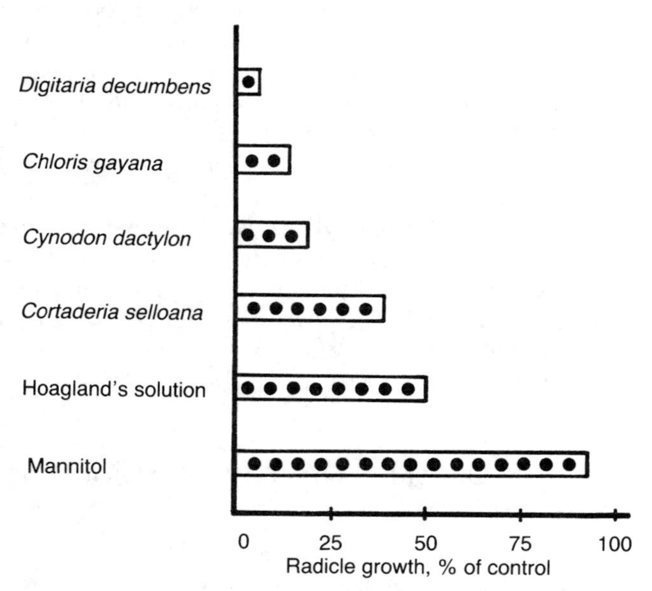

Fig. 13.4 Effects of aqueous extracts of grasses and mannitol and Hoagland's solutions at the osmotic concentration of 25 milliosmols on the radicle growth of lettuce. (After Chou and Young, 1974).

high yield when newly planted. However, several years after planting the productivity generally decreases. Chou and Young (1975) compared the phytotoxic effects of 12 subtropical grasses, including pangola grass, and found that the aqueous extracts of leaves of the 12 subtropical grasses revealed different inhibition on the plants tested. Of them, *Digitaria decumbens* showed the highest phytotoxicity of above 85% when the aqueous extract was obtained at the osmotic concentration of 25 milliosmols, which could not cause any osmotic effect on the test species (Fig. 13.4) (Chou and Young, 1974). Chou furthermore obtained aqueous leachates from four tropical grasses – namely, *Digitaria decumbens*, *Panicum repens*, *Brachiaria mutica* and *Imperata cylindrica* – and used these leachates to irrigate the seedling growth of the four species both interspecifically and intraspecifically. The results showed that the leachate of pangola grass exhibited an autotoxic effect on the growth of itself but rather stimulated the growth of *I. cylindrica* (Chou, 1989). We also conducted experiments to evaluate eight varieties of pangola grass, which were planted in farms of three experimental stations: Hsinhwa, Hengchun and Hwalien. The results showed that cultivars A84, A254 and A255 revealed the highest phytotoxicity, respectively, in the aforementioned stations, while cultivars A65, A255 and A254 respectively, showed the highest invasion ability (Liang *et al.*, 1983). The amount of phytotoxic phenolics in the leachates or extracts of the aforementioned grasses (Table 13.2) were correlated well with the degree of phytotoxicity (Chou and Young, 1975; Chou, 1977, 1987). It is thus clear that a dominant grass, such as pangola grass, can be used as forage pasture. The varieties of the dominant grass can also be used as a weed control, without using too much herbicide, to lessen environmental pollution.

13.3.2 Allelopathic effect of cover crops on weeds and orchard plants

In many fields, cover crops are often used to avoid soil erosion and to lessen the growth of weeds. Wu *et al.* (1975) conducted experiments to evaluate the phytotoxic effects of three cover crops – namely, *Centrocema*, *Indigofera* and Bahia grass – on the growth of several received crops, namely pea, mustard, cucumber, cauliflower, rape, Chinese cabbage, mungbean, water melon, tomato and rice. They found that rape was the most sensitive to the extracts of these cover crops. Among them, *Centrocema* and *Indigofera* exhibited a significant phytotoxic effect, while Bahia grass revealed the least toxicity (Fig. 13.5). In addition, the growth of banana was most inhibited by the leachate of *Centrocema*.

More recently, Chou and his student conducted experiments in the farm of Meifeng Highland Experimental Station of the National Taiwan University, located at Nantou County, to evaluate the competitive and allelopathic nature of nine cover grasses: *Bromus catharticus*, *Pennisetum clandestinum*, *Lolium multiflorum*, *L. perenne*, *Paspalum notatum*,

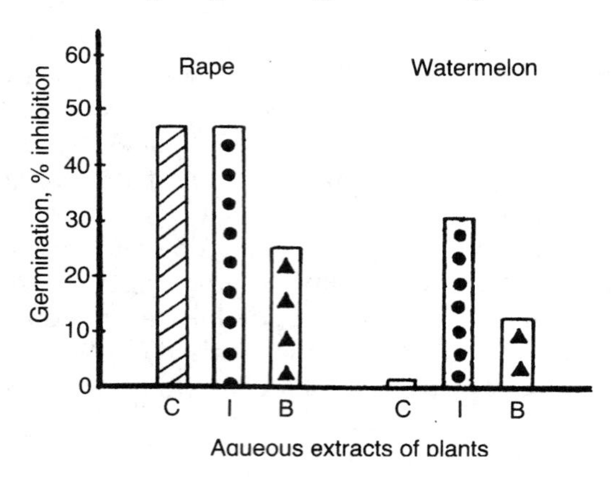

Fig. 13.5 Effects of aqueous extracts of *Centrocema* (C), *Indigofera* (I) and Bahia (B) grasses on the germination of rape and watermelon. (After Wu *et al.*, 1975).

P. dilatatum, *Dactylis glomerata*, *Eragrostis curvula* and *Triflorium repens* (white clover). Of these, *L. multiflorum*, *B. catharticus* and *D. glomerata* exhibited a high yield. *L. multiflorum*, *T. repens* and *P. clandestinum* quite quickly covered the ground, but the yield of the latter two species was comparatively lower than the first. The bioassay results of phytotoxicity showed that *T. repens* and *L. perenne* exhibited a high phytotoxic potential. However, the root exudate of the nine grasses stimulated the growth of *Pyrus lindleyi* seedlings under fertilizer dressing conditions, while *D. glomerata* and *L. multiflorum* showed a remarkably stimulatory effect without fertilizer application (Fig. 13.6). In respect to soil conservation, grasses of *P. notatum*, *T. repens*, *L. multiflorum*, *P. clandestinum* and *D. glomerata* gave a beneficial effect (Chen, 1985; Chou, 1987).

13.3.3 Allelopathic influence on crop rotation

Many monoculture fields often lead to a soil-sickness problem, presumably due to the unbalance of soil microorganisms, accumulation of soil toxins, mineral deficiency, abnormal soil pH, etc., resulting in a decrease of crop productivity. Crop rotation is thus a good control method of avoiding or eliminating the cause of the problem. In a natural ecosystem, one plant may be replaced by a following plant by means of allogenetic and autogenic succession (Daubenmire, 1968), which involves allelopathy or an auto-intoxication mechanism. It may take years to develop a stage of plant succession and may require several decades or hundreds of years to reach a stable community called climax. However, a man-made agroecosystem does need crop succession instead of natural vegetation succession, though we can shorten the time of succession by introducing a new crop to replace the preceding crop, or by rotating the crop. There

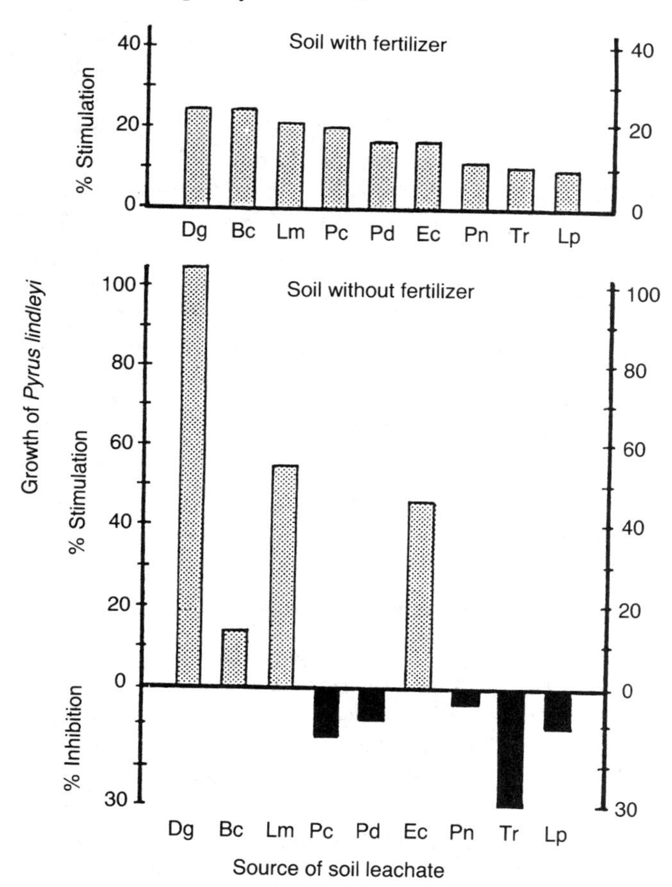

Fig. 13.6 Effects of aqueous soil leachates from nine cover grasses grown in pots with fertilizer application (upper figure) and without fertilizer application (bottom figure) on the seedling growth of *Pyrus lindleyi*. The sources of leachates were *Bromus catharticus* (Bc), *Dactylis glomerata* (Dg), *Eragrostis curvula* (Ec), *Lolium multiflorum* (Lm), *Lolium perenne* (Lp), *Paspalum notatum* (Pn), *P. dilatatum* (Pd), *Pennisetum clandestinum* (Pc) and *Triflorium repens* (Tr). (After Chen, 1985).

are many cases of crop rotation being operated in many parts of the world, and in these areas the crop productivity is always high. A unique example of crop rotation in the Ontario province of Canada is using a tobacco–rye grass–corn rotating system to solve the problem of a root rot disease by a soil-borne pathogen, *Thielaviopsis basicola* (Patrick and Koch, 1963; Chou and Patrick, 1976). After rotation, the damage of root rot would be much less. This is due to the fungitoxins produced by rye grass suppressing the germination of conidia or chlamydospore of *T. basicola*. Thus the over-population of the fungus can be controlled by allelopathic substances released by rotated crop. In Taiwan, pangola grass is a high-productivity pasture and very stable in many fields. However, after several years of plantation the productivity declines due to auto-intoxication, as mentioned earlier. The decline of productivity of this grass has been particularly pronounced in the farm of the Hengchun Experiment Station, Taiwan Livestock Research Institute. During the winter season, the Hengchun area is under a severe drought condition and pangola grass grows poorly. We therefore suggested making a crop rotation by the use of a pangola grass–water melon–pangola grass system. After water melon, the field was replanted with pangola grass, and the yield of the pangola grass was significantly increased up to 40%. Although the detailed experiments concerning the mechanism of the productivity increase has not been performed, it is generally believed that the increase of grass production could be related to the phytotoxic effect of the pangola grass. The toxic effect is apparently reduced by crop rotation. Further experiments are needed to clarify the role of allelopathy in crop rotation.

13.3.4 Allelopathic interaction in a pasture–forest intercropping system

Forests are extremely important in Taiwan for water and natural conservation, because two-thirds of the island is covered by steep mountains. Owing to the rapid population growth and economic development in the past few decades, agricultural land has become limited and therefore agriculture is moving upwards to the mountainous areas. In order to increase livestock production and to preserve the forest, a pasture–forest intercropping system in the mountains was sought to increase the pasture area for grazing. The author and his associates (Chou *et al.*, 1987, 1989) thus conducted experiments by using such a pasture–forest system. In this system, grasses were planted in the cut-over forest where tree seedlings were expected to regenerate. The selection of a suitable grass species for the intercropping system became necessary. Of the grassland species growing at high elevations in Taiwan, kikuyu grass (*Pennisetum clandestinum*) has been planted in many mountainous areas and grows well. The allelopathic interaction of a pasture–forest intercropping system was evaluated by experiments conducted in a field and greenhouse and by laboratory assays. A study site was situated in the farm of Hoshe Forest Experiment Station at Nantou County, Taiwan. After deforestation of Chinese fir (*Cunninghamia lanceolata*), a split plot design of treatments was set up. Field experiments showed that the fir litter left on the ground did not significantly suppress the growth of weeds, kikuyu grass and fir seedlings in the first four months following deforestation, while kikuyu grass significantly suppressed the growth of weeds longer than four months but did not reduce the growth of fir seedlings (Fig. 13.7). In addition, when the kikuyu grass was intercropped with three other hardwood plants, namely, *Alnus formosana, Cinnamomum camphora* and *Zelkova*

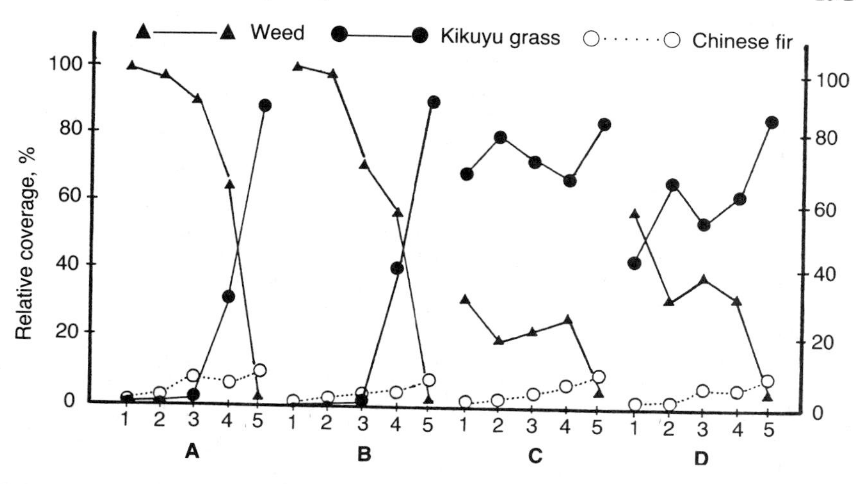

Fig. 13.7 Effect of field treatments on ground cover expressed as a percentage of relative coverage per plot of kikuyu grass, weed and Chinese fir harvest on five dates: (1) 19 April 84; (2) 21 June 1984; (3) 14 December 1984; (4) 15 May 1985; and (5) 15 December 1985. The treatments are: (A) litter removed and nothing planted; (B) litter left and nothing planted; (C) litter removed and kikuyu grass planted; (D) litter left and kikuyu grass planted. (After Chou *et al.*, 1987).

formosana, the weeds were significantly retarded but that of woody plants was not affected (Fig. 13.8). The aqueous leachate of kikuyu grass stimulated the seedling growth of *C. camphora* and *A. formosana*, but the extract stimulated the growth of *C. camphora* and inhibited that of *A. formosana*. In contrast, the aqueous extracts of three hardwood plants exhibited variable inhibition on the root initiation of kikuyu grass, and the extracts of *Z. formosana* revealed the highest phytotoxic effect. The extracts of the three hardwood plants, as well as the kikuyu grass, all suppressed the seedling growth of *Miscanthus floridulus*. The results clearly showed that the allelopathic effect was involved in controlling weed population. The responsible allelopathic substances have been identified by means of chromatography and 14 phytotoxins α-resorcyclic acid, β-resorcyclic acid, γ-resorcyclic acid, *o*-hydroxyphenylacetic acid, gallic acid, *p*-hydroxybenzoic acid, protocatechuic acid, vanillic acid, syringic acid, *p*-, *m*- and *o*-coumaric acids, cinnamic acid and ferulic acid were found (Fig. 13.9). The findings suggest that allelopathy may contribute benefits in the intercropping system to reduce the need for herbicides and to lessen the labour cost for weed control.

13.3.5 Interaction between allelopathic substances and nutrients

Wu and her co-workers (Wu *et al.*, 1976b) conducted field experiments in Lotung, an area of poor-drainage paddy soil, by applying ammonium sulphate, lime, and green manure in order to compare the eliminating

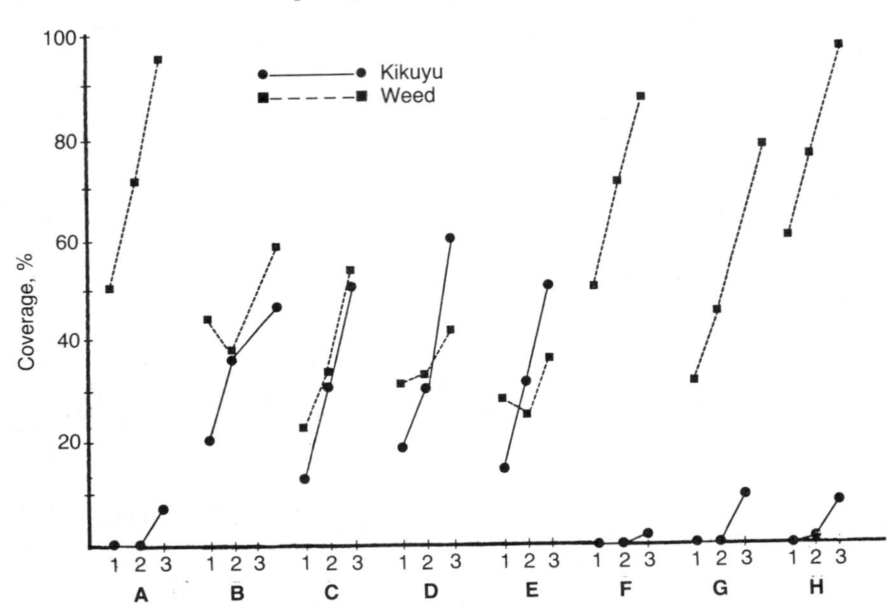

Fig. 13.8 Effects of field treatments on ground cover expressed as a percentage per plot of kikuyu grass and weed harvested on three dates: (1) 11 November 1986; (2) 29 April 1987; and (3) 28 October 1987. The treatments are: (A) open ground without planting any plant; (B) planted with kikuyu grass; (C) planted with kikuyu grass and *A. formosana*; (D) planted with kikuyu grass and *Z. formosana*; (E) planted with kikuyu grass and *C. camphora*; (F) planted with *A. formosana*; (G) planted with *Z. formosana*; and (H) planted with *C. camphora*. The data of weeds and kikuyu grass between treatments were significantly different at the 5% level. (After Chou *et al.*, 1989).

degree of phytotoxicity of rice residues decomposing in the soil. The results indicated that the paddy soil applied with lime exhibited a significantly higher yield than that with other treatments (Fig. 13.10) (Wu *et al.*, 1976b). They concluded that the increase of rice yield was simply due to a detoxification mechanism of phytotoxins, and binding of calcium with some phytotoxic substances, converting them into non-toxic forms. In addition, Chou and Chiou (1979) conducted experiments to show that different nitrogen fertilizer dressings gave different yields of rice, whereas an ammonium sulphate–nitrogen fertilizer exhibited a higher yield than that of a nitrate–nitrogen fertilizer, suggesting that the ammonium sulphate–nitrogen fertilizer may overcome the phytotoxic effect of decomposing rice residues in soil. This finding is in agreement with that of Chandrasekaran and Yoshida (1973), who concluded that ammonium sulphate effectively eliminated the injury caused by phytotoxins. We also found that the root system of rice plants was healthy and well developed under the ammonium sulphate dressing as compared with that under the nitrate fertilizer dressing. It is concluded that some nutrients may play a chelating role in detoxifying the phytotoxins in soil, and consequently give a better yield of crops.

Chou *et al.* (1977) have concluded that the more rice stubble was left in the paddy soil, the higher would be the phytotoxic phenolics and the less amount of leachable nitrogen, reflecting that the phytotoxins produced may interact with nitrogen availability in soil. They also found that the amount of leachable NH_4–N was about ten times more than that of NO_3–N (Chou and Chiou, 1979). Chou and Cheng further designed an experiment by using a ^{15}N-isotope incorporated into the soil or a soil–rice residues mixture under different temperature regimens and sequences. The results indicated that, in the absence of straw, most of the fertilizer N remained in the mineral forms. Straw enhanced N immobilization only moderately. The gradual decrease in the proportion of fertilizer N in the mineral forms was accompanied by a steady increase of fertilizer N in the amino acid fraction of organic N. Little accumulation of fertilizer N in the amino sugar or the insoluble humin fraction was found (Fig. 13.11) (Chou *et al.*, 1981). Although the experimental results were not in the distinct trend in relation to temperature variations, the temperature range of 25–30°C tended to favour N transformation activities.

During the decomposition of rice residues in soil, the amount of available minerals in soil might be affected and, consequently, could affect plant growth. Chou and Chiou (1979) conducted an experiment to see the effect of rice straw incorporated into soil on the dynamics of some cations in pot soil. The results revealed that the concentrations of cations, K, Cu and Mn, were higher in the first crop season, while those of Na, Ca, Mg and Zn were higher in the second crop season in Nankang paddy soil regardless of nitrogen fertilizer application. Most of our findings are similar to those of Patrick and Mikkelsen (1971). In the flooding soil the contents of reducible iron and manganese were relatively low. When the pot soil was mixed with rice straw and allowed to decompose the amount of K was significantly higher than that of soil alone, but those of Cu, Fe, Mn and Zn were on average significantly lower in the soil in terms of the ratio of soil to straw (Table 13.3). It is interesting to note that in several poor water-drainage areas in Taiwan – such as Changhwa, Taitung and Pingtung – the Zn deficiency is particularly pronounced during the second crop season.

13.4 PROSPECTS OF ALLELOPATHIC RESEARCH IN TAIWAN

13.4.1 Elimination of allelopathic substances by fire

Allelopathic substances can be eliminated by fire. Muller *et al.* (1968) stated that the chaparral fire cycle emerges as a sequence of events

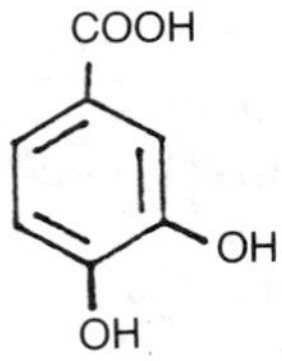

3,4-Dihydroxybenzoic acid

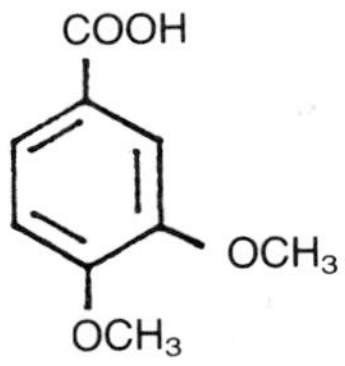

3,4-Dimethoxybenzoic acid

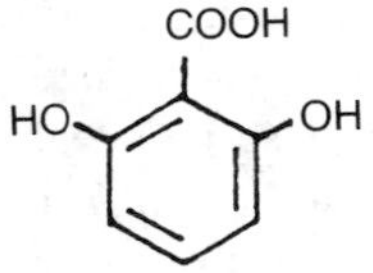

2,6-Dihydroxybenzoic acid

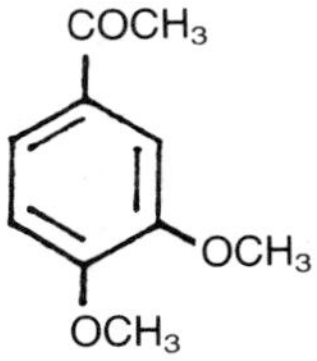

3,4-Dimethoxyacetophenone

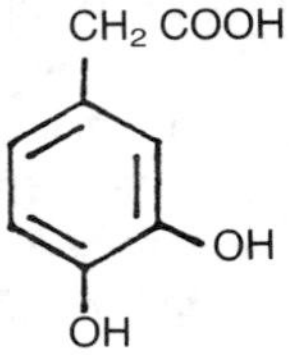

3,4-Dihydroxyphenylacetic acid

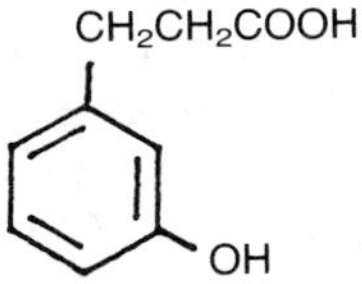

β-(*m*-Hydroxyphenyl) propionic acid

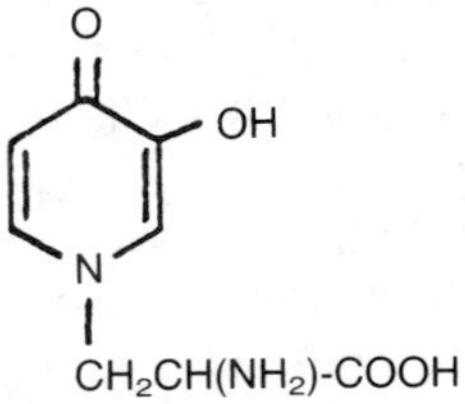

Mimosine

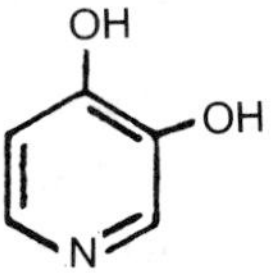

3,4-Dihydroxypyridine

Fig. 13.9 Chemical structure of allelopathic compounds isolated from plant parts mentioned in the text.

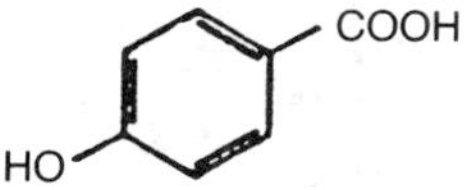

COOH
HO

p-Hydroxybenzoic acid

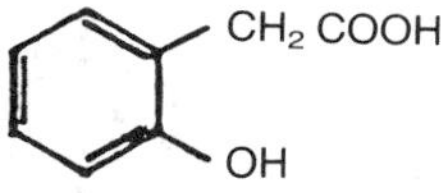

CH_2 COOH
OH

o-Hydroxyphenyl acetic acid

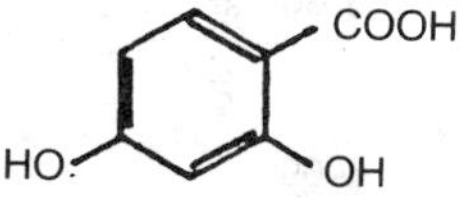

COOH
HO OH

Resorcylic acid

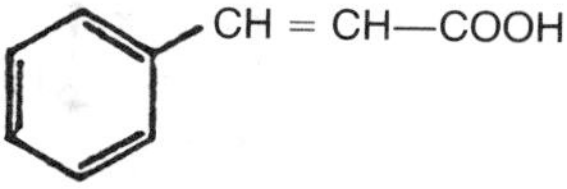

CH = CH—COOH

Cinnamic acid

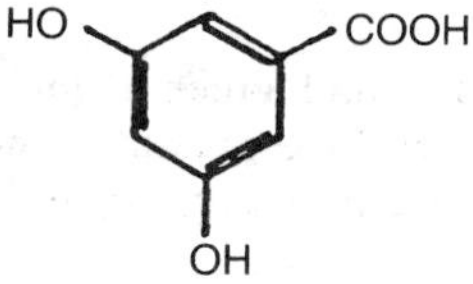

HO COOH
OH

Protocatechuic acid

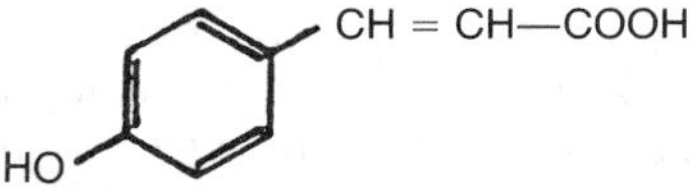

CH = CH—COOH
HO

p-Coumaric acid

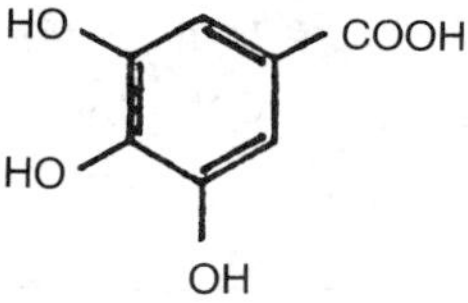

HO COOH
HO
OH

Gallic acid

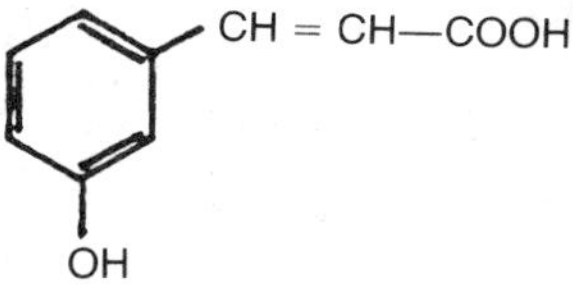

CH = CH—COOH
OH

m-Coumaric acid

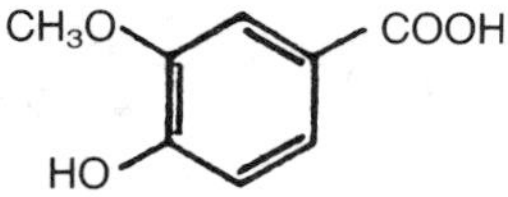

CH_3O COOH
HO

Vanillic acid

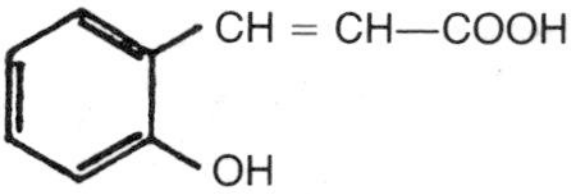

CH = CH—COOH
OH

o-Coumaric acid

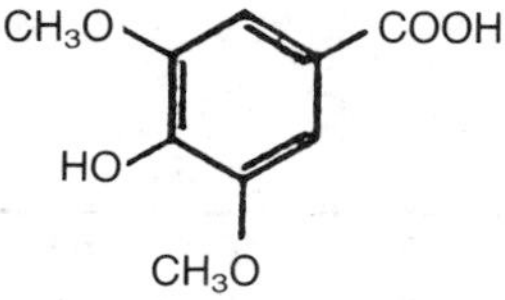

CH_3O COOH
HO
CH_3O

Syringic acid

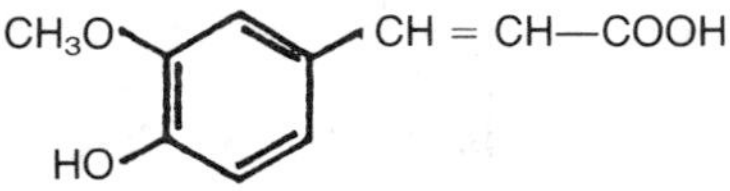

CH_3O CH = CH—COOH
HO

Ferulic acid

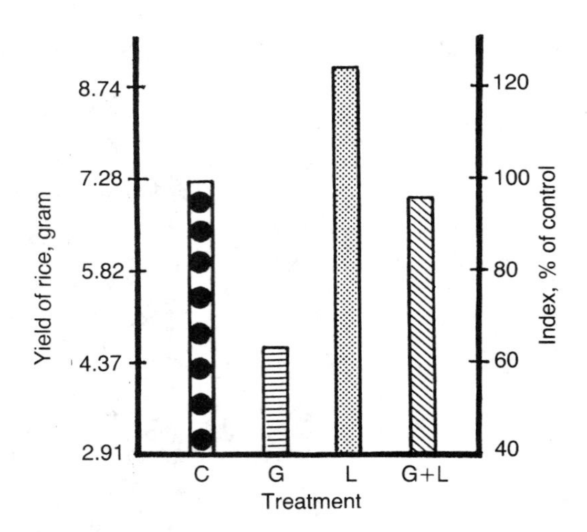

Fig. 13.10 Effects of green manure (G), liming (L) and green manure plus liming (G + L) in addition to ammonium sulphate as base dressing, and only ammonium sulphate as control (C) on the growth of rice in poorly drained paddy soil. (After Wu *et al.*, 1976b).

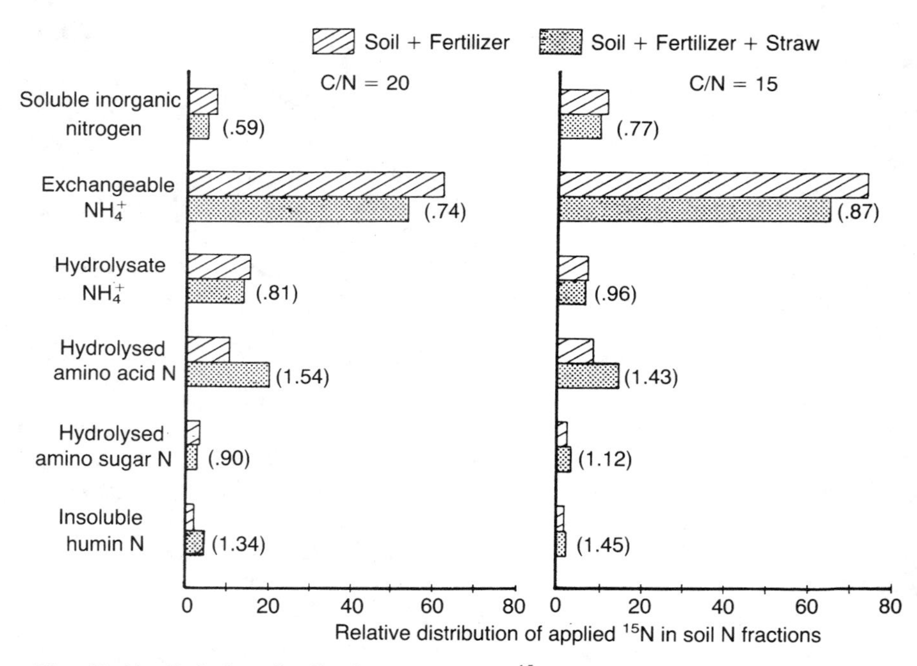

Fig. 13.11 Relative distribution of applied ^{15}N in soil nitrogen fractions. The number in parentheses indicates the ratios of AFS/AF (left figure) or BFS/BF (right figure). (After Chou *et al.*, 1981).

TABLE 13.3 *Effects of culture treatments of extractable soil cations in pot experiment (after Chou and Chiou, 1979)*[a]

	Concentration (ppm)								
	S			SR			Radio of S/SR[b]		
Cation	A[c]	B	Mean	A	B	Mean	A	B	Means
K^+	7.68	5.36	6.52	11.68	13.67	12.68	0.65[e]	0.40[e]	0.51[d]
Na^+	33.04	30.46	31.75	26.86	29.96	28.41	1.23	1.02	1.12
Ca^{2+}	52.20	50.86	51.53	34.82	37.44	36.13	1.50[d]	1.36[e]	1.43[d]
Cu^{2+}	0.05	0.04	0.05	0.07	0.02	0.05	0.71[e]	2.00[d]	1.00
Fe^{2+}	3.62	0.07	1.85	0.10	0.13	0.12	36.20[d]	0.54[d]	15.42[d]
Mg^{2+}	278.80	224.16	251.48	295.04	240.04	267.54	0.94	0.93	0.94
Mn^{2+}	11.78	16.19	13.99	14.47	7.88	11.18	0.81	2.05[d]	1.25[e]
Zn^{2+}	0.38	0.24	0.31	0.07	0.07	0.07	5.42[d]	3.43[d]	4.43[d]

[a] The data were obtained by means of five sampling times with four replications.
[b] S, SR: The pot was filled with 10 kg soil alone (S), and the soil was mixed with 100 g of chopped rice straw (SR).
[c] A, B: Top dressing of ammonium sulphate fertilizer (NH_4^+-N) and of potassium nitrate (NO_3^--N), respectively.
[d,e] Statistical significance at 5% and 1% level, respectively.

consequent to the destruction of toxins and their shrub sources by fire. Thus, a tremendous number of weeds previously suppressed by the allelopathic compound recurred in the following growing season after the fire. However, Botkin and Keller (1987) indicated that the forest fire actually plays an important role in maintaining the stability of forest community. Fire is also used to help farmers to enhance crop productivity in Taiwan. For example, farmers often burn the crop residues after harvesting not only to destroy plant pathogens but also to denature phytotoxins left in the residues. The yield reduction caused by allelopathic effects can be reduced or eliminated by ground fire. Many practices in the field have already shown that a ground fire is the best way to solve the problem of yield reduction caused by a continuous monoculture. The allelopathic substances can be denatured or evaporated by a fire, leading to an increase in crop productivity. For example, such plants as kikuyu grass, pangola grass, or rice plant have already shown that they are autotoxic. However, very few detailed experiments on the effect of fire on allelopathic compounds have been carried out in the past.

13.4.2 Polymerization of phytotoxins with humic substances in soils

Allelopathic compounds can be bound with clay mineral or other organic compounds, leading to decreased phytotoxicity (Wang *et al.*, 1971, 1978,

1983; Rice, 1984). Wang *et al.* (1978) found that protocatechuic acid, one of the phytotoxins related to trans-*p*-coumaric acid, can be polymerized into humic acid by using clay minerals as heterogeneous catalysts. In fact, the humic acid is a natural substance, which can polymerize many kinds of substances, such as amino acids, flavonoids, terpenoids, aliphatic acids, and other nitrogen-containing compounds, thus keeping the soil in a fertile state. However, it is also possible that the polymerized phytotoxins fixed into humic complex can be depolymerized under certain environmental conditions, and release free phenolic compounds that will exert a phytotoxic effect on nearby susceptible plants. If the polymerization of humic substances is a natural phenomenon, the phytotoxic substances may easily be fixed into humic complex; thus the natural device of an organo-mineral complex of humic acid would actually form a pool of detoxification for many kinds of toxic substances produced by plants.

13.4.3 Effects of environmental stresses on potential allelopathic plants

Many allelopathic compounds produced by plants are often regulated by environmental factors, such as the water potential of the environment, temperature, light intensity, soil moisture, nutrient, soil microorganisms, etc., and the compounds are released to the environment by means of volatilization, leaching, decomposition of residues and root exudation (Muller, 1966; Chou, 1983; Rice, 1984). The following paragraphs describe the phytotoxins produced and their interaction with factors under different environmental regimens.

Many aquatic plants can grow very well in a waterlogged and oxygen-deficient environment because of their adaptation mechanism. Although rice plants are not aquatic plants, they grow very well in the paddy soil. Patrick and Mikkelsen (1971) indicated that the level of oxygen was almost zero when measured at 25 cm below the soil surface in a paddy field. We obtained a similar result in paddy soil in many areas of Taiwan, namely Tsingshui (the centre of Taiwan), Chiatung and Yuanlin (the south) and Tungshan (the east coast), where the paddy fields are either poor in water drainage or have a higher water table, leading to oxygen deficiency. This is even more pronounced during the second crop season when the monsoon comes.

13.4.4 Potential herbicides from allelopathic plants

A study site of *Leucaena leucocephala* plantation was selected at the Kaoshu village of Pintung county, situated in the south of Taiwan. About 3 to 4 years after plantation, there developed an almost total lack of understorey species except *Leucaena* beneath *Leucaena* during the winter season. The absence of weeds was due to a heavy accumulation of *Leucaena* plant residues, such as leaf litter, branches, etc., which releases phytotoxic substances to suppress the growth of many understorey species, except *Leucaena* itself (Chou and Kuo, 1987). They found that the aqueous extract of leaf litter revealed a significant phytotoxicity on many plants tested, and the phytotoxins were identified as mimosine, quercetin and eight phytotoxic phenolics (Table 13.12). The seed germination and radicle growth of lettuce, rice and rye grass were significantly inhibited by aqueous mimosine at a concentration of 20 ppm, while the forest species, namely *Acacia confusa, Alnus formosana, Casuarina glauca* and *Liquidambar formosana*, were suppressed by the mimosine solution at 50 ppm or above (Chou and Kuo, 1987). However, the growth of *Miscanthus floridulus* and *Pinus taiwanensis* were retarded by mimosine solution at 200 ppm. The seedlings of *Ageratum conyzoides* died in mimosine solution at 50 ppm within 7 days and wilted at 300 ppm within 3 days. Nevertheless, both the aqueous solution of *Leucaena* leaf extract and mimosine were not toxic to the growth of *Leucaena* seedlings. It is thus concluded that the exclusion of understorey plants was evidently due to the allelopathic effects of compounds produced by *Leucaena*. It is thus suggested that mimosine and 3,4-dihydroxypyridine (Fig. 13.9) produced by *Leucaena* could possibly be used as natural herbicides.

Another case of allelochemicals produced by *Vitex negundo* was investigated. *Vitex negundo*, a dominant component of coastal vegetation, is widely distributed in the southern parts of Taiwan. Chou and Yao (1983) found that the biomass and density of its understoreys are relatively lower than in adjacent pasture. Field results showed that the natural leachate of *V. negundo* significantly retarded the growth of *Digitaria decumbens* but stimulated the growth of *Andropogon nodosus* as compared with the rainfall control. The growth of *D. decumbens* grown in pots under greenhouse conditions was significantly retarded by watering with a 1% aqueous extract of *V. negundo*, but the growth of *A. nodosus* and *Mimosa pudica* was stimulated. The allelochemicals isolated from the extracts of *V. negundo* were identified by means of chromatography, NMR and GC-mass spectrometry. The allelochemicals with allelopathic properties included phenolic compounds and ten flavonoids (Chou and Yao, 1983). The potential of allelochemicals from *V. negundo* is probably high, and their biological activity needs to be further investigated.

13.5 CONCLUSIONS AND PROSPECTS

It is concluded that the cause of yield reduction in rice plants, sugar cane, asparagus and other crops in a monoculture system is due fundamentally to adaptive auto-inhibitors produced during the decomposition of plant residues in the soil. The inhibitors accumulate in the soil to a certain extent where environmental conditions are rather unfavourable for the growth of these plants. Dominant grasses, such as pangola (*Digitaria decumbens*) and kikuyu (*Pennisetum clandestinum*) with allelopathic potential, have been used as good pasture for weed control in Taiwan. However, the pangola grass exhibits auto-intoxication about six years after the grass has been planted. A unique pasture–forest intercropping system has successfully been demonstrated by planting kikuyu grass into a deforested land for forest regeneration in order to reduce the need for herbicides and to lessen the labour costs for weed control. Continuous investigation on isolation and identification of allelopathic compounds from natural vegetation should be carried out. The molecular basis of allelopathic research, the model action of allelochemicals in plant growth regulation, and the mechanism of detoxification of allelopathic compounds in soil should be at the forefront of all future study. Further evidence to clarify the role of allelopathy in natural ecosystem and agroecosystem should therefore be revealed.

ACKNOWLEDGEMENTS

The author greatly appreciates the assistance of his former colleagues and graduate students in the compilation of this chapter. The study was supported by the Academia Sinica, the National Science Council and the Council of Agriculture of the Republic of China on Taiwan.

REFERENCES

Botkin, D. B. and Keller, E. A. (1987) *Environmental Studies*, Merrill Publishing Company, Columbus.

Chandrasekaran, S. and Yoshida, T. (1973) Effect of organic acid transformation in submerged soils on the rice plants. *Soil Sci. Plant Nutr.*, **19**, 39–45.

Chen, C. (1985) Studies on the soil conservation effect, growth and plant interactions of the nine cover crops in orchard in Meifeng area in Taiwan. MS Thesis, National Taiwan University, Taipei, Taiwan.

Chen, W. Y. (1978) Study on improvement of asparagus decline and production problem, in *Proc. 2nd Symp. Asparagus Res. Taiwan, National Chung Hsing University*, Taiwan, pp. 17–25.

Chou, C. H. (1968) Phytotoxic substances from sugar cane roots. MS Thesis. National Taiwan University, Taipei, Taiwan.

Chou, C. H. (1977) Phytotoxic substances in twelve subtropical grasses. I. Additional evidence of phytotoxicity in the aqueous fractions of grass extracts. *Bot. Bull. Acad. Sinica*, **18**, 131–41.

Chou, C. H. (1983) Allelopathy in agroecosystems in Taiwan, in *Allelochemicals and Pheromones* (eds C. H. Chou and G. R. Waller), Academia Sinica Monograph Series No. 5, Acad. Sinica, Taipei, ROC, pp. 27–64.

Chou, C. H. (1987) Allelopathy in subtropical vegetation and soils in Taiwan, in *Allelochemicals: Role in Agriculture and Forestry* (ed. G. R. Waller), ACS Symp. Ser. 330, Amer. Chem. Soc., Washington, DC, pp. 102–17.

Chou, C. H. (1989) Allelopathic research of subtropical vegetation in Taiwan. IV. Phytotoxic nature of leachate from four subtropical grasses. *J. Chem. Ecol.*, **15**, 2149–59.

Chou, C. H., Chang, H. J., Cheng, C. M. *et al.* (1989) The selective allelopathic interaction of a pasture–forest intercropping in Taiwan. II. Interaction between kikuyu grass and three hardwood plants. *Plant and Soil*, **116**, 207–15.

Chou, C. H., Chiang, Y. C. and Cheng, H. H. (1981) Autointoxication mechanism of *Oryza sativa*. III. Effect of temperature on phytotoxin production during rice straw decomposition in soil. *J. Chem. Ecol.*, **7**, 741–52.

Chou, C. H. and Chiou, S. J. (1979) Autointoxication mechanism of *Oryza sativa*. II. Effect of culture treatments on the chemical nature of paddy soil and on rice productivity. *J. Chem. Ecol.*, **5**, 839–59.

Chou, C. H., Hwang, S. Y., Peng, C. I. *et al.* (1987) The selective allelopathic interaction of a pasture–forest intercropping in Taiwan. *Plant and Soil*, **98**, 31–41.

Chou, C. H. and Kuo, Y. L. (1987) Allelopathic research of subtropical vegetation in Taiwan. III. Allelopathic exclusion of understorey by *Leucaena leucocephala* (Lam.) de Wit. *J. Chem. Ecol.*, **12**, 1431–48.

Chou, C. H., Lee, M. L. and Oka, H. I. (1984) Possible allelopathic interaction between *Oryza perennis* and *Leersia hexandia*. *Bot. Bull. Academia Sinica*, **25**, 1–19.

Chou, C. H. and Lin, H. J. (1976) Autointoxication mechanism of *Oryza sativa*, I. Phytotoxic effects of decomposing rice residues in soil. *J. Chem. Ecol.*, **2**, 353–67.

Chou, C. H., Lin, T. J. and Kao, C. I. (1977), Phytotoxins produced during decomposition of rice stubbles in paddy soil and their effect on leachable nitrogen. *Bot. Bull. Academia Sinica*, **18**, 45–60.

Chou, C. H. and Patrick, Z. A. (1976) Identification and phytotoxic activity of compounds produced during decomposition of corn and rye residues in soil. *J. Chem. Ecol.*, **2**, 369–87.

Chou, C. H. and Waller, G. R. (eds) (1983), *Allelochemicals and Pheromones*, Academia Sinica Monograph Series No. 5, Acad. Sinica, Taipei, ROC.

Chou, C. H. and Waller, G. R. (eds) (1989) *Phytochemical Ecology: Allelochemicals, Mycotoxins, and Insect Pheromones and Allomones*, Academia Sinica Monograph Series No. 9, Acad. Sinica, Taipei, ROC.

Chou, C. H. and Yao, C. (1983) Phytochemical adaptation of coastal vegetation in Taiwan I. Isolation, identification and biological activities of compounds in *Vilex negundo*. *Bot. Bull. Academia Sinica*, **24**, 155–68.

Chou, C. H. and Young, C. C. (1974) Effects of osmotic concentration and pH on plant growth. *Taiwania*, **19**, 157–65.

Chou, C. H. and Young, C. C. (1975) Phytotoxic substances in twelve subtropical grasses. *J. Chem. Ecol.*, **1**, 183–93.

Daubenmire, R. (1968). *Plant Communities*. Harper & Row, New York.

Hung, L. (1976) *Asparagus: Cultivation Technique*, Taiwan University, Taipei, p. 1.

Koeppe, D. E., Rohrbaugh, L. M. and Wender, S. H. (1971) The effect of environmental stress condition of caffeoylquinic acids and scopolin in tobacco and sunflower, in *Biochemical Interactions Among Plants*, National Academy of Science, Washington, DC, pp, 109–12.

Koeppe, D. E., Southwick, L. M. and Bittell, J. E. (1976) The relationship of tissue chlorogenic acid concentration and leaching of phenolic from sunflowers grown under varying phosphate nutrient condition. *Can. J. Bot.*, **54**, 593–9.

Liang, J. C., Sheen, S. S. and Chou, C. H. (1983) Competitive allelopathic interaction among some subtropical pastures, in *Allelochemicals and Pheromones* (eds C. H. Chou and G. R. Waller), Academia Sinica Monograph Series No. 5, Acad. Sinica, Taipei, ROC, pp. 121–33.

Muller, C. H. (1966) The role of chemical inhibition (allelopathy) in vegetational composition. *Bull. Torrey Bot. Club*, **93**, 332–51.

Muller, C. H. (1971) Phytotoxins as plant habitat variables, in *Biochemical Interactions Among Plants*, National Academy of Science, Washington, DC, pp. 64–72.

Muller, C. H. (1974) Allelopathy in the environmental complex, in *Handbook of Vegetation Science Part VI: Vegetation and Environment* (eds B. R. Strain and W. D. Billings), Dr W. Junk BV Publisher, The Hague, pp. 73–85.

Muller, C. H., Hanawatt, R. B. and McPherson, J. K. (1968) Allelopathic control of herb growth in the five cycle of California chaparral. *Bull. Torrey Bot. Club*, **95**, 225–31.

Oka, H. I. (1976) Mortality and adaptive mechanisms of *Oryza perennis* strains. *Evolution*, **30**, 380–92.

Patrick, W. H. Jr and Mikkelsen, D. S. (1971) Plant nutrient behavior in flood soil, in *Fertilizer Technology and Use*, Soil Sci. Soc. of America, Madison, Wisconsin, pp. 187–215.

Patrick, Z. A. and Koch, L. W. (1963) The adverse influence of phytotoxic substances from decomposing plant residues on the resistance of tobacco to black root rot. *Can. J. Bot.*, **41**, 747–58.

Putnam, A. R. and Duke, W. B. (1974) Biological suppression of weeds: evidence for allelopathy in accessions of cucumber. *Science*, **185**, 370–2.

Putnam, A. R. and Tang, C. S. (eds) (1986). *The Science of Allelopathy*, John Wiley, New York.

Rice, E. L. (1984) *Allelopathy*, 2nd edn, Academic Press, New York and London.

Waller, G. R. (ed.) (1987). *Allelochemicals*: *Role in Agriculture and Forestry*, ACS Symp. Ser. 330, Amer. Chem. Soc., Washington, DC.

Wang, T. S. C., Kao, M. M. and Li, S. W. (1984) The exploration and improvement of the yield decline of monoculture sugarcane in Taiwan, in *Tropical Plants* (ed. C. H. Chou), Academia Sinica Monograph Series No. 6, Acad. Sinica, Taipei, ROC., pp. 1–9.

Wang, T. S. C., Li, S. W. and Ferng, Y. L. (1978) Catalytic polymerization of phenolic compounds by clay minerals. *Soil Sci.*, **126**, 16–21.

Wang, T. S. C., Wang, M. C. and Huang, P. M. (1983) Catalytic synthesis of humic substances by using aluminas as catalysts. *Soil Sci.*, **136**, 226–30.

Wang, T. S. C., Yang, T. K. and Chaung, T. T. (1967) Soil phenolic acids as plant growth inhibitors. *Soil Sci.*, **103**, 239–46.

Wang, T. S. C., Yeh, K. L., Cheng, S. Y. and Yang, T. K. (1971) Behaviour of soil phenolic acids, in *Biochemical Interactions Among Plants*, National Academy of Sciences, Washington, DC.

Whittaker, R. H. and Fenny, P. P. (1971) Allelochemics: chemical interactions

between species. *Science*, **171**, 757–70.
Wu, H. P. and Antonovics, J. (1979) Regional distribution of rice yields in Taiwan, 1965–1976. *Bot. Bull. Acad. Sinica*, **20**, 1–8.
Wu, M. M. H., Liu, C. L. and Chao, C. C. (1976a) Identification and purification of the phytotoxin produced by *Fusarium oxysporum* Schl. in relation to yield decline of ratoon cane. *J. Chinese Agril. Chem. Soc.*, **14**, 160–5.
Wu, M. M. H., Liu, C. L., Chao, C. C. *et al.* (1976b) Microbiological and biochemical studies on the causes of low yielding in the second crop of rice. *J. Agril. Assoc. China* (New Series), **96**, 16–37.
Wu, M. M. H., Shieh, S. W., Liu, C. L. and Chao, C. C. (1975) Determination of the toxicity and phytotoxin of some cover crops inhibiting the growth of banana. *J. Agril. Assoc. China* (New Series), **90**, 54–63.
Young, C. C. (1986) Autointoxication of *Asparagus officinalis* L., in *The Science of Allelopathy* (eds A. R. Putnam and C. S. Tang), Wiley, New York, pp. 1011–12.

CHAPTER 14

Effect of pine allelochemicals on selected species in Korea

Bong-Seop Kil

14.1 INTRODUCTION

The allelopathic potential of red pine (*Pinus densiflora*), pitch pine (*Pinus rigida*) and black pine (*Pinus thunbergii*) has attracted attention of Korean botanists for the last several years. It has been a point of interest why the understorey species are sparse (Lee and Monsi, 1963) and species growing there were similar with the other pine understoreys. It was hypothesized that the similarity of floristic composition of pine stands was caused by some regulating mechanism controlled by pine through the release of certain toxic substances (allelochemicals) in the soil. When the greenhouse soil was mixed with pine leaves, the growth of the plants was suppressed. However, the toxicity of the soil was gradually reduced, and ultimately diminished after several years. This further strengthened the idea of the possible release of allelochemicals by pine. Therefore, it deemed necessary to verify experimentally whether pine is indeed producing allelochemicals which, in turn, affect neighbouring plants. For this, several experiments were performed involving various species growing inside and outside the pine forests. The results obtained from various laboratory and field studies with respect to the allelopathic effects of pine and details of isolation of allelochemicals from pine materials are reported here.

14.2 ALLELOPATHIC EFFECTS OF VARIOUS SPECIES OF PINE

14.2.1 Germination and growth test in aqueous extracts

(a) *Pinus densiflora*

Aqueous extracts of fresh leaves, fallen leaves and roots of red pine were prepared. For this, 1 litre of distilled water was added to 200 g of

Allelopathy: Basic and applied aspects Edited by S. J. H. Rizvi and V. Rizvi
Published in 1992 by Chapman & Hall, London ISBN 0 412 39400 6

each material. Each half of this material was kept at 80 and 18°C and designated A and B, respectively. pH (6–7) and osmotic pressure (not greater than 50 milliosmols) of extracts obtained after 24 hours were adjusted. Bioassay was done in both the petri dishes and plastic pots containing vermiculite having 10 ml of each extract and for control 10 ml of water. Among the test species, six showed a high germination percentage (above 90%) while 14 species gave a low germination percentage (below 50%) in the presence of extract A. On the other hand, 13 and six species showed a high and low germination percentage, respectively, in presence of extract B. Data summarized in Tables 14.1 and 14.2 clearly show that most of the differences due to treatments were significant and that extract A was more inhibitory than extract B. Further, the extent of inhibition by extracts was in the order of fresh leaf > fallen leaf > root extract. In pot studies, species having a high germination percentage were 8 and 17 in extracts A and B, respectively, but only eight and two species showed a low germination percentage in extracts A and B, respectively (Tables 14.3 and 14.4). It is clear from the results that the inside species of pine forest had significantly higher germination than the outside species. On the basis of interaction with pine, the test species were divided into three groups: (1) species which grow well inside the pine forest, (2) species which grow well outside the pine forest and (3) species which may germinate and grow equally well in either habitat. The following are some examples of the above-mentioned groups:

1. *Aster tataricus*, *Cymbopogon tortilis* var. *goeringii*, *Pencedanum terebinthaceum*, *Celastrus orbiculatus*, *Platycodon grandiflorum*, *Patrinia villosa*, *Pinus rigida*, *Atractylodes japonica*, *Lespedeza bicolor*, *Arundinella hirta*, *Pinus densiflora*, *Themeda triandra* var. *japonica*, *Leibnitzia anandria*, *Phyllanthus ussuriensis*, *Miscanthus sinensis*, *Albizzia julibrissin*, *Plantago asiatica*, *Pinus thunbergii* and *Rhododendron schlippenbachii*.
2. *Boehmeria plantanifolia*, *Cassia tora*, *Triumfetta japonica*, *Sedum erythrostichum*, *Elsholtzia ciliata*, *Oenothera odorata*, *Chenopodium album* var. *centrorubrum*, *Digitaria sanguinalis*.
3. *Setaria viridis*, *Celosia argentea*, *Erigeron annuus*, *Lactuca raddeana*, *Bidens bipinnata*, *Erigeron canadensis*, *Phytolacca americana*, *Achyranthes japonica*, *Aster yomena*, *Robinia pseudo-acacia*, *Melandryum firmum*, *Justicia procumbens* and *Cassia mimosoides* var. *nomame*.

When the extracts of fresh and fallen leaves and roots were tested against the germination of *Pinus densiflora*, *Arundinella hirta*, *Erigeron annuus* and *Leonurus sibiricus*, the extracts were found differentially inhibitory in the order of fresh leaf > fallen leaf > root extract. *L. sibiricus* was the worst affected species followed by *E. annuus*, *P. densiflora* and *A. hirta* (Fig. 14.1). Thus it is clear that phytotoxic

TABLE 14.1 *Effects of aqueous extracts of various parts of* Pinus densiflora *on germination of species inside the pine forest and species outside, in petri dishes (L, fresh leaves; FL, fallen leaves; and R, roots extracted for 24 h at 80°C). (After Kil, 1989)*

Species inside forest	Cont. (%)	L (%)	FL (%)	R (%)	Species outside forest	Cont. (%)	L (%)	FL (%)	R (%)
Setaria viridis	78.6	0.66	0.50	0.63	*Cassia tora*	56	0.70	0.31	0.33
Erigeron annuus	53	0.36	0.43	0.47	*Triumfetta japonica*	49	0.10	0.31	0.41
Aster tataricus	83.3	0.55[a]	0.95	0.75	*Sedum erythrostichum*	96.5	0.54	0.88	0.89
Cymbopogon tortilis var. *goeringii*	72	0.99	0.78	0.66	*Oenothera odorata*	50.7	0.38	0.41	0.54
Celastrus orbiculatus	83.3	0.78[a]	0.99	1.01	*Chenopodium album* var. *cen.*	61	0.37	0.60	0.52
Bidens bipinnata	93	0.44[b]	0.42[b]	0.58[a]	*Digitaria sanguinalis*	66.5	0.23	0.22	0.71
Platycodon grandiflorum	92.3	0.73[a]	1.00	0.99	*Amaranthus mangostanus*	63.3	0	0.16[b]	0.30[a]
Pinus rigida	93	1.04	0.84	0.80	*Serratula coreanata* var. *insularis*	62	0.16	0.35	0.29
Atractylodes japonica	78.8	0.71	0.94	1.04	*Persicaria hydropiper*	40	0.30	0.10	0.15
Arundinella hirta	73.3	0.86	0.87	0.95	*Leonurus sibiricus*	77.4	0.21[b]	0.28[b]	0.47[b]
Pinus densiflora	95.4	0.94	0.89	0.97	*Amorpha fruticosa*	35.6	0.57	0.66	0.74
Themeda triandra var. *japonica*	60.3	0.65	0.58	0.99	*Geranium sibiricum*	35.5	0.06	0.05	0.07
Achyranthes japonica	80.1	0.35[b]	0.33[b]	0.51[b]					
Aster yomena	50.5	0.76	0.72	0.64					
Robinia pseudo-acacia	60	0.80	0.82	0.93					
Miscanthus sinensis	58.3	0.84	0.64	0.70					
Melandryum firmum	71.2	0.11[b]	0.07[b]	0.13[b]					
Justicia procumbens	58.8	0.44	0.36	0.65					
Cassia mimosoides var. *nomame*	66.6	1.45	1.36	0.95					
Rhododendron schlippenbachii	84.8	0.73	0.91	0.81					
Lespedeza bicolor	54	1.29	0.57	1.14					
Pinus thunbergii	97	0.58	0.90	0.99					
		A	B	C			D	E	F

Superior letter 'a' indicates significantly different from control at the 0.05 level, superior 'b', at the 0.001 level.
$\bar{X}_1$: A = 0.730, B = 0.721, C = 0.786; $\bar{X}_2$: D = 0.301, E = 0.361, F = 0.452.
t values were 4.394, 3.743, 3.973 in turn, and p showed <0.001, respectively.

TABLE 14.2 *Effects of aqueous extracts of various parts of* Pinus densiflora *on germination of species inside the pine forest and species outside, in petri dishes (L, fresh leaves, FL, fallen leaves, R, roots extracted for 24 h at 18°C). (After Kil, 1989)*

Species inside forest	Cont. (%)	L (%)	FL (%)	R (%)	Species outside forest	Cont. (%)	L (%)	FL (%)	R (%)
Setaria viridis	78.6	0.80	0.71	0.88	*Cassia tora*	56	0.35	0.08	0.31
Erigeron annuus	53	0.97	0.98	0.88	*Triumfetta japonica*	49	0.37	0.51	0.53
Aster tataricus	83.3	0.98	0.86	0.82	*Sedum erythrostichum*	96.5	0.98	0.93	0.93
Cymbopogon tortilis var. *goeringii*	72	0.81	0.99	0.88	*Oenothera odorata*	50.7	0.76	0.62	0.65
Celastrus orbiculatus	83.3	1.09	0.89	1.14	*Chenopodium album* var. *centrorubrum*	61	0.67	0.70	0.63
Bidens bipinnata	93	0.86	0.81[a]	0.88	*Digitaria sanguinalis*	66.5	0.73	0.70	0.83
Platycodon grandiflorum	92.3	1.01	0.92	0.95	*Amaranthus mangostanus*	63.3	0.05[b]	0.14[b]	0.77
Pinus rigida	93	0.97	1.04	1.06	*Serratula coreanata* var. *insularis*	62	0.60	0.73	0.77
Atractylodes japonica	78.8	0.97	0.95	0.95	*Persicaria hydropiper*	40	0.75	0.40	0.10
Arundinella hirta	73.3	1.02	0.94	0.95	*Leonurus sibiricus*	77.4	0.92	0.60[a]	0.68
Pinus densiflora	95.4	1.02	0.99	0.99	*Amorpha fruticosa*	35.6	0.97	0.81	0.79
Themeda triandra var. *japonica*	60.3	0.78	0.80	0.82	*Geranium sibiricum*	35.5	0.11[a]	0.10[a]	0.89
Achyranthes japonica	80.1	0.87	0.60	0.86					
Aster yomena	50.5	0.88	0.81	0.75					
Robinia pseudo-acacia	60	0.78	0.95	0.97					
Miscanthus sinensis	58.3	1.03	0.57	0.77					
Melandryum firmum	71.2	0.39[a]	0.25[a]	0.28[a]					
Justicia procumbens	58.8	1.04	1.14	0.91					
Cassia mimosoides var. *nomame*	66.6	1.08	1.19	0.94					
Rhododendron schlippenbachii	84.8	1.01	0.87	0.87					
Lespedeza bicolor	54	1.43	1.14	1.00					
Pinus thunbergii	97	1.02	0.96	1.00					
		A	B	C			D	E	F

Superior letter 'a' indicates significantly different from control at the 0.05 level; superior 'b', at the 0.001 level.
$\bar{X}_1$: A = 0.946, B = 0.880, C = 0.912; $\bar{X}_2$: D = 0.605, E = 0.527, F = 0.657.
t values were 4.094, 4.249, 4.574 in turn, and p showed <0.001, respectively.

TABLE 14.3 *Effects of aqueous extracts of various parts of* Pinus densiflora *on germination of species inside the pine forest and species outside, sown in pots with vermiculite (L, fresh leaves; FL, fallen leaves; R, roots extracted for 24 h at 80°C)*

Species inside forest	Control (%)	L (%)	FL (%)	R (%)	Species outside forest	Control (%)	L (%)	FL (%)	R (%)
Setaria viridis	72	0.69	0.69	0.81	*Celosia argentea*	90	0.36	0.42	0.47
Aster tataricus	62	0.68^{a}	0.81	0.94	*Boehmeria platanifolia*	44	0	0.23^{a}	0.32^{a}
Peucedanum terebinthaceum	46	0.38^{b}	0.88^{a}	0.75^{a}	*Triumfetta japonica*	48	0.57^{a}	0.50^{a}	0.79^{a}
Bidens bipinnata	66	0.73^{a}	0.85^{a}	0.85^{a}	*Veronica rotunda* var. *subintegra*	56	0.50	0.47	0.83
Platycodon grandiflorum	80	0.75	0.98	0.93	*Elsholtzia ciliata*	56	0.96	0.79	0.79
Patrinia villosa	54	0.56	0.85	0.89	*Sedum erythrostichum*	86	0	0.63	0.42
Pinus rigida	96	0.96	1.04	0.96	*Oenothera odorata*	40	0.25^{a}	0.35^{a}	0.55
Phytolacca americana	66	0.88^{a}	0.73^{a}	0.67^{a}	*Digitaria sanguinalis*	58	0.17^{b}	0.55^{a}	0.55^{a}
Atractylodes japonica	58	0.90	0.90	1.10	*Amaranthus mangostanus*	88	0.73	0.75	0.59
Arundinella hirta	96	0.83	0.96	0.94	*Serratula coreanata* var. *insularis*	54	0.59	0.29	0.88
Themeda triandra var. *japonica*	78	0.77	0.92	0.95	*Aconogonum polymorphum*	78	0.21	0.69	0.44
Achyranthes japonica	92	0.96	1.00	0.91	*Leonurus sibiricus*	76	0.53^{b}	0.84^{a}	0.74^{a}
Robinia pseudo-acacia	84	0.81	0.90	0.98	*Amorpha fruticosa*	50	0.16	0.52	0.56
Miscanthus sinensis	40	0.45	0.50	0.55	*Saussurea gracilis*	74	0.03^{b}	0.57^{a}	0.19^{b}
Albizzia julibrissin	94	1.04	0.98	0.96					
Plantago asiatica	70	0.71	0.94	0.91					
Agrimonia pilosa	74	0.78	0.84	0.84					
Cassia mimosoides var. *nomame*	58	1.34	1.38	0.90					
Rhododendron schlippenbachii	50	0.72	1.08	1.48					
Pinus thunbergii	74	0.96	1.00	1.14					
		A	B	C			D	E	F

Superior letter 'a' indicates significantly different from control at the 0.05 level; superior 'b' at the 0.001 level.
$\bar{X}_1$: A = 0.795, B = 0.912, C = 0.923; $\bar{X}_2$: D = 0.361, E = 0.543, F = 0.580.
t values were 5.177, 6.133, 5.238 in turn, and p showed <0.001, respectively.

TABLE 14.4 *Effects of aqueous extracts of various parts of* Pinus densiflora *on germination of species inside the pine forest and species outside, sown in pots with vermiculite (L, fresh leaves; FL, fallen leaves; R, roots extracted for 24 h at 18°C)*

Species inside forest	Cont. (%)	L (%)	FL (%)	R (%)	Species outside forest	Cont. (%)	L (%)	FL (%)	R (%)
Setaria viridis	72	0.78	1.11	1.17	*Celosia argentea*	90	0.47	0.69	1.07
Aster tataricus	62	0.98	0.90	0.81	*Boehmeria platanifolia*	44	0.86	1.14	1.18
Peucedanum terebinthaceum	46	1.63	1.50	0.88	*Triumfetta japonica*	48	0.43[a]	0.57[a]	0.57[a]
Bidens bipinnata	66	0.39[a]	0.45[a]	0.94	*Veronica rotunda* var. *subintegra*	56	0.83	0.89	0.56
Platycodon grandiflorum	80	0.98	1.03	0.35	*Elsholtzia ciliata*	56	1.07	1.09	0.93
Patrinia villosa	54	0.93	1.20	1.02	*Sedum erythrostichum*	86	0.93	0.93	0.65
Pinus rigida	96	0.98	1.02	1.04	*Oenothera odorata*	40	1.23	0.93	0.95
Phytolacca americana	66	0.55[a]	0.70	0.73	*Digitaria sanguinalis*	58	0.59	1.00	0.90
Atractylodes japonica	58	0.93	0.86	0.83	*Amaranthus mangostanus*	88	0.82	0.84	0.77
Arundinella hirta	96	0.88	1.00	1.00	*Serratula coreanata* var. *insularis*	54	0.35	1.35	0.65
Themeda triandra var. *japonica*	78	1.03	1.08	1.12	*Aconogonum polymorphum*	78	0.90	0.77	0.78
Achyranthes japonica	92	1.04	0.83	0.96	*Leonurus sibiricus*	76	0.71	0.58[a]	0.47[a]
Robinia pseudo-acacia	84	1.02	0.98	1.02	*Amorpha fruticosa*	50	0.32	0.28	0.28
Miscanthus sinensis	40	0.70	0.88	0.90	*Saussurea gracilis*	74	0.51[a]	0.49[a]	0.08[b]
Albizzia julibrissin	94	0.82	0.91	0.87					
Plantago asiatica	70	1.06	1.14	0.91					
Agrimonia pilosa	74	1.00	1.11	0.86					
Cassia mimosoides var. *nomame*	58	1.10	1.21	0.93					
Rhododendron schlippenbachii	50	1.24	1.32	1.68					
Pinus thunbergii	74	0.82	0.91	1.00					
		A	B	C			D	E	F

Superior letter ‘a’ indicates significantly different from control at the 0.05 level; superior ‘b’ at the 0.001 level.
$\bar{X}_1$: A = 0.943, B = 1.007, C = 0.951; $\bar{X}_2$: D = 0.716, E = 0.825, F = 0.703.
t values were 2.814, 2.142, 2.748 in turn, and p showed <0.05, respectively.

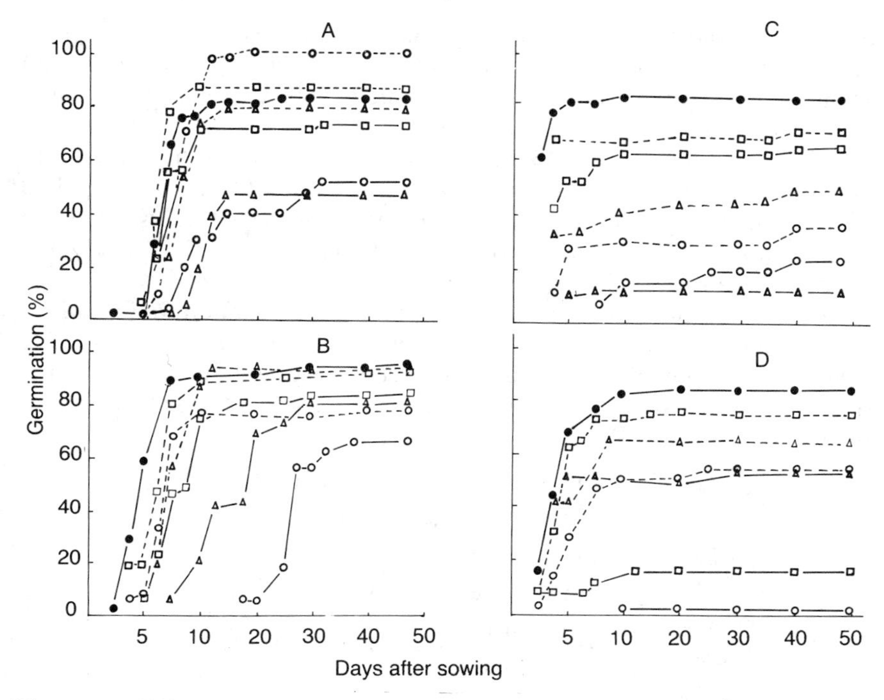

Fig. 14.1 Effects of pine extracts on the germination percentage of *Pinus densiflora* (A), *Arundinella hirta* (B), *Leonurus sibiricus* (C) and *Erigeron annuus* (D). ●—●, control; ○—○, fresh leaves, extracted at 80°C; ○---○, fresh leaves, at 18°C; △—△, fallen leaves, at 80°C; △---△, fallen leaves, at 18°C; □—□, roots, at 80°C; □---□, roots, at 18°C. (After Kil, 1989. Reproduced with permission, Kil, B. S. and Academia Sinica, Taipei, ROC.)

substances secreted from pine tree differentially inhibit seed germination of various species and the extent of inhibition varies with a change in source and extraction procedures. These findings are in agreement with the earlier reports of Guenzi and McCalla (1962), Quienn (1974), Ashraf and Sen (1978), Khosh-Khui and Bassiri (1979) and Datta and Chatterjee (1980).

(b) *Pinus thunbergii*

Two hundred grams of fresh leaves, fallen leaves and roots were extracted in 1000 ml of distilled water under various temperatures and duration. The aqueous extract, obtained after immersing 200 g of black pine leaves in 1000 ml of water for 48 h at 18°C, was adjusted to pH 3–9 as and when needed (Kim and Kil, 1984). Germination tests were carried out in petri dishes. For growth studies, the seeds of experimental species were sown in plastic pots (depth 12 cm, height 8 cm) with holes at the bottom. The holes were covered by gauze and the pots were filled with vermiculite. These pots were kept in a large plastic pot (depth 20 cm) and 15 ml of the extract or water was supplied daily to each of the large supporting pots.

Thinning was done after 15 days of sowing and five healthy plants were maintained in each pot. Hoagland's nutrient solution (25 ml) was supplied to each pot every week and they were so arranged as to allow them to receive sufficient light. The harvest was gathered after 50 days of sowing, washed in water, dried in an oven and dry weight was recorded.

Relative germination ratio (RGR), relative elongation ratio (RER) and relative dry weight ratio (RDR) were calculated for quantitative evaluation of the effect of phytotoxic substances on the experimental species. The following formula were employed for the calculation of RGR, RER and RDR.

$$\mathrm{RGR} = \frac{\text{Germination percentage of test plant}}{\text{Germination percentage of control}} \times 100$$

$$\mathrm{RER} = \frac{\text{Elongation percentage of test plant}}{\text{Elongation percentage of control}} \times 100$$

$$\mathrm{RDR} = \frac{\text{Dry weight of test plant}}{\text{Dry weight of control}} \times 100$$

Rumex japonicus, *Lactuca sativa*, *Perilla frutescens* and *Matricaria chamomilla* were used for germination tests and *R. japonicus* and *L. sativa* were concurrently used for growth tests. The results of the germination test showed an inhibitory tendency, on the whole, compared with the control.

As observed in the germination test, growth (RDR) was inhibited in direct proportion to the time used in preparing the extracts (Fig. 14.2). This result coincides with that of an earlier study that was carried out using radish and sorghum (Einhellig and Rasmussen, 1978). *M. chamomilla*, *Salsola kamarovi*, *Hibiscus syriacus* and *Allium fistulosum* were tested for germination (RGR) in petri dishes in the aqueous extracts obtained from black pine at different temperatures; the results are shown in Fig. 14.3.

The germination and elongation of four species were tested in the aqueous extracts (pH 3 to pH 9). The germination rates were generally low as compared with the control, though they varied according to experimental species. The elongation was also low at pH 3 and pH 9, respectively (Table 14.5). This result is similar to that of a preceding germination using extracts with pH below 3 and above 9 (Chou and Young, 1974).

(c) *Pinus rigida*

Aqueous extracts were obtained by soaking 200 g fresh leaves in 1000 ml and the rain leachates were collected at 1, 3 and 5 leaved stages. Seeds

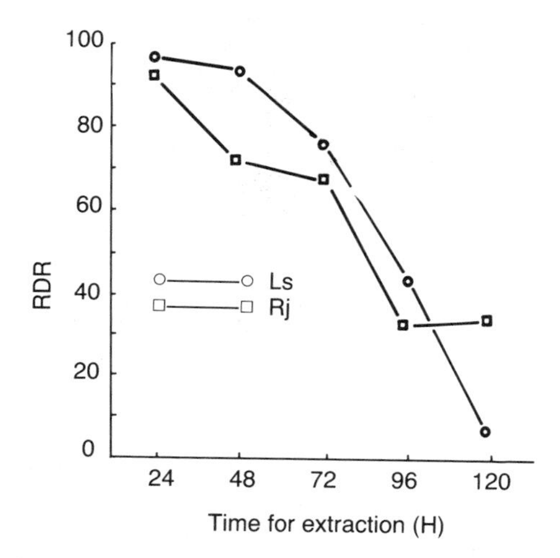

Fig. 14.2 Dry weights of different species tested in aqueous extracts of black pine leaves treated for various times at 18°C. Ls, *Lactuca sativa*; Rj, *Rumex japonica*. (After Kil, B. S., 1989. Reproduced with permission, Kil, B. S. and Academia Sinica, Taipei, ROC.)

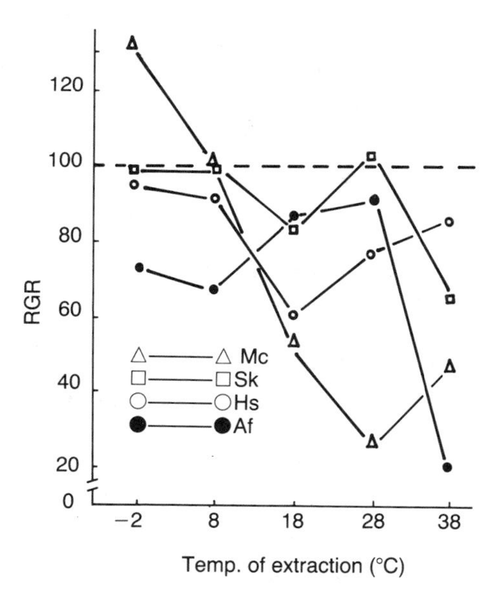

Fig. 14.3 Relative germination ratio of four species tested in aqueous solutions of black pine leaves extracted at different temperatures for 48h. Mc, *Matricaria chamomilla*; Sk, *Salsola kamarovi*; Hs, *Hibiscus syriacus*; Af, *Allium fistulosum*. (After Kil, B. S., 1989. Reproduced with permission, Kil, B. S. and Academia Sinica, Taipei, ROC.)

TABLE 14.5 *Relative germination ratio (RGR) and relative elongation ratio (RER) of selected species tested in aqueous extracts of black pine (*Pinus thunbergii*) at different pH values*

Species	pH (extracts)						
	3	4	5	6	7	8	9
RGR							
Setaria italica	92.9	85.7	57	78.6	92.9	85.7	71.4
Rumex japonicus	65.2	60.9	65.2	47.8	69.6	56.5	56.5
Raphanus sativus var. *hortensis* for. *acanthiformis*	76.5	41.1	82.4	70.6	64.7	58.8	76.5
Hibiscus syriacus	0	66.7	44.4	38.9	50.0	44.4	66.7
Lycopersicon esculentum	57.9	105.3	94.7	105.3	100	84.2	73.7
Carpesium abrotanoides	76	76	80	80	76	80	84
Allium fistulosum	57	50	71	64	28	64	86
RER							
Setaria italica	30.8	51.6	58.5	55.7	55.3	49	36.4
Rumex japonicus	52.6	153.7	180	104.4	147.2	102	46
Raphanus sativus var. *hortensis* for. *acanthiformis*	44.2	93.4	96.0	115.2	108.9	115.9	43
Hibiscus syriacus	0	120	94.8	120.6	111.4	99.4	64.5
Lycopersicon esculentum	53.7	169.2	120.1	137.5	129.5	109.8	66.1
Carpesium abrotanoides	53.8	115.7	132.9	139	165.2	116.7	55.7
Allium fistulosum	138	105	165	158	175	128	69

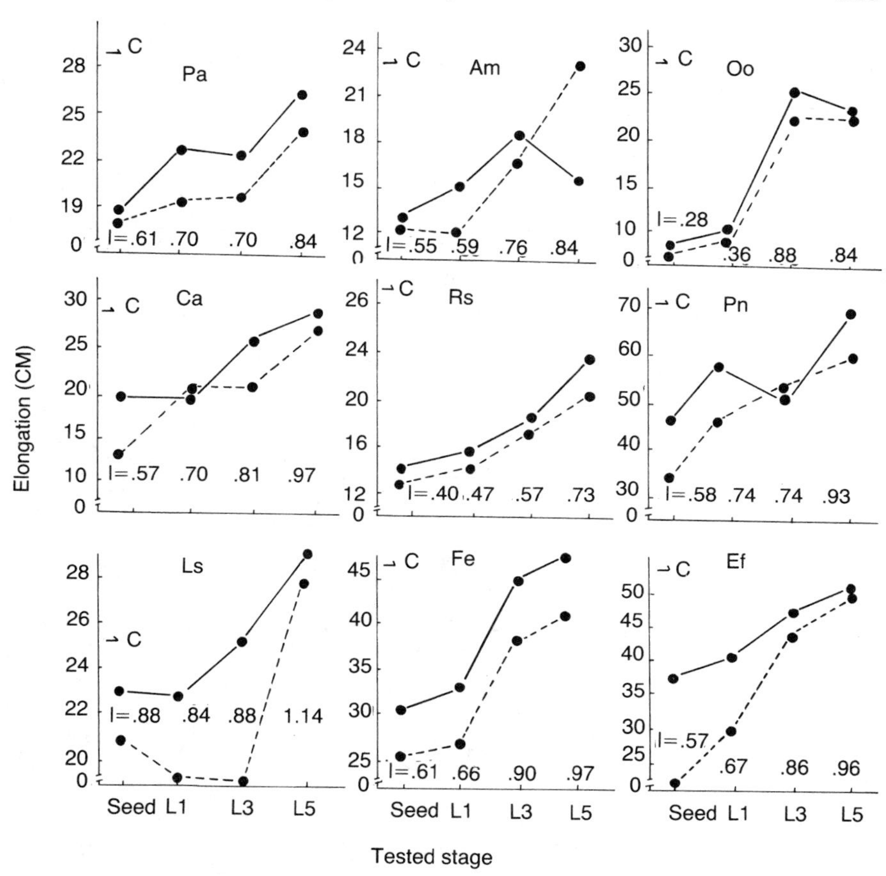

Fig. 14.4 Seedling elongation of selected species tested in aqueous extracts (●---●) and rain leachates (●—●) of *Pinus rigida*. L1, treatment time of 1-leaf seedling; L3, of 3-leaves; L5, of 5-leaves. Pa, *Poa annua*; Am, *Amaranthus mangostanus*; Oo, *Oenothera odorata*; Ca, *Celosia argentea*; Rs, *Raphanus sativus* var. *hortensis* for. *acanthiformis*; Pn, *Pharbitis nil*; Ls, *Lactuca sativa*; Fe, *Fagopyrum esculentum*; Ef, *Eragrostis ferruginea*. (After Kil, 1989. Reproduced with permission, Kil, B. S. and Academia Sinica, Taipei, ROC.)

sown in vermiculite-containing pots were treated with the aqueous extract and rain leachates. The seedling length of harvested species was measured and their fresh and dry weight were recorded (Rho and Kil, 1986). In the test involving developmental stages the 'I' value was used to represent the mean inhibition of growth by extracts and leachates in comparison to the control. 'I' was calculated by the formula:

$$I = \frac{\text{Effect of aqueous extract + effect of rain leachate}}{\text{Effect of control} \times 2}$$

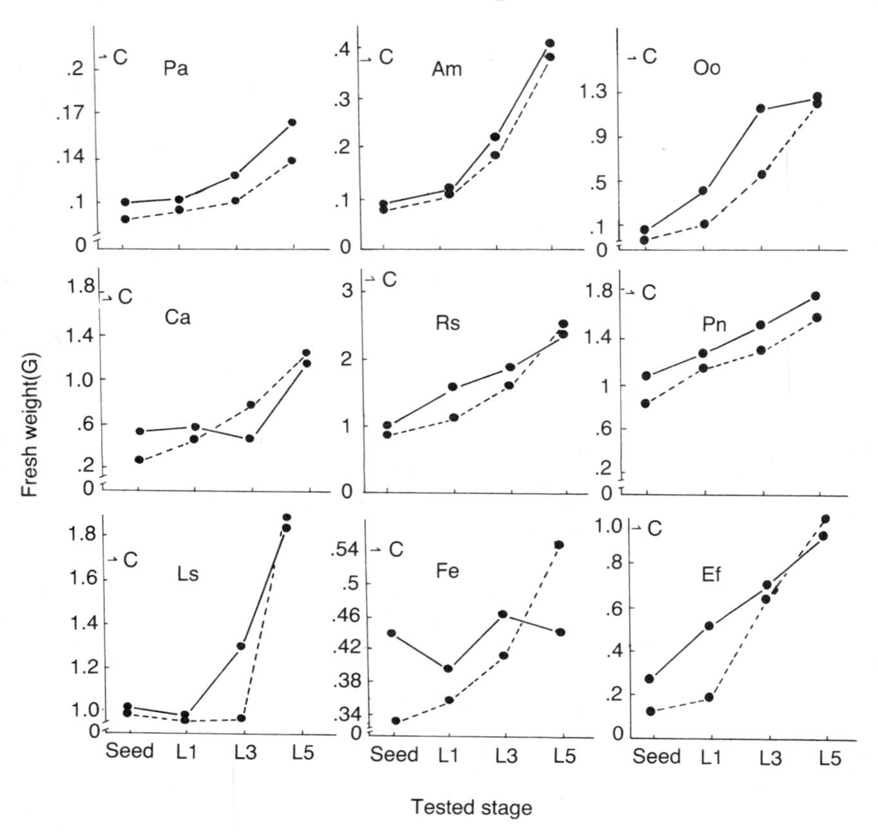

Fig. 14.5 Fresh weights of selected species tested in aqueous extracts (●---●) and rain leachates (●—●) of *Pinus rigida*. Keys to legends and species as in Fig. 14.4. (After Kil, 1989. Reproduced with permission, Kil, B. S. and Academia Sinica, Taipei, ROC.)

The results of these experiments are summarized in Figs 14.4, 14.5 and 14.6. The growth of experimental species was more severely inhibited when they were treated with an extract of pitch pine at their early growth stage. The elongation rates of all potted species other than *Fagopyrum esculentum* and *Eragrostis ferruginea* were relatively low and those of *Raphanus sativus* var. *hortensis* for. *acanthiformis* and *Poa annua* showed particularly low values.

The growth of nine experimental test species was measured in terms of dry weight. Among them, except *Lactuca sativa* and *Eragrostis ferruginea*, other species showed considerable inhibition. In the case of *Metaplexis japonica* there was a correlation between inhibition, concentration of the extract and the developmental stage of the extract source.

The aqueous extract of pitch pine was also adjusted to pH 1 to 14 and

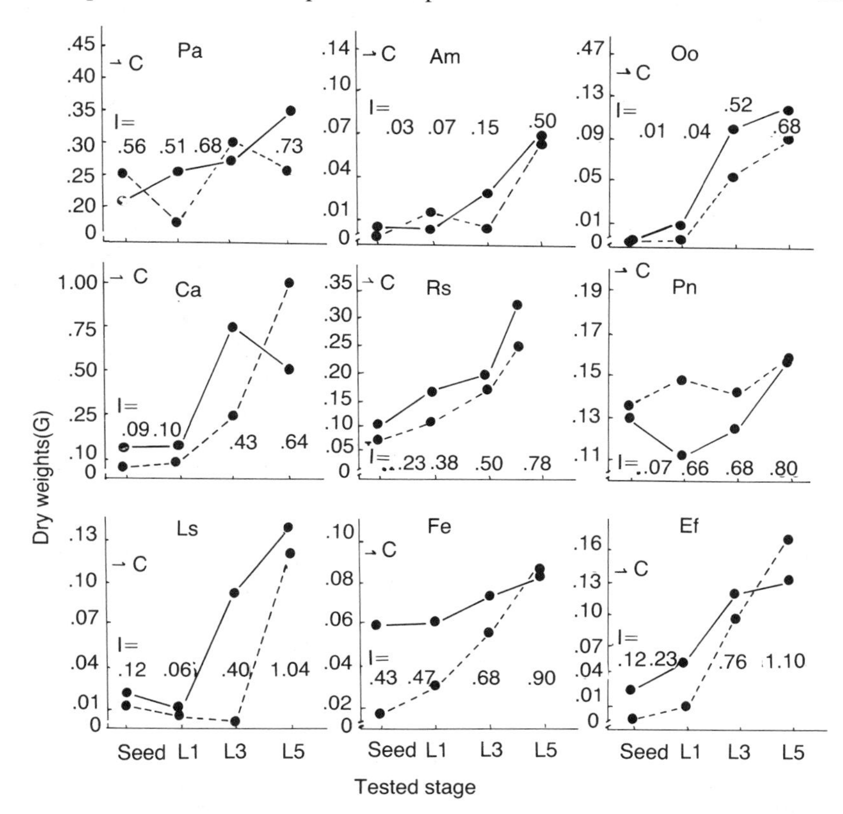

Fig. 14.6 Dry weights of selected species tested in aqueous extracts (●---●) and rain leachates (●—●) of *Pinus rigida*. Key to legends and species as in Fig. 14.4. (After Kil, 1989. Reproduced with permission, Kil, B. S. and Academia Sinica, Taipei, ROC.)

its effect was studied on germination, seedling elongation and dry weight of lettuce (Table 14.6). The species did not germinate at all in extracts of pH 1, 13 and 14, while the highest germination rate was recorded at pH 7. The elongation rate of seedling was inhibited by approximately 59% at pH 7 and 13, and 45.9% at pH 3. The inhibitory effect on dry weight was in the sequence of fresh leaves > roots > fallen leaves extract. When seeds of 25 wild plants were germinated in the presence of aqueous extracts obtained at 18°C from leaves, fallen leaves and roots of pitch pine, eight species showed a relatively higher germination ratio (Table 14.7). These species grow on the hills and plains where pitch pine is planted. However, five species which showed a low germination rate commonly grow on the plains or around dwelling houses. The effects of these extracts were also recorded on the germination rate of the test species, which were severely inhibited in all the species but minimum in *D. sanguinalis* and *T. triandra* var. *japonica* when treated with the extract

TABLE 14.6 *Germination and growth of* Lactuca sativa *in aqueous extracts of* Pinus rigida *at different pH values. (L, fresh leaves; FL, fallen leaves; R, roots extracted for 24 h at 18°C) (after Kil, 1989)*

	Germination (%)			Seedling length (cm)			Dry weight (g)		
Solution	L	FL	R	L	FL	R	L	FL	R
Control	100	100	100	13.7	13.7	13.7	0.88	0.88	0.88
pH 1	0	0	0	0	0	0	0	0	0
pH 3	61.2	33.2	88.0	6.3	7.7	7.5	0.25	0.34	0.26
pH 5	80.0	34.6	90.0	7.7	7.9	9.1	0.31	0.41	0.33
pH 7	92.0	43.2	96.6	8.1	7.6	8.9	0.28	0.41	0.37
pH 9	86.6	48.6	94.6	7.9	7.2	9.3	0.27	0.37	0.37
pH 11	86.6	36.0	92.0	8.1	7.2	9.6	0.23	0.37	0.31
pH 13	0	0	0	0	0	0	0	0	0
pH 14	0	0	0	0	0	0	0	0	0

TABLE 14.7 *Effects of aqueous extracts of* Pinus rigida *on the relative germination ratio of different species (L, fresh leaves; FL, fallen leaves; and R, roots extracted for 24 h at 18°C)*

Species	L	FL	R
Pinus rigida	100	100	91.1
Pinus densiflora	100	100	100
Pinus thunbergii	100	100	100
Arundinella hirta	82.9	100	97.9
Setaria viridis	81.8	69.1	90.9
Digitaria sanguinalis	74.5	53.1	87.1
Miscanthus sinensis	82.9	20.0	54.3
Cymbopogon tortilis var. *goeringii*	84.2	100	100
Themeda triandra var. *japonica*	85.0	32.5	82.5
Chenopodium album var. *centrorubrum*	100	50.0	50.0
Amaranthus mangostanus	6.1	8.5	15.1
Achyranthes japonica	97.7	92.3	88.5
Phytolacca americana	8.6	5.3	46.6
Melandryum firmum	10.0	22.9	30.8
Cassia tora	42.9	14.3	57.1
Lespeza bicolor	78.9	63.2	94.7
Phyllanthus ussuriensis	41.7	38.9	52.8
Oenothera odorata	78.6	42.9	35.7
Leonurus sibiricus	43.9	14.6	73.9
Justicia procumbens	100	100	71.0
Erigeron annuus	85.7	89.3	82.1
Erigeron canadensis	64.3	76.2	88.1
Artemisia princeps var. *orientalis*	100	68.3	41.5
Bidens bipinnata	88.0	94.0	74.0
Youngia sonchifolia	100	77.5	85.0

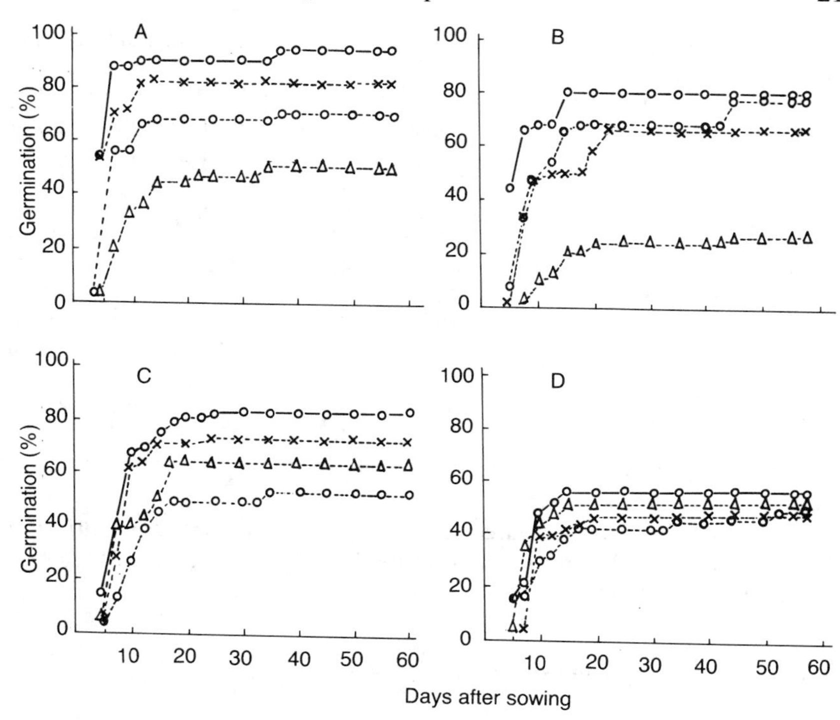

Fig. 14.7 Effects of *Pinus rigida* leaves extracts on the germination percentage of *Digitaria sanguinalis* (A), *Themeda triandra* var. *japonica* (B), *Erigeron canadensis* (C) and *Erigeron annuus* (D). △---△ Fallen leaves, at 18°C; ○---○ Fresh leaves, at 18°C; X---X Roots, at 18°C; ○—○ Control.

of fallen leaves. *E. canadensis* and *E. annuus* also showed a similar tendency with the leaf extract (Fig. 14.7).

When *Pinus densiflora* and *Oenothera odorata* were grown in the presence of the above extracts, their RDR was relatively higher among all the species tested. However, *D. sanguinalis*, *Amaranthus mangostanus* and *Triumfetta japonica* were severely inhibited. These results are similar to those of the germination test discussed previously. It is clearly indicated by the results (Tables 14.7 and 14.8) that even though the toxic matter of pitch pine has no significant effect on seed germination, the dry weight of the species continuously treated with the extract is markedly inhibited. Similar results have also been recorded with *Setaria glauca* which inhibited the dry weight of bean and corn (Bhowmik and Doll, 1983).

When lettuce was treated with various concentrations of leaf extract of pitch pine obtained at an interval of 24 h, the growth of the plant was significantly inhibited by the extracts obtained at 24 to 72 h (Fig. 14.8). This suggests that possibly the toxic substances of pitch pine exude con-

TABLE 14.8 *Dry weights (RDR) of different plants grown in pine leachate made at room temperature*

Species	Leaching time (h)				
	12	24	48	72	96
Erigeron annuus	87	1	0	0	0
Arundinella hirta	45	43	24	10	9
Achyranthes japonica	36	25	11	6	2
Plantago asiatica	66	24	10	1	0
Celosia argentea	25	10	7	0	0
Aconogonum polymorphum	51	32	16	2	0
Leonurus sibiricus	50	22	7	4	2
Saussurea gracilis	104	16	12	7	0

tinuously and dissolve in water leading to an increased concentration. These results coincide with those reported by Le Tourneau and Heggeness (1975) and Khan (1982).

14.2.2 Germination and growth test in leachates

(a) *Pinus densiflora*
Samples of fresh and fallen pine leaves and pine forest soil were put in plastic pipes (diameter 15 cm, length 150 cm) fixed on strong supports and each was connected to a funnel at the bottom. Each funnel was connected to a rubber tube closed by a pinch cock. Every 200 g sample was leached with 1000 ml of water and the leachates thus obtained were stored in a refrigerator. These were tested for their effects on seed germination and seedling growth. Results showed that the maximum inhibition of germination and growth was caused by leachates of fresh leaves. The effect of pine leaf leachates, obtained at 12, 24, 48 and 72 h, was tested on seed germination (Fig. 14.9). In 12 and 24 h leachates, seeds germinated well but they were strongly inhibited by 48 and 72 h leachates. The germination of the inside species of the pine forest was much greater than that of the outside species, i.e. the outside species of pine forest were more inhibited in pine leachates than the inside species. As compared with the control, dry weights of most of the species were strongly inhibited by 24, 48, 72 and 96 h leachates (Table 14.8).

(b) *Pinus thunbergii*
Leachates of black pine leaves inhibited the germination and seedling growth of all the test species but the difference was not significant (Table 14.9, 14.10). The relative growth ratio and relative dry weight ratio were also inhibited by rain leachate of *P. thunbergii* (Fig. 14.10).

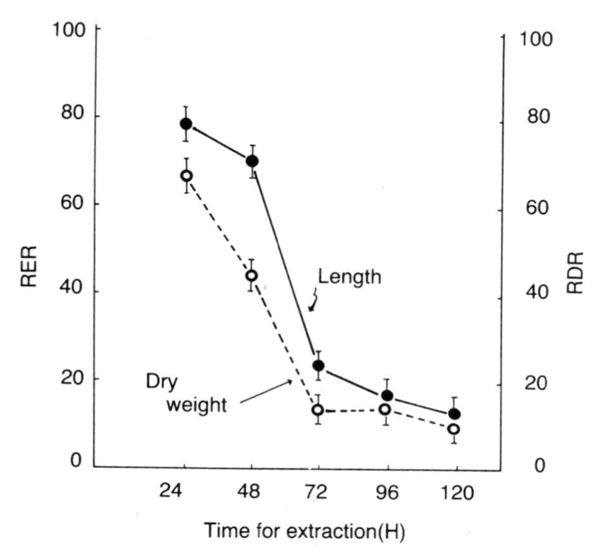

Fig. 14.8 Effect of aqueous extracts of *Pinus rigida* leaves on lettuce radicle elongation (relative elongation ratio, RER) and growth (relative dry weight ratio, RDR). (After Kil, 1989. Reproduced with permission, Kil, B. S. and Academia Sinica, Taipei, ROC.)

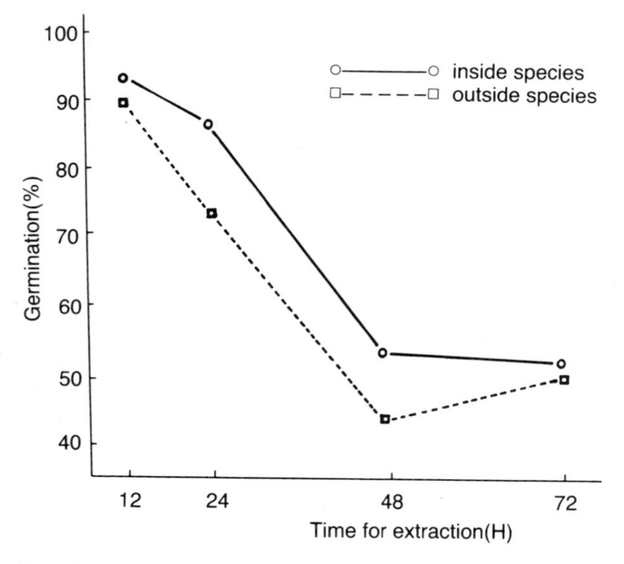

Fig. 14.9 Germination percentage of inside species and outside species of pine forest in presence of leachates obtained at different intervals.

TABLE 14.9 *Germination percentage of different species tested in petri dish supplied with aqueous leachate of black pine needles*

Species	Control	Leachate
Perilla frutescens var. *japonica*	97.5a	91.5a
Lactuca sativa	88.0a	70.5a
Artemisia princeps var. *orientalis*	94.0a	79.0a
Carpesium abrotanoides	86.5a	57.0a
Setaria viridis	60.5a	53.5a
Raphanus sativus var. *hortensis* for. *acanthiformis*	89.5a	84.0a

Means within rows followed by same letter do not differ significantly at the 5% level of probability.

TABLE 14.10 *Elongation (mm) of different species tested in petri dish supplied with aqueous leachate of black pine needles*

Species	Control	Leachate
Perilla frutescens var. *japonica*	63.3a	74.4b
Lactuca sativa	80.8a	100.9b
Artemisia princeps var. *orientalis*	24.5a	24.6a
Capesium abrotanoides	23.9a	16.6b
Setaria viridis	63.2a	63.3a
Raphanus sativus var. *hortensis* for. *acanthiformis*	149.1a	170.5a

Means within rows followed by same letter do not differ significantly at the 5% level of probability.

(c) *Pinus rigida*

The germination and growth of seven species were tested using leachates obtained from the fresh, fallen leaves and roots of pitch pine (Fig. 14.11). The germination rates of these species were relatively high but their dry weights were extremely low. This could mean that, although plants may germinate, their growth is doubtful in the presence of the leachate of pitch pine. These findings are in agreement with earlier reports that plant and rain leachate had inhibitory effect on the germination and growth of plants (McPherson and Muller, 1969; Lodhi and Rice, 1971).

14.2.3 Germination and growth test in soil

(a) *Pinus densiflora*

Pine forest soils and field soils (B-horizon) were put into germination boxes (1 m × 1 m) and seeds of species growing inside and outside the

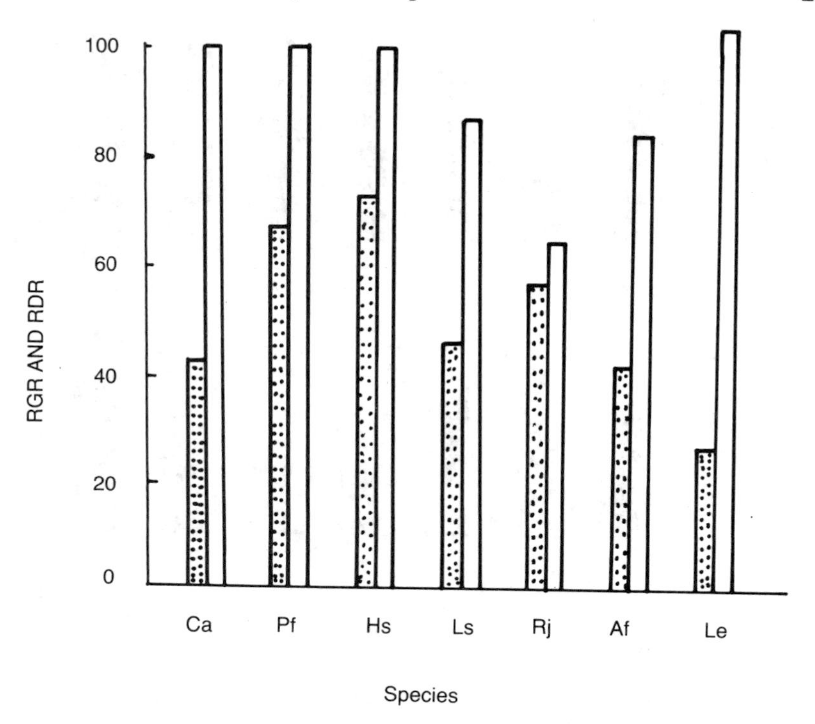

Fig. 14.10 RGR (▭) and RDR (▨) of different species tested in rain leachates of black pine. Ca, *Carpesium abrotanoides*; Pf, *Perilla frutescens*; Hs, *Hibiscus syriacus*; Ls, *Lactuca sativa*; Rj, *Rumex japonica*; Af, *Allium fistulosum*; Le, *Lycopersicon esculentum*.

pine forest were sown at 10 cm intervals, watered and protected with a vinyl cover. An analysis of both types of soils revealed that there were marked differences with regards to pH and some other parameters considered (Table 14.11).

Results showed that inside species germinated well in the pine forest soil while outside species showed better germination in the field soil. Similarly, the growth (dry weights) of inside and outside species were better in the forest and field soils, respectively (Table 14.12).

Based on these findings it can be concluded that the inside species of pine forest have developed a resistance for the pine toxins and therefore they grew well in comparison with the outside species.

(b) *Pinus thunbergii*

Plastic pots (12 cm × 7 cm) were filled with the soil of black pine forest and outside soil. Fifty seeds each of the test species were planted in each pot. After 10 days of germination 200 ml of Boysen Jensen nutrient

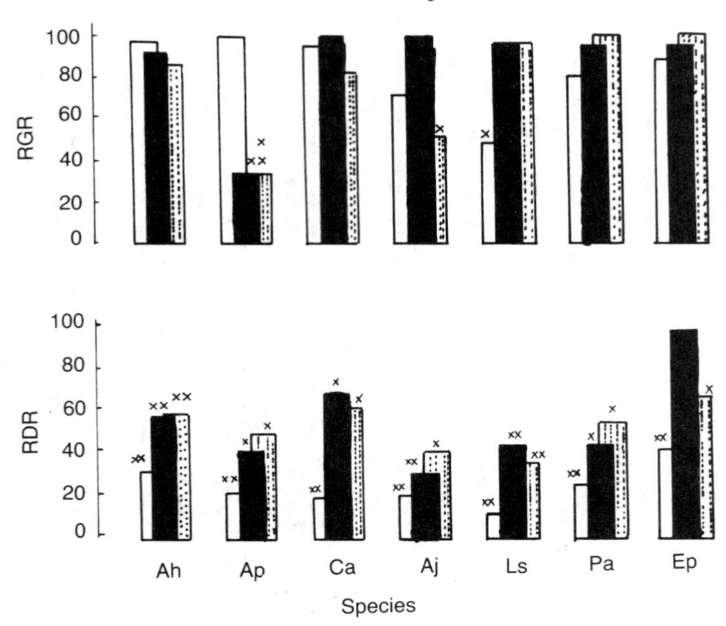

Fig. 14.11 Relative germination ratio (RGR) and relative dry weight ratio (RDR) of selected species seedlings tested in leachates of *Pinus rigida*. (Fresh leaves (white bars), fallen leaves (dotted bars) and roots (black bars) of *P. rigida* leachates. X, significant difference from control at 0.05 level; XX, at 0.001 level. Ah, *Arundinella hirta*; Ap, *Aconogonum polymorphum*; Ca, *Celosia argentea*; Aj, *Achyranthes japonica*; Ls, *Leonurus sibiricus*; Pa, *Plantago asiatica*; Ep, *Eclipta prostrata*.

solution was supplied once every 4 days. The plants were harvested 55 days after sowing, washed and dried in an oven at 80°C.

Inside species such as *Pinus rigida*, *Achyranthes japonica*, *Arundinella hirta*, *Themeda triandra* var. *japonica*, *Phytolacca americana*, *Albizzia julibrissin* showed a higher germination percentage and dry weight of seedling in the soil of black pine forest than outside species like *Setaria*

TABLE 14.11 *Soil analysis of the studied sites*

Soils	pH	P_2O_5 (ppm)	SiO_2 (ppm)	OM (%)	Lime req. (kg/10a)	Ca	Mg	K	H	C.E.C
Inside of pine forest	4.5	25	43	5.4	196	1.1	1.0	0.24	1.76	4.10
Outside of pine forest	6.0	21	67	1.4	98	5.4	1.3	0.16	+	6.89

TABLE 14.12 *Dry weight (mg) of species inside pine forest and species outside*

Inside species	PS (mg)	FS (mg)	PS/FS	Outside species	PS (mg)	FS (mg)	PS/FS
Aster tataricus	0.22	0.16	1.34	*Celosia argentea*	0.04	0.72	0.06
Peucedanum terebinthaceum	0.16	0.14	1.14	*Boehmeria platanifolia*	0.94	0.97	0.97
Bidens bipinnata	0.29	0.21	1.38	*Calystegia japonica*	0.18	0.15	1.20
Platycodon grandiflorum	0.48	0.29	1.66	*Veronica rotunda* var. *subintegra*	0.02	0.06	0.33
Patrinia villosa	0.39	0.28	1.39	*Elsholtzia ciliata*	0.51	1.26	0.40
Pinus rigida	0.51	0.31	1.65	*Oenothera odorata*	0.09	0.03	3.00
Arundinella hirta	0.24	0.47	0.51	*Aconogonum polymorphum*	0.57	0.33	1.73
Leibnitzia anandria	0.38	0.36	1.06	*Leonurus sibiricus*	0.09	1.38	0.07
Achyranthes japonica	0.53	0.10	5.30	*Amorpha fruticosa*	0.19	0.40	0.48
Ablizzia julibrissin	0.59	0.83	0.71	*Saussurea gracilis*	0.52	1.17	0.44
Melandryum firmum	0.28	0.29	0.97	*Triumfetta japonica*	0.61	0.93	0.66
Justicia procumbens	0.71	0.43	1.65	*Amaranthus mangostanus*	0.08	0.23	0.35
Plantago asiatica	0.23	0.11	2.09	*Serratula coreanata* var. *insularis*	0.61	0.89	0.69
Agrimonia pilosa	0.97	0.50	1.94	*Sedum erythrostichum*	0.41	0.39	1.05
Cassia mimosoides var. *nomame*	0.52	0.89	0.58				
Rhododendron schlippenbachii	0.36	0.38	0.95				
Pinus thunbergii	0.37	0.29	1.28				
			$\bar{X}_1 = 1.505$				$\bar{X}_2 = 0.816$

$\bar{X}$ significantly different at the 0.05 probability level. $t = 2.064$.
PS, pine stand soil; FS, outside soil.

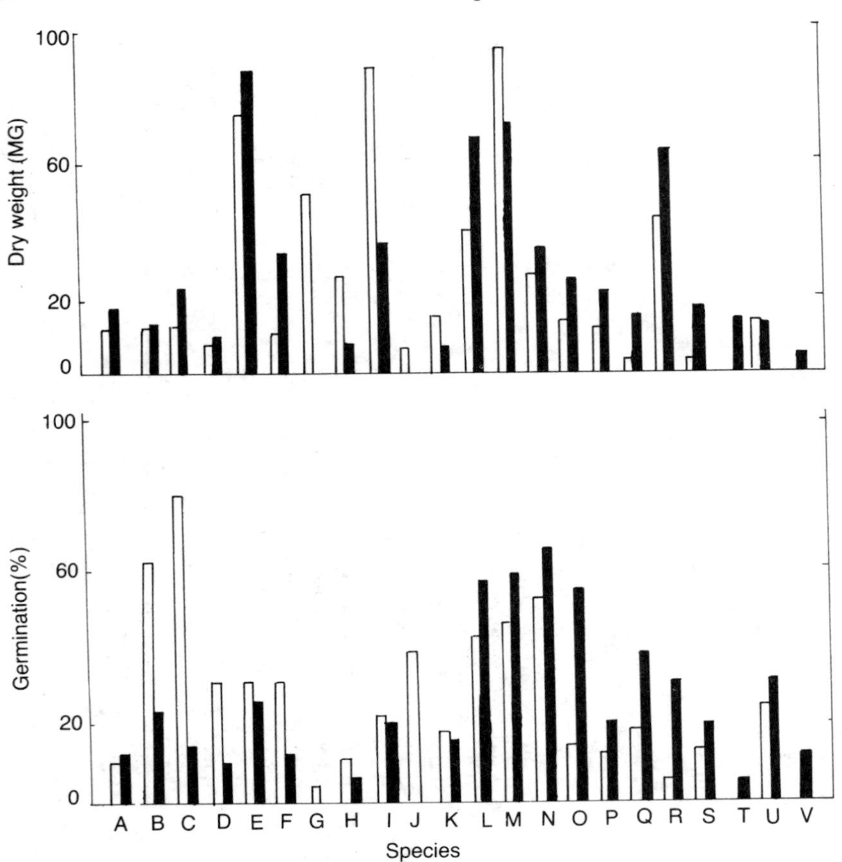

Fig. 14.12 Bar graph showing germination percentage and dry weight of seedling tested in black pine forest soil (□) and farm soils (■). A, *Pinus thunbergii*; B, *Pinus rigida*; C, *Achyranthes japonica*; D, *Arundinella hirta*; E, *Themeda triandra* var. *japonica*; F, *Phytolacca americana*; G, *Albizzia julibrissin*; H, *Justicia procumbens*; I, *Robinia pseudo-acacia*; J, *Miscanthus sinensis*; K, *Lespedeza bicolor*; L, *Setaria viridis*; M, *Cassia tora*; N, *Digitaria sanguinalis*; O, *Amaranthus tricolor*; P, *Triumfetta japonica*; Q, *Oenothera odorata*; R, *Bidens bipinnata*; S, *Phyllanthus ussuriensis*; T, *Melandryum firmum*; U, *Amorpha fruticosa*; V, *Youngia sonchifolia*.

viridis, *Digitaria sanguinalis*, *Amaranthus mangostanus*, *Triumfetta japonica*, *Oenothera odorata* (Fig. 14.12).

(c) *Pinus rigida*

The germination and dry weight of the inside and outside species were variously affected when grown in pitch pine and outside soil. The results are summarized in Table 14.13.

TABLE 14.13 *Germination percentage and oven-dry weight (mg) of selected species grown in the soil of* Pinus rigida *forest*

Species	Germination (%)			Dry weight (mg)		
	A	B	B/A	C	D	D/C
Pinus rigida	64	62	0.97	44.50	40.81	0.92
Pinus thunbergii	92	92	1.00	34.17	35.00	1.02
Arundinella hirta	90	64	0.71	23.71	47.03	1.98
Aconogonum polymorphum	80	64	0.80	136.33	178.50	1.31
Celosia argentea	88	74	0.84	219.02	311.08	1.42
Achyranthes japonica	96	74	0.77	433.71	513.47	1.18
Plantago asiatica	40	24	0.60	119.80	135.17	1.13
Eclipta prostrata	44	62	1.41	72.50	163.84	2.26
Lactuca sativa	36	96	2.67	35.00	78.12	2.23

A and C, in-forest soil; B and D, out-forest soil.

Thus germination and growth tests in aqueous extracts, leachates and soils provided interesting data regarding the behaviour of inside and outside species and their interactions with various pine species.

14.3 ISOLATION AND IDENTIFICATION OF ALLELOCHEMICALS

14.3.1 By paper chromatography

The procedures used to isolate the compounds from pine materials were modified from those of Lodhi and Rice (1971) (originally from Rice, 1965; Guenzi and McCalla, 1966). Fresh pine leaves and fallen leaves (25 g) were added to 500 ml water at 60°C for 2–3 h. It was then filtered, acidified to pH 2 using 2 N HCl, and extracted with two half-volumes of diethyl ether. Ether and water fractions were evaporated to dryness and were taken up in 3 ml of 85% ethanol and 10 ml of distilled water, respectively.

Also, 10 g each of fresh pine leaves and fallen leaves were ground to pass through a 10 mesh screen and hydrolysed in an autoclave for 45 min with 150 ml of 2 N NaOH. The extracts were filtered and acidified to pH 2.0 with 1 N HCl and extracted with two half-volumes of 5% $NaHCO_3$. It was acidified again to pH 2.0 and extracted with two half-volumes of ether. The ether fraction was evaporated to dryness and the residue was taken up in 3 ml of 95% ethanol.

These fractions were chromatographed in two dimensions on Whatman

TABLE 14.14 *Results of paper chromatography showing Rf value of chemical substances isolated from black pine leaves (after Kil, 1989)*

Substances	Standard		Black pine leaves	
	BAW[a]	AA[b]	BAW[a]	AA[b]
Gallic acid	0.62	0.54	0.62	0.52
Salicylic acid	0.92	0.69	0.92	0.64
Cinnamic acid	0.85	0.35	0.85	0.37
p-Hydroxybenzoic acid	0.87	0.66	0.87	0.68
Ferulic acid	0.85	0.52	0.86	0.52
Syringic acid	0.80	0.50	0.80	0.50
p-Coumaric acid	0.82	0.35	0.82	0.32
Benzoic acid	0.88	0.54	0.88	0.54
Vanillic acid	0.88	0.59	0.88	0.60
Sulphosalicylic acid	0.90	0.68	0.91	0.67
Catechol	0.90	0.64	0.90	0.60
Gentisic acid	0.80	0.72	0.80	0.72
Chlorogenic acid	0.65	0.68	0.64	0.66

[a] *n*-Butanol–acetic acid–water (63:10:27, v/v/v).
[b] 6% aqueous acetic acid.

TABLE 14.15 *Rf value of phenolic compounds (standard) and components of* Pinus rigida *leaves isolated by paper chromatography*

No	Chemical substance	Standard		*Pinus rigida*	
		BAW[a]	AA[b]	BAW[a]	AA[b]
1	Cinnamic acid	0.85	0.35	0.85	0.39
2	Chlorogenic acid	0.65	0.68	0.62	0.76
3	Syringic acid	0.80	0.50	0.78	0.55
4	Gentisic acid	0.80	0.72	0.80	0.68
5	*p*-Coumaric acid	0.82	0.35	0.82	0.27
6	Ferulic acid	0.85	0.52	0.84	0.58
7	Benzoic acid	0.88	0.54	0.83	0.63
8	*p*-Hydroxybenzoic acid	0.87	0.66	0.83	0.66
9	Vanillic acid	0.88	0.59	0.87	0.58

[a] *n*-Butanol–acetic acid–water (63:10:27, v/v/v).
[b] 6% aqueous acetic acid.

paper with *n*-butanol-acetic acid-water (BAW 63:10:27, v/v/v), followed by 6% aqueous acetic acid (AA 6%). The chromatograms were inspected under ultraviolet light (2537 Å) and compounds were marked. Further, the chromatograms were dipped in various reagents (Rice, 1965). Thirteen and nine different chemicals were identified in black pine (Table 14.14) and pitch pine (Table 14.15), respectively.

14.3.2 By thin layer chromatography

To isolate chemicals from an extract of fresh pine leaves, 20 g of fresh pine leaves were sliced into small pieces, added to 150 ml of ether and kept at 0°C for 20 h. The ether extract was treated with a sodium bicarbonate solution, acidified with 15% tartaric acid, and then re-extracted with ether. Ether-soluble acidic substances (acid fraction) were concentrated in a vacuum evaporator and 1 ml of absolute methanol was added. The ether fraction was evaporated to 10 ml, distilled water was added and the ether was completely removed by shaking. The aqueous solution was filtered to remove the impurities such as resin and green pigments and re-extracted with ether. The neutral fraction (having ether-soluble neutral substances) was concentrated in a vacuum evaporator and 1 ml of absolute methanol was added to it. All procedures were done in total darkness under a safe lamp. The concentrated sample solution was stored in a refrigerator until used for chromatography. For thin layer chromatography 0.1 ml of the sample was streaked on a thin layer plate (20 cm × 20 cm) and was developed using the solvent system: Isopropyl alcohol:25% NH_4OH:distilled water (10:1:1, v/v/v). In order to detect the substances, dried chromatogram was subjected to the following treatment:

(a) Ferric chloride test – 1 g of ferric chloride was dissolved in 100 ml of distilled water and sprayed on the chromatogram.
(b) Ehrlich reagent test – 1 g of *p*-dimethylaminobenzaldehyde was dissolved in N–HCl solution and distilled water was added to make 100 ml solution. This was sprayed on the chromatogram.
(c) Ultra-violet fluorescent lamp detection – UV-fluorescent lamp (2537 Å) was used.

The results obtained from above tests are summarized in Table 14.16.

14.3.3 By high-pressure liquid chromatography

HPLC (Waters Co. 440 type, column u – Bondapak C-18, detector UV absorbance detector 254 nm 0.05 aufs, flow rate 0.4 ml/min and chart speed 1.00 min^{-1}) was employed in order to detect allelochemicals.

Finely smashed experimental material (20 g) was put in 400 ml of water and filtered after 24 h, then 75 g NaCl was added to 250 ml of filtered

TABLE 14.16 *Thin layer chromatographic characteristics of chemical substances extracted from fresh leaves of pine*

Fractions	Rf values	UV-F1 253 nm	Ehrich reaction	$FeCl_3$
Acid	0.17	Pale blue	Light orange	Blue
	0.32	Pale blue	Brown	Blue
	0.37	Blue	Pink	–
	0.42	Silvery white	–	–
	0.47	Pale blue	–	Blue
	0.52	Dark blue	Light pink	–
	0.57	Pale blue	Light pink	–
	0.67	Dark blue	Pink	–
	0.72	Blue	Pink	Light blue
	0.77	Blue	Pink	–
	0.87	Blue	Yellow	–
	0.92	Milky white	Yellow	–
Neutral	0.27	Silvery white	Pink	–
	0.32	Silvery white	–	–
	0.47	Light blue	–	–
	0.52	Dark blue	Orange	Green
	0.57	Light blue	–	Blue
	0.67	Purplish blue	–	–
	0.72	Purplish blue	Brown	Dark blue
	0.77	Brownish blue	–	–
	0.82	Blue	Yellow	Blue
	0.87	Purplish blue	Silvery white	Blue
	0.92	Silvery white	Orange	–
	0.97	Dark blue	Orange	–

solution to get it saturated. HCl was added to the filtered solution and the pH of the solution was adjusted to 1. The filtered solution was extracted repeatedly with 25 ml of ether. To this, 25 ml of 0.1 M pyrophosphate solution (pH 10) was added and thoroughly shaken before extraction. This was acidified to pH 1 with HCl and again extracted with 10 ml of ether. This ether solution was concentrated in a rotary evaporator and the concentrate was analysed using HPLC. The results obtained are summarized in Table 14.17.

14.3.4 By gas chromatography

(a) Gas chromatographic conditions and materials

The analysis of various extracts was performed on a Varian 2440 gas

TABLE 14.17 *Chromatographic identification of chemical compounds from* Pinus rigida *leaves (after Kil, 1989)*

Compounds	PC[a]	HPLC	GC
Cinnamic acid	+[b]	–[c]	+
Chlorogenic acid	+	–	–
Syringic acid	+	–	+
Gentisic acid	+	–	+
p-Coumaric acid	+	–	–
Ferulic acid	+	–	–
Benzoic acid	+	–	–
p-Hydroxybenzoic acid	+	–	+
Vanillic acid	+	–	+
Tannic acid	–	+	–
Hydroquinone	–	+	–
Catechol	–	+	–
Caffeic acid	–	–	+
Gallic acid	–	–	+

[a] PC, paper chromatography; HPLC, high-performance liquid chromatography; GC, gas chromatography.
[b] +, present.
[c] –, absent.

chromatograph which was equipped with a 6 ft × $\frac{1}{8}$ inch i.d. stainless steel column, packed with 2.5% SE-30 on ABS (90/100 mesh). The carrier gas (helium) flow was 40 ml min^{-1}. The flame ionization detection was set at a sensitivity of 2 × 10. Injection port and detector temperatures were held at 230 and 205°C, respectively. The column temperature was programmed at the rate of 6°C min^{-1} from 100 to 250°C. One microlitre samples were injected with a 5 ml SGE syringe, employing the solvent flush technique. Materials analysed were fresh pine leaves, fallen leaves and pine rain leachate.

(b) Procedure

Twenty grams of whole fresh leaves were placed in one litre amber bottles with water (400 ml), and shaken on a mechanical shaker for the periods of 1 and 24 h. Fresh pine leaves were also soaked in 0.1 M pyrophosphate solution at pH 10 for two weeks at room temperature. Samples of fresh and fallen leaves were extracted for different time periods. After the respective extraction time had elapsed, the leaves were removed, the aqueous extract was saturated with NaCl and its pH was adjusted to 1 with HCl. This solution was extracted three times with 50 ml of distilled diethyl ether. In some cases ether extract at this point contained a brown gelatinous substance which, on standing, precipitated out of ether. In such cases ether was decanted off from the precipitate, the precipitate was washed twice with 25 ml of ether, and the washings were combined with ether extract. To avoid the possibility that non-polar compounds may be co-extracted with phenolic acids into ether extract, the acidic compounds in ether were back extracted into a pH 10, 1 M Na_3PO_4 buffer (3 × 25 ml). The pyrophosphate extract was saturated with NaCl and the pH was adjusted to 1. This solution was then extracted three times with 10 ml of distilled ether.

Ether extract was transferred to a 250 ml recovery conical flask. A three-ball Snyder distillation column Kontes was attached to the recovery flask and ether was distilled off to a volume of approximately 2 ml. The concentrate was then transferred to a centrifuge tube and evaporated to dryness by blowing a gentle stream of N_2. Residue on the walls were washed with 100 ml of spectrograde acetonitrile. To this solution, 150 ml of BSTFA was added and allowed to stand for 5 min. The prepared sample was then chromatographed. Five hundred millilitres of pine rain leachate were also analysed using the above-mentioned method.

The retention times of provided standards and leaf extracts were compared in order tentatively to identify the phenolic acids present in the samples.

As shown in Table 14.18, 11 phenolic acids and benzoic acid were detected. Relative amount and relative retention time indicated that gentisic, gallic and caffeic acids were of less occurrence in comparison to the other phenolics identified.

It has also been reported that rain drops or dew contain phytotoxic substances which regulate plant–plant interactions. Therefore, pine rain leachates were analysed by gas chromatography and benzoic acid along with one unknown substance were detected. Phenolics are known to exude from plants or be produced during decay of plant residues (Shindo and Kuwatsuka, 1977) and have earlier been identified by many workers (Wang *et al.*, 1967; De Bell, 1971; McPherson *et al.*, 1971; Rice and Pancholy, 1973a,b; Shibakusa, 1973; Tames *et al.*, 1973; Al-Mousawi and Al-Naib, 1975; Chou and Patrick, 1976; Thakur, 1977; Ballester *et al.*, 1979; Carballeira, 1980). Most of these phenolic compounds have been implicated in allelopathy and are known to affect germination and seedling growth of plants (Olmsted and Rice, 1970; Tinnin and Muller, 1972; Kapustka and Rice, 1976; Rasmussen and Einhellig, 1977; Rice *et al.*, 1980). It is interesting to note that the allelopathic actions of these compounds were more uniform and significant when the availability of nutrient was poor (Stowe and Osborn, 1980).

The results obtained from studies with soaking of fallen leaves revealed that when the leaves were soaked in water for 1 h, only five phenolic acids were isolated but after 24 h soaking, the number of isolated compounds increased to 8. The additional phenolics were salicylic, cinnamic and *p*-coumaric acids. Also, there was a change in the yield of vanillic and

TABLE 14.18 *Identification of chemical compounds from pine leaves by gas chromatography (after Kil and Yim, 1983)*

No.	Chemical compounds	Relative retention time	Fresh leaves[a]			Fallen leaves[a]			Rain leachate
			W1	W24	P2	W1	W24	P2	
1	Benzoic acid	0.28	+++	+++	+++	+	+	++	+
2	Salicylic acid	0.75		+++			+		
3	Cinnamic acid	0.78	+	+			+		
4	*p*-Hydroxybenzoic acid	1.0		+++	+	+	+	+	
5	Gentisic acid	1.38		+					
6	Protocatechuic acid	1.48		+	+	+	+	+	
7	Syringic acid	1.63		+		+	+		
8	*p*-Coumaric acid	1.70		+++			+	+	
9	Gallic acid	1.80						+	
10	Ferulic acid	2.03		+					
11	Caffeic acid	2.15		+					
12	Vanillic acid	1.33			+	+	++	+	

[a] W1, W24 mean soaked in water for 1 and 24 h, P2 means soaked in pyrophosphate solution for 2 weeks. Quantitative comparison was made by the order +++ > ++ > +.

protocatechuic acid. These findings suggest that the quality and quantity of phenolics are variable, which at least partially can be attributed to the microbial action on the plant tissue.

Horsley (1976) pointed out that phenolic compounds, particularly benzoic acid, cinnamic acid and coumarins, are among the most important phytotoxins in temperate ecosystems. Besides phenolics, there are other substances which may have an important effect on germination and seedling growth. However, there is a possibility that phenolic acids present in pine leaves may play a crucial role with respect to its allelopathic effect on understorey species, which results into a special floristic composition. Analyses of pine leaf have shown that pitch pine has seven phenolic compounds (Table 14.17), and most of the 14 isolated chemicals from black pine leaves were also of a phenolic nature (Fig. 14.13, Table 14.19).

Different concentrations of chemicals identified from pine were prepared and used for lettuce seed germination bioassay. The RGR, RER, RFR and RDR of lettuce were variously affected by concentrations of different chemicals (Figs 14.14 and 14.15).

14.4 LABORATORY AND GREENHOUSE BIOASSAY

Working with other allelochemicals many researchers have reported similar findings. When various concentrations of juglone were tested against 16 test species, 10^{-3} M and 10^{-4} M concentrations caused death and severe inhibition, respectively. But there was no harmful effect from

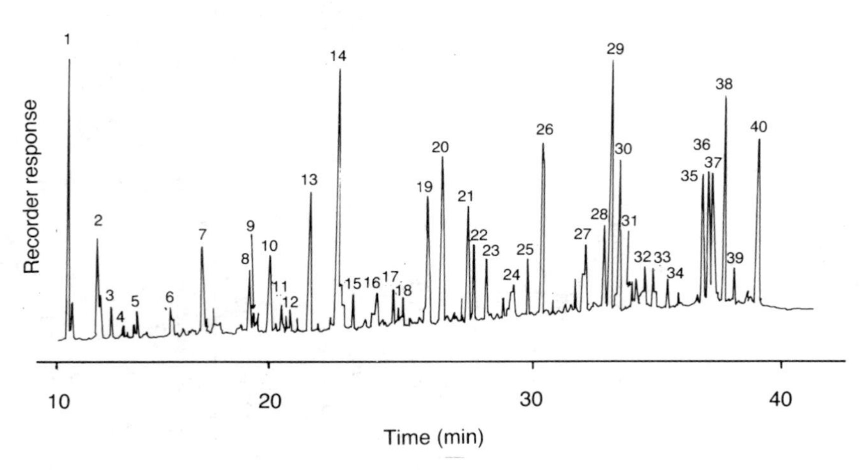

Fig. 14.13 Gas chromatograms obtained by using a sample of the combined ether extracts of *Pinus thunbergii* leaves. Key to numbers as in Table 14.19.

TABLE 14.19 *Chemical substances of* Pinus thunbergii *leaves identified from chromatograms in Fig. 14.13*

Peak number	Structural identification	Retention time
1	Benzoic acid	11.12
2	Phenylacetic acid	12.50
3	Unidentified	
4	Catechol	13.43
5, 6, 7, 8	Unidentified	
9	Salicylic acid	18.87
10, 11	Unidentified	
12	*m*-Hydroxybenzoic acid	20.49
13	Unidentified	
14	*p*-Hydroxybenzoic acid	22.38
15	Phloroglucinol	22.98
16	Phthalic acid + *p*-phenylbenzaldehyde	24.04
17, 18	Unidentified	
19	Vanillic acid	26.04
20, 21	Unidentified	
22	Protocatechuic acid	27.98
23	Unidentified	
24	Syringic acid	29.51
25	Unidentified	
26	*p*-Coumaric acid	30.65
27, 28	Unidentified	
29, 30	Unidentified	
31	Ferulic acid	34.37
32, 33	Unidentified	
34	Caffeic acid	35.83
35–40	Unidentified	

10^{-5} M and 10^{-6} M concentrations (Rietveld, 1983). Formation of root and plant elongation have been found to be inhibited by sprinkling leaf surface with 0.001 M and 0.01 M of salicylic acid, cinnamic acid and catechol (Sitaramaiah and Pathak, 1981). Thus, findings reported here are in agreement with earlier reports.

The effective concentrations of substances determined in the laboratory are usually higher than those found in natural conditions. Therefore, to confirm the laboratory results and their utility, field bioassay is a prerequisite. For this, an experimental plot (2.5 × 2.5 m) was set in the forest of pitch pine and seeds of several species were sown. A control plot of the same size was also established outside the forest. The brightness of sunlight was properly adjusted using shade. In addition, two germination boxes, each of 1.2 × 1.2 × 0.2 m, containing vermiculite were prepared, one of which was placed so as to be exposed to rain leachate of pitch pine

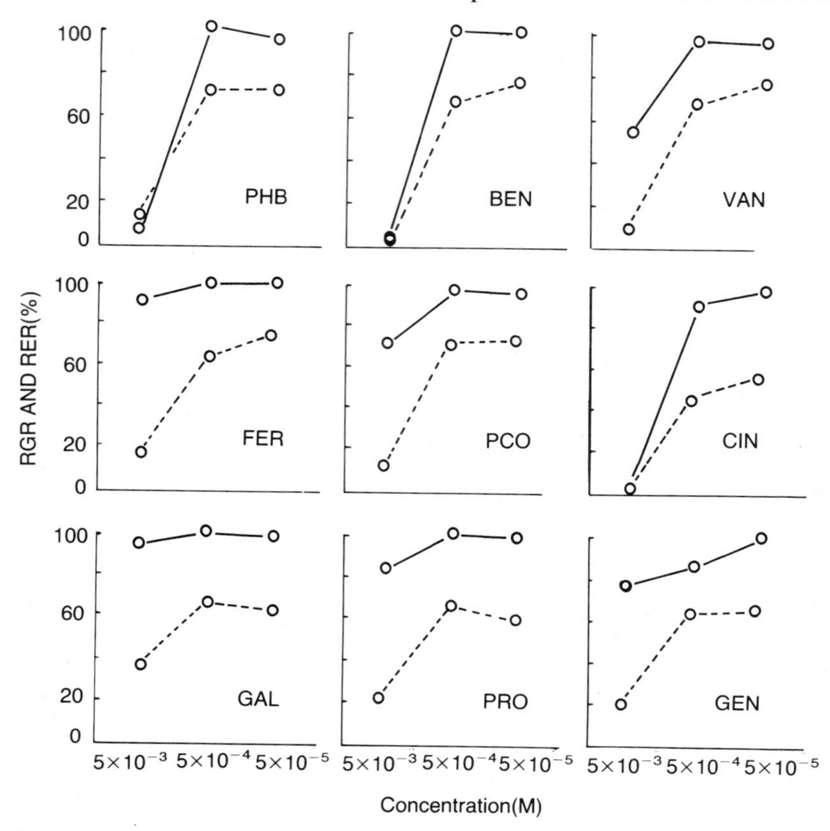

Fig. 14.14 RGR (○–○) and RER (○--○) of lettuce tested in different concentration of chemicals. PHB, *p*-hydroxybenzoic acid; BEN, benzoic acid; VAN, vanillic acid; FER, ferulic acid; PCO, *p*-coumaric acid; CIN, cinnamic acid; GAL, gallic acid; PRO, protocatechuic acid; GEN, gentisic acid.

and the other one (control) was watered and placed outside the forest. The light condition of both boxes was equally adjusted. The results, summarized in Tables 14.20 and 14.21, reveal that RFR and RDR of the test species were variously affected.

14.5 CONCLUSIONS AND PROSPECTS

In the field survey the author found that the floristic composition in the forest floor of pine stands were similar to each other but very different from those of nearby fields. It was hypothesized that this kind of floristic composition was caused by some regulating mechanism controlled by toxins produced by pine. Accordingly, pine material was extracted under

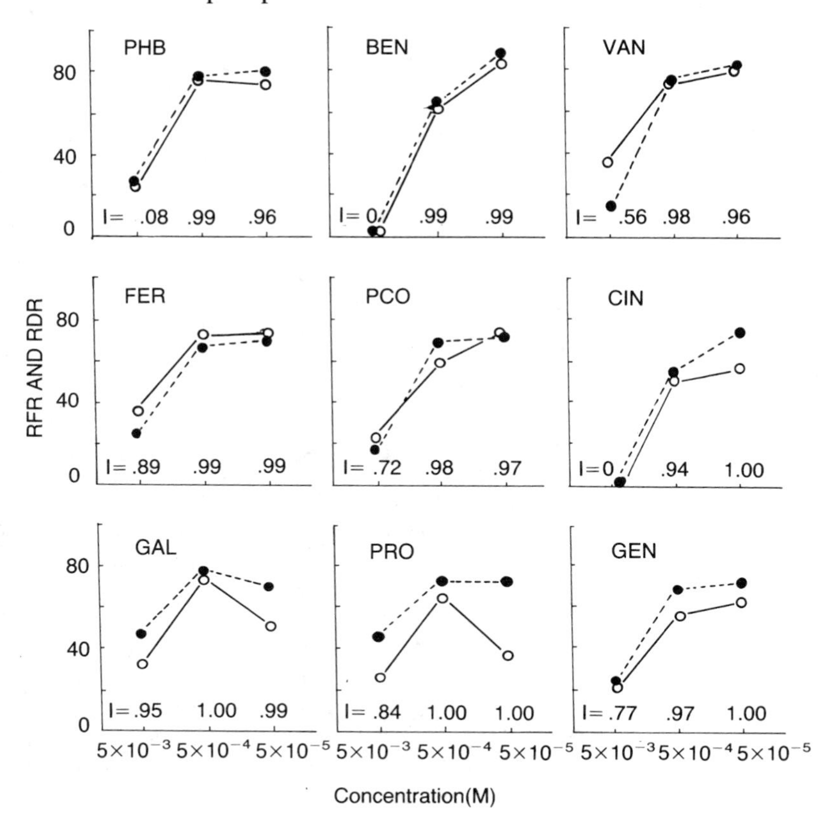

Fig. 14.15 RFR (●---●) and RDR (○–○) of lettuce tested in different concentrations of chemicals. Keys to chemicals as in Fig. 14.14 and I, inhibition index. RFR means relative fresh weight ratio.

different conditions for various durations. Results showed that extracts made at 80°C were more toxic than those made at 18°C and the inhibitory activity of different plant parts was in the order of fresh leaves > fallen leaves > roots. The results of germination experiments conducted in forest and field soils did not confirm these findings, but data obtained from growth studies did so. In other words, the inside species of pine grew best in the inside soils of the pine forest while the outside species grew better in the outside soil.

Thin layer and gas chromatographic studies revealed that pine leaves contain several toxins, the majority of which were phenolic acids like salicylic, cinnamic, *p*-hydroxybenzoic, gentisic, protocatechuic, syringic, *p*-coumaric, gallic, ferulic, caffeic and vanillic acids. However, paper chromatography and HPLC confirmed that the phytotoxic substances implicated in allelopathy are largely contained in black pine leaves.

TABLE 14.20 *Relative fresh weight ratio (RFR) and relative dry weight ratio (RDR) of selected species seedling grown in vermiculite under crown of* Pinus rigida *forest*

Species	Lp	Cm	Cb	Bb	So	Rs	Ss	Ef	Ap	Oo	Ah	Co	Ph
RFR	65	80	79.7	183.3	100	112.4	83.3	83	26	62.5	467.8	65.6	65
RDR	88.6	97.6	87.7	56.6	130.2	106.4	145.8	16.4	83.3	80	156.3	6.53	61.3

Keys to species: Lp, *Liriope platyphylla*; Cm, *Cassia mimosoides* var. *nomame*; Cb, *Cosmos bipinnatus*; Bb, *Bidens bipinnata*; So, *Sanguisorba officinalis*; Rs, *Raphanus sativus* var. *hortensis* for. *acanthiformis*; Ss, *Scilla scilloides*; Ef, *Eragrostis ferruginea*; Ap, *Artemisia princeps* var. *orientalis*; Oo, *Oenothera odorata*; Ah, *Arundinella hirta*; Co, *Cassia occidentalis*; Ph, *Phaseolus angularis*.

TABLE 14.21 *Relative fresh weight ratio (RFR) and relative dry weight ratio (RDR) of selected species seedling grown in soil under the crown of* Pinus rigida *forest*

Species	Lp	Rs	So	Pav	Pn	Ef	Oo	Cac	Ph	Pa
RFR	70.1	88.2	110	23.3	38	58	400	100	77.2	5
RDR	93.1	71.4	91	14	60	96	24	60	81.7	80

Keys to species: Lp, Rs, So, Ef, Oo, Ph, as in Table 14.20; Pav, *Polygonum aviculare*; Pn, *Pharbitis nil*; Cac, *Chenopodium album* var. *centrorubrum*; Pa, *Poa annua*.

The experimental findings presented here clearly demonstrated that these phenolic compounds may at least partially be responsible for the allelopathic effects of pine on other species.

It is emphasized that similar studies should be undertaken in other plant communities which would provide a better insight into the natural ecosystem leading to an improved management and optimum exploitation of the natural resources. Besides these aspects, the allelochemicals produced by pine or other forest trees may also be evaluated for their plant growth regulatory and pest-controlling properties. Considering the diverse chemical nature of allelochemicals, possibilities are fair that some of the allelochemicals reported here may either directly be used as agrochemicals or may provide 'molecular models' for creating some potent and safer agrochemicals. However, a multidisciplinary approach would be essential to exploit fully the allelopathic potential of plants.

REFERENCES

Al-Mousawi, A. H. and Al-Naib, F. A. G. (1975) Allelopathic effects of *Eucalyptus microtheca* F. Muell. *J. Univ. Kuwait Sci.*, **2**, 59–60.

Ashraf, N. and Sen, D. N. (1978) Allelopathic potential of *Celosia argentea* in arid land crop fields. *Oecol. Plant.*, **13**, 331–8.

Ballester, A., Vieitez, A. M. and Vieitez, E. (1979) The allelopathic potential of *Erica australis* and *Erica arborea*. *Bot. Gaz.*, **140**, 433–6.

Bhowmik, P. C. and Doll, J. D. (1983) Growth analysis of corn and soybean response to allelopathic effects of weed residues at various temperatures and photosynthetic photon flux densities. *J. Chem. Ecol.*, **9**, 1263–80.

Carballeira, A. (1980) Phenolic inhibitors in *Erica australis* L. and the associated soil. *J. Chem. Ecol.*, **6**, 593–6.

Chou, C. H. and Young, D. C. (1974) Effects of osmotic concentration and pH on plant growth. *Taiwania*, **19**, 157–65.

Chou, C. H. and Patrick, Z. A. (1976) Identification and phytotoxic activity of compounds produced during decomposition of corn and rye residues in soil. *J. Chem. Ecol.*, **2**, 369–87.

Datta, S. C. and Chatterjee, A. K. (1980) Allelopathic potential of *Polygonum orientale* in relation to germination and seedling growth of weeds. *Flora (Jena)*, **169**, 456–65.

De Bell, D. S. (1971) Phytotoxic effects of cherrybark oak. *Forest Sci.*, **17**, 180–5.

Einhellig, F. A. and Rasmussen, J. A. (1978) Synergistic inhibitory effects of vanillic and *p*-hydroxybenzoic acids on radish and grain sorghum. *J. Chem. Ecol.*, **4**, 425–36.

Guenzi, W. D. and McCalla, T. M. (1962) Inhibition of germination and seedling development by crop residues. *Soil Sci. Soc. Amer. Proc.*, **26**, 456–8.

Guenzi, W. D. and McCalla, T. M. (1966) Phytotoxic substances extracted from soil. *Soil Sci. Soc. Amer. Proc.*, **30**, 214–16.

Horsley, S. B. (1976) Allelopathic interference among plants. II. Physiological modes of action. *Proc. Fourth North Amer. For. Biol. Workshop*, pp. 93–136.

Kapustka, L. A. and Rice, E. L. (1976) Acetylene reduction (N_2-fixation) in soil and old succession in central Oklahoma. *Soil Biol. Biochem.*, **8**, 497–503.

Khan, M. I. (1982) Allelopathic potential of dry fruits of *Washingtonia filifera* inhibition of seed germination. *Physiol. Plant*, **54**, 323–8.

Khosh-Khui, M. and Bassiri, A. (1979) Inhibition of seedling growth by wild myrtle. *Weed Res.*, **19**, 45–50.

Kil, B. S. (1989) Allelopathic effects of five pine species in Korea, in *Phytochemical Ecology: Allelochemicals, Mycotoxins, and Insect Pheromones and Allelomones* (eds C. H. Chou and G. R. Waller), Academia Sinica, Monograph Ser. No. 9, Acad. Sinica, Taipei, pp. 81–100.

Kil, B. S. and Yim, Y. J. (1983) Allelopathic effects of *Pinus densiflora* on the floristic composition of undergrowth of red pine forest. *J. Chem. Ecol.*, **9**, 1135–51.

Kim, G. S. and Kil, B. S. (1984) Allelopathic effects of aqueous black pine extracts on the selected species. *J. Basic Natu. Sci. Wonkwang Univ.*, **3**, 38–45.

Lee, I. K. and Monsi, M. (1963) Ecological studies on *Pinus densiflora* forest. I. Effects of plant substances on the floristic composition of the undergrowth. *Bot. Mag. (Tokyo)*, **76**, 400–13.

Le Tourneau, D. and Heggeness, H. G. (1975) Germination and growth inhibitors in leafy spurge foliage and quackgrass rhizomes. *Weeds*, **5**, 12–19.

Lodhi, M. A. K. and Rice, E. L. (1971) Allelopathic effects of *Celtis laevigata*. *Bull. Torrey Bot. Club*, **98**, 83–9.

McPherson, J. K., Chou, C. H. and Muller, C. H. (1971) Allelopathic constituents of the chaparral shrub *Adenostoma fasciculatum*. *Phytochemistry*, **10**, 2925–33.

McPherson, J. K. and Muller, C. H. (1969) Allelopathic effects of *Adenostoma fasciculatum* 'chamise', in the California chaparral. *Ecol. Monogr.*, **39**, 177–98.

Olmsted, C. E. and Rice, E. L. (1970) Relative effects of known plant inhibitors or species from first two stages of old-field succession. *Southwestern Nat.*, **15**, 165–73.

Quinn, J. A. (1974) *Convolvulus sepium* in old field succession on the New Jersey Piedmont. *Bull. Torrey Bot. Club*, **101**, 89–95.

Rasmussen, J. A. and Einhellig, F. A. (1977) Synergistic inhibitory effects of *p*-coumaric and ferulic acids on germination and growth of grain sorghum. *J. Chem. Ecol.*, **3**, 197–205.

Rho, B. J. and Kil, B. S. (1986) Influence of phytotoxin from *Pinus rigida* on the selected plants. *J. Natu. Sci. Wonkwang Univ.*, **5**, 19–27.

Rice, E. L. (1965) Inhibition of nitrogen-fixing and nitrifying bacteria by seed plant. II. Characterization and identification of inhibitors. *Physiol. Plant*, **18**, 255–68.

Rice E. L. and Pancholy, S. K. (1973a) Inhibition of nitrification by climax ecosystems. II. Additional evidence and possible role of tannins. *Amer. J. Bot.*, **60**, 691–702.

Rice, E. L. and Pancholy, S. K. (1973b) Inhibition of nitrification by climax ecosystems. III. Inhibitors other than tannins. *Amer. J. Bot.*, **61**, 1095–103.

Rice, E. L., Lin, C. H. and Huang, C. Y. (1980) Effects of decaying rice straw on growth and nitrogen fixation of a bluegreen alga. *Bot. Bull. Acad. Sinica*, **21**, 111–17.

Rietveld, W. J. (1983) Allelopathic effects of juglone on germination and growth of several herbaceous and woody species. *J. Chem. Ecol.*, **9**, 295–308.

Shibakusa, R. (1973) Growth inhibitors in leaves of *Abies sachalinensis* Masters in the dormant period(II). *J. Japan For. Soc.*, **55**, 91–4.

Shindo, H. and Kuwatsuka, S. (1977) Behavior of phenolic substances in the decaying process of plants. VI. Changes in quality and quantity of phenolic substances in the decaying process of rice straw in a soil. *Soil Sci. Plant Nutr.*, **23**, 319–32.

Sitaramaiah, K. and Pathak, K. N. (1981) Effect of growth regulators, phenolics and aromatic acid on root-knot severity (*Meloidogyne incognita* and *M. javanica*) on tomato. *J. Plant. Diseases and Protection*, **88**, 651–4.

Stowe, G. L. and Osborn, A. (1980) The influence of nitrogen and phosphorus levels on the phytotoxicity of phenolic compounds. *Can. J. Bot.*, **58**, 1149–53.

Tames, R. S., Gesto, M. D. V. and Vieitez, E. (1973) Growth substances isolated from tubers of *Cyperus esculentus* var. *aureus*. *Physiol. Plant*, **28**, 195–200.

Thakur, M. L. (1977) Phenolic growth inhibitors isolated from dormant buds of sugar maple. *J. Exp. Bot.*, **28**, 795–803.

Tinnin, R. O. and Muller, C. H. (1972) The allelopathic influence of *Avena fatua*: The allelopathic mechanism. *Bull. Torrey Bot. Club*, **99**, 287–92.

Wang, T. S. C., Yang, T. K. and Chuang, T. T. (1967) Soil phenolic acids as plant growth inhibitors. *Soil Sci.*, **103**, 239–46.

CHAPTER 15

Allelopathy in Brazil

A. G. Ferreira, M. E. A. Aquila, U. S. Jacobi and V. Rizvi

15.1 INTRODUCTION

Considering the large tropical wet region and a rich flora, the amount of work done with respect to allelopathic interaction among Brazilian plant species seems negligible. Reports on the allelopathic effect of *Mangifera indica* and *Cyperus rotundus* on seed germination and early seedling growth of various crop plants and microorganisms are among some of the pioneer work done in this direction (Teles, 1968; Meguro, 1969a, b; Meguro and Bonomi, 1969). This was followed by evaluation of a number of plant species for their allelopathic potential. Crude extracts of *Solanum chloranthum* (Moraes and Vicente, 1970), *Andira humilis*, *A. anthelmia*, *A. vermifuga*, *A. fraxinifolia* (Rizzini, 1971), *Calea cuneifolia* (Coutinho and Hashimoto, 1971), *Wedelia paludosa*, *Anacardium occidentalis* (Barbosa, 1972), *Solanum paniculatum*, *Artemisia absinthicum*, *Achryrocline alata*, *Athaeae rosea* (Dionello-Basta and Basta, 1984), *Sorghum halepense* (Castro *et al.*, 1984), *Brassica napus* (Vilhordo *et al.*, 1985; Burin and Vilhordo, 1986), *Phytolacca thyrsyflora* (Nagao *et al.*, 1986), *Beta vulgaris*, *Sorghum bicolor* and *Ilex domestica* (Santos *et al.*, 1987) have been found to be allelopathic to several crop plants. However, isolation and characterization of allelochemicals from these species is yet to gain a momentum in Brazil and also in other Latin American countries.

15.2 ALLELOPATHIC ACTIVITY OF SOME BRAZILIAN SPECIES

15.2.1 *Mimosa bimucronata* (DC) OK

This is a native bush of southern Brazil and is often found in pure stands on slopes and on low lands (Ferreira, 1976). The leaves and stalks of the plant are often used as a green manure in Brazil. However, in some cases

Allelopathy: Basic and applied aspects Edited by S. J. H. Rizvi and V. Rizvi
Published in 1992 by Chapman & Hall, London ISBN 0 412 39400 6

TABLE 15.1 *Radicle lengths (mm) of various species as affected by different concentrations of* Mimosa bimucronata *extract*

	Concentration			
Species	1:2	1:4	1:8	Control
Lettuce	8.2	11.4	16.4	28.0
Rice	14.0	18.4	28.1	42.7
Carrot	11.0	17.3	22.2	38.3
Chicory	9.2	12.1	17.4	26.9
Kale	8.5	12.9	21.7	44.3
Cucumber	11.0	20.7	31.5	67.7
Cabbage	6.5	10.8	21.0	42.2
Tomato	12.0	25.1	33.1	82.0

an inhibition of cultivated species was noticed after manuring with *M. bimucronata*. Such an inhibition might be the result of the release of allelochemicals from the leaves and stalks of *M. bimucronata*. Therefore, the possibility was tested by evaluating its aqueous extract against a number of species. For this, leaves of *M. bimucronata* were collected from natural stands, dried in an oven at 70°C, and stored in desiccators till further use. Dried plant material was extracted in distilled water for 24 h at 4°C. After filtration and centrifugation the extract was diluted to several strengths and tested for its allelopathic activity against a number of species. This was done by seed germination bioassay at 25 ± 2°C in a germination cabinet with a photoperiod of 16 h at 1800–2000 lux. The length of radicle was recorded after 5 days of germination. All the concentrations of extract were found ineffective with respect to seed germination, however, these significantly inhibited the radicle length in most of the cases (Table 15.1).

Members of legume genera, *Leucaena* and *Mimosa*, are known to contain mimosine, a non-protein amino acid. The allelopathic activity of mimosine is well demonstrated in the case of *Leucaena leucocephala* (Kuo *et al.*, 1982). To check the presence of mimosine in *M. bimucronata*, its extract was co-chromatographed with authentic mimosine and 3,4-dihydroxypyridine, the end product of mimosine degradation. Two prominent spots were noted on the chromatogram at Rf 0.5 and 0.7 corresponding to the Rf values of authentic mimosine and 3,4-dihydroxypyridine, respectively. The chromatogram was cut into ten strips representing each of the Rf values and were eluted in distilled water. Solutions thus obtained were then subjected to lettuce seed germination bioassay. None of the eluates could inhibit the seed germination, but the radicle growth was inhibited by all the fractions.

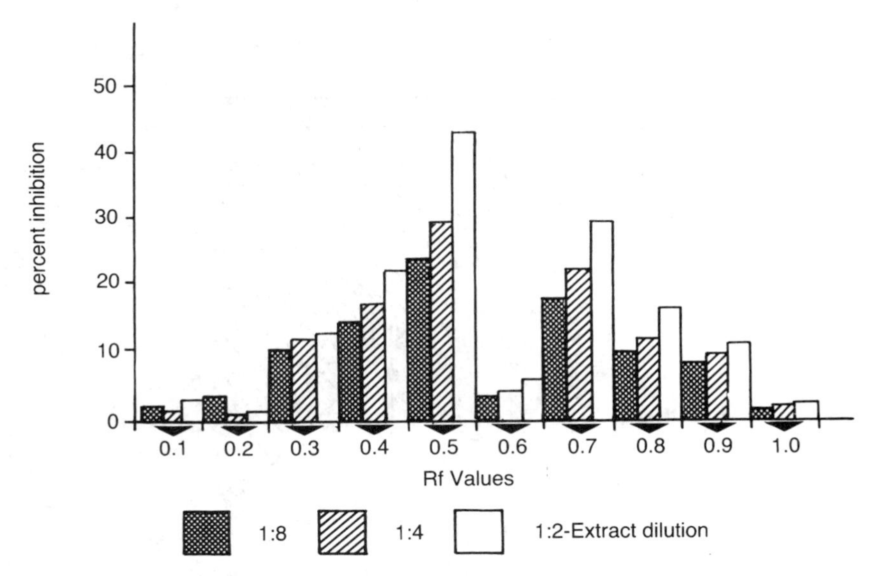

Fig. 15.1 Percentage of growth inhibition of lettuce seedling radicle by *Mimosa bimucronata* extracts. The controls are plotted as the base line (zero inhibition).

However, eluate obtained from the strip representing the 0.5 Rf zone showed maximum inhibition, followed by the eluate from 0.7 Rf zone (Fig. 15.1). Radicle length was also inhibited variously by different concentrations of authentic mimosine (Fig. 15.2). Thus, the Rf values of active substances and those of mimosine and 3,4-dihydroxypyridine were found to be the same. This clearly indicated that the allelopathic activity of *M. bimucronata* is at least partly due to the presence of these aforementioned substances. These findings are substantiated from the reports of Smith and Fowden (1966), Chou and Kuo (1986) and Jacobi and Ferreira (1987).

15.2.2 *Ilex paraquariensis* St Hil.

This is a native tree in the southern region of Brazil, north-east of Argentina and Paraguay. Several xanthines, namely theobromine, theophylline and caffeine, have been found to be present in various parts of the plant (Ferreira Filho, 1957; Leitao, 1969; Filip *et al.*, 1983). However, no or little attention has been paid on its possible allelopathic activity. Therefore, young top and bottom leaves along with the stalks were tested for their allelopathic activity. Plant parts were separately extracted for 10 min at 70°C in water. Crude extracts thus obtained were directly subjected to the maize seed germination bioassay. After 4 days of germination, length and dry weight of radicle and plumule were recorded.

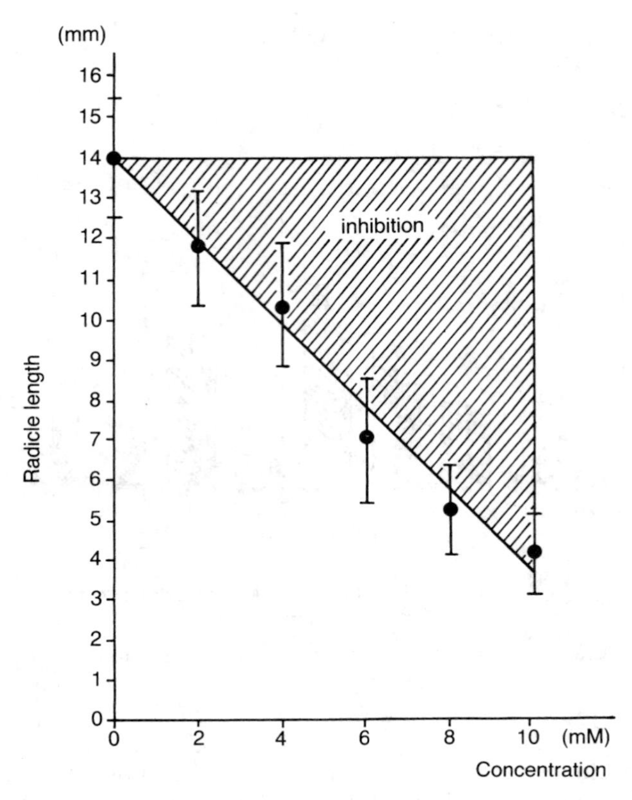

Fig. 15.2 The inhibitory effects of authentic mimosine on radicle growth of lettuce. Mean (n = 40) and S.D. are shown at each concentration.

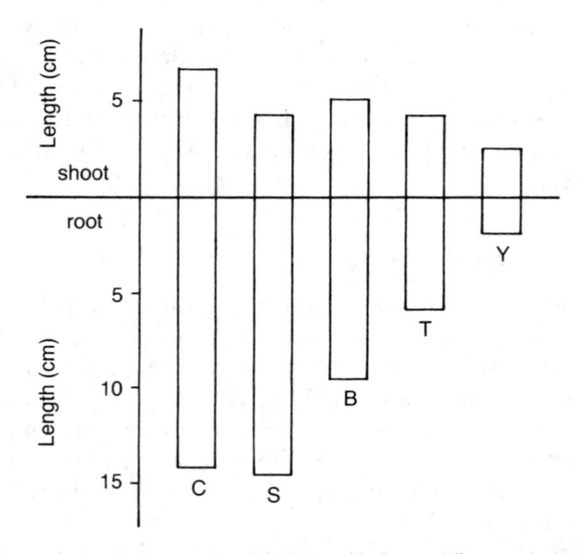

Fig. 15.3 Dry weight of *Zea mays* seedlings (4 days old) germinated in presence of aqueous extract of *Ilex paraguariensis*. C = control, S = stalks, B = bottom leaves, T = top leaves, Y = young leaves.

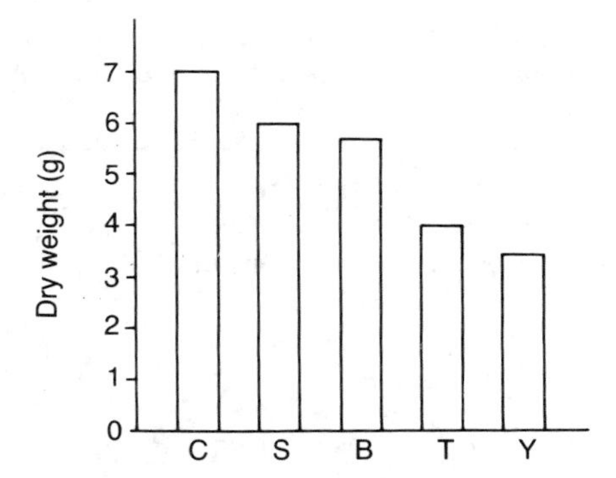

Fig. 15.4 Plantlets size of *Zea mays* (4 days old) germinated in presence of aqueous extracts of *Ilex paraquariensis*. C = control, S = stalks, B = bottom leaves; T = top leaves; Y = young leaves.

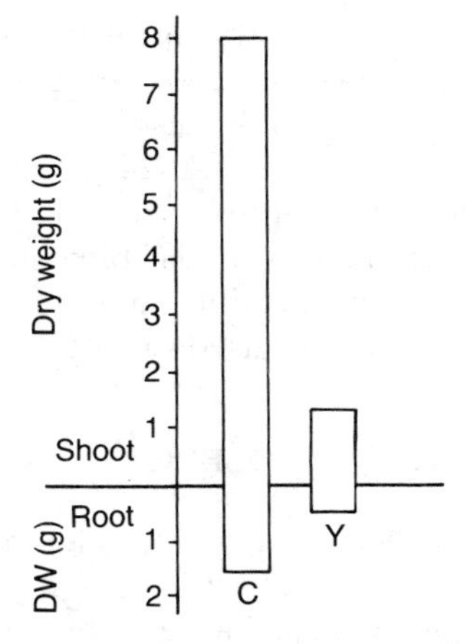

Fig. 15.5 Dry weight of *Zea mays* plants after 42 days of germination. C = control, Y = *Ilex paraquariensis* green manure added to the soil.

The results obtained showed that young leaf extract caused maximum inhibition of radicle, plumule length and seedling dry weight, while stalk extracts caused minimum inhibition (Figs 15.3 and 15.4). In addition to maize, the extracts of various parts of the plant were also found to be inhibitory to *Lycopersicon esculentum*, *Daucus carota*, *Beta vulgaris*, *Raphanus sativus* and *Brassica oleracea*. When the leaves of *Ilex paraquariensis* were mixed in soil as green manure, it significantly reduced the dry weight of the plants (Fig. 15.5).

TABLE 15.2 Lactuca sativa*: Effect of various concentrations of* Baccharis trimera *extract on germination and hypocotyl length of Boston and Honson varieties, at 48 h*

	Germination (%)		Hypocotyl length (mm)	
Concentrations	Boston	Honson	Boston	Honson
Control	94.5a	85.5a	48.6a	44.0a
1:32	16.0b	30.5b	37.1a	40.4a
1:16	2.5c	4.5c	23.5b	27.1b
1:8	3.0c	1.5c	23.6b	17.0c
1:4	2.0c	1.5c	22.0b	19.0c

15.2.3 *Baccharis trimera* (Less) A. P. de Candole

Allelopathic activity of *B. trimera* was measured in terms of inhibition of germination and radicle growth of two varieties of *Lactuca sativa*. For this, fresh leaves were extracted in distilled water at 70°C for 10 min followed by an incubation at 5°C for 24 h. Various concentrations of the extract thus obtained, were prepared and subjected to lettuce seed germination bioassay. The results of the bioassay demonstrated that all the dilutions of extract significantly inhibited the germination and hypocotyl length of both varieties used (Table 15.2).

15.3 CONCLUSIONS AND PROSPECTS

The data presented here demonstrate the allelopathic potential of some of the Brazilian species. The inhibitory activity of plants indicate the presence of some allelochemicals which may prove significant in the context of agriculture and forestry. The importance of allelopathy/ allelochemicals in crop rotation and agroforestry systems (Waller, 1987; Rizvi *et al.*, 1989a, b), controlling weeds, plant pathogenic fungi and insects (Rizvi *et al.*, 1980a, b; Waller, 1987; Rizvi *et al.*, 1988) and in many other aspects of agriculture, forestry and horticulture (Rice, 1984) has already been established in several countries. Therefore, comprehensive allelopathic investigations of Brazilian flora is imperative and needs immediate attention.

REFERENCES

Barbosa, D. C. A. (1972) Efeito inibidor do extrato aquoso de folhas de *Anacardium occidentale* L. sobre o crescimento de plantas de *Phaseolus vulgaris L. An. Soc. Bot. Brasil* (Garanhuns), pp. 189–92.

Burin, M. E. and Vilhordo, B. W. (1986) Efeito alelopatico do extrato de colza (*Brassica napus* var. *oleifera* Metzg) sobre a germinaçaõ de sementes de trigo, soja e tomate. *Agronomia Sulriograndense* (Porto Alegre) **22**, 35–54.

Castro, P. R. C., Rodrigues, J. D., Maimoni-Rodella, R. C. S. *et al.* (1984) Açaõ alelopatica de alguns extratos de plantas daninhas na germinaçaõ de arroz (*Oryza sativa* L. cv. IAC-165). *An. E. S. A. Luiz de Queiroz.*, **41**, 369–81.

Chou, C. H. and Kuo, Y. L. (1986) Allelopathic research of subtropical vegetation in Taiwan. III. Allelopathic exclusion of understorey by *Leucaena leucocephala* (Lam.) de Wit. *J. Chem. Ecol.*, **12**, 1431–48.

Coutinho, L. M. and Hashimoto, F. (1971) Sobre o efeito inibitorio da germanaçaõ de sementes produzido por folhas de *Calea cuneifolia* DC. *Cien. Cult.* (São Paulo), **23**, 759–64.

Dionello-Basta, S. B. and Basta, F. (1984) Inibidores de germinaçaõ e de crescimento em plantas usadas em medicina popular. *Cien. Cult.* (São Paulo), **36**, 1602–6.

Ferreira, A. G. (1976) Germinaçaõ de sementes de *Mimosa bimucronata* (DC.) OK (Marica) I – Efeito da escarificoçoã e do pH. *Cien. Cult.* (São Paulo), **28**, 1200–4.

Ferreira Filho, J. C. (1957) *Cultura e Preparo de Erva-mate*, Serv. Inf. Agric. Ministerio da Agricultura, Brasil.

Filip, R., Iglesias, D. I. A., Rondina, R. V. D. and Coussio, J. D. (1983) Analysis of the leaves and stems of *Ilex argentina* l. Xanthines. *Acta Farm. Bonaerense* (Buenos Aires), **2**, 87–90.

Jacobi, U. S. and Ferreira, A. G. (1987) Allelopathic effects of *Mimosa bimucronata* (DC.) O. Kuntze. *Abstracts, 14th Int. Botanical Congr.* (Berlin), p. 413.

Kuo, Y. L., Chou, C. H. and Hu, T. W. (1983) Allelopathic potential of *Leucaena leucocephala*, in *Allelochemicals and Pheromones* (eds C. H. Chou and G. R. Waller), Academia Sinica, Monograph Ser. No. 5, Acad. Sinica, Taipei, ROC., pp. 107–19.

Leitao, E. L. (1969) Quimica do mate. *A Lavoura.* 72(9/10), 25–32.

Meguro, M. (1969a) Substancias reguladoras de crescimento em rizoma de *Cyperus rotundus* L. I. Efeito do extrato de rizoma na germinaçaõ e crescimento de plantas superiores. *Bol. Fac. Fil. Cien. Letras USP Botanica* (São Paulo), pp. 127–44.

Meguro, M. (1969b) Substancias reguladoras de crescimento em rizoma de *Cyperus rotundus* L. II. Natureza e propriedades do inibidor. *Bol. Fac. Fil. Cien. Letras USP Botanica* (São Paulo), pp. 127–72.

Meguro, M. and Bonomi, M. V. (1969) Acao inibidora do extrato de rizoma de *Cyperus rotundus* L. no desenvolvimento de alguns fungos. *Bol. Fac. Fil. Cien. Letras USP Botanica* (São Paulo), pp. 173–94.

Moraes, W. B. C. and Vicente, M. (1970) Estudo sobre um inibidor da germinaçaõ de sementes presente em frutos de *Solanum chlorathum* DC. *An. Acad. bras. Cien.*, **42** (suppl.), 367–9.

Nagao, E. D., Garbe, M., Utimada, H. H. *et al.* (1986) Dados preliminares do efeito do extrato aquoso de sementes de *Phytolacca thyrsiflora* na germinaçaõ e crescimento de milho e pepino. *Resumos 38th Meeting SBPC* (Curitiba), p. 1000.

Rice, E. L. (1984) *Allelopathy*, Academic Press, New York.

Rizvi, S. J. H., Jaiswal, V., Mukerji, D. and Mathur, S. N. (1980a) Antifungal properties of 1,3,7-trimethylxanthine isolated from *Coffea arabica. Naturwiss.*, **67**, 459–60.

Rizvi, S. J. H., Pandey, S. K., Mukerji, D. and Mathur, S. N. (1980b) 1,3,7-Trimethylxanthine, a new chemosterilant for stored grain pest *Callosobruchus chinensis. Z. Angew Entomol.*, **90**, 378–80.

Rizvi, S. J. H., Singh, V. K., Rizvi, V. and Waller, G. R. (1988) Geraniol an allelochemical of possible use in integrated pest management. *Plant Protect. Qly*, **3**, 112–14.

Rizvi, S. J. H., Mishra, G. P. and Rizvi, V. (1989a) Allelopathic effects of nicotine on maize. I. Its potential application in crop rotation. *Plant and Soil*, **116**, 289–91.

Rizvi, S. J. H., Mishra, G. P. and Rizvi, V. (1989b) Allelopathic effects of nicotine on maize. II. Some aspects of its mode of action. *Plant and Soil*, **116**, 292–4.

Rizzini, C. T. (1971) Germinaçaõ e inibiçaõ em cinco especies de Andira. *Rev.bras.Biol.* (Rio de Janeiro), **31**, 209–18.

Santos, D. S. B., Mello, V. D. C., Tillman, M. A. A. and Buitrago, I. (1987) Efeito alelopatico de extratos de sementes no desenvolvimento de plantulas de olericolas. *Resumos 39th Meeting SBPC* (Brasilia), p. 825.

Smith, I. K. and Fowden, L. (1966) A study of mimosine toxicity in plants. *Exp. Bot.*, **17**, 750–61.

Teles, F. F. F. (1968) Evidencia do efeito herbicida em mangueira (*Mangifera indica* L.) sobre leguminosa. *An. Soc. Bot. Brasil* (Fortaleza), 129–32.

Vilhordo, B. W., Burin, M. E., Gandolfi, V. H. *et al.* (1985) Efeito alelopatico da colza (*Brassica napus* L. var. *oleifera* Metzg.) na rotaçaõ e suceçaõ trigo e soja. *Agronomia Sulriograndense* (Porto Alegre), **21**, 55–64.

Waller, G. R. (ed.) (1987) *Allelochemicals: Role in Agriculture and Forestry*. ACS Symp. Ser. 330, Amer. Chem. Soc., Washington, DC.

CHAPTER 16

Allelopathic research activity in Iraq

I. S. Alsaadawi

16.1 INTRODUCTION

During the past three decades, extensive work in the field of allelopathy has been done. Rice (1984) reviewed most of the research works associated with the role of allelopathy in agricultural and ecological ecosystems. This field of research has drawn the attention of several scientists since there is a strong feeling that allelopathic research can be applied to so many current problems such as weed control (Putnam *et al.*, 1983; Putnam and Duke, 1974) and regulation of nitrogen cycle in soil (Alsaadawi, 1987; Alsaadawi and Rice, 1982a, b; Alsaadawi *et al.*, 1983; Putnam and Duke, 1974). Accordingly a large volume of research work has been done and published in various approaches of allelopathy.

In Iraq, considerable attention has been given to allelopathic research works with the aim of understanding the nature of some ecological and agricultural problems such as vegetational patterning, orchards replanting, crops–weeds interaction, and inhibition of nitrogen fixing and nitrifying bacteria. Most of these works have been conducted by former graduate students of Dr E. L. Rice, or by researchers influenced by him. There are also some research works on the isolation of natural products of antibacterial, fungal, nematodal and viral activities, but these will not be included in this chapter.

16.2 ROLE OF ALLELOPATHY IN FORESTRY AND HORTICULTURE

Al-Mousawi and Al-Naib (1975) found that several herbaceous species were not able to grow well and complete their life cycles under the canopy of *Eucalyptus microtheca*, although the same species grow normally under the canopy of the adjacent trees of *Casuarina cunninghamiana*. These workers, therefore, decided to determine the causative agent responsible for the pattern of herbs associated with *E. microtheca*.

Allelopathy: Basic and applied aspects Edited by S. J. H. Rizvi and V. Rizvi
Published in 1992 by Chapman & Hall, London ISBN 0 412 39400 6

Initial experiments were designed to determine if there were significant differences in soil moisture, selected minerals, light intensity, soil texture or soil pH under the canopies of *E. microtheca* and *C. cunninghamiana*. Analyses of total nitrogen, P, K, Ca, Fe, Mn and Zn indicated no differences in the two stands. The soil pH and texture were virtually the same in the two stands. The soil moisture and light intensity were significantly higher under the canopy of *E. microtheca* than under *C. cunninghamiana*. Thus, the growth reduction of the associated herbs under *E. microtheca* was not due to a deficiency in soil moisture and light intensity. Subsequent work showed that soil collected under the canopy of *E. microtheca* inhibited seed germination and growth of the test species. Aqueous extracts and decaying materials of senescent and non-senescent *E. microtheca* leaves significantly inhibited seed germination and seedling growth of the herbaceous species. Therefore, allelopathy appeared to be the basic factor responsible for the reduction of plant growth (Halligan, 1976). Additional work by Al-Naib and Al-Mousawi (1976) and Al-Mousawi and Al-Naib (1976) showed that the inhibitory compounds α-pinene, camphene, cineole, isochlorogenic, ferulic, *p*-coumaric and caffeic acids were produced by fresh leaves of *E. microtheca*. Apparently, these investigators did not attempt to isolate the inhibitors from senescent leaves or from soil under the canopy of *E. microtheca*. However, some of these inhibitors were previously isolated from *E. globulus* by Del-Moral and Muller (1969), and the importance of allelopathy as a factor in ecological process has been pointed out (Muller, 1969).

Sour orange, *Citrus aurantium* L., grows in gardens and some commercial orchards in central Iraq. Preliminary observations indicated that several herbaceous plants were not able to grow under sour orange even with repeated attempts and adequate irrigation and fertilization. The same observations were also made in a citrus orchard left for more than three years without any agricultural practices. However, these herbaceous species grow normally under the *Phoenix dactylifera* L. Therefore, we decided to determine the possible factor responsible for the failure of the herbs to grow under sour orange trees (Alsaadawi and Al-Rubeaa, 1985).

Light intensities were measured 50 times in different regions under sour orange and date palm trees, and no significant differences were evident in light intensity under both species. Thus, this factor could not be responsible for the differences in growth of the test species under sour orange and date palm. Soil moisture, pH, soil texture and selected minerals were analysed to see if sour orange causes changes in some of these soil factors which could account for the change in vegetational patterning. The soil moisture was determined from March to August 1983. The mineral analyses were made for total N, total C, and available P, K, Mg, Fe and Zn. The amounts of NH_4–N and NO_3–N were also determined. There were no significant differences in all the above

TABLE 16.1 *Effect of field soil from under sour orange trees on germination and growth of test species*

Species	Oven-dry weight of seedlings (mg)[a]		Germination (% of control)
	Control	Test	
Avena sativa	55.7	59.6	15.1
Amaranthus retroflexus	29.7	6.8b	11.6
Cynodon dactylon	37.0	5.4b	38.5
Chenopodium album	33.6	11.0b	9.1

[a] Average of at least 20 seedlings.
[b] Dry weight significantly different from control at 0.05 level (after Alsaadawi *et al.*, 1985).

test parameters under sour orange and date palm trees, and thus the reduction in growth of the herbs under sour orange was apparently not due to competition for the above test factors.

Allelopathy was therefore thought to be the remaining mechanism for the vegetational patterning, and we started several experiments to investigate the biological activities of soil under sour orange, leaf extracts and decaying leaves against seed germination and growth of the test species.

Soil collected under sour orange drastically reduced seed germination of all test species (Table 16.1). Seedling growth of all test species except *Avena sativa* was significantly inhibited by soil collected under sour orange trees. This indicates that the soil under sour orange contains allelopathic agents. Subsequent experiments showed that aqueous extracts, decaying residues and volatile materials of both senescent and non-senescent leaves were inhibitory to seed germination and/or growth of all test species. Further work in the laboratory (Alsaadawi *et al.*, 1985a) revealed that sour orange senescent and non-senescent leaves contain four phenolic phytotoxins and five volatile inhibitors, four of which were terpenes. All volatile inhibitors significantly reduced seed germination and/or seedling growth of *Amaranthus retroflexus* (Table 16.2).

The problem of replanting various fruit trees such as apple, grape, cherries, plums and others following the removal of old orchards has been recognized to be important in North America and Europe for many years. Rice (1984) pointed out that the problem is usually species specific and there was considerable involvement of allelopathy in creating the replant problem for peach and apple. Recently, we observed that the replant problem of citrus in old citrus orchards occurred even with repeated plantation, proper fertilization and other agricultural management. This problem has also been recorded and investigated in the other regions of the world. Martin (1948) suggested that biotic factors, par-

TABLE 16.2 *Germination and growth of* Amaranthus retroflexus *in atmosphere containing volatile compounds isolated from sour orange leaves*

Compound	Amount (μl)	Mean length (% of control)[a]			Seed germination (% of control)
		Radicle	Hypocotyl	Seedling	
α-Pinene	10	52.53	49.07	50.79	87.5
	20	21.21	34.25	27.73	95.8
β-Pinene	10	31.22	42.84	37.01	67.5
	20				0.0
(+)-Limonene	10	95.35b	79.20	87.28	87.5
	20	47.90	35.28	41.58	95.8
(−)-Limonene	10	68.51	31.17	49.80	100.0
	20	42.92	32.54	37.72	93.5
Octanol	2.5	24.71	30.95	27.83	87.5
	5				0.0
Citronellal	10	86.17	68.84	77.50	66.6
	20	64.69	36.71	50.70	13.3

[a] Average of at least 30 seedlings.
[b] Mean lengths were not significantly different from control at 0.05 level (after Alsaadawi *et al.*, 1985a).

ticularly soil fungi, are responsible for the citrus replant problem in North America. Burger and Small (1983) worked on this problem in South Africa and found that allelopathy is partly involved. Rice (1984) reviewed the works of these authors and concluded that many questions need to be answered. Some of these questions are concerned with the interactive effects of allelopathy and soil pathogens and the indirect effect of allelopathic agents of plant and/or microbial origins on the development of citrus replant problem. Moreover, preliminary investigation conducted in our laboratories showed results of considerable differences from those obtained by Martin. Therefore, due to the lack of actual magnitude of the mechanisms involved, we initiated several experiments to determine the contribution of allelopathy, soil microorganisms and the interaction of both in the development of the citrus replant problem.

The possible role of soil fungi and nematodes in the citrus replant problem was firstly investigated. Several pathogenic and non-pathogenic fungi and the nematode *Tylenchulus semipentrans* were isolated from soils of both declined and non-declined citrus orchards in about equal abundance (Hassan *et al.*, 1989b). However, in almost all declined trees, the pathogen *Phytophthora citrophthora* was found to have advanced into the roots until it reached the crown of the trees.

Fumigation of old citrus soil with methyl bromide caused a dramatic increase in sour orange growth. To test the effect of each fungus group and nematode on growth of sour orange, the selective pesticides nemacure, benlate and Ridomil 5G were applied to old citrus soil separately or in combination. It was found that the elimination of *Phytophthora citrophthora* increased the growth of sour orange seedling significantly. Subsequent experiment revealed that the inoculation of sterilized soil with the isolated pathogenic and non-pathogenic fungal groups separately or together caused a significant reduction in root and top growth in most cases. When the isolated pathogenic fungi were added to sterilized soil, it was found that both *P. citrophthora* and *Fusarium solani* caused a significant reduction in growth of sour orange seedlings. Martin *et al.* (1956) found that inoculation of fumigated soil with selected pathogenic fungi including *P. citrophthora* and *F. solani* did not have much influence on the growth of citrus in most cases. The one exception was inoculation with *Thielaviopsis basicola* which greatly reduced the growth of orange seedlings.

Separate experiments were made to evaluate the possible role of allelopathy in the citrus replant problem (Hassan *et al.*, 1989). The results revealed that extracts of soils collected from old citrus orchards significantly reduced the growth of sour orange seedlings (Table 16.3). This suggests that the old citrus soil contains inhibitory compounds of plants and/or microbial origin. Subsequent experiments showed that extracts of non-senescent sour orange leaves and decaying senescent leaves markedly inhibited sour orange growth. Thus, we believe that allelopathy is at least involved directly in the citrus replant problem. We did not attempt to isolate the phytotoxins from sour orange plants. However, Alsaadawi *et al.* (1985a) were able to isolate or identify several phytotoxic compounds from sour orange senescent and non-senescent leaves. Burger and Small (1983) isolated homovanillic acid, seseline, xanthyletin and two unidentified inhibitors from decomposed rough lemon roots. The first

TABLE 16.3 *Effect of extracts of citrus and non-citrus soils on growth of sour orange seedlings*

Extract	Concentration (%)	Dry weight (mg/plant)[a]			Seedling weight (% of control)
		Root	Top	Seedling	
Control		1.67	0.38	2.05	100
Old citrus soil	10	1.12	0.23	1.35	66
	15	1.30	0.28	1.58	77
Non-citrus soil	10	1.83	0.30	2.13	104
	15	2.02	0.41	2.43	119
LSD $p < 0.5$		0.53	0.08	0.59	

[a] Average of five replicates (after Hassan *et al.*, 1989a).

TABLE 16.4 *Growth and root disease index of* Citrus aurantium *treated with combination of pathogenic and non-pathogenic fungi and citrus root residues*

Treatments	Dry weight (g/plant)[a]			Root disease index
	Root	Top	Seedling	
Control	24.60	31.53	56.13	0.3
Root residue	18.90	26.20	45.10	1.5
Root residue + pathogenic fungi	17.90	23.50	41.23	1.5
Root residue + non-pathogenic fungi	20.15	27.70	47.85	1.5
Root residue + pathogenic and non-pathogenic fungi	21.30	32.10	53.40	1.8
LSD $p < 0.05$	0.59	0.85	1.25	

[a] Average of five replicates (after Hassan *et al.*, 1989c).

compound was the major growth inhibitor which stopped root cell elongation and caused root tip swelling.

Additional experiments were also conducted to investigate the interactive effects of allelopathic agents from root residues and soil fungi on the development of the citrus replant problem (Hassan *et al.*, 1989c). The results revealed that incorporation of root residues with the pathogenic and non-pathogenic fungal groups separately or together in soil caused more reduction to sour orange growth than did the root residues alone (Table 16.4). However, the highest reduction was found when the pathogenic fungal group was incorporated with root residues in soil. This suggests the occurrence of some sort of interaction. Subsequent experiments indicated that extracts of different parts of sour orange and lechates of the isolated soil fungi increased the disease index of sour orange roots grown *in vitro*. On the other hand, the citrus extracts did not affect growth of the test fungi. These results suggest that allelopathic compounds which build up in old citrus soil could also cause citrus decline by rendering the root systems more susceptible to soil pathogens as Rice hypothesized previously.

16.3 ROLE OF ALLELOPATHY IN CROPS–WEEDS INTERACTION

Recently, the recognized importance of allelopathy in agricultural practices has been accelerated with the main goal of using this phenomenon in biological control (Alsaadawi and Rice, 1982a; Putnam *et al.*,

1983; Leather, 1983; Rice, 1984). One approach for utilizing this phenomenon is to screen various accessions of allelopathic crops for their ability to reduce weeds, and a few crops have been evaluated in this aspect (Putnam and Duke, 1974; Fay and Duke, 1977; Leather, 1982).

Sorghum bicolor has been known to provide a great weed-killing capacity (Putnam *et al.*, 1983) and some research works showed that this crop is autotoxic and should be rotated with other crops to maximize crop productivity (Leon, 1976; Hussain and Gadoon, 1981). Recently, more than 100 cultivars have been introduced to our centre for various purposes, and it was hypothesized that genetic variations for allelopathic traits might occur among these cultivars. Therefore, we started a programme to screen 100 cultivars of grain sorghum introduced from Bulgaria for their allelopathic activity against weeds.

Screening experiment was conducted to examine the activity of root exudates of the test cultivars to inhibit the germination and seedling growth of *Amaranthus retroflexus*. The experiment was run under controlled conditions using sand culture medium (Alsaadawi *et al.*, 1985b). The results exhibited a great variability in the ability of the test cultivars to alter seed germination and/or seedling growth of the test species. Seed germination was reduced by approximately 70% of the control in 11 cultivars and 82% of the control in 25 cultivars. The other cultivars caused no significant reduction in seed germination. The growth range was 77–113% of control. Ten cultivars, some of them economically important (CV. 219, 260), inhibited *A. retroflexus* growth by more than 79% of control (Table 16.5), and 25 cultivars by approximately 85% of control. The others did not show significant inhibition or stimulation in seedling growth. Additional experiment was conducted to collect and characterize the root exudates of one of the most phytotoxic cultivars. The system used for collection and characterization of root exudates was

TABLE 16.5 *Allelopathic effect of root exudates of selected* Sorghum bicolor *cultivars on seed germination and seedling growth of* Amaranthus retroflexus

	Mean length (mm)[a]			
Cultivar	Radicle	Hypocotyle	Seedling	Seed germination (% of control)
Control	15.8a	20.6b	36.4b	100
219	10.8c	18.8c	29.6c	70
260	11.9c	17.5c	29.4c	72
177	13.8b	25.1a	38.9a	104
264	14.4b	25.4a	39.8a	98

[a] Average of at least 20 seedlings. Means within column followed by the same letter are not significantly different according to Duncan's multiple-range test (after Alsaadawi *et al.*, 1985b).

TABLE 16.6 *Inhibitory effects of different fractions of root exudates from* Sorghum bicolor *cv. 219 on seedling growth of* Amaranthus retroflexus

Extract fraction	Radicle length (mm)			Hypocotyl length (mm)			Seed germination (% of funnel control)
	Distilled water	Funnel control	Root exudates	Distilled water	Funnel control	Root exudates	
Ethanol							
Acidic, pH 2.0	22.35	21.64	10.87b	7.54	6.64a	6.96	104
Neutral, pH 6.8	22.35	15.65a	8.29b	7.54	5.50a	4.47b	81
Basic, pH 12.0	22.35	22.36	22.32	7.54	6.71a	6.13	110
Acetone	22.10	19.43a	19.12	8.76	7.04a	5.60b	96

[a] Significantly less than water control at 0.05 level by Student's *t* test.
[b] Significance compared with pot control at 0.05 level by Student's *t* test (after Alsaadawi *et al.*, 1985b).

basically that of Tang and Young (1982) with slight modification. It was found that neutral fraction of root exudates was more inhibitory than the acidic and basic fractions (Table 16.6). This result is particularly important since the pH used is approximately close to the pH of the field soil. Tang and Young (1982) found that the neutral fraction of *Hemarthria altissima* root exudates showed greater toxicity than the basic and acidic fractions.

Additional investigation demonstrated that cultivars, which were found to be non-toxic through root exudation, exhibited considerable toxicity through aqueous extracts and decaying materials. However, the degree of phytotoxicity is less than those of toxic root exudates. Thus, it appears that some cultivars of *Sorghum bicolor* has a potential inhibitory activity against weeds and such a phenomenon could significantly reduce the requirement for commercial herbicides.

Subsequently, research activity was carried out to see if the low doses of gamma radiation can increase the allelopathic potential of sorghum against weeds (Alsaadawi *et al.*, 1986). From previous studies (Riov *et al.*, 1969; Koeppe *et al.*, 1970) radiation was known to increase phytotoxin production in plants. Our results showed that plants originated from seeds exposed to 0.5, 1.0 and 1.0 krad of gamma rays have more inhibitory activity in their root exudates, aqueous extracts, and decaying residues against growth of *Amaranthus retroflexus*. This result may open a new possible approach for increasing the allelopathic potential of some crops without causing genetic damages.

Based on the above-mentioned information, *S. bicolor* has a potential inhibitory activity against weed growth. The inhibitory compounds occurring in sorghum plants are mostly phenols (Guenzi and McCalla, 1966). The mechanisms of action of some of these phenolic compounds are known (Rice, 1984), while the mechanisms of action of the others such as syringic, caffeic and protocatechuic acids have not been investigated. Experiments were, therefore, started to investigate the possible effect of these phenolic acids on chlorophyll content and ion uptake as suggested mechanisms in plants (Alsaadawi *et al.*, 1985b). These two parameters are known to limit plant growth (Colton and Einhellig, 1980; Epestein, 1976). Three concentrations (10^{-4}, 5×10^{-4}, 10^{-3} M) of each acid were used to test their effect on cowpea (*Vigna sinensis*) seedlings using sand culture medium. Seedling growth, chlorophyll a and b, total chlorophyll, chlorophyll a/b ratio, and the uptake of N, P, K, Fe and Mo were significantly reduced by most of the test concentrations of the phenolic acids. However, chlorophyll b content and Mg uptake were not significantly inhibited by 5×10^{-4} M and 10^{-3} M of caffeic acid and 5×10^{-4} M of protocatechuic acid. In most cases, the reduction in dry weight was parallel to the reduction in chlorophyll content and ion uptake, and the reduction in chlorophyll was also parallel to the reduction in ion uptake.

Euphorbia prostrata L., which is locally named as cancer of bermudagrass (*Cynodon dactylon*), is a perennial weed in bermudagrass lawns in Iraq. The weed often expanded the size of its stand from a few plants to a large area in a few months. The expansion occurred in spite of a heavy stand of bermudagrass. As *E. prostrata* spreads, it almost eliminates the bermudagrass and occurred virtually in pure stands.

The rapid spread of *E. prostrata* into bermudagrass sod suggested the possibility of an allelopathic mechanism in addition to competition. Therefore, we initiated several experiments to test whether *E. prostrata* is allelopathic to bermudagrass and selected crop and weed species, and to obtain preliminary information about the possibility of using *E. prostrata* in biological control of weeds in some crops.

The results indicated that the rapid invasion of *E. prostrata* into heavy sods of bermudagrass is apparently not due to changes in pH, electrical conductivity, minerals and soil texture (Alsaadawi *et al.*, 1989). Also, the observed vegetational patterning could not be attributed to soil moisture since it commonly occurs in bermudagrass lawns and gardens that have been customarily irrigated. Light could not account for the vegetational patterning either because *E. prostrata* is very tiny and does not shade light appreciably. However, soil collected from *E. prostrata* stands at the end of its life cycle and stored until the beginning of the growing season significantly inhibited germination and growth of bermudagrass. This suggests that toxins are released from *E. prostrata* and remain stable for enough time to affect growth of bermudagrass, and also it probably explains why bermudagrass plants are not observed the following season in the area occupied by *E. prostrata*. Subsequent experiments indicated that toxins are water soluble and can be released from *E. prostrata* by decomposition of plant residues and root exudations. It is significant that only a small amount of *E. prostrata* residue is quite enough to affect the growth of bermudagrass significantly. The accumulation of these small amounts with time may have more detrimental effect on bermudagrass growth at the end of the season, and these amounts apparently remain active to prevent seed germination and growth of bermudagrass in *E. prostrata* stands the following season.

Thus, it appears from the previous experiments that allelopathy is the major factor responsible for the patterning of vegetation observed in this study. However, once germination and growth of bermudagrass are inhibited by allelopathic compounds released from *E. prostrata*, competition undoubtedly accentuates the growth inhibition. Preliminary work indicated that the aqueous extract of *E. prostrata* contained some inhibitory compounds phenolic in nature.

Our results showed that *E. prostrata* is more allelopathic to bermudagrass and *A. retroflexus* than to *M. sativum* and cotton. This suggests that residues or living plants of *E. prostrata* may prove useful to control bermudagrass and *A. retroflexus* in certain crops. It is too early to suggest a specific methodology; however, such results may add additional information about the possibility of using allelopathic plant species in weed control.

Dodder (*Cuscuta* spp.) are important parasitic weeds that attack many legumes including alfalfa. The close attachment of this weed to host and its ability to produce a large number of viable seeds make its control difficult. Ashton (1976) indicated that mechanical, cultural and chemical methods are used to control dodder. However, herbicides are the most promising but their usage depends on whether alfalfa is cultivated for seeds or for animal feed.

Recently, Abdul-Rahman and Habib (1986) conducted a lathhouse experiment to investigate the phytotoxic activity of extracts of bermudagrass *Cynodon dactylon*, Johnson grass *Sorghum halepense*, thumble pigweed *Amaranthus albus* against doddor *Cuscuta campestris* on alfalfa after attachment. Different concentrations of each of the herbicides dimethyltetrachloroterephthalate (DCPA), glyphosate and metribuzin were also included in the study for comparison. It was found that the test concentrations of all weeds controlled dodder plant significantly. The percent of control increased with the increase in extract concentration (Table 16.7), while, the increase in percent of control was parallel with the increase of alfalfa growth. The potential of weed extracts in controlling dodder parasite was the same as DCPA and glyphosate did. The highest concentration of bermudagrass extract was as effective as the higher dose rate of metribuzin but showed less alfalfa injury. In most tests, the weed extracts exhibited less alfalfa injury than the other test herbicides. Subsequent experiment revealed that the extracts of the test weeds contained several phenolic compounds, and apparently bermudagrass contained the highest quantity of total phenols while the tumble pigweed contained the lowest.

Habib and Abdul-Rahman (1988) extended this work by testing the phytotoxic potential of the previous weeds plus wall goosefoot weed *Chenopodium murale* in controlling the dodder parasite under both lathhouse and field conditions. The test concentration of each weed was applied after the dodder stem and tendril covered the alfalfa plants. The test concentrations of extracts of all weeds significantly increased the control of dodder parasite (Table 16.8). Bermudagrass and wall goosefoot weed extracts showed the highest ability to control dodder plant. Meanwhile, extracts of these weeds exhibited a high percentage of alfalfa injury when compared with the extracts of the other two weeds. The data of lathhouse and field experiments exhibited a higher degree of consistency. Analyses of total phenols indicated that the higher toxicity of bermudagrass was correlated with a higher amount of phenols produced by this weed. However, this is not the case with the other weeds. It is possible that the differential amounts of particular phytotoxin(s) may be responsible for such potential phytotoxicity. Thus, it would be fruitful to

TABLE 16.7 *Dodder control and alfalfa injury and dry weight as affected by different herbicidal and plant extract treatments*

Treatments	Cencentrations[a,b]	Dodder control (%)	Alfalfa injury (%)	Alfalfa dry weight (g plant^{-1})
DCPA	6	68.85bcd	10e	0.749a
	8	68.85bcd	10e	0.754a
	10	83.85ab	10e	0.794a
Glyphosate	0.075	54.05de	10.00e	0.377d
	0.100	76.86abc	30.00c	0.507c
	0.150	79.54abc	45.25c	0.761a
Metribuzin	0.15	90a	90a	0.585bc
	0.25	90a	90a	0.567bc
	0.50	90a	90a	0.545bc
Bermudagrass	0.05	10f	6.61e	0.17e
	0.25	75abc	10.00e	0.60bc
	0.50	90a	20.00d	0.80a
Johnsongrass	0.05	66.0bcd	10.00e	0.237e
	0.25	76.6abc	6.66e	0.552bc
	0.50	83.3ab	10.00e	0.745a
Tumblepigweed	0.05	40e	3.3f	0.224e
	0.25	71bc	10.0e	0.554bc
	0.50	80abc	10.0e	0.633ab
Control	0	0f	0f	0.204e

[a] Rate of herbicide in kg a.i. ha^{-1}.
[b] Rate of plant extract in gram crude plant material cm^{-3} water (after Abdul-Rahman and Habib, 1986).

evaluate the phytotoxic ability of each phenolic compound separately and in combination, in order to determine the limiting factor(s) responsible for such important inhibitory action. The phenolic compounds isolated in this study have been recognized as major phytotoxins in allelopathic plant species. The way in which these extracts enter the dodder plant remained unknown.

The allelopathic effects of weed species on crop plants has also been studied for various reasons. One of these reasons is to understand the causative factors responsible for the strong interference of some weeds with various crops in agricultural lands and with weeds under natural conditions. For example, *Imperata cylindrica* is an important weed in Iraq. Khalil (1973) observed that this weed appeared in pure stands

TABLE 16.8 *Effect of weed extracts on field dodder and alfalfa under field conditions (after Habib and Abdul-Rahman, 1988)*

Weed extract	Concentrations (g ml^{-1} water)	Dodder control (%)	Dodder dry weight (g plant^{-1})	Alfalfa injury (%)	Alfalfa dry weight (g plant^{-1})
Tumble pigweed	0.00	00.0	0.437	0.00	0.256
	0.05	23.3	0.100	0.00	0.542
	0.25	60.00	0.859	3.30	0.369
	0.50	83.30	0.015	6.60	0.885
Wall goosefoot	0.00	00.00	0.413	00.00	0.212
	0.05	50.00	0.070	20.00	0.365
	0.25	73.3	0.050	20.00	0.674
	0.50	96.60	0.013	43.3	0.937
Bermudagrass	0.00	00.00	0.416	00.00	0.237
	0.05	23.30	0.072	00.00	0.234
	0.25	73.30	0.034	35.00	0.527
	0.50	96.60	0.014	43.30	0.898
Johnsongrass	0.00	00.00	0.408	0.00	0.208
	0.05	26.60	0.074	10.0	0.335
	0.25	36.60	0.051	10.0	0.797
	0.50	76.60	0.016	10.0	0.797
LSD ($p < 0.05$)		6.9	0.036	4.07	0.051

shortly after invasion of the fields or natural areas and almost no plant could grow within its stand. It was found that aqueous extracts of rhizomes significantly inhibited the growth of *Amaranthus viridis*, *Beta vulgaris*, *Gossypium hirsutum*, *Medicago sativa* and *Lycopersicon esculentum*. Chromatographic analysis showed that the rhizome extract contains chlorogenic acid, catechol, *p*-coumaric acid and *p*-hydroxybenzaldehyde. Apparently, no attempt was made to evaluate the phytotoxicity of the top plant materials. However, Abdul-Wahab and Al-Naib (1972) isolated scopolin, scopoletin, chlorogenic and isochlorogenic acids from leaves and clums of *Imperata cylindrica*. The allelopathic effects of this weed against some crops have also been confirmed by Eussen (1978) and Eussen and Soerjani (1978).

Abdul-Rahman and Al-Naib (1986) observed that bermudagrass strongly interfered with cotton plants in the field. The presence of such interference led these workers to hypothesize that allelopathy may be involved. Therefore, they conducted several experiments to test this hypothesis. Their results showed that aqueous extracts and root exudates of bermudagrass significantly inhibited germination and growth of cotton and the weeds prosopis *Lagongchium farctum*, Johnson grass and cocklebur *Xanthium strumarium* which also occurred in the cotton field. Chemical analyses indicated the presence of numerous inhibitory phenolic compounds.

16.4 ALLELOPATHIC SUPPRESSION OF NITRIFICATION BY SELECTED CROPS

The nitrification process is important for the regulation of nitrogen leaching in soils. Such leaching may severely reduce the productivity of plants, cause great losses in the energy requirement for growing crops and increase the water pollution and disease incidence of plants (Huber *et al.*, 1977). In an attempt to enhance the utilization of nitrogen fertilizers, synthetic nitrification inhibitors have been used extensively (Alexander, 1977). The inhibition of nitrification by allelopathic plant species has also been reported in natural and agricultural ecosystems (Rice and Pancholy, 1972, 1973, 1974; Lodhi, 1981; Santhi *et al.*, 1986). The presence of a potential nitrification inhibitor crop may be useful to regulate the leaching of fertilizer nitrogen and thus increase the productivity of plants. Therefore, we started to evaluate four cultivars of *Sorghum bicolor* and six cultivars of *Helianthus annuus* against nitrification. These plant species are known to be allelopathic to several weed species (Leather, 1982; Hussain and Gadoon, 1981). The effects of aqueous extracts and decaying residues of both cultivars on nitrification were evaluated using soil incubation methods. Aqueous extracts and decaying residues of the test sorghum cultivars significantly inhibited the nitrification process (Alsaadawi *et al.*, 1985b). Some of the cultivars that were more toxic to weeds exhibited more toxicity to the nitrification process. The potential inhibition of these cultivars to nitrification was found to increase when the seeds of these cultivars were exposed to low doses of gamma irradiation (Alsaadawi *et al.*, 1986b). In further experiments, water extracts, and decaying residues of roots and shoots of the test sunflower cultivars were also found to reduce nitrification significantly (Alsaadawi, 1988). When the allelopathic potential of root and shoot bioassay experiments were pooled and presented against incubation periods to reflect the allelopathic activity of each sunflower cultivar, it was found that cultivars citosol and local showed more inhibition to nitrification than the others (Table 16.9).

The presence of allelopathic potential in sorghum and sunflower against nitrification may help to augment nitrogen use efficiency of the added fertilizer. The data provide information on the possible uses of allelopathic crops in no-tillage cropping systems to control weeds and regulate the nitrification process. Although it is too early to make any definite conclusions before field evaluation on this aspect is made, such findings should be taken into consideration since the nitrification inhibition by

TABLE 16.9 *Allelopathic potential of different sunflower cultivars against nitrification (number of observations = 96[a])*

Cultivar	Intercept (a)	Regression coefficient (b)[b]	Correlation coefficient of determination (r^2)[c]
Control	1.45	5.93	0.973
Cheremn	−0.68	5.33	0.970
Peredovik	−0.25	5.26	0.973
Record	−0.05	5.23	0.966
Citosol	−0.05	5.07	0.968
Local	−0.23	5.03	0.963

[a] Nitrification and incubation periods are considered as dependent and independent variables respectively
[b] Significant at $p < 0.01$ level according to t test.
[c] Significant at $p < 0.01$ level (after Alsaadawi, 1988).

allelopathic plants such as neem has been previously found to be fruitful and practical under field conditions (Sahrawat and Parmar, 1975; Santhi *et al.*, 1986).

16.5 CONCLUSIONS AND PROSPECTS

The allelopathic phenomenon has been previously demonstrated as a causative agent of several ecological phenomena such as species distribution pattern, succession and others. Recently, this phenomenon has been a subject of many scientists world wide since there is a strong belief that it offers unlimited opportunities to solve several agricultural problems.

There is a common feeling that we are on the threshold of some exciting discoveries in using the allelopathic idea in weed management. I believe that in the near future we shall be able to breed crop plants for allelopathic potential against important weeds just as we breed plants for disease resistance. This can be achieved if more screening programmes are carried out on the germplasm collection of crops and their wild type relatives.

I also predict we shall soon be using mulches consisting of residues of allelopathic weeds or crop plants to control weeds and inhibit soil nitrification. The research activities in this approach will rapidly develop and may add other advantages to the use of no-tillage practices in agriculture.

There will also be a special consideration for researches concerning the redesigning of crop rotation on the basis of allelopathic phenomenon.

This type of research would be carried out to minimize the negative impact of allelopathic mechanisms on crops and exploit it to control weeds and other pests.

The future looks bright for identifying new classes of pesticides and growth regulators from allelopathic plants to replace the synthetic dangerous chemicals used at present. The cooperation of chemists, plant physiologists, microbiologists, ecologists and perhaps others is necessary to achieve maximum progress in this important field of research.

ACKNOWLEDGEMENTS

I wish to thank Drs H. A. Jaddou, F. A. Fattah, and J. K. Al-Uqaili for their helpful reviews of an earlier draft of this chapter. Special thanks to Nidhal S. Al-Dilimi for her rapid typing of the manuscript.

REFERENCES

Abdul-Rahman, A. A. S. and Al-Naib, F. A. G. (1986) The effects of bermudagrass *Cynodon dactylon* (L.) Pers. on the germination and seedling growth of cotton and three weed species. *JAWRR*, **5**, 115–28.

Abdul-Rahman, A. A. S. and Habib, S. A. (1986) Effectiveness of herbicides and some plant extracts in controlling dodder on alfalfa. *JAWRR*, **5**, 53–64.

Abdul-Wahab, A. S. and Al-Naib, F. A. G. (1972) Inhibitional effects of *Imperata cylindrica* (L.). *Bull. Iraq Nat. Hist. Mus.*, **5**, 17–24.

Alexander, M. (1977) *Soil Microbiology*, Wiley, New York.

Al-Mousawi, A. H. and Al-Naib, F. A. G. (1975) Allelopathic effects of *Eucalyptus microtheca* F. Muell. *J. Univ. Kuwait*, **2**, 59–66.

Al-Mousawi, A. H. and Al-Naib, F. A. G. (1976) Volatile growth inhibitors produced by *Eucalyptus microtheca*. *Bull. Biol. Res. Center*, **7**, 17–23.

Al-Naib, F. A. G. and Al-Mousawi, A. H. (1976) Allelopathic effects of *Eucalyptus microtheca*: Identification and characterization of phenolic compounds in *Eucalyptus microtheca*. *J. Univ. Kuwait*, **3**, 83–8.

Alsaadawi, I. S. (1987) Phytotoxin resistance by *Rhizobium meliloti* mutants. *Plant and Soil*, **102**, 279–82.

Alsaadawi, I. S. (1988) Biological suppression of nitrification by selected cultivars of *Helianthus annuus* L. *J. Chem. Ecol.*, **14**, 722–42.

Alsaadawi, I. S. and Al-Rubeaa, A. J. (1985) Allelopathic effects of *Citrus aurantium* L. Vegetational patterning. *J. Chem. Ecol.*, **11**, 1515–33.

Alsaadawi, I. S. and Rice, E. L. (1982a) Allelopathic effects of *Polygonum aviculare* L. I. Vegetational patterning. *J. Chem. Ecol.*, **8**, 993–1009.

Alsaadawi, I. S. and Rice, E. L. (1982b) Allelopathic effects of *Polygonum aviculare* L. II. Isolation, characterization and biological activities of phytotoxins. *J. Chem. Ecol.*, **8**, 1011–23.

Alsaadawi, I. S., Rice, E. L. and Karns, T. K. B. (1983) Allelopathic effects of *Polygonum aviculare* L. III. Isolation, characterization and biological activities of phytotoxins other than phenols. *J. Chem. Ecol.*, **9**, 761–74.

Alsaadawi, I. S., Arif, M. B. and Al-Rubeaa, A. J. (1985a) Allelopathic effects of *Citrus aurantium* L. II. Isolation, characterization and biological activities of phytotoxins. *J. Chem. Ecol.*, **11**, 1527–34.

Alsaadawi, I. S., Al-Uqaili, J. K., Al-Rubeaa, A. J. and Al-Hadithy, S. M. (1985b) Effect of gamma irradiation on allelopathic potential of *Sorghum bicolor* against weeds and nitrification. *J. Chem. Ecol.*, **12**, 1737–45.

Alsaadawi, I. S., Al-Hadithy S. M. and Arif, M. B. (1986a) Effects of three phenolic acids on chlorophyll content and ion uptake in cowpea seedlings. *J. Chem. Ecol.*, **12**, 221–7.

Alsaadawi, I. S., Al-Uqaili, J. K., Al-Rubeaa, A. J. and Al-Hadithy S. M. (1986b) Allelopathic suppression of weed and nitrification by selected cultivars of *Sorghum bicolor* (L.) Moench. *J. Chem. Ecol.*, **12**, 209–19.

Alsaadawi, I. S., Sakeri, F. A. and Al-Dilimi, S. M. (1990) Allelopathic inhibition of *Cynodon dactylon* (L.) Pers. and other plant species by *Euphorbia prostrata* L. (submitted).

Ashton, F. M. (1976) *Cuscuta* spp. (dodder): A literature review of its biology and control. *University of California Bulletin No. 1880*, Davis, California, USA, p. 22.

Burger, W. P. and Small, J. G. C. (1983) Allelopathy in citrus orchards. *Scientia Horticulturae*, **20**, 361–75.

Colton, C. E. and Einhellig, F. A. (1980) Allelopathic mechanisms of velvetleaf (*Abutilon theophrasti* Medic. Malvaceae) on soybean. *Amer. J. Bot.*, **67**, 1407–13.

Del-Moral, R. and Muller, C. H. (1969) Fog drip: A mechanism of toxin transport from *Eucalyptus globulus*. *Bull. Torrey Bot. Club*, **96**, 467–75.

Epestein, E. (1976) *Mineral Nutrition of Plants: Principle and Perspectives*, Wiley, New York.

Eussen, J. H. H. (1978) Isolation of growth inhibiting substances from alang-alang [*Imperata cylindrica* (L.) Beauv.], in *Studies on Tropical Weed* Imperata cylindrica (L.) *Beauv. var.* major (ed. J. H. H. Eussen), Durkerij Elinkwijk BV, Utrecht.

Eussen, J. H. H. and Soerjani, M. (1978) Allelopathic activity of alang-alang [*Imperata cylindrica* (L.) Beauv.], in *Studies on Tropical Weed* Imperata cylindrica *(L.) Beauv. var.* major (ed. J. H. H. Eussen), Durkeij Elinkwijk BV, Utrecht.

Fay, P. K. and Duke, W. B. (1977) An assessment of allelopathic potential in *Avena g=rmplfism. 6f2W·ed S?i* b325, 224–8.

Guenzi, W. and McCalla, T. M. (1966) Phenolic acids in oat, wheat, sorghum, and corn residues and their phytotoxicity. *Agron. J.*, **58**, 303–4.

Habib, S. A. and Abdul-Rahman, A. A. (1988) Evaluation of some weed extracts against field dodder on alfalfa (*Medicago sativa*). *J. Chem. Ecol.*, **14**, 443–52.

Halligan, J. P. (1976) Toxicity of *Artemisia californica* to four associated herb species. *Amer. Midl. Nat.*, **95**, 406–21.

Hassan, M. S., Alsaadawi, I. S. and El-Behadli, A. H. (1989a) Citrus replant problem in Iraq. II. Possible role of allelopathy. *Plant and Soil*, **116**, 157–60.

Hassan, M. S., El-Behadli, A. H. and Alsaadawi, I. S. (1989b) Citrus replant problem in Iraq. I. Possible role of some soil microorganisms. *Plant and Soil*, **116**, 151–5.

Hassan, M. S., El-Behadli, A. H. and Alsaadawi, I. S. (1989c) Citrus replant problem in Iraq. III. Interactive effect of soil fungi and allelopathy. *Plant and Soil*, **116**, 161–6.

Huber, D. W., Warren, H. L., Nelson, D. W. and Tasi, C. Y. (1977) Nitrification inhibitors: New tools for food production. *Bioscience*, **27**, 523–9.

Hussain, F. and Gadoon, M. A. (1981) Allelopathic effects of *Sorghum vulgare* Pers. *Oecologia (Berl.)*, **51**, 284–8.

Khalil, A. I. (1973) Allelopathic effects of *Imperata cylindrica* rhizomes on some cultivated and wild Iraqi plants. MSc Thesis, College of Science, Univ. of Baghdad, Iraq.

Koeppe, D. E., Rohrbaugh, L. M., Rice, E. L. and Wender, S. H. (1970) The effect of X-radiation on the concentration of scopolin and caffeoylquinic acids in tobacco. *Radia. Bot.*, **10**, 261–5.

Leather, G. R. (1982) Sunflowers (*Helianthus annuus*) are allelopathic to weeds. *Weed Sci.*, **31**, 37–42.

Leather, G. R. (1983) Weed control using allelopathic crop plants. *J. Chem. Ecol.*, **9**, 983–90.

Leon, W. B. (1976) Phytotoxicité induite par les résidus de recolte de *Sorghum vulgare* dans les sols sableux d'ouest African. *Doctoral thesis, Université de Nancy, France*.

Lodhi, M. A. K. (1981) Accelerated soil mineralization, nitrification, and revegetation of abandoned fields due to the removal of crop–soil phytotoxicity. *J. Chem. Ecol.*, **7**, 685–94.

Martin, J. P. (1948) Fungus flora of some California soils in relation to slow decline of citrus trees. *Proc. Soil Sci. Soc. Amer.*, **12**, 209–14.

Martin, J. P., Klotz, L. J., De Wolfe, J. A. and Ervins, J. O. (1956) Influence of some common soil fungi on growth of citrus seedlings. *Soil Sci.*, **81**, 259–67.

Muller, C. H. (1969) Allelopathy as a factor in ecological process. *Vegetatio. Haag*, **18**, 348–57.

Putnam, A. R., Defrank, J. and Barnes, J. P. (1983) Exploitation of allelopathy for weed control in annual and perennial cropping systems. *J. Chem. Ecol.*, **9**, 1001–11.

Putnam, A. R. and Duke, W. B. (1974) Biological suppression of weeds: evidence for allelopathy in accessions of cucumber. *Science*, **185**, 370–2.

Rice, E. L. (1984) *Allelopathy*, Academic Press, New York.

Rice, E. L. and Pancholy, S. K. (1972) Inhibition of nitrification by climax ecosystems. *Amer. J. Bot.*, **59**, 1033–40.

Rice, E. L. and Pancholy, S. K. (1973) Inhibition of nitrification by climax ecosystems. II. Additional evidence and possible role of tannins. *Amer. J. Bot.*, **60**, 691–702.

Rice, E. L. and Pancholy, S. K. (1974) Inhibition of nitrification by climax ecosystems. III. Inhibitors other than tannins. *Amer. J. Bot.*, **61**, 1095–103.

Riov, J., Monselise, S. P. and Kahan, R. S. (1969) Ethylene-controlled induction of phenylalanine ammonia-lyase in citrus fruit peel. *Plant Physiol.*, **44**, 631–5.

Sahrawat, K. L. and Parmar, B. S. (1975) Alcohol extract of neem (*Azadirachta indica* L.) seed as nitrification inhibitor. *J. Indian Soc. Soil Sci.*, **23**, 131–4.

Santhi, S. R., Palaniappan, S. P. and Purushothaman, D. (1986) Influence of neem leaf on nitrification in a lowland rice soil. *Plant and Soil*, **93**, 133–5.

Tang, C. S. and Young, C. C. (1982) Collection and identification of allelopathic compounds from undisturbed root system of Bigalta limpograss (*Hermarthria altissima*). *Plant Physiol.*, **69**, 155–60.

CHAPTER 17

Impact of allelopathy in the traditional management of agroecosystems in Mexico

A. L. Anaya, R. C. Ortega and V. Nava Rodriguez

17.1 INTRODUCTION

Some of the consequences of extensive agriculture are the impoverishment of the cultivated and wild germplasm, and the disappearance of hundreds of species from natural ecosystems that have been destroyed to open agricultural spaces. On the other hand, the modernization of this kind of agriculture, with the use of machinery, herbicides, fungicides and insecticides, that are frequently and indiscriminately applied in different environments, has contributed to the uncontrolled advance of pollution and erosion.

Considering all this, research on the traditional management of resources is, in fact, a rediscovery of the rational use of nature and the search for ideal practices of organic agriculture that can give us the chance to find appropriate technologies for sustainable production without damage to the environment.

The management of resources in traditional agroecosystems represents the sum of multiple experiences and numerous assays realized day by day in an empirical way in the field. These practices have originated from different techniques adapted to local conditions and customs. In Mexico, for example, corn has been cultivated for thousands of years and has been one of the main food sources for the indigenous people and for all Mexicans up to the present time. In many parts of the country, agricultural practices are based on the combination of corn with other crops, each one with different needs along different periods of time. Due to the topographic diversity in Mexico and the consequent ecological diversity, it is possible to imagine all the various agricultural practices that arose around corn farming, for instance in such magnificent cultures as Maya, Teotihuacan, Purepecha and Aztec. When the Spanish conquerors came to Mexico, all the agricultural techniques of fifteenth-century Spain were

Allelopathy: Basic and applied aspects Edited by S. J. H. Rizvi and V. Rizvi
Published in 1992 by Chapman & Hall, London ISBN 0 412 39400 6

brought to Mexico. Since that time, agriculture in Mexico has developed through the combination of both indigenous and Spanish techniques, which were improved by their mutual contact and which constitute the current traditional agriculture in Mexico.

A great part of the fundamental 'genetic belt' (the term applied to describe the genetic richness that surrounds the Earth between the Tropics of Cancer and Capricorn) is located in Mexico. Mexico, having an area of 1 972 544 km^2, is the fourteenth largest country of the world, and the fourth in biotic diversity. There are more than 2000 genera of angiosperms, and 30 000 species with a high degree of endemism; there exist 8416 species of vertebrates and thousands of invertebrates which have not yet been completely censused (data of the Instituto de Biologia of UNAM).

All this reinforces the need to look for new options of production, congruent to local, ecological, economical and cultural conditions, to preserve the extraordinary biotic richness that characterizes Mexico.

Nowadays most of the studies on allelopathy are focused on agricultural problems, and, in our country – especially when dealing with traditional agriculture, because of its obvious applications in the management of biotic resources – control of weeds, insects and pathogens. So, the useful considerations of allelopathy are justified not only because of the potential industrial use of secondary metabolites, but also because of the facility to apply biological control in agroecosystems, through the promotion or elimination of those plants, animals and microorganisms whose ecological role is well known and has been studied in depth (Farnsworth, 1977).

Allelopathy within this context can contribute to the particular management of species based on the knowledge of potential chemical interactions among organisms and of the chemical compounds that mediate them.

17.2 ALLELOPATHIC RESEARCH IN MEXICO

17.2.1 Tropical zones

(a) Studies in Los Tuxtlas, Veracruz

The first studies in allelopathy in Mexico were carried out in natural ecosystems, mainly in the secondary communities that replace the rain forest when it is destroyed. In 1970, the studies on allelopathy in the tropical zones were scarce, particularly in Mexico. The contributions of Müller and Müller (1956), Marinero (1962), Müller (1966, 1969, 1970), Frei and Dodson (1972), Webb *et al.* (1967), Gliessman and Müller (1972) and Quaterman (1973), provided an important background for the study that was carried out concerning the allelopathic potential of the secondary vegetation in Veracruz, Mexico (Anaya, 1976), at the biological station of the National Autonomous University of Mexico (UNAM), located at Los Tuxtlas, Veracruz, in the coastal plateau of the Gulf of Mexico. Almost all species studied inhibited the growth of test seedlings. The species with higher allelopathic potential were: *Piper auritum*, *P. hispidum*, *Croton pyramidalis* and *Siparuna nicaraguensis*. The essential oils of the Piperaceae were highly inhibitory to seed germination and growth of seedlings, while that of *Croton pyramidalis* was less inhibitory and also produced stimulations. Safrol constitutes 60 or 70% of the essential oil of *P. auritum* (Collera, 1956), but α- and β-pinene, scarcely present in this oil, exhibited a major allelopathic activity on the test seeds.

As part of the same project, a study of one of the most common weeds in some disturbed habitats, also in Los Tuxtlas, was carried out by Anaya and del Amo (1978). *Ambrosia cumanensis* was found as an important component of the ruderal vegetation. It grows vigorously and in almost pure stands, so we decided to assess its allelopathic potential in order to evaluate this phenomenon as a determining factor for the structure of the community as well as in the secondary succession process.

Root and leaf aqueous leachates of *Ambrosia cumanensis* produced a strong inhibition on the growth of some weed species. The aqueous solutions of soil collected under *A. cumanensis* in July (during its flowering) were strongly allelopathic to weed growth. Decomposition of leaves and roots in pots also caused an inhibition to some weeds. Microorganisms have a major role in this process, as shown by results from sterile and non-sterile soils (Table 17.1).

Bioassays with several sesquiterpenic lactones from *A. cumanensis* showed that these compounds produced different effects (stimulatory and inhibitory) on the germination and growth of several species of the secondary vegetation (del Amo and Anaya, 1978). Therefore, it is possible that the allelopathic potential of *A. cumanensis* contributes to the self-control of its population by preventing the growth of seedlings of its own species.

The information obtained from this project at Los Tuxtlas shows that some species from the secondary vegetation produce one or more substances, mainly in leaves or through the decomposition of their organic matter, that can inhibit the growth or cause other damages on other plants (Ramos Prado, 1991), and may cause parallel effects that are related to the tropisms, role of auxins, pigments synthesis and to other metabolic processes.

In relation with other types of compounds found in Piperaceae, Holyoke and Reese (1987) mention that unsaturated isobutylamides offer protection to the plants which produce them at relatively low concentrations, also they are the most appealing natural product model for development of new insecticides from pyrethroids.

TABLE 17.1 *Percent of inhibition or stimulation produced by leaves and roots of* Ambrosia cumanensis *incorporated into sterile and non-sterile soil in pots, on the growth of some secondary species of a tropical zone of Mexico (after Anaya and del Amo, 1978).*

Part of *Ambrosia* in soil	Test species													
	Ambrosia cumanensis		*Mimosa pudica*		*Bidens pilosa*		*Ochroma lagopus*		*Mimosa somnians*		*Cassia jalapensis*		*Crotalaria spectabilis*	
	Root	Stem	Root	Stem	Root	Stem	Root	Stem	Root	Stem	Root	Stem	Root	Stem
Leaves:														
Sterile soil	24[b]	(3)	(48)[b]	(8)	(7)	(148)[b]	100[b]	100[b]	100[b]	100[b]	83[b]	64[b]	26[b]	(8)
Non-sterile soil	61[b]	(75)[b]	(21)[b]	(50)[b]	(38)[b]	(120)[b]	(347)[b]	(38)[b]	6	(94)[b]	66[b]	62[b]	35[b]	24[b]
Roots:														
Sterile soil	(1)	33[b]	(32)[b]	(24)[b]	(52)[b]	(49)[b]	61[a]	0	52[b]	94[b]	42[b]	46[b]	(34)[b]	(50)[b]
Non-sterile soil	60[b]	44[b]	23[b]	17[a]	33[b]	51[b]	77[b]	(2)	100[b]	100[b]	47[b]	27[b]	39[b]	9

Numbers in parentheses are stimulations.
[a] $p < 0.005$.
[b] $p < 0.001$.

The production of allelopathic compounds in tropical zones, particularly if they are continuously released into the environment, may contribute to the elimination of secondary, already established species and to the selection of those which are beginning to get established in the community (Anaya, 1976; Anaya *et al.*, 1987a).

(b) Studies in Coatepec, Veracruz

In 1979, Anaya and co-workers initiated the study of allelopathy in relation with agroecosystems. One of them was carried out in shaded coffee plantations that have great economical importance in Mexico because coffee is one of the main export products. This research was made in Coatepec, Veracruz, in a traditional coffee plantation which is characterized by the presence of shade trees resembling the structure of the deciduous temperate forests, with three well-defined strata: the herbaceous layer, the shrub layer represented by coffee plants, and the tree layer. The main goal of this study was to assess the allelopathic interactions among the species that constitute this community, in particular the coffee plants (Anaya *et al.*, 1982). Figure 17.1 shows the effect of the aqueous solutions of the soils of two varieties of coffee: Typica and Bourbon. Both soils contain inhibitory compounds that significantly reduce the radicle and stem growth of *Bidens pilosa* and *Mimosa pudica*.

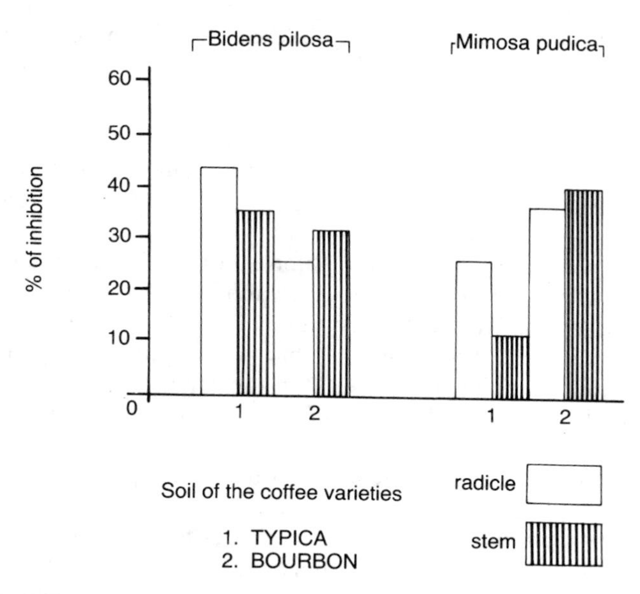

Fig. 17.1 Effect of aqueous leachates of coffee soil in Coatepec, on radicle and stem growth of *Bidens pilosa* and *Mimosa pudica*. All data are significant $p < 0.001$. (After Anaya *et al.*, 1987a. Reproduced with permission, Anaya, A.L. and Amer. Chem. Soc., Washington, DC.)

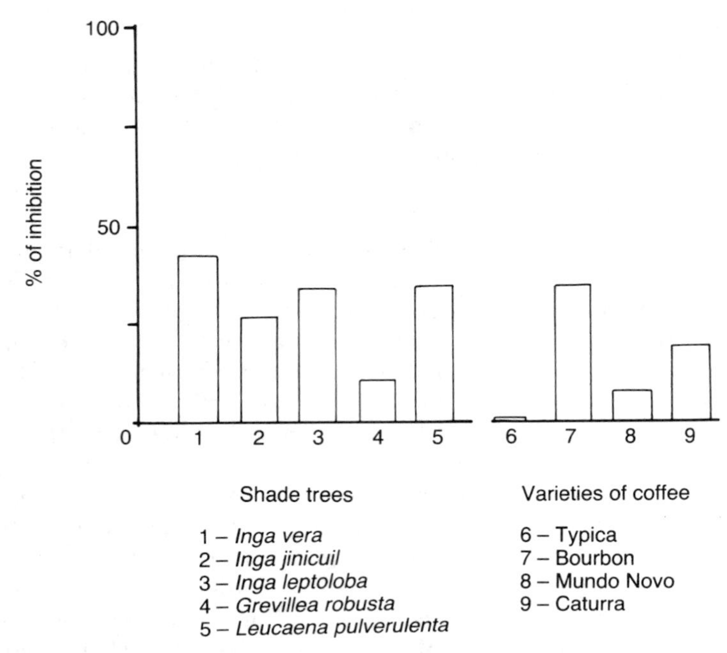

Fig. 17.2 Effect of aqueous leachates of dry leaves of some shade trees of coffee and some varieties of Coatepec coffee plants, on the radicle growth of *Rumex crispus*. Effect by Typica and Mundo Novo leachates are non-significant. (After Anaya *et al.*, 1987b. Reproduced with permission, Anaya, A.L. and Amer. Chem. Soc. Washington, DC.)

Waller *et al.* (1986) mention that these effects might be explained by the accumulation of caffeine and other alkaloids in the soil of old coffee plantations.

Figure 17.2 shows the allelopathic potential of dry leaves of shade trees and four varieties of coffee. *Inga* species and *Leucaena pulverulenta* are the shade trees species that exhibited the most inhibitory effect on *Rumex* growth. A very interesting result is that of the absence of inhibitory effects of dry leaves of Typica coffee. On the other hand, the soil of this variety showed a strong inhibitory effect on the growth of two other weeds: *Bidens pilosa* and *Mimosa pudica*.

Table 17.2 shows the greatest allelopathic effect produced by plants from the herbaceous layer. Various species of Commelinaceae and the fern *Pteridium aquilinum* have exhibited this potential in different types of climatic zones (Gliessman and Muller, 1972; Gliessman, 1978).

Leachates of coffee leaves and roots, and extracts from soil, were shown to contain (among other compounds): (1) from soil – caffeine and methyl esters of myristic through docosanoic acids (which have allelopathic activity), and (2) from leaves, stems and roots – furfural, *N*-phenyl-1-

TABLE 17.2 *Effect of aqueous leachates and acetone extracts of four herbaceous species of a coffee plantation in Coatepec, Veracruz, on the growth of* Bidens pilosa *and* Mimosa pudica *(after Anaya* et al., *1982).*

Herbs species (leachates/extracts)	Inhibition/Stimulation (%)			
	Bidens pilosa		*Mimosa pudica*	
	Radicle	Stem	Radicle	Stem
Commelina diffusa				
aqueous leachate	48[a]	18[a]	42[a]	16[a]
acetone extract	9	6	24[a]	19[a]
Commelina erecta				
aqueous leachate	62[a]	8	24[a]	16[a]
acetone extract	(58)[a]	(20)[a]	22[a]	13
Tripogandra serrulata				
aqueous leachate	(20)[a]	(2)	29	19[a]
acetone extract	(38)[a]	(30)[a]	16[a]	(5)
Pteridium aquilinum				
aqueous leachate	56[a]	24[a]	40[a]	29[a]
acetone extract	19[a]	16[a]	17[a]	6

Numbers in parentheses are stimulations.
[a] $p < 0.001$.

naphthylamine, *N*-alkanes and caffeine (Rizvi *et al.*, 1980; Waller *et al.*, 1986). Caffeine from the Coatepec coffee plantation soil indicated that around old coffee shrubs, considerable amounts of these alkaloids may be released from the litter and accumulate in the vicinity of roots over the years (1–2 g of caffeine m^2 per year) (Waller *et al.*, 1986). This fact may provide an explanation for the worldwide phenomenon of early degeneration of coffee plantations at 10–25 years of age (Wellman, 1961).

Continuing with the study of allelopathic potential of herbs in coffee plantations, Ramos *et al.* (1983) made a study of three of the most abundant species in coffee orchards: *Commelina diffusa*, *Tripogandra serrulata* and *Zebrina* sp. All species, fresh, dried, chopped, as well as their litter, exert a strong growth inhibition upon some weed species. It seems that the allelopathic compounds of comelinas are mainly produced during the rainy season, and those in April, having no means of being released into the environment, are probably produced in lesser proportions. Figure 17.3 shows a hypothetical relation between the relative percentage of phytotoxicity of the comelinas and the month of the year

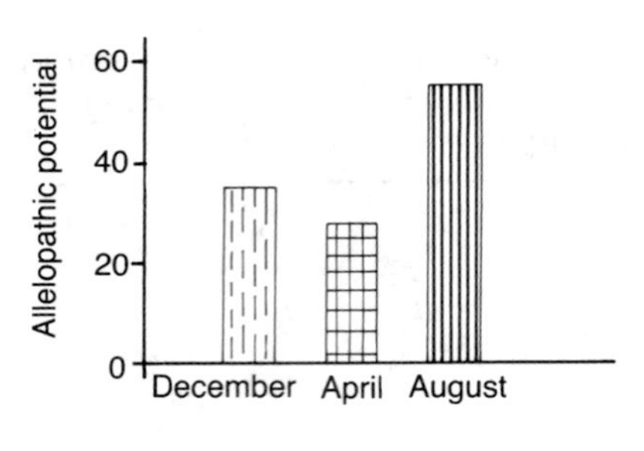

Fig. 17.3 Hypothetical relation between the allelopathic potential (relative phytotoxicity) of the Commelinas and the month of the year when plants were collected and leached. (After Ramos *et al.*, 1983. Reproduced with permission, Anaya, A.L. and Plenum Publishing Corporation, New York.)

when plants were collected and made to leach. Wellman (1961) states that in some cases comelinas are encouraged to grow in lightly shaded coffee, presumably to protect the soil and keep a natural layer of mulch in perpetual production. These shallow-rooted perennial weeds keep up a pronounced dominance which seems to contribute towards regulating the establishment of the other weeds. In Coatepec, comelinas are promoted by the coffee growers to cover the soil over the year as a live and/or mulch cover.

The study of the allelopathic potential of *Pteridium aquilinum*, that was clearly manifested in the coffee plantation of Coatepec as we mentioned, was initiated by Gliessman and Müller (1972). This fern can quickly invade abandoned cultivated fields and it also develops in the environmental conditions created after a fire in the tropics. In the tropical regions, the green fronds of *Pteridium* are the most effective source of allelopathic toxins. New fronds and consequently toxins are formed throughout the whole year (Gliessman, 1978).

(c) Studies in Tabasco

Gliessman and his co-workers at the Colegio Superior de Agricultura Tropical in Cárdenas, Tabasco, Mexico, carried out different studies of allelopathy in agroecosystems, especially at The Chontalpa Region, between 1979 and 1982.

Alejos (1980) carried out a study in this region of Tabasco, in some corn fields where he controlled different densities of *Bidens pilosa* ('cadillo'), a weed considered by local people as a bad weed, 'mal monte', that causes a harmful effect to corn. Alejos confirmed this fact; he observed that *B. pilosa*, when dominant, has an injurious effect on corn growth and productivity in the first 30 days after the sowing of corn. This can be partially explained because of the allelopathic potential exhibited by *B. pilosa* in the laboratory and the greenhouse.

The interplanting of squash (*Cucurbita pepo*) by peasants in corn/bean polycultures or in corn or bean monoculture stands during the proper season in southeastern Mexico contributes to weed control (Amador, 1980). Squash is planted mainly because of its ability to control weeds for improved corn and bean production, and any harvest of fruit is mostly an added bonus. The squash forms a continuous plant cover over the low-growing weedy species, eliminating them as well as restoring productivity to the soil by what appears to be large quantities of organic matter (8 to 10 tonnes dry matter per hectare). Laboratory bioassays implicate the role of allelopathy as one of the possible mechanisms of this inhibition.

Martínez Becerra (1982) in Cárdenas, Tabasco, tested ten common weed species with a high importance index, incorporated into soil of pots in the greenhouse on the growth of corn, bean, radish and squash as test species. Corn exhibited a great inhibition of its growth especially by the effect of *Bidens pilosa*, *Digitaria horizontalis*, *Scleria pterota* and *Spilanthes americana*. On the other hand, *Cyperus odoratus*, *Lagasea mollis*, *Commelina diffusa* and *Paspalum conjugatum* stimulated corn growth. Chacon and Gliessman (1982) revealed that peasants in tropical zones of Mexico possess a great empirical knowledge about the role that weeds (*monte*) play in agroecosystems. Peasants classify them as good (*buen monte*) and bad (*mal monte*) weeds, as was mentioned by Alejos (1980). It seems that, in general, good weeds appear late in crop fields, because they do not belong to the first stages of secondary succession. On the other hand, bad weeds appear early in the crop fields.

During the agricultural management, peasants remove bad weeds promptly and totally, and they promote or simply do not cut good weeds because of their multiple beneficial effect to crops, soil and man. The same practice was observed with comelinas in coffee plantations, and we can assume that these weeds are a very good example of *buen monte* (Ramos *et al.*, 1983). However some weeds behave differentially and are both good and bad. In the southeast of Mexico weeds are considered to be an integral part of agroecosystems.

An interesting and different aspect of the dynamics of the corn–bean–squash polyculture system is the frequency of reduction in pest damages. Studies have shown that this occurs through a combination of effects such as increased population of beneficial predators and parasites of the pest insects (Le Tourneau, 1983).

Gliessman and Altieri (1982) have found a reduction in plant-feeding insects when they interplant vegetable crops with other crops that are not attractive to the pest, or when they allow limited weed growth around the crop plants.

Rosado-May *et al.* (1985, 1986) stated that in the Mexican tropics some weeds in the agroecosystems are interrupting the life cycle of some plant pathogens and in this way are reducing their populations; the mechanism that determines this could be allelopathy. The presence of *Cyperus rotundus* can reduce the incidence of *Tanatephorus cucumeris*, which causes the 'web blight' in common bean (*Phaseolus vulgaris*); on the other hand, *Bidens pilosa* can control the population of several genera of plant parasitic nematodes which severely affect the corn.

(d) Studies in Uxpanapa, Veracruz

As part of a project on the management of different stages of secondary succession in Uxpanapa, Veracruz, Mexico, a comparative study of the dynamics of weed and crop production in a monoculture of corn and a 14-crop polyculture was made by Caamal and del Amo (1986, 1987). The results indicated that crop production was twice as high in the polycropped plots. Polycultures represent an efficient management system that deals with a natural control of weeds, particularly on grass species, very aggressive as weeds in tropical zones. Sweet potato (*Ipomoea batatas*) is very efficient in controlling weeds, in the same way as *Stizolobium pruriens* and other species of *Ipomoea*, by an interference where a very aggressive competition potential is complemented with a high allelopathic effect (Anaya *et al.*, 1990; Peterson and Harrison, 1991).

In the same project Nava *et al.* (1987) carried out different bioassays with the aim of detecting the allelopathic effect of green fronds of *P. aquilium* on the germination and radicle growth of seeds of some cultivated plants and weeds, on the mycelial growth of phytopathogenic fungi, and on phytopathogenic bacteria.

The fronds incorporated into soil inhibited the growth of four of the five weeds, and that of peanut (*Arachis hypogaea*), crystal jicama (*Pachyrrhizus erosus*) and mustard (*Brassica nigra*), used as test species (Table 17.3).

The fungal growth was strongly inhibited by the aqueous extract (Fig. 17.4). *P. aquilium* offers a wide range of potential uses as a natural herbicide and fungicide. One of the goals of the study was to suggest the incorporation of some of the common species of secondary plants into the integrated management of resources (del Amo, 1986).

In the southeastern part of Mexico, many peasants plant legumes as *Stizolobium pruriens* (Gliessman and Garcia, 1979) intercropped with corn. They use it as a cover crop, protection against soil erosion and as an aid to weed control, mainly in the case of grasses such as *Paspalum conjugatum*. *Stizolobium* incorporates nitrogen into the soil, improving the yield of corn, and it has shown a significant ability to reduce both the number and biomass of weeds when it is planted in rotation with corn (Gliessman, 1982). This legume had shown an allelopathic potential in laboratory bioassays; when it was tested on corn, bean and cabbage

TABLE 17.3 *Effect of macerated green fronds of* Pteridium aquilinum *incorporated to soil in pots, on the germination and radicle growth of some cultivated plants used in a multiple cropping system in Uxpanapa, Veracruz (after Nava* et al., *1987)*

	Mean no. germinated seeds						Mean radicle growth (mm)	
Species	Control	Fronds placed on the soil	Control	Fronds mixed with the soil	Control	Fronds placed on the soil	Control	Fronds mixed with the soil
Corn	18	16	10	10	8.46	4.15	4.07	8.15
Jicama	23	19	3	13[a]	2.10	0.90[a]	2.46	3.07
Peanut R.R.	17	16	–	–	4.02	2.11[a]	–	–
Peanut T.II.	16	16	–	–	2.62	2.51	–	–
Tomato	7	4	8	4	3.39	3.56	1.08	0.63
Mustard	11	2[a]	15	7[a]	4.39	3.83	3.88	2.57
Sesame seed	18	17	12	14	5.02	4.35	5.02	4.44
Chile ancho	11	13[a]	6	12[a]	1.00	1.49	0.51	1.31

[a] $p < 0.05$.

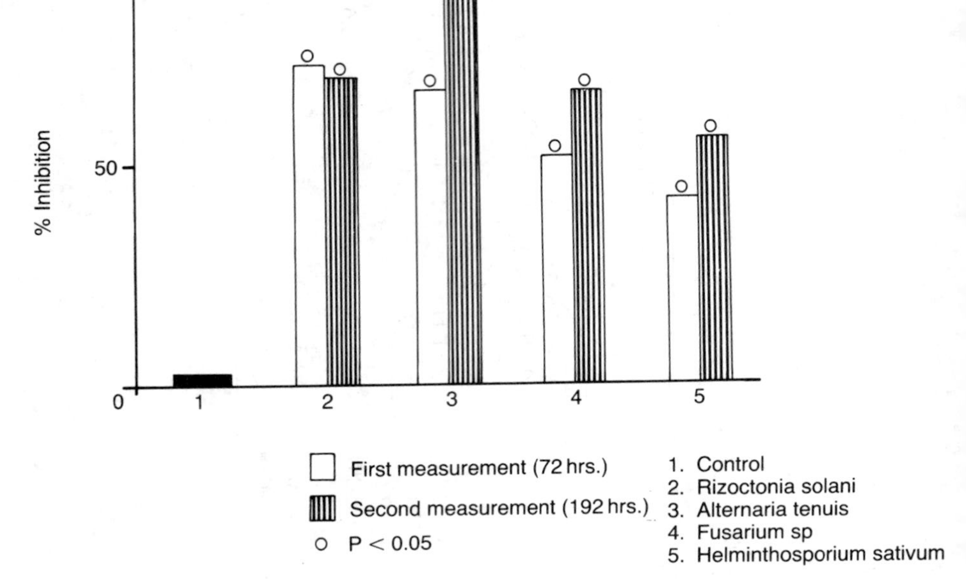

Fig. 17.4 Effect of aqueous extract of green fronds of *Pteridium aquilinum* on the radial growth of some phytopathogenic fungi. (After Nava *et al.*, 1987. Reproduced with permission, Anaya, A.L. and Elsevier Science Publishers, Netherlands.)

seeds, the latter species was the most susceptible (Chacon and Gliessman, 1982).

Stizolobium or velvet bean is also used in other states of Mexico (Puebla, Oaxaca, Chiapas and Veracruz), where the peasants not only use it as an aid to weed control, but they also use it as forage or they eat the seeds boiled or toasted and prepare a drink that is similar to coffee. For this reason peasants call the velvet bean 'nescafe'.

Arevalo and Jiménez (1988) carried out a survey in Uxpanapa, Veracruz, and found that peasants of this zone also use 'nescafe' to eliminate the vegetation more easily in zones of secondary communities before the agricultural cycle. They intercrop 'nescafe' with corn to eliminate bad weeds and as cover crop. The use of this legume represents an alternative for weed control, because here hand weeding, or the use of herbicides, is not easy.

Escárzaga (1987), in laboratory bioassays, determined the phenological stage in which *Stizolobium* presents the highest concentrations of allelochemicals. Vegetative, flowering and fructification stages of *Stizolobium* (dry leaves and roots leachates) were tested upon some cultivated and weed seeds. In general, the dicotyledonous weeds are more inhibited by

TABLE 17.4 *Effect of dry leachates of three* Stizolobium pruriens *phenological stages on the radicle growth of some cultivated and weed species (after Escárzaga, 1987)*

Phenological stage	Vegetative		Flowering		Fructification	
Leachates of Species	Leaves	Roots	Leaves	Roots	Leaves	Roots
			Radicle growth (mm)			
Crops:						
Zea mays	29.2[a]	(27.3)	11.2	–	(16.7)[b]	(17.0)[b]
Sorghum	(18.3)[b]	(9.8)	(11.5)	–	(38.8)[a]	(3.4)
Phaseolus sp.	(16.8)[a]	80.6[a]	(20.4)[a]	70.0[a]	10.5	(8.7)
Medicago	9.8	52.0[a]	(47.1)[a]	76.9[a]	14.3	22.4
Lycopersicum	(4.3)	(5.6)	(32.9)[a]	(50.3)[a]	(50.1)[a]	(39.1)[a]
Weeds:						
Amaranthus	29.4[a]	15.8[b]	25.6[a]	(3.2)	49.9[a]	17.1[b]
Lolium	(16.2)[b]	(22.8)[a]	63.3[a]	(16.1)[b]	(35.4)[a]	(14.4)
Portulaca	20.0[a]	9.3	(32.9)[a]	(3.8)	84.6[a]	47.6[a]

Numbers in parentheses are stimulations.
[a] $p < 0.001$
[b] $p < 0.005$

leaves leachates of flowering plants and the monocotyledonous weeds by leaves leachates of plants in fructification (Table 17.4).

The effects on fungal growth are stimulatory or inhibitory, depending on the test species. Leachates of dry leaves inhibit (55%) the radial growth of *Helminthosporium sativum* and stimulate (50%) the radial growth of *Fusarium* (Fig. 17.5).

Results from qualitative and quantitative evaluation of some chemical constituents, particularly phenols and volatile terpenes, from juvenile stages of three primary species belonging to the perennial rain forest, enabled del Amo *et al.* (1986) to confirm significant differences in the chemical composition of the studied species *Nectandra ambigens*, *Omphalea oleifera* and *Licaria alata*, depending on the season of the year and the degree of natural fungal infection. Among these three species, the essential oil from the latter plant showed the highest inhibitory activity against all the fungal species tested. Essential oil from undamaged plants showed the highest inhibitory activity, especially against *Colletotrichum* sp. Qualitative and quantitative differences related to the degree of infection in plants confirm the statement of Stoesl (1970) in relation to those compounds that remain constant and belong to the so-called constitutive components, and others that are only found in damaged plants: a fact that places them among the group of phytoalexins.

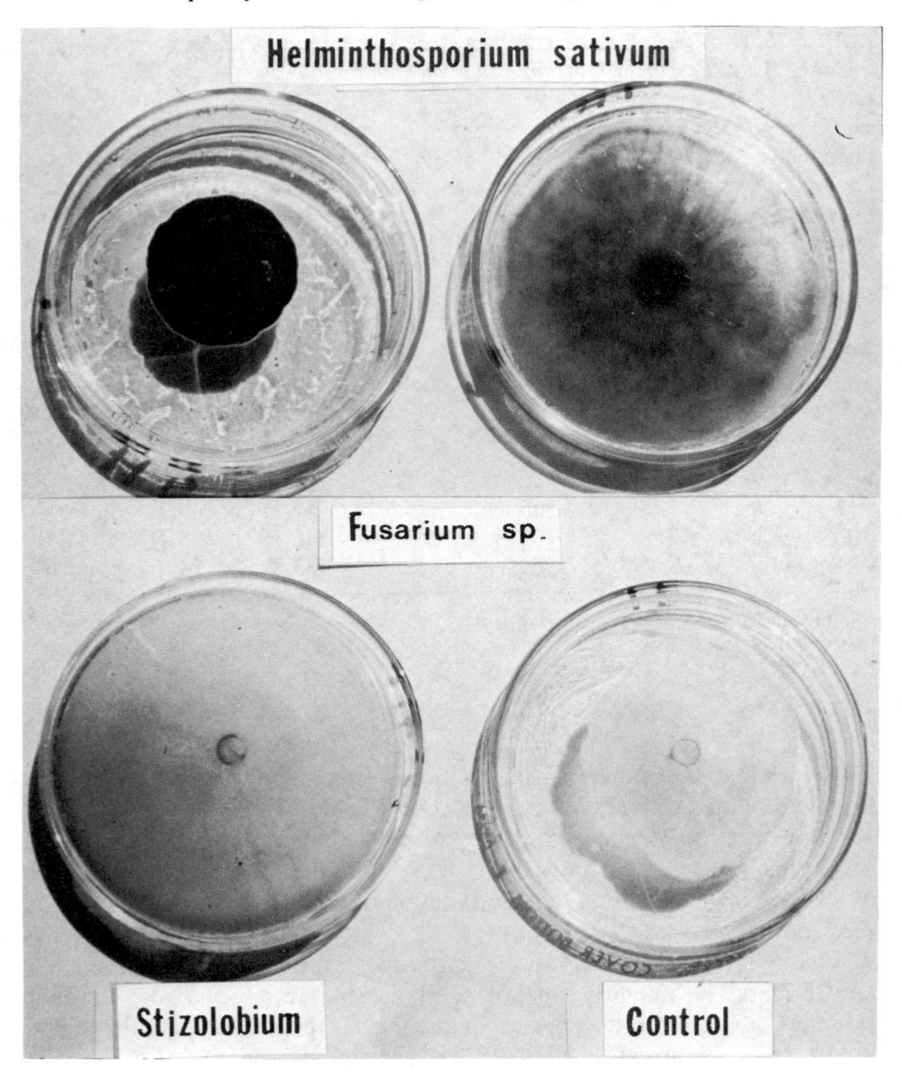

Fig. 17.5 Effect of dry aqueous leachate (8%) of *Stizolobium pruriens* on the radial growth of fungus.

17.2.2 Temperate zones

(a) Studies in Xochimilco and Mixquic

Simultaneous to the studies of the coffee plantation was the study of the allelopathic potential of the water hyacinth (*Eicchornia crassipes*) (Anaya *et al.*, 1987a). This aquatic plant, introduced in Mexico at the beginning of this century, invades many of the water reservoirs and streams. It is

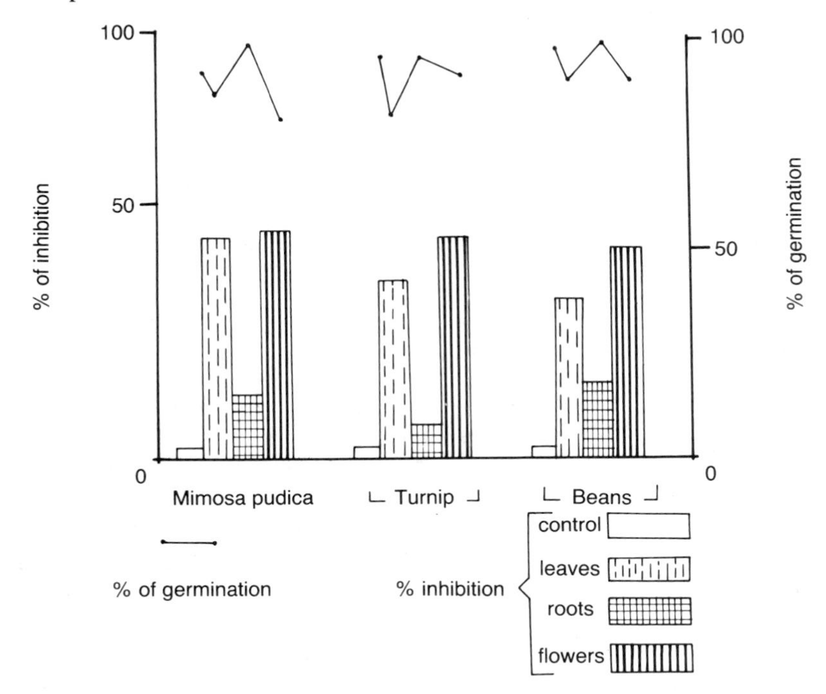

Fig. 17.6 Effect of aqueous leachates of leaves, roots and flowers of water hyacinth on germination and radicle growth of *Mimosa pudica*, turnip and beans. All effects on radicle growth are significant, except that of roots leachates of water hyacinth. (After Anaya *et al.*, 1987b. Reproduced with permission, Anaya, A.L. and Amer. Chem. Soc., Washington, DC.)

considered to be one of the worst aquatic weeds in our country. Its capacity to establish in several water habitats and its extensive vegetative growth suggests an aggressive mechanism of invasion in which allelopathy could be important. Several bioassays were done by testing aqueous extracts from leaves, tops, roots and flowers, upon weeds and cultivated plants. Results showed a strong inhibitory effect especially from top and flower leachates (Fig. 17.6). Water hyacinth is widely used as a green fertilizer in the Valley of Mexico, mainly in the ancestral traditional agroecosystems known as 'chinampas' – long narrow strips of land surrounded on at least three sides by water. Water hyacinth improves the physical and chemical properties of soil and it exerts a certain control of weeds through its decomposition in the soil.

Once the allelopathic potential of water hyacinth was demonstrated in laboratory bioassays, we decided to study the effect of this plant on the agricultural production and growth of weeds in a chinampa where turnip, radish, lettuce and cabbage were cultivated.

TABLE 17.5 *Vegetable crop production (tonnes ha^{-1}) in a traditional 'chinampa' in Xochimilco, under different treatments (after Anaya* et al., *1987a)*

	Crops				
Treatments	Turnip	Radish	Lettuce	Cabbage	Total
Control (soil only)	0.111	0.109	0.565	0.354	1.139
Soil with:					
Chemical fertilizers	1.435[a]	0.190	1.075	0.633	3.233[a]
Cow manure	2.303[a]	0.901[a]	3.22[a]	1.766[a]	8.19[a]
Water hyacinth manure:					
Whole plant	3.344[a]	1.156[a]	4.523[a]	2.193[a]	11.212[a]
Roots	2.864[a]	1.058[a]	4.934[a]	3.541[a]	12.397[a]
Top	3.062[a]	1.913[a]	5.369[a]	2.693[a]	13.07[a]

[a] $p < 0.05$.

The soil in the chinampa was prepared in the traditional way at Xochimilco, by making a seed bed with mud from the bottom of the channels that surround the chinampa. When the mud is dry, it is cut in small cubes where the seeds are planted. The seedbed is then covered with soil and twigs. Once the seedlings reached 10–15 cm they were transplanted to a plot previously weeded, ploughed and fertilized with water hyacinth and manure. Treatments were placed randomly and covered with mud. These were: (a) control (soil only); (b) chemical fertilizers; (c) cow manure; and (d) water hyacinth fertilizer (whole plant, roots or tops). Table 17.5 shows that all plots with water hyacinth exhibited a higher crop yield than the other treatments. This was mainly due to the improvement of the soil texture and a decrease in the salinity of soil owing to a reduced water evaporation and salt deposition on the soil surface. We also found that the whole plant and the top of water hyacinth added to the soil caused the greatest significant inhibition on the biomass of the weeds present in turnip, radish and cabbage plots (Table 17.6). These studies opened new and wide research alternatives not only related to allelopathy, but to other agroecological aspects such as crop–weed interactions (Chou and Lin, 1976; Trenbath, 1976). Within this context, Jiménez and Schultz (1981) conducted a study in a chinampa at San Andres Mixquic, D. F., in the southeastern Valley of Mexico. In this study they found that corn crops are less affected by the presence of weeds when they are associated with *Cucurbita ficifolia*; however, the latter is strongly affected by corn and to a lesser degree by weeds. Also they found that leachates from cultivated plants (corn, squash and beans) stimulated the growth of their own seedlings and inhibited that of weeds. Finally it was demonstrated that corn pollen has a strong allelopathic

TABLE 17.6 *Dry weight of total weeds (tonnes ha^{-1}) present in the plots of some crops in a traditional 'chinampa' at Xochimilco, under different treatments (after Anaya* et al., *1987a)*

Treatments	Crops				
	Turnip	Radish	Lettuce	Cabbage	Total
Control (soil only)	0.1733	2.24	2.8	3.33	8.54
Soil with:					
Chemical fertilizer	0.1733	2.24	3.2	6.0	11.61
Cow manure	0.1733	1.30[a]	1.95	4.33	7.75
Water hyacinth manure:					
Whole plant	0.08[a]	1.14[a]	2.8	1.8[a]	5.82[a]
Roots	0.2	1.3[a]	3.4	2.66	7.56
Top	0.033[a]	1.1[a]	2.4	1.4[a]	4.93[a]

[a] $p < 0.05$.

potential; the radicle growth of some test weed species was significantly inhibited by different concentrations of aqueous leachates of it. *Bidens pilosa* was the most susceptible species, while *Rumex crispus* was less affected; also, the radicle growth of another weed, *Cassia jalapensis*, was inhibited by corn pollen sprinkled on its seeds growing in different substrates (Jiménez *et al.*, 1983) (Figs 17.7 and 17.8).

(b) Studies in Tlaxcala

Altieri and Trujillo (1987) described the traditional ways in which Tlaxcalan farmers grow corn, mentioning the array of polyculture and agroforestry designs that result in a series of ecological processes important for insect pest and soil fertility management. Some measurements show that trees integrated into cropping systems modify the aerial and soil environment of associated understorey corn plants, influencing their growth and yields. With decreasing distance from trees, surface concentrations of most soil nutrients increase. Certain tree species affect corn yields more than others. Arthropod abundance also varies depending on their degree of association with one or more of the vegetational components of the system. Densities of predators and the corn pest *Macrodactylus* sp. depend greatly on the presence and phenology of adjacent alfalfa strips.

In the traditional agroecosystems, the biological heterogeneity constitutes a self-protection that arises naturally (Janzen, 1973). Altieri *et al.* (1983) mentioned that different plants and their volatile compounds can improve and make more attractive the habitat and the selection of pest host. When *Amaranthus* sp. extracts are sprayed on corn crops, *Trichogramma* sp. wasp increases its parasitic attack on eggs of the corn worm.

Fig. 17.7 Geotropism inversion of seedling roots of *Cassia jalapensis* by the effect of corn pollen.

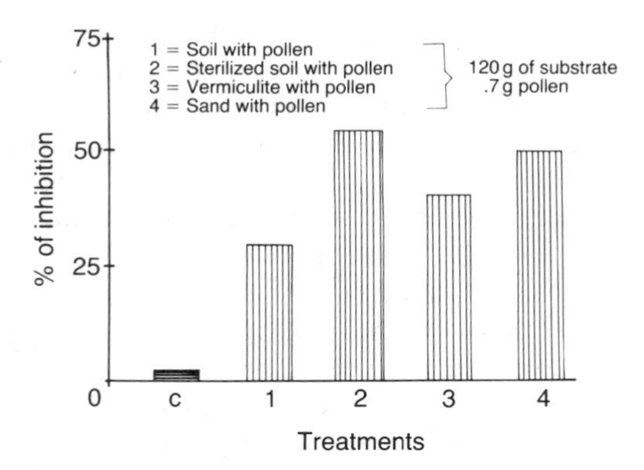

Fig. 17.8 Effect of sprinkled pollen of corn on the radicle growth of *Cassia jalapensis* growing in different substrates. All data are significant $p < 0.001$. (After Jiménez *et al.*, 1983. Reproduced with permission, Anaya, A.L. and Plenum Publishing Corporation, NY.)

Although the data of Altieri and Trujillo (1987) were derived from non-replicated fields, they point out some important trends and information that can be used to design new crop associations that will achieve sustained soil fertility and low pest attack potentials.

As part of the Chemical Ecology Project of the Institute of Cell Physiology (IFC) of UNAM (sponsored by CONACYT. PCEC-BNA-001079), Anaya *et al.* (1987a) carried out a field experiment in a traditional 'camellon' in Tlaxcala, where the systems of agricultural management are similar to those of 'chinampas' in the Valley of Mexico. 'Camellones' are very complex agroecosystems, which include the management of trees, shrubs and herbs in the same way and with the same proposal as in the chinampas. In fact, as Altieri and Trujillo (1987) confirmed, they are real agroforestry systems. 'Camellones' are surrounded by water and trees, mainly *Alnus firmifolia* that constitutes one of the main components in the agricultural landscape. Peasants cultivate a great variety of plants there, and use different plant and animal manures: leaves of maguey, capulin, sabine and alder, or the decomposing organic matter from the plants accumulated in the water channels that surround some of the fields (González, 1984; Anaya *et al.*, 1987b). The channels in 'camellones' represent the rest of an ancient hydrological system comprising lagoons, rivers and channels. This system has, for hundreds of years, represented a very specialized and efficient form of water and plant management for agricultural production (Armillas, 1984).

In an experiment (Anaya *et al.*, 1987b), a plot was sown with 'pinto' corn combined with bean and squash, in a design that allowed a comparison of the results of the three crops under the different treatments of (1) monoculture and polyculture, and (2) the effect of the use of plant covers (leaves of *Alnus firmifolia*, *Berula erecta* and *Juncus* sp.). The experiment also allowed us to record the growth of weeds in the different plots and the development of *Rhizobium* nodules in bean roots. The detection of the allelopathic potential of crops and weeds, and the consequences of these in relationships between different plants, were also studied (Fig. 17.9).

The number of weeds in the plots with *Alnus*, *Berula* and *Juncus* covers was significantly reduced compared with the number of weeds in plots without cover. Also the number of weeds in plots with *Berula* and *Juncus* cover was significantly smaller than with *Alnus* cover. This suggests that the decomposition of both kinds of cover (especially *Berula* and *Juncus*) may release some allelopathic compounds that contribute to the control of weeds. In the same way the total number of weeds in the corn–squash plots was significantly smaller than in the corn–bean and monoculture corn plots (Table 17.7). This is probably due to the allelopathic potential of squash that has also been observed in other agroecosystems (Chacon and Gliessman, 1982; Jiménez and Schultz, 1981).

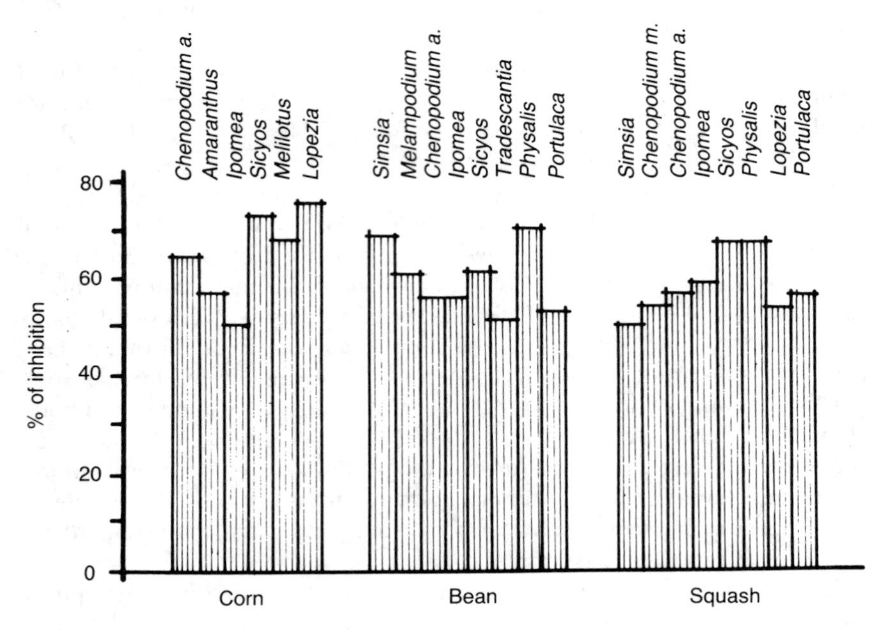

Fig. 17.9 Effect of lixiviates of dry weeds on radicle growth of corn, bean and squash. All data are significant at $p < 0.001$.

TABLE 17.7 *Total number of weeds present in plots of mono- and polyculture, with different plant cover treatments in a traditional 'camellon' in Tlaxcala (after Anaya* et al., *1987a)*

	Plant cover treatment			
Crops	Without cover	Leaves of *Alnus*	Leaves of *Berula* and *Juncus*	Total
Corn	1782	1723	1215	4720
Corn–bean	1976	1527	1430	4933
Corn–squash	1787	1346	1411	4544[a]
Total	5545	4596[b]	4056[b]	

[a] $p < 0.001$.
[b] $p < 0.01$.

TABLE 17.8 *Production of crops (kg m^{-2}) with different plant cover treatments, in a traditional 'camellon' in Tlaxcala (after Anaya* et al., *1987b)*

Plant cover treatment	Crop			
	Corn	Bean	Squash	Total
Without cover	198	1.7	11.9	211.6
With leaves of *Alnus*	203	2.6[a]	14.8	220.4
With leaves of *Berula* and *Juncus*	202	2.7[a]	25.5	230.2
Total	603	7.0	52.2	

[a]$p < 0.05$.

TABLE 17.9 *Number of root nodules in beans inoculated and uninoculated with* Rhizobium, *under different treatments in a traditional 'camellon' in Tlaxcala (after Anaya* et al., *1987b)*

Treatments	Number of nodules		
	Beans inoculated	Beans not inoculated	Total
Plots without weeding	134	72	206
Plots with weeding	68	78	146
Plots covered with leaves of *Alnus*	180	111	291
Plots covered with leaves of *Berula* and *Juncus*	273	96	369
Total	655	357	1012

All numbers are significant $p < 0.001$

The production of corn was similar in the three treatments whether in mono- or polyculture. Production of beans was significantly higher in the plots with both plant covers; and that of squash was higher only in the *Berula* and *Juncus* treatment (Table 17.8). It is observed that the latter cover had the most inhibitory effect on the growth of weeds; on the other hand, it was the only one that improved squash production. Finally, squash seemed to be the only crop with an inhibitory effect on weed growth in this field experiment.

In relation to the number of nodules on the roots of beans, the results showed that there were significant differences between inoculated and non-inoculated beans at the 60th day after they were sown (Table 17.9). Both plant covers (*Alnus* and *Berula*) caused an increases in the number of nodules. In the plots without weeding, the number of nodules in inoculated beans was also increased. Pena Cabriales and Alexander (1983) suggested that appreciable *Rhizobium* growth in unamended soil occurs only in the presence of germinating seeds, growing roots and decomposing nodules. In the case of the experiment in camellones, it is possible that the presence of a great variety of compounds in the rich rhizosphere of the plots without weeding can be a factor that increases the development of *Rhizobium*. It is also possible that these compounds are inhibitory to other microorganisms or even plants, because the effects are different, depending on the receptor organism, which is a general rule in allelopathy. Nevertheless, in the case of camellones, there are various nitrogen-fixing microorganisms in the different species of cultivated and non-cultivated plants (*Alnus* is one of the most important) whose dynamics are determined to a great extent by the special management the peasants give to the agroecosystem. Microorganisms associated with plant roots are sensitive and are affected by various factors (water stress, water-flooding, soil aeration, temperature, pH, agricultural chemicals and naturally occurring plant materials). The nature of plant exudates plays a key role in rhizosphere development and in turn rhizosphere organisms have either a direct or indirect effect on the host plant growth and development. All nutrients and/or compounds a plant obtains from the soil must pass through the rhizosphere and be subjected to biological and chemical transformation in this zone (Hoagland and Williams, 1985).

Anaya *et al.* (1987b) also mentioned the role of *Alnus* trees and a dominant shrub, *Baccharis glutinosa*, as 'insect traps'; both species trapped *Macrodactylus* sp., one of the main pests of corn and other cultivars, in the summer.

Nava (in preparation) studied the allelopathic potential of some cultivated legumes in this traditional camellones, emphasizing the *Rhizobium*–legume relationship and the significance of allelopathy on the crop management. Different species and a great number of varieties of legumes are cultivated in camellones as one of the more popular practices of natural nitrogen fertilization. In this place all kinds of beans, alfalfa, faba beans and vetch are cultivated. Each kind of crop is sown by peasants in monoculture, associated, intercropped or in rotation. Some of them, e.g. beans and faba beans, are commonly sown in the summer rainy season; and others, such as alfalfa and vetch, are sown in the winter.

In this study (Nava, in preparation), the roots of alfalfa and vetch were incorporated into soil in pots where corn and bean were sown in association. This resembles the incorporation of some residues of the former crops to the soil, before the summer agricultural cycle, in the alfalfa–corn or vetch–corn rotation of common practices. Soil that had not been cultivated for six months was taken as the control in the greenhouse experiment. The results demonstrated that the roots of alfalfa significantly stimulated the number of weed grasses in the pots with this treatment (Table 17.10). On the other hand, roots and tops of alfalfa, and

TABLE 17.10 *Mean number and mean dry weight (g) of mono- and dicotyledonous plants present in soil of pots with a corn–bean association under different treatments, in a 45 days' greenhouse experiment (after Nava, in preparation).*

Treatments	Monocotyledonous		Dicotyledonous	
	Number	Dry weight	Number	Dry weight
Control	0.5	0.0242	22.8	1.9238
Soil + alfalfa root	62[a]	2.0814[a]	21.7	1.0488
Soil + alfalfa root and top	12.2	0.9371	4.5	0.4739
Soil + vetch root	1.2	0.0486	16.2[a]	1.1162
Soil + vetch root and top	0.33	0.0026	10.0[a]	2.8942[a]

Numbers are the mean of six replicates.
[a] $p < 0.001$.

TABLE 17.11 *Mean dry weight (g) of corn and bean, and mean number of nodules of bean associated with corn in pots, under different treatments in a 45 days' greenhouse experiment (after Nava, in preparation)*

Treatment	Corn		Bean		Number of nodules
	Top	Root	Top	Root	
Control	4.1412	4.5672	2.5079	1.2265	500.8
Soil + alfalfa root	3.6661	3.3526	3.6613	1.9515	469.7
Soil + alfalfa root and top	8.1897[a]	5.8496	4.4779[a]	2.3375	437.8
Soil + vetch root	5.8214	3.9028	2.7983	2.6640	612.2
Soil + vetch root and top	7.2510[a]	4.5136	4.6313[a]	1.8118	427.3

Numbers are mean of six replicates.
[a] $p < .001$

roots of vetch diminished the number of dicotyledonous weeds in pots. Roots and tops of alfalfa and vetch significantly stimulated the growth of stems and leaves of corn and bean. It seems that the residues of alfalfa and vetch have a double effect when they are incorporated into soil: (1) they can inhibit some weeds, and (2) they can, at the same time, promote the growth of corn and bean. There is also some evidence that these treatments also affect root nodule development (Table 17.11)

These studies pointed out the complex interdependence between management practices (rotation, incorporation of crop residues to the soil, weeding, etc.) and multiple biotic interactions, some undoubtedly mediated by allelochemicals.

(c) Studies in Santa Cruz, California and Texcoco, Mexico

A cooperative research between the Agroecology Program of the University of California in Santa Cruz, and our Chemical Ecology Laboratory of the IFC, UNAM, was carried out in both places (Anaya *et al.*, 1988). Seven of the main weeds common in Santa Cruz and Texcoco were collected in each locality and added as dry plant cover to soils under different management conditions (soil recently ploughed, soil with a cultivar, soil undisturbed and soil of an annual grassland). The weeds used were: *Chenopodium* sp., *Amaranthus* sp., *Raphanus raphanistrum*, *Brassica campestris*, *Sonchus oleraceus*, *Malva parviflora* and *Polygonum aviculare*. These greenhouse experiments were carried out to observe the effects of different kinds of management of agricultural soils and, at the same time, to observe the effects of dried weeds mixed with soils, on the floristic potential, and the growth of corn seedlings. Simultaneously, laboratory bioassays were performed with leachates of fresh and dry weeds to elucidate the mechanism involved in the responses of the greenhouse experiment.

The results showed that *Raphanus*, *Malva* and *Brassica* strongly inhibited a number of weeds in pots in the soils of Santa Cruz. *Amaranthus*, *Raphanus* and *Brassica* were the most inhibitory treatments to weed growth in the soils of Texcoco (Table 17.12). The growth of Santa Cruz 'dry corn' was inhibited by *Brassica*, *Chenopodium* and *Raphanus* and that of Texcoco 'criollo corn' by *Chenopodium*, *Amaranthus* and *Malva*.

The results showed that latitude, altitude and climate can affect the production of different allelochemicals and, in some circumstances, can have deleterious effects on plants and microorganisms. The result of this study also points out the need for a better understanding of agroecosystems in the context of their regional particularities and an awareness that transplanting technologies from one region to another may not be appropriate.

Weeds must be considered as active components of the agroecosystem in developing regionally appropriate technologies. The inhibitory potential of weeds can be useful for production, and conservation goals of farmers if they are used in the appropriate way, either as intercropped plants, as cover crops or green manure, depending on the sensitivity of the crops that are going to be exposed to their effects.

17.2.3 Arid zones

(a) Studies in Monterrey, Nuevo León

The dominance and spatial distribution of *Larrea divaricata* in Mexican deserts, suggest the allelopathic mechanism as an important element of the adaptative traits of this shrub. The aqueous leachates of the top part of plants, collected in the surroundings of Monterrey, Nuevo Léon, showed some inhibitory and also stimulatory effects both on radicle and

TABLE 17.12 *Total number of weeds and dry weight of corn (g) in soil of pots with different weed species used as cover crop in a 30 days' greenhouse experiment in Santa Cruz, California and Texcoco, Mexico (after Anaya* et al., *1988)*

	Total number of weeds		Dry weight (g)	
Cover crop species	Santa Cruz, California	Texcoco, Mexico	Santa Cruz, California 'dry corn'	Texcoco, Mexico 'criollo corn'
Without cover	1080	230	4.3	2.2
Polygonum aviculare	1050	220	4.1	2.0
Sonchus oleraceus	1250[a]	490[a]	3.6	1.8
Chenopodium[b]	730[a]	160[a]	2.8[a]	1.6[a]
Amaranthus[c]	680[a]	100[a]	3.4	1.7[a]
Brassica campestris	570[a]	157[a]	2.6[a]	2.3
Raphanus raphanistrum	480[a]	150[a]	3.0[a]	1.9
Malva parviflora	530[a]	170[a]	4.6	1.7[a]

[a] $p < 0.001$.
[b] *Chenopodium* species for Santa Cruz was *C. album* and for Texcoco was *C. murale*.
[c] *Amaranthus* species for Santa Cruz was *A. retroflexus* and for Texcoco was *A. hybridus*.

stem growth of test seeds. The inhibitory effects were more frequent on dicotyledonous than on monocotyledonous species (Rovalo, 1978).

Some evidence is given for the allelopathic potential of *Helietta parvifolia* (Rutaceae) from the submontane scrub in Nuevo Léon, Mexico. Aqueous leachates of young and mature leaves inhibited growth of root and stem of bean, corn and tomato. The presence of alkaloids in this aqueous extract was demonstrated. The essential oil of the leaves significantly inhibited the growth of tomato, corn, onion, radish, squash and bermudagrass. Also, it exhibited a strong fungicidal and bactericidal action on *Aspergillus* and other Fungi Imperfecti, and on some bacteria isolated from soil. On the other hand, essential oil stimulated, very significantly, the growth of bean. Vegetation analysis indicates that *H. parvifolia* is the species with the greatest importance in the community. In the vicinity of this plant, saplings and seedlings are numerous and show great diversity, but never reach the shrub stage. It is possible that some allelochemicals of *H. parvifolia* contribute to regulate their growth (Graue and Rovalo, 1982).

17.3 CHEMICAL AND PHYSIOLOGICAL STUDIES

Along with the studies of allelopathy in different agroecosystems, we deal with the evidence of the production and ecological activity of different secondary plant compounds. As a result of this, and also as a logical research complement, we made some studies on the isolation and identification of allelopathic compounds, and tried to establish the most probable mode of action of these compounds on the affected organisms that have provided more information about the inhibition or stimulation of the growth of plants and microorganisms, caused by the allelopathic compounds. Examples of this kind of study are numerous (Waller, 1983; Rice, 1984). González de la Parra *et al.* (1981) identified two monoterpenes diastereoisomers, piquerol A and B, from *Piqueria trinervia*, a common weed that grows in tropical and temperate zones in Mexico. Both show strong biological activity, but piquerol A is the more active. In general, piquerol A was more inhibitory to the growth of roots of the test species, and piquerol B to the growth of stems.

Uribe *et al.* (1984, 1985) demonstrated that one of the active component of essential oil of *Piper auritum*, β-pinene, uncouples the oxidative phosphorylation process and inhibits respiration in isolated rat liver mitochondria. These effects can be explained by the interaction of β-pinene with the mitochondrial membrane. They also found that this terpene inhibited respiration of yeast cells, and that this inhibition depends on the ratio between the terpene concentration and the amount of yeast cells; for a fixed concentration of β-pinene, inhibition decreased as the amount of yeast cells increased. In relation with this, Weidenhamer *et al.* (1987) demonstrated that lower phytotoxin concentrations can produce equivalent or greater inhibitory effects than higher concentrations when the amount available for uptake per seed is greater. They germinated cucumber seeds under various combinations of solution volume and seed number with a range of ferulic acid concentrations.

Cruz Ortega *et al.* (1988) studied two physiological effects caused by the allelopathic compounds of the ethanolic extract of corn pollen on isolated watermelon hypocotyl mitochondria and on meristematic cells of the radicle of this plant. They found that the extract acts as an inhibitor of the electron pathway, decreasing oxygen consumption in state 3, with malate and succinate substrate. The specific inhibition site is probably located before the cytochrome-c. The ethanolic extract of corn pollen also decreased the mitotic activity and produced irregular pycnotic nuclei. At present the isolation and identification of the chemical nature of the allelopathic compounds of corn pollen is being carried out.

17.4 CONCLUSIONS AND PROSPECTS

Perspectives of allelopathy on the study of traditional agroecosystems are very wide, from a holistic point of view. The biotic variety, the abundance of niches, the richness of agricultural techniques, the management of different plant species, rotation of crops, special combination of them, use of natural fertilizer, cover crops, green manures, control of weeds, insects and pathogens allow to the ecologist, agronomist, phytopathologist, physiologist, chemist, etc., to evaluate the importance of secondary compounds and the present and future impact of their management, as well as many other scientific aspects of this kind of agroecosystem. These we consider as the point of departure for the future production of food, raw materials, fuel, medicine, etc. The need for a holistic vision is urgent because of the rapid and alarming damage to the environment, and the accelerated destruction of natural resources, especially in developing countries like Mexico. It is necessary to think about the new alternatives that this kind of research gives us in favour of a more rational, less destructive and more efficient management of resources. Scientific research in this field of chemical ecology and allelopathy will allow us to conciliate quantity of production with quality of natural resources management. By searching between these two apparent extremes, we could find the equilibrium point that favours the sustainable use of resources.

REFERENCES

Alejos, P. M. A (1980) El potencial alelopático del 'cadillo' (*Bidens pilosa* L.) en relación con su manejo en el cultivo de maíz, Thesis. Colegio Superior de Agricultura Tropical, Cárdenas, Tabasco, México.

Altieri, M. A. and Trujillo, J. (1987) The agroecology of corn production in Tlaxcala, Mexico, *Human Ecol.*, **15**, 189–220

Altieri, M. A., Martin, P. B. and Lewis, W. J. (1983) A quest for ecologically based pest management systems, *Environ. Management*, **7**, 91–100.

Amador, M. F. (1980) Comportamiento de tres especies (maíz, frijol, calabaza) en policultivos en La Chontalpa, Tabasco, México. Thesis, Escuela Superior de Agricultura Tropical, Cárdenas, Tabasco, México.

Amo, R. S. del (1986) Management of secondary succession for the creation of useful rainforest in Uxpanapa, Veracruz, Mexico. A case study. An intermediary alternative between transformation and modification. *Proc. Int. Workshop of Rainforest Regeneration and Management*, Gury, Venezuela. MAB–UNESCO, pp. 24–8.

Amo, R. S. del and Anaya, A. L. (1978) Effect of some sesquiterpenic lactones on the growth of certain secondary tropical species, *J. Chem. Ecol.*, **4**, 305–13.

Amo, R. S. del, Ramírez, J. G. and Espejo, O. (1986) Variation of some secondary metabolites in juvenile stages of three plant species from tropical rain forest. *J. Chem. Ecol.*, **12**, 2021–38.

Anaya, A. L. (1976) Estudio sobre el potencial alelopático de algunas plantas secundarias en una zonas cálido-húmeda de México, Doctoral Thesis, Science Faculty, UNAM, México.

Anaya, A. L. and Amo, R. S. del (1978) Allelopathic potential of *Ambrosia cumanensis* H.B.K. (Compositae) in a tropical zone of México, *J. Chem. Ecol.*, **4**, 289–304.

Anaya, A. L., Roy Ocotla, G., Ortíz, L. M. and Ramos, L. (1982) Potencial alelopático de las principales plantas de un cafetal, in *Estudios ecológicos en el agroecosistema cafetalero* (eds A.E. Jiménez and A. Gómez Pompa), Compañia Editorial Continental, S.A. México, DF, pp. 85–94.

Anaya, A. L., Ramos, L., Hernández, J. G. and Cruz Ortega, R. (1987a) Allelopathy in Mexico, in *Allelochemicals: Role in Agriculture and Forestry* (ed. G. R. Waller), ACS, Symp. Ser. 330, Amer. Chem. Soc., Washington, DC, pp. 89–101.

Anaya, A. L., Ramos, L., Cruz Ortega, R. *et al.* (1987b) Perspectives on allelopathy in Mexican traditional agroecosystems: a case study in Tlaxcala. *J. Chem. Ecol.*, **13**, 2083–101.

Anaya, A. L., Gliessman, S., Cruz Ortega, R. *et al.* (1988) Comparative effects of allelopathic weeds used as cover crop on the floristic potential of soils, in *Global Perspectives on Agroecology and Sustainable Agriculture, Proc. Sixth Int. Scientific Conf. of the International Federation of Organic Movements* (IFOAM), Vol. I, University of California, Santa Cruz, pp. 607–18.

Anaya, A. L., Calera, M. R., Mata, R. and Pereda-Miranda, R. (1990) Allelopathic potential of *Ipomoea tricolor* Cav. [Convolvulaceae]. *J. Chem. Ecol.*, **16**, 2145–57.

Arevalo, R. J. and Jiménez, O. J. J. (1988) Uso y Manejo del 'nescafé' (*Stizolobium pruriens* var. *utilis*) entre los pobladores del Valle de Uxpanapa, Veracruz, in *Cuatro Estudios sobre Sistemas Tradicionales* (ed. R. S. del Amo), Instituto Nacional Indigenista, Serie de Investigaciones Sociales No. 17, México, DF.

Armillas, P. (1984) Notas sobre sistemas de cultivo en Mesoamérica. *Cuicuilco*, Año IV (13), México.

Caamal, A. and Amo, R. S. del (1986) Comparación de la dinámica de arvenses en sistemas de policultivo y monocultivo. *Biotica* (México), **11**, 127–36.

Caamal, A. and Amo, R. S. del (1987) La milpa múltiple como punto de partida del manejo de la sucesión secundaria. *Turrialba*, **37**, 195–210.

Chacon, J. C. and Gliessman, S. R. (1982) The use of the non weed concept in traditional tropical agroecosystems of Southeastern Mexico. *Agroecosystems*, **8**, 1–11.

Chou, C. H. and Lin, H. J. (1976) Autointoxication mechanism of *Oryza sativa*, I. Phytotoxic effect of decomposing rice residues in soil. *J. Chem. Ecol.*, **2**, 353–67.

Collera Zúñiga, O. (1956) Estudio del aceite esencial de *Piper auritum*. Thesis, Chemistry Faculty, UNAM, México.

Crossley, D. A. Jr, House, G. J., Snider, R. M. *et al.* (1984) The positive interactions in agroecosystems, in *Agricultural Ecosystems* (eds R. Lowrance, B. R. Stinner and G. J. House), Wiley, New York, pp. 73–81.

Cruz Ortega, R., Anaya, A. L. and Ramos, L. (1988) Effects of allelopathic compounds of corn pollen on respiration and cell division of watermelon. *J. Chem. Ecol.*, **14**, 71–86.

Escárzaga, G. E. (1987) Determinación del potencial alelopático del 'nescafé' (*Stizolobium pruriens* (L.) Medec var. *utilis* Wallex Wight) sobre cinco cultivos y tres malezas. Thesis, Instituto Tecnológico y de Estudios Superiores de Monterrey, Campus Querétaro, México.

Farnsworth, N. R. (1977) The current importance of plants as a source of drugs, in *Crop Resources* (ed. D. S. Seigler), Academic Press, New York, pp. 61–73.

Frei Sister, J. K. and Dodson, C. H. (1972) The chemical effect of certain bark substrates on the germination and early growth of epiphytic orchids. *Bull. Torrey. Bot. Club*, **99**, 301–7.

Gliessman, S. R. (1978) The establishment of bracken following fire in tropical habitats. *Amer. Fern. J.*, **68**, 41–4.

Gliessman, S. R. (1982) Allelopathy and biological weed control in agroecosystems, in *Allelochemicals and Pheromones* (eds C. H. Chou, and G. R. Waller), Academia Sinica, Monograph Series No. 5, Acad. Sinica, Taipei, ROC, pp. 77–86.

Gliessman, S. R. and Altieri, M. A. (1982) Advantages of polyculture cropping. *California Agriculture*, 14–17.

Gliessman, S. R. and Garcia, E. R. (1979) The use of some tropical legumes in accelerating the recovery of productivity of soils in the lowland humid tropics of Mexico, in *Tropical Legumes: Resources for the Future*, NAS Publ. No. 27, Washington, DC, pp. 292–3.

Gliessman, S. R. and Müller, C. H. (1972) The phytotoxic potential of bracken, *Pteridium aquilinum* (L.) Kuhn. *Madroño*, **21**, 299–303.

González-Jácome, A. (1984) Agroecosistemas en las tierras altas de México. *Boletín de la Universidad Iberoamericana*, **146**, 26–33.

González de la Parra. M., Anaya, A. L., Espinoza, F., Jiménez, M. and Castillo, R. (1981) Allelopathic potential of *Piqueria trinervia* (Compositae) and piquerols A and B. *J. Chem. Ecol.*, **7**, 509–15.

Graue, W. B. and Rovalo, M. M. (1982) Potencial alelopático y microbicida de *Helieta parvifolia. Biótica* (México), **7**, 405–16.

Hoagland, R. E. and Williams, R. D. (1985) The influence of secondary plant compounds on the associations of soil microorganisms and plant roots, in *The Chemistry of Allelopathy* (ed. A. C. Thompson), ACS Symp. Ser. 268, Amer. Chem. Soc., Washington, DC, pp. 302–25.

Holyoke, C. W. and Reese, J. C. (1987) Acute insect toxicants from plants, in *Handbook of Natural Pestricides* (eds E. O. Morgan and N. B. Mandava), CRC Press, Boca Raton, FL, pp. 67–118.

Janzen, D. H. (1973) Community structure of secondary compounds in plants. *Pure. Appl. Chem.*, **34**, 529–38.

Jiménez-Osornio, J. J. and Schultz, C. K. (1981) Relaciones cultivo-arvenses en una chinampa. Thesis, Sciences Faculty, UNAM, México.

Jiménez-Osornio, J. J., Anaya, A. L., Shultz, K., Hernández, J. and Espejo, O. (1983) Allelopathic potential of corn pollen. *J. Chem. Ecol.*, **9**, 1011–25.

Le Tourneau, D. (1983) Population dynamics of insect pests and natural control in traditional agroecosystems in tropical Mexico. PhD Dissertation, Univ. California, Berkeley.

Marinero, R. N. (1962) Influencia de *Melinis minutiflora* Beauv en el crecimiento de *Cordia alliodora* (R. and P.) Cham. MS Thesis, Instituto de Ciencias Agrícolas, Turrialba, Costa Rica.

Martínez Becerra, A. (1982) La alelopatía como un factor ecológico en una comunidad de arvenses para la Planicie Tropical de Tabasco. MS Thesis, Colegio Superior de Agricultura Tropical, Cárdenas, Tabasco, México.

Müller, C. H. (1966) The role of chemical inhibition (allelopathy) in vegetational composition. *Bull. Torrey Bot. Club*, **93**, 332–51.

Müller, C. H. (1969) Allelopathy as a factor in ecological process. *Vegetatio Haag*, **18**, 348–57.

Müller, C. H. (1970) Phytotoxins as plant habitat variables. *Recent Adv. Phytochem.*, **3**, 105–27.

Müller, W. H. and Müller, C. H. (1956) Association patterns involving desert plant that contain toxic products. *Amer. J. Bot.*, **43**, 354–61.

Nava, R. V., Fernández, L. E. and Amo, R. S. del (1987) Allelopathic effect of green fronds of *Pteridium aquilinum* on cultivated plants, weeds, phytopathogenic fungi and bacteria. *Agric. Ecosyst. Env.*, **18**, 357–79.

Nava, R. V. (in preparation) El potencial alelopático de algunas leguminosas cultivadas y su importancia en el manejo agrícola en los 'camellones' de Tlaxcala, PhD Thesis, Science Faculty, UNAM, México.

Pena Cabriales, J. J. and Alexander, M. (1983) Growth of *Rhizobium* in unamended soils. *Soil Sci. Soc. Amer. J.*, **47**, 81–4.

Peterson, J. K. and Harrison, H. F. Jr (1991) Isolation of a substance from sweet potato *Ipomoca batatas* periderm tissue that inhibits seed germination. *J. Chem. Ecol.*, **17**, 943–51.

Quaterman, E. (1973) Allelopathy in cedar glade plant communities. *J. Tennessee Assoc. Sci.*, **48**, 147–50.

Ramos, L., Anaya, A. L. and Nieto de Pascual, J. (1983) Evaluation of allelopathic potential of dominant herbaceous species in a coffee plantation. *J. Chem. Ecol.*, **9**, 1079–97.

Ramos-Prado, J. (1941) Interacciones gulmicas entre especies primarias y secundarias y selva aita peremifolia. Thesis, Science Faculty, UNAM, México.

Rice, E. L. (1984) *Allelopathy*, Academic Press, London.

Rizvi, S. J. H., Jaiswal, V., Mukerji, D. and Mathur, S. N. (1980) Antifungal properties of 1,3,7 trimethylxanthine isolated from *Coffea arabica*. *Naturwiss.*, **67**, 459–60.

Rosado-May, F. J. and García, R. E. (1985) Incidencia de la 'mustia hilachoza' (*Tanatephorus cucumeris*) en frijol común como resultado del manejo del suelo. *Rev. Mex. Fitopatol.* (México), **3**, 92–9.

Rosado-May, F. J., Gliessman, S. R. and Alejos, M. P. (1986) El 'cadillo' (*Bidens pilosa*), arvense asociada al maíz en la Chontalpa Tabasco: Planta con propiedades nematicidas, *Rev. Mex. Fitopatol.* (México), **4**.

Rovalo, M. M. (1978) Bioensayos con la 'gobernadora' (*Larrea divaricata* Cav.) como posible planta alelopática y/o herbicida. Primera Reunión sobre los Reguladores de las Plantas y los Insectos, Consejo Nacional de Ciencia y Tecnología, Monterrey, México.

Stoesl, A. (1970) Antifungal compounds produced by higher plants, in *Recent Advances in Phytochemistry* (ed. C.C. Stuluik), Appleton Century Crofts, New York, pp. 143–80.

Trenbath, B. R. (1976) Plant interactions in mixed crop communities, in *Multiple Cropping*, ASA Special Publ. No. 27, American Society of Agronomy, CSSA and SSSA, Madison, Wisconsin.

Uribe, S., Alvarez, R. and Peña, A. (1984) Effects of β-pinene, a non-substituted monoterpene, on rat liver mitochondria, *Pest. Biochem. Physiol.*, **22**, 43–50.

Uribe, S., Ramírez, J. and Peña, A. (1985) Effects of β-pinene on yeast membranes functions. *J. Bacteriol.*, **161**, 1195–200.

Waller, G. R. (1983) Frontiers of allelochemicals research, in *Allelochemicals and Pheromones* (eds C. H. Chou and G. R. Waller), Academia Sinica Monograph Series No. 5, Acad. Sinica, Taipei, ROC., pp. 1–25.

Waller, G. R., Kumari, D., Friedman, J. *et al.* (1986) Caffeine autotoxicity in *Coffea arabica* L, in *The Science of Allelopathy* (eds A. R. Putnam and C. S. Tang), Wiley, New York, pp. 243–69.

Webb, L. J., Tracey, J. G. and Haydock, K. P. (1967) A factor toxic to seedlings of the same species associated with living roots of the non-gregarious subtropical rain forest tree *Grevillea robusta. J. Appl. Ecol.*, **4**, 13–25.

Weidenhamer, J. D., Morton, T. C. and Romeo, J. T. (1987) Solution volume and seed number. *J. Chem. Ecol.* **13**, 1481–91.

Wellman, F. L. (1961) *Coffee, Botany, Cultivation and Utilization*, World Crops Series, Wiley-Interscience, New York.

CHAPTER 18

The allelopathic potential of aromatic shrubs in phryganic (east Mediterranean) ecosystems

D. Vokou

18.1 INTRODUCTION

One of the characteristic types of mediterranean ecosystems is phrygana, occurring not only around the Mediterranean Basin, but in all other regions of the Earth with a mediterranean-type climate (synonyms of phrygana: batha – Israel, tomillares – Spain, renosterbos – S. Africa, gariga – Italy, coastal sage – California). Particularly in Greece, this ecosystem type occupies approximately 13% of the surface of the country (Diamantopoulos, 1983). According to Aschmann's definition of a mediterranean climate (1973), these open and low shrub communities occur at the dry end of the precipitation gradient, whereas, at the wet end, dense, evergreen sclerophyll communities (maquis) develop. The dominant life-form in phrygana is that of therophytes (>40%), but their physiognomy is determined by woody plants, adapted to cope with the summer drought of the mediterranean climate through the mechanism of seasonal dimorphism (Margaris, 1981).

Another characteristic of phryganic ecosystems is the abundance of aromatic plants among their components. The characteristic differentiating these plants from all others, in spite of the fact that they belong to many different families, is the production of chemically related secondary compounds, the low molecular weight isoprenoids. These compounds form diverse mixtures of volatile oils, giving the producing plant its characteristic fragrance.

This remarkable contribution of aromatic plants, both in species number and in biomass, in phrygana, led us to undertake research aiming to understand the function of volatile oils in the frame of this type of ecosystem. Allelopathic effects were, therefore, under investigation.

The subjects treated in this chapter, focusing on aromatic shrubs from

Allelopathy: Basic and applied aspects Edited by S. J. H. Rizvi and V. Rizvi
Published in 1992 by Chapman & Hall, London ISBN 0 412 39400 6

phryganic ecosystems, include (i) allelopathic effects on herbaceous plants, (ii) autopathy,* and (iii) effects on soil microorganisms from the same area, where these aromatic shrubs occur.

18.2 EARLY EVIDENCE OF ALLELOPATHIC INTERACTIONS INDUCED BY AROMATIC SHRUBS IN THE RELATED ECOSYSTEM TYPE OF CALIFORNIAN COASTAL SAGE COMMUNITIES

The triggering factor that initiated a long series of research activities in the field of allelopathy, induced by aromatic shrubs, was the discovery of striking patterns of vegetation in and around patches of *Salvia* and *Artemisia* spp., in the coastal sage communities of California. They were considered to be the outcome of inhibitory effects exerted by the volatile oils that these aromatic shrubs produce (Muller *et al.*, 1964). Intensive work followed; inhibitory terpenes were identified (Muller, 1965), and mechanisms of action were suggested (Muller and del Moral, 1966; Muller and Hauge, 1967; Muller *et al.*, 1968a, b). Many other chaparral species were proved toxic in assay tests (Muller, 1966). However, later on, Halligan (1973, 1975, 1976), investigating possible causes for the formation of bare areas around stands of *Artemisia californica* Less., concluded that the activity of small mammals was the most important factor affecting the pattern of herb distribution, and criticized Muller for having overrated the role of allelopathy. Animal activity had been suggested to be the primary factor responsible for the zonation pattern even earlier by Bartholomew (1970), but this view was initially criticized by Muller and del Moral (1971). The relevant literature on the whole matter is extensively reviewed by Rice (1974, 1978).

In my opinion this outcome had, to a certain extent, influenced research in the area. The problem with plant derived toxins and antibiotics (proved as such *in vitro*) is that they may be quickly inactivated in the soil by adsorption to soil colloids, by chemical or microbial degradation (Elad and Misaghi, 1985). The fact that one of the best-known examples of allelopathy acting in the field (not only under laboratory conditions), when put under re-examination, led to contradictory conclusions resulted in a retreat of the previous enthusiastic activity. Research in the field of allelopathy, particularly due to aromatic plants, has been fairly slow during the last years.

* On this occasion, I would suggest replacement of the term autoallelopathy by autopathy. Since these terms have Greek origins I believe that their English counterparts should be meaningful; autoallelopathy consists of the roots of three words: auto (meaning itself), allelo (meaning reciprocally) and pathy (meaning suffering). No sense could be made out of their combination.

18.3 ALLELOCHEMICAL EFFECTS OF AROMATIC PLANTS FROM PHRYGANIC ECOSYSTEMS

Aromatic plants are very common in phrygana. In Greece, in particular, *Coridothymus capitatus* (L.) Reichenb. fil. is the most prominent aromatic plant in phryganic ecosystems, whereas *Satureja thymbra* L., *Teucrium polium* L., *Origanum onites* L., *Origanum vulgare* L. ssp. *hirtum* (Link) Ietswaart, *Lavandula stoechas* L., *Salvia fruticosa* Miller, *S. pomifera* L. have a very wide distribution. The volatile oils of the first three, as well as that of *Rosmarinus officinalis* L. (not native in the phrygana of Greece), were analysed both quantitatively and qualitatively (Vokou, 1983; Vokou and Bessière, 1985; Vokou and Margaris, 1986a), and their effect on seed germination and seedling growth was tested (Vokou and Margaris, 1982). The experimental material were seeds of annual legumes, growing wild in the same ecosystem type, viz. *Astragalus hamosus* L., *Hymenocarpos circinnatus* (L.) Savi and *Medicago minima* (L.) Bartal. Seeds and volatile oils were always only in aerial contact, in petri dishes hermetically closed.

Mediterranean ecosystems are subject to frequent fires. The early post-fire period is characterized by the high participation of annuals, especially legumes, a result of enhanced seed germination of these plants and subsequent establishment of their seedlings in the field. Among the explanations proposed for this increased germinability is the destruction by fire of germination inhibitors (like volatile oils) present in the soil (Muller *et al.*, 1968; Christensen and Muller, 1975). In order to test this hypothesis we have selected legume seeds as experimental material.

It was found that germination was generally not affected (Table 18.1) with the quantities tested (1, 3, 6, 10 and 20 μl, which, taking into account the volume of the petri dishes used, correspond to concentrations of 10, 30, 60, 100 and 200 ppm). On the contrary, seedling growth and particularly radicle elongation was inhibited (Figs 18.1(a),(b),(c)). Of the four volatile oils tested, that of *R. officinalis* proved to be always the most toxic. However, *R. officinalis* is not an element of this ecosystem type; it grows in more mesic habitats and was used in that study for the sake of comparison. It could be thus argued that these legumes, growing in proximity to aromatic plants, are less susceptible to the inhibitory effect of their volatile oils. Therefore, destruction by fire of the residual volatile oils in the soil is not the primary reason for the increased germinability of annual legumes in the initial post-fire period. The high percentages of inhibition, resulting after addition of relatively high quantities of volatile oils, are not to be expected in the field, as such concentrations hardly exist under natural conditions.

Most striking are the results concerning seed germination and seedling elongation of *C. capitatus* in the presence either of its volatile oil or of its leaves (Vokou and Margaris, 1986b). Both these processes are severely

TABLE 18.1 *Percent germination of* Hymenocarpos circinnatus, Astragalus hamosus *and* Medicago minima *seeds treated with 1, 3, 6, 10 and 20 μl of* Rosmarinus officinalis *(R),* Coridothymus capitatus *(C),* Satureja thymbra *(S) and* Teucrium polium *(T) volatile oils (after Vokou and Margaris, 1982)*

	Germination (%)		
Treatment	*Hymenocarpus circinnatus*	*Astragalus hamosus*	*Medicago minima*
Control	93	92	98
R1	90	93	92
R3	97	76	93
R6	95	67	95
R10	93	62	76
R20	48	56	63
C1	92	96	98
C3	88	84	92
C6	85	72	90
C10	92	72	95
C20	93	48	94
S1	89	85	99
S3	95	82	89
S6	87	69	86
S10	71	83	83
S20	61	78	69
T1	97	98	96
T3	88	93	90
T6	91	96	100
T10	94	96	100
T20	91	91	90

inhibited (Table 18.2). Concerning the inhibitory effect exerted by intact leaves, it should be taken into account that their volatile oil content at the time of collection (June) is ≃4%. Accordingly, the quantities of 0.5 and 1 g tested correspond to 20 and 40 μl, respectively. However, the drastic quantities should be much lower because of the limitations to emanation that the plant tissues impose. The volatile oil quantities of 1 and 3 μl, strongly inhibitory to *C. capitatus* seed germination, had no effect on the three legumes mentioned before (Table 18.1). Additionally, the inhibitory effect on seedling elongation was again much more pronounced in the case of *C. capitatus* volatile oil and its own seedlings. It is remarkable to note, therefore, that the volatile oil of *C. capitatus*, as an allelopathic agent, is most potent against its own seeds.

Taking into account the chemical composition of *C. capitatus* volatile

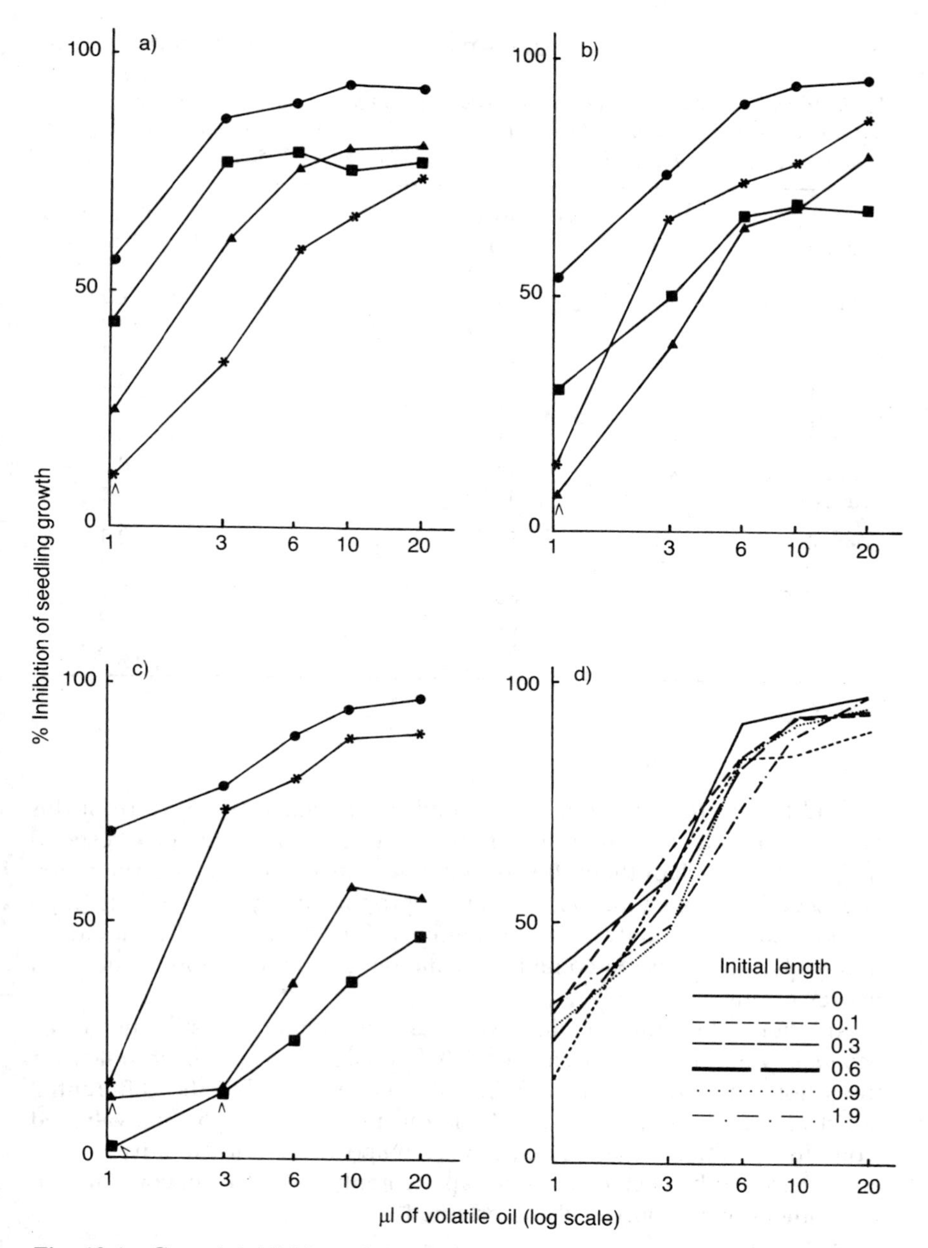

Fig. 18.1 Growth inhibition of *Astragalus hamososus* (a), *Medicago minima* (b), and *Hymenocarpos circinnatus* (c) seedlings induced by 1, 3, 6, 10 and 20 μl of *Rosmarinus officinalis* (●), *Coridothymus capitatus* (■), *Satureja thymbra* (▲) and *Teucrium polium* (*) volatile oils, and comparative seedling growth inhibition (d) of non- and pregerminated *Cucumis sativus* seeds treated with *R. officinalis* volatile oil. Pregerminated seed were separated into five groups according to their initial seedling length; each group was examined separately; ∧ Designates non-significant difference at $p = 0.05$. (After Vokou and Margaris, 1982. Reproduced with permission, Vokou, D. and Kluwer Academic Publishers, The Netherlands.)

TABLE 18.2 *Seed germination and seedling elongation of* Coridothymus capitatus *in presence of its volatile oil, leaves (containing 4% volatile oil) and litter (containing 0.5% volatile oil) (after Vokou and Margaris, 1986a)*

Treatment	Germination (%)	Seedling length (cm)
Control	73	1.7
VO 1 μl	64	0.2
VO 3 μl	24	0.1
VO 6 μl	25	0.1
VO 10 μl	1	0.1
VO 20 μl	1	0.1
Leaves 0.5 g	44	0.2
Leaves 1 g	46	0.1
Control	80	2.1
Litter: 1 g	86	2.1
1.5 g	88	2.1
3 g	87	2.3
5 g	80	2.3

VO = volatile oil.

oil and testing the individual compounds which make the mixture of this volatile oil (Table 18.3), it was found that seeds and seedlings respond differently to each of them. It was found that, in general, seed germination and seedling elongation values varied proportionally, either both high or both low, with only a few exceptions, e.g. camphene. It should be noted also that oxygen-containing compounds are far more toxic than hydrocarbons.

Another interesting finding was that the major constituent of *C. capitatus* volatile oil, carvacrol (>70%), is highly toxic, which holds for the whole mixture of this volatile oil. On the basis of the differential effect of the constituents of a volatile oil, its toxicity might be evaluated from its qualitative and quantitative composition (the identity of the effect imposed by each compound, applied either alone or in combination with others, being previously ascertained).

18.3.1 Susceptible stages to allelochemic inhibition and reversibility of the process

Experiments were conducted in order (i) to define the stages at which the toxic effects are exerted and (ii) to check for the reversibility of the inhibition. In this case, the experimental material was the volatile oil of *R. officinalis*, proved to be the most toxic against other plant species, and

TABLE 18.3 *Seed germination and seedling elongation of* Coridothymus capitatus *in presence of standard isoprenoid compounds, constituents of its volatile oil. The quantity tested was 10 μl (after Vokou and Margaris, 1982)*

Isoprenoid compounds	Germination (%)	% of control	Seedling length (cm)	% of control
Control	82		1.7	
Hydrocarbons:				
$C_{10}H_{16}$: myrcene	80	98	1.6	94
$C_{10}H_{16}$: limonene	61	74	0.9	53
$C_{10}H_{16}$: γ-terpinene	81	99	1.6	94
$C_{10}H_{16}$: α-pinene	45	55	0.2	12
$C_{10}H_{16}$: β-pinene	70	85	1.1	65
$C_{10}H_{14}$: *p*-cymene	52	63	1.5	88
$C_{15}H_{24}$: caryophyllene	80	98	1.6	94
$C_{20}H_{32}$: camphene	66	80	0.6	35
Alcohols:				
$C_8H_{16}O$: 1-octen-3-ol	0	0	0	0
$C_{10}H_{18}O$: linalol	0	0	0	0
Phenols:				
$C_{10}H_{14}O$: carvacrol	8	10	0.1	6

seeds of *Cucumis sativus* L., used repeatedly as experimental material in related works because of its rapid and uniform germination (Muller, 1966; McCahon *et al.*, 1973; Weaver and Klarich, 1977). As in the case of wild legumes, *R. officinalis* volatile oil in the quantities tested (1, 3, 6, 10 and 20 μl) did not affect seed germination but severely inhibited seedling elongation (by 14, 31, 80, 94 and 96%, respectively).

The first test was to discover if any toxic effect was exerted on the seeds at the dry state. For this reason seeds of *C. sativus* were hermetically closed in petri dishes, in only aerial contact with *R. officinalis* volatile oil (quantities applied: 6, 10 and 20 μl). After 25 days the seeds were transferred into clean petri dishes, water was added and measurements were made after another 5 days. There was no significant difference in seed germination and seedling length among the control and the other three treatments. This means that volatile oils are unlikely to exert any inhibitory effect on the seeds at the dry state.

The next test was to discover if there was any difference in the toxic effect (as manifested in seedling elongation), dependent on the addition of the inhibitory agent before or after germination (radicle emergence taken as criterion of germination). To this aim *C. sativus* seeds were first allowed to germinate and were then exposed to *R. officinalis* volatile oil.

According to their initial radicle length, they were divided into five groups, viz. of mean initial length of 0.1 cm, which corresponds to mere radicle emergence, of 0.3, 0.6, 0.9 and 1.8 cm, and were treated separately with different quantities of *R. officinalis* volatile oil. In parallel, there was the treatment of non-pregerminated seeds, put in the atmosphere of *R. officinalis* volatile oil at the same time with water addition, and the control treatments without any addition of volatile oil. The level of inhibition for each group and quantity added was estimated according to the following formula

$$\% \text{ Inhibition} = \frac{\Delta L_C - \Delta L_R}{\Delta L_C} \times 100$$

where ΔL_C = difference of mean initial seedling length from mean final seedling length of the respective control treatment.
ΔL_R = difference of mean initial seedling length from mean final length of seedlings exposed to *R. officinalis* volatile oils.

The bundle of curves (Fig. 18.1(d)), each one corresponding to a different initial seedling length, shows that there is no difference among the levels of inhibition imposed by each quantity of *R. officinalis* volatile oil on the different groups of *C. sativus* seedlings and seeds. These results again suggest that the effect of volatile oils is exerted on water-imbibed seeds and it is equally strong even if the seedling is fairly elongated.

While the tissues of inhibited seedlings – especially those grown in the atmosphere of 10 and 20 µl of volatile oil – seem to be seriously damaged (radicles in particular), the potential of recovery is retained and is evidently expressed when the toxic agent is removed. In Table 18.4 it can be seen that the inhibited *C. sativus* seedlings of the treatments 6 and

TABLE 18.4 *Potential of recovery of inhibited* Cucumis sativus *seedlings pretreated with the most toxic quantities of 6, 10 and 20 µl of* Rosmarinus officinalis *(R) volatile oil*

Stages	Treatment		Seedling length (cm)
	Control		7.1 ± 0.5
1. *Cucumis sativus* seeds treated with R6, R10, R20. Seedling length after 5 days: 0.1–0.8 cm	R6	→	7.7 ± 0.5
2. Transfer to clean atmosphere	R10	→	7.0 ± 0.4
3. Seedling length measurement (7 days)	R20	→	3.3 ± 0.3

→ designates treatment imposed before transfer far from the toxic agent.
Control = control of seven days.

10 μl *R. officinalis* volatile oil have totally recovered after 7 days, with their mean length not differing from the untreated 7-day-old seedlings; those of the treatment 20 μl, though differing significantly, have started to grow with obvious signs of quick recovery, e.g. with the emergence of lateral rootlets taking the place of the main radicle, almost completely destroyed.

This reversibility of the inhibitory effect was also proved in the case of *C. capitatus*. In the 1-octen-3-ol (one of its volatile oil constituents) treatment, for example, seed germination was completely inhibited (Table 18.3). After 8 days of the seeds transfer to clean petri dishes, the percentage of germination rose to 55% and the seedling length to 2.3 cm.

18.3.2 Evidence of allelopathy in the field

Toxicity of volatile oils is a fact. The real problem, however, is whether concentrations that naturally occur in the field might induce such inhibitory effects.

To test this hypothesis experiments were carried with *C. capitatus* litter, because it is with this material that seeds are in immediate contact. The procedure followed was the same as in the case of leaves. As can be seen from Table 18.2, no inhibitory effect was exerted with any of the quantities applied. The absence of phytotoxicity could not be attributed to alteration of the composition of the volatile oil because of physical or biological factors, since both the phenolic as well as other oxygen-containing compounds proved to be very toxic, contribute in the litter volatile oil with high percentages (Vokou, 1983). If we compare the inhibition of seed germination and seedling elongation exerted by the leaves and the volatile oil of *C. capitatus*, we find that the quantities of leaves tested correspond in fact to 1–6 μl. This means that the drastic quantity in the leaves is ≃1/10 to 1/20 of that estimated by hydro-distillation. Therefore, the drastic quantities in the litter for all up to 5 g treatments seem to be lower than 1 μl. The absence of any inhibitory effect – though expected – at the 5 g treatment might be due to differential emanation and diffusion characteristics of the material used in the quantities applied (high compactness of litter in the 5 g treatment should not be disregarded).

Parallel to that, the spatial distribution of *C. capitatus* mature individuals was studied in Velestino, Thessaly, an area where it is the only dominant woody plant. Two 25 m surfaces, differing in the density of coverage, were mapped. Estimation of pattern was conducted by using the nearest-neighbour technique. It was found (Vokou and Margaris, 1986b) that in the more densely covered site the distribution is normal, whereas in the less densely covered it is random. Absence of clumps suggest functioning of factors impeding aggregation. Random distributional patterns were also found in homogeneous stands of the aromatic shrubs *Artemisia tridentata* and *A. cana* (Hazlett and Hoffman, 1975).

Working in another phryganic ecosystem (Mt Hymettus, Athens), Argyris (1977) studied the seed ecology of the dominant woody plants. He found that *C. capitatus* produces a large number of seeds (524 M^{-2}) in comparison to the other phryganic woody species. These seeds are dormancy-free (proved under laboratory conditions), and readily germinate with high percentages (80–95%) in a short period (2–3 days) under a fairly wide range of favourable temperatures (10–25°C). In the field, however, during the period October–May (1974–1975) that measurements were made, only 67 seedlings were counted in an area of 135 M^2. Additionally, in spite of their quick and high germinability, and contrary to what happened in the case of other, non-aromatic phryganic species studied, thyme seedlings had not appeared in October, though the conditions prevailing (rainfalls, temperature) were very favourable. The distribution of seedlings around the mother plant was also estimated. Four zones were defined; the litter zone, the adjacent 10 cm zone (0–10 cm), a 20 cm zone further (10–30 cm) and, beyond that, another 20 cm zone (30–50 cm). Thyme seedlings, in each zone, accounted for 14, 41, 39 and 5% of the total, respectively. Relative patterns of seedling emergence were reported for the aromatic shrub *Artemisia herba-alba* in the Negev desert of Israel (Friedman and Orshan, 1975). The low percentage of seedlings in the litter zone was attributed (Argyris, 1977) to inhibitory effects of *C. capitatus* litter. In contrast, seedlings of most of the other woody phryganic species are most frequent in this zone.

Both the distributional patterns of mature plants and seedlings, as well as the germination characteristics of *C. capitatus*, suggest mediation of a factor regulating the density of the mature plants and controlling seed germination and seedling establishment. A firm conclusion could not be drawn concerning the identity of this factor on the basis of data available. Nevertheless, allelopathy is a suspect.

It should be noted that the newly fell leaf litter (August) is very rich in volatile oils, ≃3.7% (Vokou and Margaris, 1986a). This concentration gradually falls down to the value of 0.2% at the end of January. In late October, after the first rainfalls which initiate seed germination of phryganic species, it is 0.8%. Though low, comparatively to the time of leaf fall, it might be sufficient to impede seedling emergence, observed in the field by Argyris (1977). It was argued in the past that if volatile oils were to exert any effect, they should get primarily adsorbed on the soil particles (Muller and del Moral, 1966; Kelsey *et al.*, 1978). However, at least in the case of *C. capitatus*, it does not seem to be an absolute prerequisite. *C. capitatus* drops its leaves in August along with the mature seeds. This material is held at the base of the plant, taking the form of a cone. It is there, and not deeply in the soil, that seed germination takes place. However, this microenvironment seems to be hostile for seed germination and seedling emergence. It could be argued that litter should first discharge (at least partly) the toxic volatile oils to allow for these processes to take place. Furthermore, it might be expected that seeds escaping the influence of the mother plant by being transported to other sites could germinate and establish easily, which is evident by the increased seedling emergence beyond the litter zone.

In a recent study on seed germination of *C. capitatus*, it is suggested that autotoxicity of this plant is an important ecophysiological mechanism regulating the germination process (Thanos and Skarou, 1988). Its seeds, which holds for many species of the Lamiaceae family, are dispersed – probably through zoochory – as one-seeded nutlets enclosed in groups of four within the desiccated fruiting calyx. The calyx opens only in high humidity (hygrochasy). Nevertheless, in the high majority of the instances observed, the seeds germinated inside the calyx, which thus serves as a kind of water-filled pot. On the other hand, calyces (like leaves) bear glands with volatile oil. Therefore, seed germination and seedling elongation could be manifested only after a considerable amount of this inhibitor is eliminated – washed out by rain or through microbial processes (Vokou *et al.*, 1984), initiated by rain. This rain-measuring device seems to function as a delay-of-germination mechanism. Since *C. capitatus* seeds are completely non-dormant when ripe, and their germination is quite fast, this delay mechanism would be ecologically advantageous, especially in the rapid alternation of dry and wet conditions, met at the beginning of the rainy season in a mediterranean-type climate. Furthermore, *C. capitatus* seed germination, though optimal in darkness or daylight, can be inhibited by Far Red irradiation and this fact implies that germination might not be manifested under a dense canopy. The authors conclude that thanks to both a shade avoidance and a rain gauge mechanism seed germination of this plant and subsequent seedling establishment is secured in open, illuminated sites, well into the rainy period.

The fact that *C. capitatus* volatile oil toxicity is best manifested against its own seeds and seedlings and less against the nearby growing annual legumes, might be explained on the basis of requirements and adaptations that plants have evolved to survive in an environment with limited resources (Vokou and Margaris, 1986b). The hard period in the mediterranean climate is summer, characterized by prolonged drought. *C. capitatus* is a seasonal dimorphic shrub with a shallow root system with a high capacity to absorb water. Water availability is extremely low in summer. Unless there is sufficient space among plants drawing water from the same resource pool, and with the same mechanism, survivorship to the adult stage would be very low. During that period, annuals are in the form of seeds, not competing for the limited water resources.

Avoidance of competition with annuals during the hard period might be a reason why *C. capitatus* has evolved such a mechanism of suppressing mostly its own seeds and seedlings and much less those of annuals.

However, in a related study, in Israel, Katz *et al.* (1987) found that *C. capitatus* inhibits in the field seedling emergence of the neighbouring non-legume annuals, *Plantago psyllium* and *Erucaria hispanica*. These authors suggest the existence of two types of this aromatic shrub; the so-called 'aggressive' and 'non-aggressive' type. This differentiation is made on the basis of formation of annuals-free belts around the shrubs. The emergence of deliberately planted annuals in the vicinity of the 'aggressive' shrubs was considerably reduced both when compared with those planted as controls (10 M away from the shrub) and those planted close to 'non-aggressive' shrubs. Experiments under laboratory conditions showed that seed germination of both *P. psyllium* and *E. hispanica* was inhibited by diffusates or volatiles derived from blended shoots of *C. capitatus*, as well as by its volatile oil. The variability of annuals suppression by *C. capitatus* shrubs could not be explained. Nonetheless, the authors suggest, as possible components determining this variability, water stress dependent changes in the amount or in the composition of volatile oils, as well as the intraspecific chemical variability demonstrated in various members of Lamiaceae.

As far as the autotoxicity of *C. capitatus* is concerned, it results in lowering the reproductive rate of the population. By this mechanism, intraspecific competition is avoided and the seed deposit is not wasted. Such density-dependent regulations are beneficient when populations reach or exceed optimal densities by preventing them from overtaxing and eventually overexploiting or destroying the resources of their habitats. As reported by Friedman and Orshan (1975), density regulation in perennial plant populations may prevent overcrowding and enhance emergence and survival of seedlings following catastrophic loss of adult plants.

18.4 AROMATIC SHRUBS AFFECTING MICROBES

The antimicrobial activity of volatile oils is well documented. A lot of information and literature references on that matter can be found in the relevant chapters of Rice's *Allelopathy* (1974) and the updated review (1978), in Belaiche's treatise (1979), and in the review by Janssen *et al.* (1987).

The fact that many different volatile oils have proved to be potent antimicrobial agents against both bacteria and fungi, should not lead to generalizations of universal toxicity, taking into account the diversity and complexity both of their constituents and of the microorganisms themselves. It could also be argued that microorganisms from the environ-

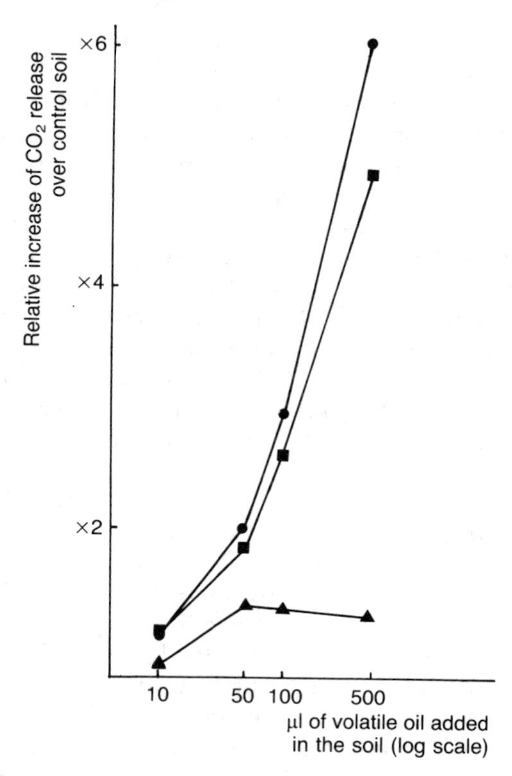

Fig. 18.2 CO_2 evolved from soil seven days after addition of increasing quantities of *Satureja thymbra* (●), *Coridothymus capitatus* (■) and *Rosmarinus officinalis* (▲) volatile oils. Results are expressed as relative increase in comparison to the control. (After Vokou and Margaris, 1984. Reproduced with permission, Vokou, D. and Pergamon Press Ltd, UK.)

ment where aromatic plants grow or that live on their various parts should have been adapted to cope with the inhibitory effects of their constituents, or even take advantage of them. Additionally, we should remark that one of the characteristics of mediterranean-type ecosystems is that the decomposition process and , therefore, carbon and nutrients recycling, is impeded by the peculiarities of a mediterranean climate. It is restricted mostly to autumn and spring, when low temperatures and drought are no longer limiting factors. Taking into account the remarkable contribution of aromatic plants in these systems, the inhibition of microbial activity by their volatile oils would retard even more the decomposition process.

Therefore, experiments were carried out in order to study the effect of volatile oils on soil microbial activity. The total soil respiration of samples, taken from the area where aromatic plants grow, was determined. Addition of *C. capitatus* and *S. thymbra* volatile oils in the soil samples significantly activated soil respiration (Fig. 18.2). This increase of soil respiration, however, might have been a primary or a secondary effect of volatile oils: primary, by directly activating soil microbes; secondary, by killing some to provide substrate easily decomposable to others. Experiments dealing with the specific effects of these compounds, separately on the fungi and bacteria, might give the answer.

The fungi tested were *Mucor hiemalis* and *Penicillium citrinum*. *M. hiemalis* was selected because it is a ubiquitous soil fungus with rapid growth; it shows a preference for alkaline soils (characteristic of the phryganic ecosystems of Greece), predominates in the upper soil layers (where any effect of volatile oils is expected to be exerted) and colonizes freshly-added substrates, such as litter (charged with volatile oils in the case of aromatic plants). *P. citrinum* was chosen on account of the initial high percentage germination of its conidia on water agar and its sensitivity to soil fungistasis (Lynch, 1982).

S. thymbra proved to exert a strong allelopathic effect on the mycelial growth of *M. hiemalis*. Concentrations as low as $0.25\,\mu l\ ml^{-1}$ in culture medium were sufficient to inhibit completely any growth of the fungus.

P. citrinum spore germination was suppressed after addition of various quantities of *S. thymbra* and *R. officinalis* volatile oils to soil samples. The percentages of spore germination after 18 h were 25, 6.8 and 0.8 for the filter paper, the control, and 1 μl *S. thymbra* oil per gram of soil, respectively (Vokou *et al.*, 1984). When the period of incubation was prolonged to 42 h, the number of spores germinating in the filter paper and control treatments increased, but remained unchanged in the soil treated with *S. thymbra* volatile oil.

Therefore, a strongly inhibitory allelopathic effect is exerted on the fungi tested. This is not the case, though, with bacteria, the number of which is raised more than 80 times, when *S. thymbra* volatile oil is applied at 1 μl per gram of soil. Additionally, bacterial numbers increased with time after addition of volatile oils to soil samples (Vokou *et al.*, 1984). The increase in number of bacteria implies provision of a carbon and energy source, which, as proved, were the volatile oils themselves (Vokou *et al.*, 1984). As seen in Fig. 18.3, soil respiration could be maintained significantly higher than that of the control even during a period of approximately one year, by the mere addition of 0.1 ml of volatile oil, every 7 (initial phase of 21 weeks) or 15 days (subsequent 27 weeks) in 150 g soil samples (Vokou and Margaris, 1988). The ability of bacteria to decompose volatile oils and individual terpenes, constituting them, was also proved by Gibbon and Pirt (1971), Gibbon *et al.* (1972) and Gunsalus and Marshall (1971).

While volatile oils of the aromatic shrubs tested inhibited fungal spore germination and mycelial growth of the species studied, they activated some bacteria that could use them as a substrate for growth. Presumably these bacteria have been adapted to coexist and take benefit from other-

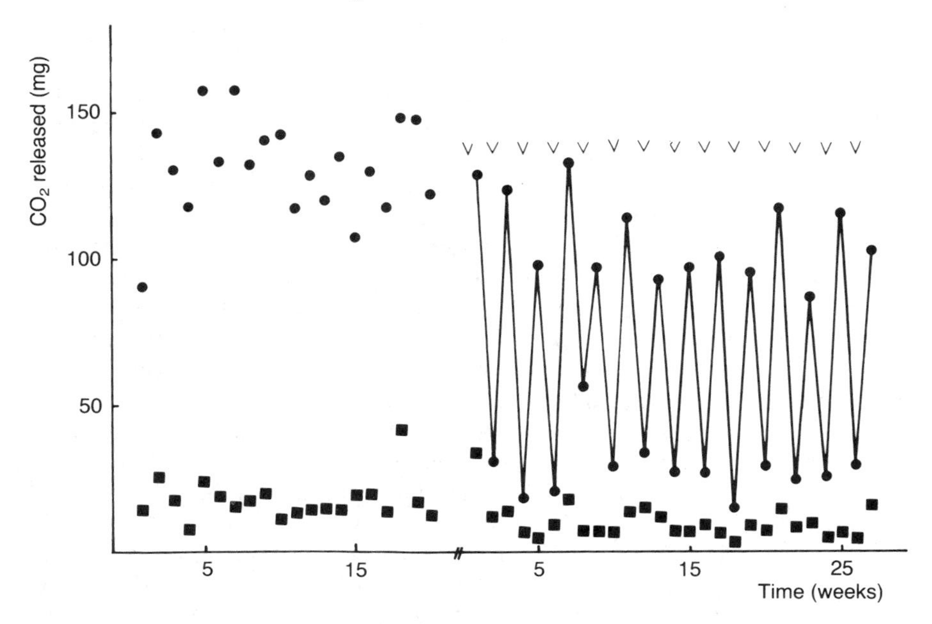

Fig. 18.3 Weekly values of CO_2 release from 150 g soil samples treated with 100 µl of *Satureja thymbra* volatile oil, added every seven days for a period of 21 weeks and every 15 days for a period of 27 weeks; V represents time of addition of the volatile oil. (After Vokou and Margaris, 1988. Reproduced with permission, Vokou, D., and Veb Gustav Fischer Verlag, Jena, Germany.)

wise antimicrobial compounds. Since the litter of aromatic plants contains large quantities of volatile oils, it might be expected that the decomposition process in phryganic ecosystems is mediated by allelopathic interactions, causing shifting of the population balance in soil from fungi to bacteria, particularly in the surface horizons. This activation of bacteria by volatile oils might be considered as an adaptive mechanism, where phryganic systems develop.

Katz *et al.* (1987), in their study concerning the allelopathic potential of *C. capitatus*, have also found that incubation of green shoots of this aromatic shrub in soil for 7 days induced a considerable increase in the population levels of actinomycetes; by 9.5-fold in soils sampled from 'non-aggressive' and by 36.2-fold in samples from 'aggressive' ones. Since various isolates of actinomycetes inhibited germination of test plants and subsequent seedling development *in vitro*, the authors suggest that annual suppression around the canopies of *C. capitatus* might be the consequence of a direct alleopathic effect of diffusates and/or volatiles from this shrub, associated with a synergistic interaction of an indirect inhibitory effect of secondary metabolites produced by soil-borne actinomycetes, enhanced by the shrubs emanated metabolites.

18.5 CONCLUSIONS AND PROSPECTS

Volatile oils are a group of natural products chemically well studied; the reason being their use in cosmetics, pharmaceutical and alimentary industries. However, their function in biological systems is hardly known. Only their antimicrobial activity, particularly against pathogens, is well documented, but mostly in a purely descriptive way.

As with most of the products of secondary metabolism, they have no apparent physiological role for the producing plant. Nevertheless, they may play important roles in the ecological context. Mediating various processes in the frame of an ecosystem, they become indirectly beneficial to plants. Although there is fairly adequate information concerning the involvement of volatile oils in processes of adaptive character in certain ecosystem types, such as the mediterranean, this is not true for most of them; this is a gap to be filled. Research areas in this field that ask for more experimental evidence and stronger argumentation concern (i) the contribution of allelopathy, induced by volatile oils, into the vegetation patterning, (ii) the rates of their production and emanation as well as their persistence in the soil environment, (iii) the possible synergistic or antagonistic effects with other compounds present, (iv) their effects on mineral absorption and particularly on water relations, given that aromatic plants which are rich in volatile oils occur mostly in xeric habitats and (v) their interference with microorganisms, which have either positive or negative effects to plant growth, through which they would indirectly affect the higher plants themselves.

REFERENCES

Argyris, J. (1977) Seed ecology of some phryganic species. PhD Thesis, University of Athens, Greece.

Aschmann, H. (1973) Distribution and peculiarity of Mediterranean ecosystems, in *Mediterranean-Type Ecosystems* (eds F. di Castri and H. Mooney), Springer-Verlag, Berlin, pp. 11–19.

Bartholomew, B. (1970) Bare zone between California shrub and grassland communities: the role of animals. *Science*, **170**, 1210–12.

Belaiche, P. (1979) *Traité de Phytothérapie et d'Aromathérapie I. L'Aromatogramme*, Maloine S. A., Paris.

Christensen, N. L. and Muller, C. H. (1975) Effects of fire on factors controlling plant growth in *Adenostoma* chaparral. *Ecol. Monogr.*, **45**, 29–55.

Diamantopoulos, J. (1983) Structure and distribution of phryganic ecosystems of Greece. PhD Thesis, Univ. of Thessaloniki, Greece.

Elad, Y. and Misaghi, I. J. (1985) Chemically mediated interactions between plants and other organisms, in *Recent Advances in Phytochemistry* (eds G. A. Cooper-Driver, T. Swain, and E. E. Conn), Plenum Press, New York and London, pp. 21–45.

Friedman, J. and Orshan, G. (1975) The distribution, emergence and survival of seedling of *Artemisia herba-alba* Asso. in the Negev desert of Israel in relation to distance from the adult plants. *J. Ecol.*, **63**, 627–32.

Gibbon, G. H. and Pirt, S. J. (1971) The degradation of α-pinene by *Pseudomonas* PX1. *FEBS Letters*, **18**, 103–5.

Gibbon, G. H., Millis, N. F. and Pirt S. J. (1972) Degradation of α-pinene by bacteria. *Proc. IVth Int. Fermentation Symp: Fermentation Technology Today*, pp. 609–12.

Gunsalus, I. C. and Marshall, V. P. (1971) Monoterpene dissimilation: chemical and genetic models. *CRC Crit. Rev. Microbiol.*, 291–310.

Halligan, J. P. (1973) Bare areas associated with shrub stands in grasslands: the case of *Artemisia californica. Bioscience*, **23**, 429–32.

Halligan, J. P. (1975) Toxic terpenes from *Artemisia californica. Ecology*, **56**, 990–1003.

Halligan, J. P. (1976) Toxicity of *Artemisia californica* to four associated herb species. *Amer. Midl. Nat.*, **95**, 406–21.

Hazlett, D. L. and Hoffman, G. R. (1975) Plant species distributional patterns in *Artemisia tridentata* and *Artemisia cana* dominated vegetation in western North Dakota. *Bot. Gaz.*, **136**, 72–7.

Janssen, A. M., Scheffer, J. J. C. and Baerheim-Svendsen, A. (1987) Antimicrobial activity of essential oils: a 1976–1986 literature review. Aspects of the test methods. *Planta Medica*, **53**, 395–8.

Katz, D. A., Sneh, B. and Friedman, J. (1987) The allelopathic potential of *Coridothymus capitatus* L. (Labiatae). Preliminary studies on the roles of the shrub in the inhibition of annuals germination and/or to promote allelopathically active actinomycetes. *Plant and Soil*, **98**, 53–66.

Kelsey, R. G., Stevenson, T. T., Scholl J. P. *et al.* (1978) The chemical composition of the litter and soil in a community of *Artemisia tridentata* spp. *vaseyana. Biochem. Syst. Ecol.*, **6**, 193–200.

Lynch, J. M. (1982) Limits to microbial growth in soil. *J. Gen. Microbiol.*, **128**, 405–10.

Margaris, N. S. (1981) Adaptive strategies in plants dominating Mediterranean-type ecosystems, in *Mediterranean-Type Shrublands* (eds F. di Castri, D. W. Goodal, and R. L. Specht), Elsevier, Amsterdam, pp. 309–15.

McCahon, C. B., Kelsey R. G., Sheridan R. P. and Shafizadeh, F. (1973) Physiological effects of compounds extracted from sagebrush, *Bull. Torrey Bot. Club*, **100**, 23–8.

Muller, C. H. (1965) Inhibitory terpenes volatilized from *Salvia* shrubs. *Bull. Torrey Bot. Club*, **92**, 38–45.

Muller, C. H. (1966) The role of chemical inhibition (allelopathy) in vegetational composition. *Bull. Torrey Bot. Club*, **93**, 332–51.

Muller, C. H. and del Moral, R. (1966) Soil toxicity induced by terpenes from *Salvia leucophylla. Bull. Torrey Bot. Club*, **93**, 130–7.

Muller, C. H. and del Moral, R. (1971) Role of animals in suppression of herbs by shrubs. *Science*, **173**, 462–3.

Muller, C. H., Muller, W. H. and Haines, B. L. (1964) Volatile growth inhibitors produced by shrubs. *Science*, **143**, 471–3.

Muller, C. H., Hanawalt, R. B. and McPherson, J. K. (1968a) Allelopathic control of herb growth in the fire cycle of California chaparral. *Bull. Torrey Bot. Club*, **95**, 225–31.

Muller, W. H. and Hauge, R. (1967) Volatile growth inhibitors produced by *Salvia leucophylla*: effect on seedling anatomy. *Bull. Torrey Bot. Club*, **94**, 182–91.

Muller, W. H., Lorber, P. and Haley, B. (1968b) Volatile growth inhibitors produced by *Salvia leucophylla*: effect on seedling growth and respiration. *Bull. Torrey Bot. Club*, **95**, 415–22.

Rice, E. (1974) *Allelopathy*, Academic Press, New York.

Rice, E. (1978) Allelopathy – an update. *Bot. Rev.*, **45**, 15–109.

Thanos, C. A. and Skarou, F. (1988) The ecophysiology of seed germination in thyme (*Coridothymus capitatus*). *Book of Abstracts, 6th Congress of FESPP*, Split (Yugoslavia).

Vokou, D. (1983) Volatile oils and their role in phryganic ecosystems. PhD Thesis, Univ. of Thessaloniki Greece.

Vokou, D. and Bessière, J. -M. (1985) Volatile constituents of *Teucrium polium. J. Nat. Products (Lloydia)*, **48**, 498–9.

Vokou, D. and Margaris, N. S. (1982) Volatile oils as allelopathic agents, in *Aromatic Plants: Basic and Applied Aspects* (eds N. S. Margaris, A. Koedam, and D. Vokou), Martinus Nijhoff, The Hague, pp. 59–72.

Vokou, D. and Margaris, N. S. (1986a) Variation of volatile oil concentration of Mediterranean aromatic shrubs *Thymus capitatus* Hoffmag et Link, *Satureja thymbra* L., *Teucrium polium* L., and *Rosmarinus officinalis* L. *Int. J. Biometeor.*, **30**, 147–55.

Vokou, D. and Margaris, N. S. (1986b) Autoallelopathy of *Thymus capitatus. Oecol. Plant.*, **7**, 157–63.

Vokou, D. and Margaris, N. S. (1988) Decomposition of terpenes by soil microorganisms. *Pedobiologia*, **31**, 413–19.

Vokou, D., Margaris, N. S. and Lynch, J. M. (1984) Effects of volatile oils from aromatic shrubs on soil microorganisms. *Soil Biol. Biochem.*, **16**, 509–13.

Weaver, T. W. and Klarich, D. (1977) Toxic effects of volatile exudates from *Artemisia tridentata* Nutt. on soil microbes. *Amer. Midl. Nat.*, **97**, 508–12.

CHAPTER 19

Possible role of allelopathy in growth inhibition of softwood seedlings in Newfoundland

A. U. Mallik

19.1 INTRODUCTION

Failure of natural regeneration on harvested and burned areas and very poor growth of planted seedlings of softwood species, particularly black spruce (*Picea mariana*),[1] on many sites of central Newfoundland is a major silvicultural problem on the island. Some sites initially are sufficiently 'stocked' but one year later the site may experience 40% seedling mortality (W. J. Meades, pers. comm.). It is often the case that even if seedlings survive, their growth rate is extremely poor (cf. Botwood, Newfoundland). In the forests of central Newfoundland, a native ericaceous dwarf shrub called *Kalmia angustifolia* (hereafter mentioned as *Kalmia*) occurs most abundantly as an understorey species (Ryan, 1978). This ericaceous weed is particularly dominant in more open stands of black spruce (Vincent, 1965; Richardson and Peter Hall, 1973a, b; Richardson, 1974a, b, 1981), where it may constitute 80–95% cover under the forest canopy. *Kalmia* also dominates extensive areas of heathlands in Newfoundland (Meades, 1983a, b). It is also an important understorey component in the forests of other Maritime Provinces of Canada and of the eastern United States (Hall *et al.*, 1973). In Newfoundland, large areas of productive forest land are being invaded by this vigorously growing plant, particularly following fire and logging (Page, 1970; van Nostrand, 1971; Richardson, 1975). In the presence of *Kalmia*, the growth of black spruce seedlings is dramatically reduced (Candy, 1951; Wall, 1977; Richardson, 1979). Growth of Jack pine (*Pinus banksiana*) trees in New Brunswick plantations is also reported to be inhibited by increasing densities of *Kalmia* (Krause, 1986). It has been argued that such growth inhibition is attributable to allelopathy and competition from

[1] Nomenclature for vascular plants follows Scoggan (1979).

Allelopathy: Basic and applied aspects Edited by S. J. H. Rizvi and V. Rizvi
Published in 1992 by Chapman & Hall, London ISBN 0 412 39400 6

the vigorous vegetative regrowth of *Kalmia*. It has also been suggested that *Kalmia* may accumulate phytotoxic substances in the leaves which, when leached into soil through precipitation, may make the habitat inhospitable to softwood tree seed germination and seedling growth (Peterson, 1965). Mallik (1987) suggested that the allelopathic compounds involved in the growth inhibition of softwood seedlings are probably the decomposition product(s) of *Kalmia* as well as black spruce organic matter (Mallik and Newton, 1988). Rice (1984) updated the various aspects of allelopathy in relation to natural plant communities, crop plants and plants important in forestry and horticulture.

While it has been widely recognized that *Kalmia* presents a serious silvicultural problem in Newfoundland and eastern North America it is not until very recently that the allelopathic growth inhibition of softwood seedlings due to *Kalmia* – black spruce organic matter received attention in terms of scientific research (Mallik, 1987; Mallik and Newton, 1988). The present chapter summarizes the results obtained so far from various laboratory, greenhouse and field experiments and observations in Newfoundland. Since the study of *Kalmia* allelopathy has been initiated very recently, many of the results reported here are preliminary in nature. Opportunity for future research and its practical implications for forest management is discussed.

19.2 INHIBITORY EFFECTS OF *KALMIA* WATER EXTRACTS

A dramatic inhibition of primary root growth of black spruce was obtained compared with controls when water extracts (at room temperature) of *Kalmia* leaf, root, litter and soil was applied on petri dishes keeping the seeds moist with the extracts (Table 19.1). Root growth was totally inhibited due to *Kalmia* litter extracts (Fig. 19.1) Bioassays performed with *Kalmia* humus samples from different sites in central and eastern Newfoundland showed large variations in the extent of root inhibition of

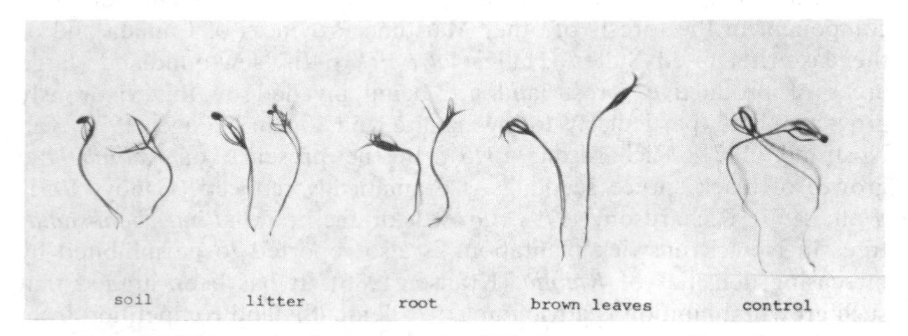

Fig. 19.1 Inhibition of primary growth of black spruce seedlings due to water extract of *Kalmia*. Note the total root growth inhibition in the *Kalmia* litter and soil extract treated seedlings.

TABLE 19.1 *Effect of water extracts of* Kalmia *leaves, root, litter and soil on germination of black spruce seeds and length of primary root and shoot of the seedlings (after Mallik, 1987)*

Site	Treatment	Germination (%) ±SE	Length (mm) Root ±SE	Length (mm) Shoot ±SE
Harbour Grace	Control	92 ± 2.5	19 ± 1.0a	31 ± 0.4a
	Green leaf	94 ± 0.8	15 ± 0.6b	28 ± 0.5b
	Brown leaf	90 ± 1.0	13 ± 0.8b	29 ± 1.5bc
	Root	90 ± 1.6	7 ± 1.1c	27 ± 0.8bc
	Litter	94 ± 0.8	0 ± 0.0d	26 ± 1.1cd
	Soil	91 ± 1.9	3 ± 1.2c	26 ± 0.9d

Each value represents the mean ±standard error of 50 seedlings distributed in groups of ten per petri dish. Treatments with the same letter are not significantly different; $p \leq 0.05$ with one-way ANOVA.

black spruce seedlings. Root extracts and soil extracts of *Kalmia* were also very inhibitory. Leaf extracts of *Kalmia* showed some inhibitory effects but not as dramatic as in other treatments (Table 19.1). Percentage germination due to treatments were not affected significantly. Although shoot length was significantly less in treated plants, it was not as dramatic a decrease as was observed in the root growth.

Light microscopy of the root–shoot transition region of the affected seedlings revealed that they did not produce normal radical growth. *Kalmia* litter and soil extract treated seedlings were incapable of producing a normal primary root with a root cap and root hairs, instead the root–shoot transition region was extended, swollen and curved and had a blunt tip (Figs 19.2(a)–(c)).

When black spruce, red pine (*Pinus resinosa*) and balsam fir (*Abies balsamea*) seeds were planted directly on *Kalmia* soil (mostly humus) and kept soaked in petri dishes with distilled water, germination percentages again did not differ significantly. However, length of primary roots and shoots was significantly less as compared with controls for all species. Growth inhibition due to *Kalmia* humus was most pronounced in black spruce and red pine rather than in balsam fir. This result is consistent with the field observations where natural regeneration was more affected in black spruce and red pine stands infested with *Kalmia* as compared with the balsam fir stands. Balsam fir seedlings however, are much more robust than black spruce and red pine. *Kalmia* humus is considerably acidic (pH 3.0–3.5) and results in Table 19.2 show that the root growth of balsam fir was much better than that of red pine and black spruce on *Kalmia* humus of similar acidity.

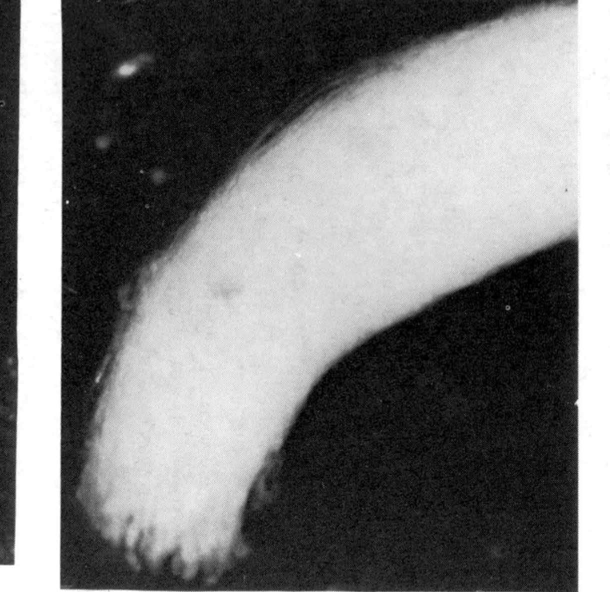

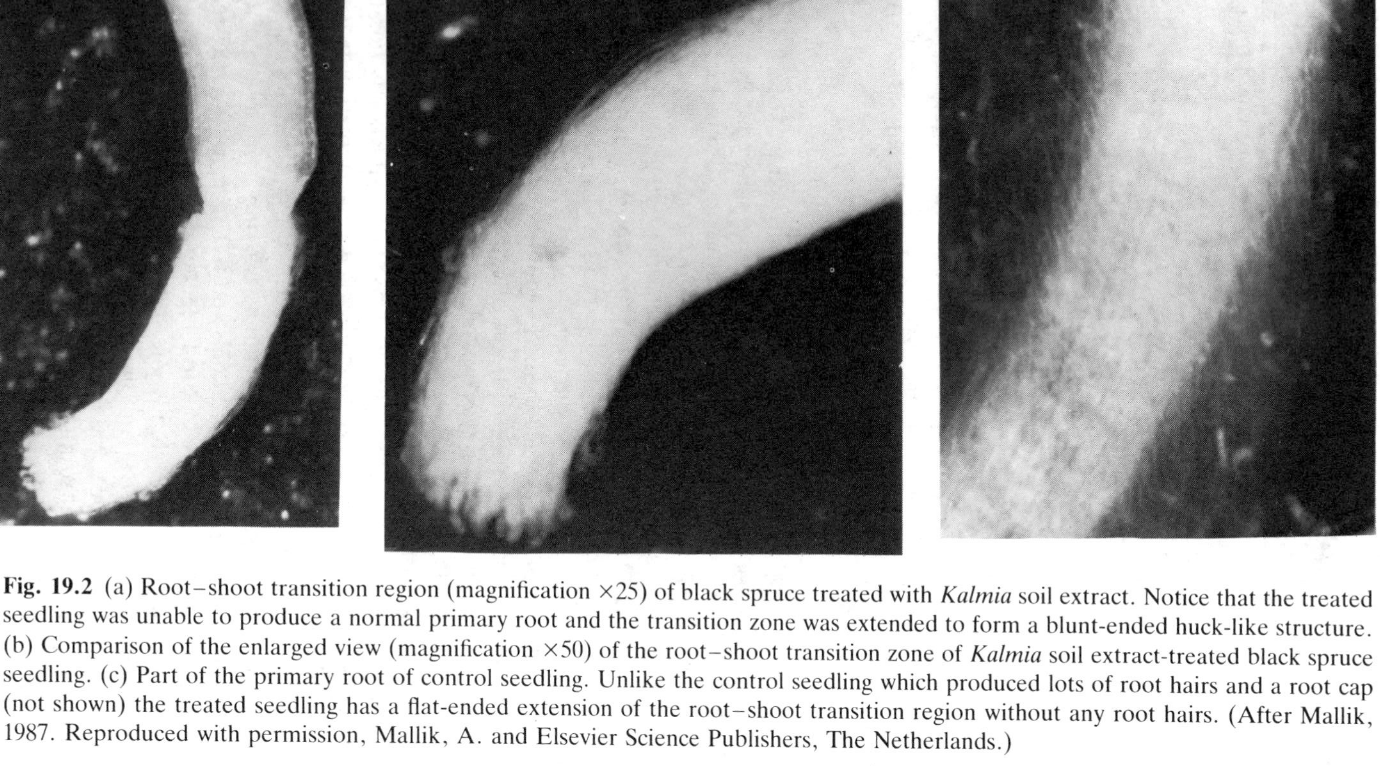

Fig. 19.2 (a) Root–shoot transition region (magnification ×25) of black spruce treated with *Kalmia* soil extract. Notice that the treated seedling was unable to produce a normal primary root and the transition zone was extended to form a blunt-ended huck-like structure. (b) Comparison of the enlarged view (magnification ×50) of the root–shoot transition zone of *Kalmia* soil extract-treated black spruce seedling. (c) Part of the primary root of control seedling. Unlike the control seedling which produced lots of root hairs and a root cap (not shown) the treated seedling has a flat-ended extension of the root–shoot transition region without any root hairs. (After Mallik, 1987. Reproduced with permission, Mallik, A. and Elsevier Science Publishers, The Netherlands.)

TABLE 19.2 *Effect of forest floor substrates of central Newfoundland on germination and growth of primary root and shoot of black spruce, red pine and balsam fir (modified after Thompson and Mallik, 1989)*

			Primary root		Shoot		
Species	Treatment	Germination (%)	Length (mm)	Fresh weight (mg)	Length (mm)	Fresh weight (mg)	pH of medium
Black spruce	Control (sand)	96a	11.4a	1.0a	28.3a	5.1a	6.3a
	Kalmia soil	96a	6.0b	0.7b	17.9b	3.8b	3.0b
Red pine	Control (sand)	84a	13.6a	3.6a	44.5a	25.5a	7.1a
	Kalmia soil	96a	3.9b	1.2b	30.3b	21.2b	3.2b
Balsam fir	Control (sand)	86a	27.4a	6.3a	39.0a	29.0a	6.4a
	Kalmia soil	83a	25.1b	3.8b	37.7a	29.5a	3.5b

Mean value of 50 observations; ten seeds in each observation (balsam fir seeds were given 5 weeks of scarification treatment at 4°C prior to commencement of germination experiment). Values in column with the same letter are not significantly different; $p \leqslant 0.05$ with one-way ANOVA.

19.2.1 Hospitability of seedbed substrates under different canopy cover

Black spruce seedling bioassay described above, on seedbed substrates of a range of habitat types with *Kalmia* present, was used as an indication of seedbed hospitability. Since allelopathic substances are known to concentrate in the top soil, humus samples were collected from the top 10 cm of the forest floor where *Kalmia* was an abundant undergrowth. The habitat types considered were mature black spruce and red pine forest, cut-over and burnt-over black spruce forest, and open *Kalmia* heath that originated from a previously forested area that was repeatedly burned in the past 50–80 years where no tree regeneration had been taking place since the last fire. Results presented in Table 19.3 show that with similar types of *Kalmia* cover, mature red pine forest and black spruce forest cut-overs were most inhibitory to primary root growth of black spruce followed by mature black spruce forest. Burnt-over forest floor is relatively more hospitable in terms of primary root growth of black spruce although *Kalmia* spreads very vigorously following fire by vegetative regeneration as was observed by the higher cover value of *Kalmia* (Table 19.3).

Once a medium- to poor-quality site became dominated by *Kalmia* for over 8–10 years, conifer regeneration became practically impossible. Planted trees did not die but experienced 'growth check' and remained stunted for a number of years (4–6 years) then turned yellowish with eventual death. During that time *Kalmia* became vigorous and a thick layer of duff accumulated from the litter, fine roots, course roots and rhizomes of *Kalmia*, possibly adding more allelopathic substances in the habitat. It is not uncommon to see patches of vigorous *Kalmia* in forest openings surrounded by seed bearing black spruce trees but no seedling establishment in the *Kalmia* patch.

19.2.2 Stair–step structure for allelopathic experiment

In order to separate the growth inhibition of *Kalmia* allelopathy from competition, a stair–step pot-culture experiment was performed in the greenhouse following the principle of Bell and Koeppe (1972). The stair–step structure was built on a movable wooden platform (2 × 1.5 m; 2.5 m high). The 2.5 m high structure made up of 7.6 × 10.2 cm wood had four shelves on either side made of plywood (36 cm wide; 2 cm thick). Vertical distance between the shelves was 36 cm and on each shelf six large pots (30 × 30 cm; 30 cm high) were placed in such a way that the pots on the upper shelves corresponded to those of the lower shelves in vertical rows (Fig. 19.3). Circular holes (14.5 cm dia.) were cut in the shelves corresponding to the bottom of the pot to fit a 15 cm diameter plastic funnel. Each pot had five holes (0.5 cm dia.) at the bottom to allow drainage of excess water that was collected by the funnel. The stem of the funnel was fitted with a plastic elbow connected to a hose through which leachates of one pot was passed on to the surface of the next lower pot. The lower most

TABLE 19.3 *Seed germination and seedling growth of black spruce on substrates of different forest and* Kalmia *cover types in central and eastern Newfoundland*

Site types with *Kalmia* understorey	*Kalmia* cover (%)	Black spruce seed germination (%)	Primary root		Shoot	
			Length (mm)	Fresh weight (mg)	Length (mm)	Fresh weight (mg)
Control, washed sand	–	91.0a	14.8a	9.7a	31.0a	53.3a
Mature black spruce forest, Wooddale	57 ± 6	90.8a	5.7b	4.2b	30.4a	51.9a
Mature red pine forest, Terra Nova	58 ± 4	71.0b	1.9c	1.2c	32.8a	40.5b
Black spruce forest cutover, Wooddale	54 ± 5	92.4a	1.5c	4.1b	27.4a	54.0a
Burnt-over black spruce forest, Botwood	72 ± 5	92.0a	6.7b	4.5b	30.2a	54.2a
Open *Kalmia* heath previously covered with black spruce–balsam fir forest burnt in 1975 & possibly once before, Pouch Cove	68 ± 5	88.0a	6.0b	–	30.0a	–

For *Kalmia* cover, mean value of ten 1 m^2 quadrats; for germination and seedling growth, mean value of 50 seedlings, ten seedlings in each petri dish with *Kalmia* humus was used. Washed sand was used as control. Values in columns with the same letter are not significantly different; $p \leq 0.05$ with one-way ANOVA.

a

b

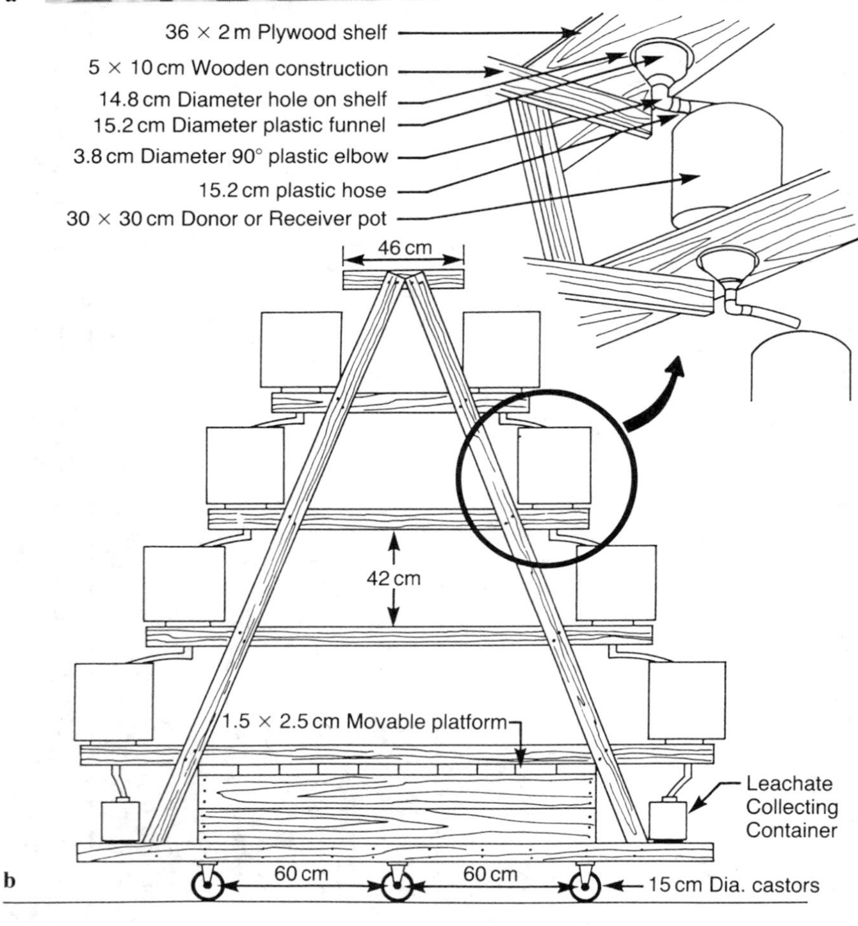

Fig. 19.3 (a) Front view of the stair–step structure for allelopathic experiment in the greenhouse. (b) Side view of the stair–step structure.

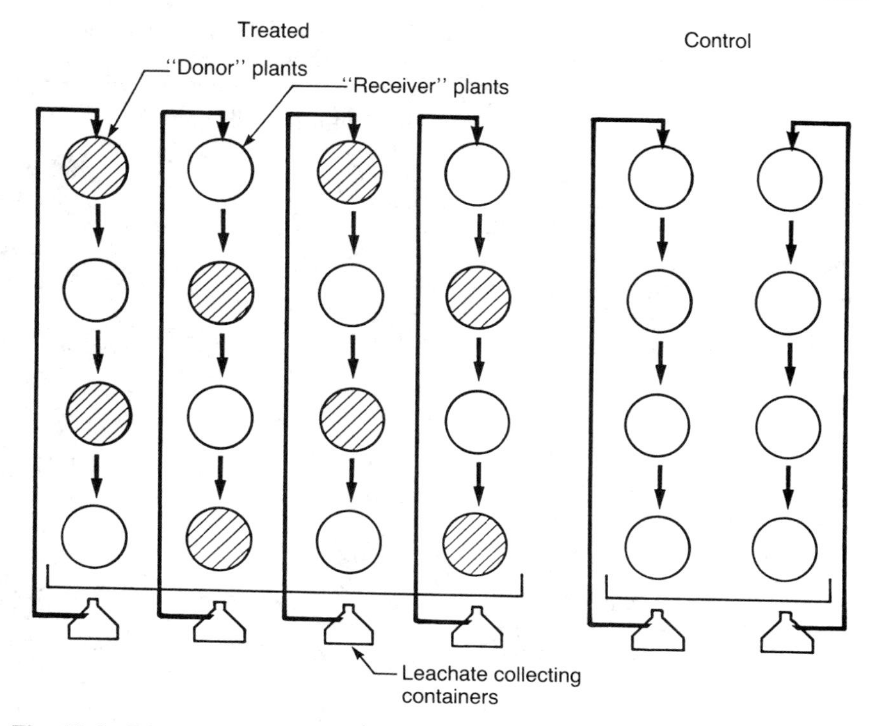

Fig. 19.4 Diagramatic sketch of the stair–step–pot allelopathic experiment.

pot of each of the vertical rows was connected to a collecting container. Watering was done on the top pot of each vertical row after rinsing the content of the collecting bottle. This allowed circulation of the same water leachate through the individual vertical row of pots.

While performing experiments, black spruce, larch and red pine seedlings were grown in sand culture in the pots of vertical rows. To find out the inhibitory effects of water-soluble allelopathic compounds the treated pots were interrupted with the potentially allelopathic plants (in this case *Kalmia*) grown in its own soil monolith in pots. Control plants were kept uninterrupted (Fig. 19.4). In this way, direct competition from the 'donor' plants in the allelopathic experiment was removed by growing them separately. Thus, any growth inhibition of the treated plants would be due to the water leachates coming through the canopy and soil monolith of *Kalmia*. To average the local environmental variability, the whole structure was rotated 180° at a regular (weekly) interval.

19.2.3 Growth inhibition of two-year-old softwood seedlings

Experiments performed in the stair–step–pot set-up with black spruce, larch and red pine indicate that plants receiving the interruptions from

TABLE 19.4 *Inhibitory effects of water leachates passed through* Kalmia *soil monolith on plant height and oven dried biomass of softwood seedlings*

Plants	Treatment	Plant height 4 months after treatment (cm)	Oven dried biomass (g)	
			Above ground	Below ground
Black spruce	Control (sand culture)	25.3a	2.08a	3.61a
	Kalmia interrupted	14.7b	1.43b	2.44b
Larch	Control (sand culture)	21.6a	1.86a	2.12a
	Kalmia interrupted	16.2a	1.31b	1.74b
Red pine	Control (sand culture)	11.7a	0.70a	1.35a
	Kalmia interrupted	9.6a	0.44b	0.95b

Values in column with the same letters are not significantly different; $p \leq 0.05$ with one-way ANOVA.

Kalmia soil monolith experienced significantly less plant height growth and biomass (Table 19.4).

The biomass difference of control (sand culture grown) and treated (sand culture grown but interrupted with *Kalmia* soil monolith) black spruce seedlings, four months after the stair–step–pot experiment began, is depicted in Fig. 19.5. It is evident from Table 19.4 that growth inhibition was most pronounced in black spruce followed by larch and red pine. Nutrient analyses of the seedlings at the end of the experiment indicated that the inhibition of growth in the affected softwood seedlings was not due to lack of nutrients. In fact, the NPK concentrations were rather higher in the stunted, treated seedlings than in the controls (Mallik and Roberts, unpublished data).

19.2.4 Inhibitory effects of softwood forest floor substrates

In a recent study by Mallik and Newton (1988) it was discovered that the forest floor substrates of mature black spruce (where *Kalmia* was not present as an understorey species), is inhibitory to primary root growth and development of black spruce. The morphology and anatomy of the affected seedlings showed similar symptoms as was observed in the *Kalmia* soil water extract-induced primary root inhibition (Mallik, 1987). This

Fig. 19.5 Biomass difference of black spruce seedlings four months after commencement of stair–step–pot allelopathic experiment. Note the dramatic reduction of biomass due to the water leachate of *Kalmia*–soil monolith.

TABLE 19.5 *Effects of black spruce forest floor substrates of central Newfoundland (where no* Kalmia *was present) on seed germination and seedling growth of black spruce (after Mallik and Newton, 1988)*

			Primary root		Shoot	
Black spruce stand age (years)	Substrate types	Germination (%)	Length (mm)	Fresh weight (mg)	Length (mm)	Fresh weight (mg)
Control		90	14.6	1.0	28.3	5.1
15	LFH	91	5.8	0.6	28.7	3.8
	Ae	72	6.3	0.7	23.7	4.1
	Bf	76	7.8	0.9	24.1	4.8
30	LFH	84	4.1	0.4	23.5	3.4
	Ae	76	4.6	0.6	21.9	4.2
	Bf	78	9.8	1.0	30.8	5.5
45	LFH	82	4.1	0.5	25.2	4.3
	Ae	81	1.4	0.3	17.1	3.7
	Bf	81	9.6	1.0	32.0	5.8
60	LFH	80	3.9	0.4	23.8	3.5
	Ae	78	1.6	0.2	18.9	4.6
	Bf	79	8.1	1.2	31.6	5.9

leads the author to assume that the compound(s) responsible for this kind of root growth inhibition may be the decomposition product(s) of the organic substances of both *Kalmia* and black spruce. The root growth inhibiting property was more pronounced in the top LFH and Ae soil horizons as compared with the Bf horizon (mineral) soil (Table 19.5). It is also evident from the results that the degree of root growth inhibition was increased in the top soil with increasing age of the stand. Further experimentation was conducted by planting 2 + 0 seedlings of black spruce, larch and red pine in their own soils in pot culture (30 × 30 cm) in the greenhouse. A soil sample from a burnt-over red pine stand was also used as another growing medium. Soils of all those forest sites had thick (10–15 cm) organic matter except in the burnt-over soil, where 4–8 cm of organic matter was consumed by the forest fire and the soil samples contained ash and black charred material on top. Four months after commencement of the experiment all seedlings, on all but the burnt soils, showed very poor growth in terms of height and above- and below-ground biomass. The growth of red pine seedling on burnt soil, however, was significantly higher as compared with those grown on the unburnt soil (Table 19.6).

TABLE 19.6 *Effect of growing media on plant height and oven dry biomass of softwood seedlings*

Plant	Growing Medium	Plant height 4 months after pot culture (cm)	Oven dry biomass (g)	
			Above ground	Below ground
Black spruce	Black spruce–larch forest soil	15.3	1.58	1.21
Larch	Black spruce–larch forest soil	16.7	1.59	1.28
Red pine[a]	Red pine forest soil	11.3a	0.54a	0.79a
	Burnt over red pine forest soil	14.1b	1.10b	0.93b

[a] Means within column followed by different letters indicate significant difference at 0.05 level.

TABLE 19.7 *Effects of burnt and unburnt black spruce forest humus on seed germination and seedling growth of black spruce*

Treatment	Germination (%)	Length	
		Primary root (mm)	Shoot (mm)
Control (washed sand)	90a	14.6a	28.6a
Burnt humus	86a	9.7b	25.2a
Unburnt humus	86a	7.0c	26.5a

Values in column with the same letter are not significantly different, $p \leq 0.05$ with one-way ANOVA.

19.3 FOREST FIRE AND SEEDBED HOSPITABILITY

Root inhibitory property of black spruce due to *Kalmia* and/or black spruce organic matter is found to be considerably reduced after a forest fire. Seed germination and seedling growth of black spruce were tested on burnt and unburnt *Kalmia*–black spruce humus. Tests were also conducted using water extracts of the burnt and unburnt forest floor organic

TABLE 19.8 *Effects of water extracts of burnt and unburnt black spruce forest humus on seed germination and seedling growth of black spruce*

Treatment	Germination (%)	Length	
		Primary root (mm)	Shoot (mm)
Control	84b	16.5ab	27.7a
Burnt humus (no *Kalmia*)	92a	17.1a	31.1a
Unburnt humus (no *Kalmia*)	91a	3.5c	20.8b

Values in column with the same letter are not significantly different, $p \leq 0.05$ with one-way ANOVA.

TABLE 19.9 *Nature of red pine seedling establishment around the seed bearing trees on burnt and unburnt sites in central Newfoundland (after Mallik and Roberts, submitted)*

Treatment and parameters	Distance from the seedbearing trees[a]			
	0–1 m	1–2 m	2–3 m	3–5 m
Burnt:				
Seedlings	19	45	28	8
Height (cm)	18.9	14.9	12.5	16.6
Organic matter depth (cm)	2.0	2.1	2.6	2.8
Unburnt:				
Seedlings	0	2	2	7
Height (cm)	–	13.5	19.0	16.8
Organic matter depth (cm)	–	4.0	5.0	2.7

[a] Total number of seedlings around seven trees and mean value of seedling heights and mean value of organic matter depth beside the seedlings.

substances. Tables 19.7 and 19.8 show that length of primary roots of black spruce was significantly higher in burnt organic matter compared with the unburnt humus. Root growth in the water extract of burnt organic matter was very similar to that of control. However, seedlings treated with water extracts of unburnt organic matter were severely inhibited (Table 19.8). Response of red pine seedling establishment and growth in the field as to the burnt and unburnt organic matter was similar to the above findings.

Studies conducted on regeneration of red pine on burnt and unburnt stands indicate that a burnt stand was more conducive to seedling regeneration than the unburnt. Table 19.9 clearly shows that very few seedlings occurred around the seedbearing trees on an unburnt stand compared with the burnt stand. Fire removed the organic substances, making the habitat more hospitable for seedling establishment and growth.

19.3.1 Fire temperature, pH change and root growth

Organic matter of mature black spruce and red pine forest soil with *Kalmia* understorey (70–80% cover) was heated in a muffle furnace for 30 minutes at each of the following temperatures: 200, 400, 600 and 800°C. Black spruce and red pine seed germination and seedling growth (length of primary root and shoot) was estimated in the samples using washed sand as control. pH of the germination media were also noted. Results presented in Fig. 19.6 show that pH of the organic matter was increased with increasing fire temperature. A concomitant increase in the primary root length of black spruce up to 600°C was also observed. Above that temperature, although there was a further increase in pH of the burnt organic matter, root growth in that substance was less as compared with

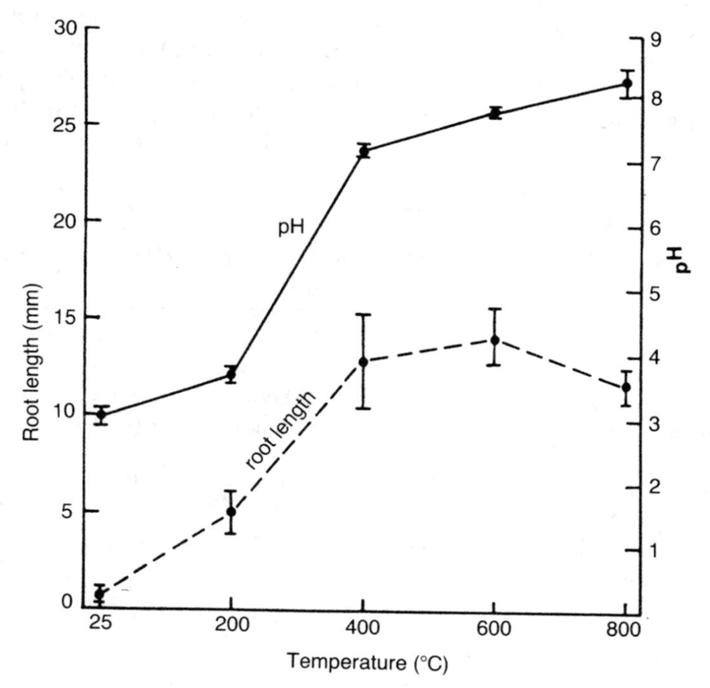

Fig. 19.6 Effect of burning temperature of organic matter of black spruce with *Kalmia* on pH and root length of black spruce germinants.

the 600°C treatment. Thus, up to 600°C burning temperature there was an exponential increase in root length and pH. This result supports the view that moderate fire makes the habitat more conducive to seedling establishment and growth. Whether this was achieved by increasing pH or removing allelopathic compounds or both was the subject for subsequent experiments. The pH of the *Kalmia* humus and the water extract of the humus was raised to neutral (pH 7) by adding NaOH and the black spruce seedling bioassay was repeated in those two media.

19.4 NATURE AND STABILITY OF GROWTH-INHIBITING COMPOUNDS IN THE WATER EXTRACTS

Results presented in Table 19.10 indicate that it was not the acidity factor that caused primary root inhibition since neutral pH of the medium did not help to improve the root growth. It was rather more inhibitory. Chemistry of the compound(s) in the water extracts of *Kalmia*–black spruce soil responsible for primary root growth inhibition of softwood is not yet known. Research is underway to identify and characterize the compound(s) by using solvent partition methods followed by seedling bioassay as an indication to know which part of the extract contains the active compound(s) responsible for root growth inhibition. Preliminary results indicate that the root inhibitory compound(s) are polar in nature. Among other properties, it seems that the root inhibitory compound(s) are not sparingly soluble in water since different replicates of the same extract yielded various degrees of inhibition (Mallik, 1987). Addition

TABLE 19.10 *Effects of* Kalmia *organic matter, higher pH and PVP on length and fresh weight of primary root and shoot of black spruce*

	Length		Fresh weight	
Treatments	Root (mm)	Shoot (mm)	Root (mg)	Shoot (mg)
Control (washed sand)	9.8	26.1	0.6	5.6
Kalmia O.M. pH 3.5	0.2	25.1	0.04	5.1
Kalmia O.M. + PVP	2.3	31.6	0.9	6.9
Control (distilled water)	16.4	22.3	1.6	5.5
Kalmia humus extract	7.1	25.7	0.8	5.2
Kalmia humus extract (pH 7)	3.4	23.9	0.4	5.3

Average value of five replicates; ten seeds in each replicate.

TABLE 19.11 *Effects of treated water extracts of* Kalmia *organic matter (with autoclaving, centrifugation, millipore filtration and δ-radiation) on germination and seedling growth of black spruce*

Treatment	Germination (%)	Length of primary root (mm)	Length of shoot (mm)
Control (distilled water)	90	15.3a	23.4c
Water extract	80	7.1e	26.5ab
Water extract centrifuged	86	6.7ef	23.3c
Water extract autoclaved (30 minutes)	92	9.2d	24.7bc
Water extract autoclaved and centrifuged	92	9.8d	24.6bc
Water extract centrifuged and millipore filtered	88	11.8c	25.1bc
Water extract δ-ray, 75 kr	86	13.2bc	26.3ab
Water extract δ-ray, 100 kr	86	12.6bc	25.5bc
Water extract δ-ray, 125 kr	90	12.0bc	25.1bc
Water extract δ-ray, 150 kr	84	13.3b	25.1bc

Treatments in column with the same letter are not significantly different at $p \leq 0.05$ with nested ANOVA (after Sokal and Rohlf, 1981).

of polyvinyl pyrollidone, a polymer which is supposed to dissolve the phenolic compound(s) (D. Boyle, pers. com.) from the organic matter, did not help to remove the root growth inhibition, suggesting that the growth inhibitory factor(s) may not be phenolic in nature (Table 19.10).

High temperatures seem to make the organic matter more conducive to seedling growth which may mean that the allelopathic compounds are inactivated by high temperature. Autoclaving, centrifugation and millipore filtration and a high dose of δ-radiation also removed the root growth inhibitory effects of the water extracts of *Kalmia* organic matter (Table 19.11). The growth-inhibitory property of the water extract was also diminished by storing the extract for five weeks at 4°C followed by shaking, prior to use for the bioassay test (Table 19.12).

19.5 BRACKEN FERN ALLELOPATHY

It has been observed at several sites in western and central Newfoundland where forest openings were dominated by bracken fern (*Pteridium aquilinum*) that conifer regeneration in those patches was poor or sometimes lacking. Since bracken fern is known to have allelopathic effects on the habitat (Stewart, 1975; Glass, 1976; Gliessman and Muller, 1972,

TABLE 19.12 *Effects of storage (5 weeks at 4°C), followed by shaking prior to application, of water extracts of burnt and unburnt black spruce forest humus on seed germination and seedling growth of black spruce*

Treatment	Germination (%)	Length of primary root (mm)	Length of shoot (mm)
Control	89 ± 2	16.7a	26.2a
Burnt	87 ± 4	18.0a	25.4a
Unburnt	91 ± 3	14.5ab	26.7a

Treatments with the same letter in the column are not significantly different at $p \leq 0.05$ with one-way ANOVA.

1978; Rice, 1984; Ferguson and Boyd, 1988), it is suspected that lack of tree regeneration in bracken-dominated patches may be due to allelopathy. However, this hypothesis needs to be tested experimentally.

19.6 FUTURE RESEARCH: IMPLICATIONS FOR FOREST RENEWAL IN NEWFOUNDLAND

Research is underway to identify and characterize the compound(s) responsible for primary root growth inhibition of black spruce in fresh leaves and partially decomposed organic matter of *Kalmia*. It is possible that the allelopathic compound(s) present in *Kalmia* humus may interfere with the mycorrhizal fungi of black spruce which eventually lead to the 'growth check' of black spruce seedlings. Studies in *Calluna* heathlands where this kind of growth inhibition was observed in Sitka spruce (*Picea sitchensis*) seem to suggest allelochemical interference of seedling–mycorrhizal relationships (Handley, 1961, 1963; Robinson, 1971, 1972). A similar phenomenon may be involved in *Kalmia*–black spruce 'growth check' in Newfoundland. Research along this line is likely to resolve the silvicultural problems of growth inhibition of planted black spruce. A specific mycorrhizae–black spruce association may be able to overcome this kind of allelopathic growth inhibition experienced in the field. Research is underway in selecting the 'appropriate' mycorrhizae capable of forming a symbiotic association with black spruce and, at the same time, being able to withstand the allelopathic compound(s) present in the *Kalmia*-infested sites. Inoculation of those fungi in the root system of black spruce may help the seedlings grow better in the inhospitable habitats of central Newfoundland.

ACKNOWLEDGEMENTS

The research was conducted during the tenures of an NSERC Visiting Fellowship and a Canada–Newfoundland Forest Resources Development Agreement Research Contract awarded to the author. Logistic support of Forestry Canada, Newfoundland and Labrador Region is gratefully acknowledged. The author is grateful to Dr A. Rahimtula of Memorial University of Newfoundland and Mr R. S. van Nostrand of Forestry Canada for their comments on the draft manuscript.

REFERENCES

Bell, D. T. and Koeppe, D. E. (1972) Noncompetitive effects of giant foxtail on growth of corn. *Agron. J.*, **64**, 321–5.

Candy, R. H. (1951) Reproduction on cutover and burned-over land in Canada. *Can. Dept Res. and Dev., For. Res. Div., Silv. Res. Note No. 92*, p. 224.

Ferguson, D. E. and Boyd, R. J. (1988) Bracken fern inhibition of conifer regeneration in northern Idaho. *USDA For. Serv. Res. Paper INT-388*, pp. 1–11.

Glass, A. D. M. (1976) The allelopathic potential of phenolic acids associated with the rhizosphere of *Pteridium aquilinum*. *Can. J. Bot.*, **54**, 2440–4.

Gliessman, S. R. and Muller, C. H. (1972) The phytotoxic potential of bracken, *Pteridium aquilinum* (L.) Kuhn. *Madrono*, **21**, 299–304.

Gliessman, S. R. and Muller C. H. (1978) The allelopathic mechanisms of dominance in bracken (*Pteridium aquilinum*) in southern California. *J. Chron. Ecol.*, **4**, 337–62.

Hall, I. V., Jackson, L. P. and Everett, C. F. (1973) The biology of Canadian weeds. 1. *Kalmia angustifolia* L. *Can. J. Plant Sci.*, **53**, 865–73.

Handley, W. R. C. (1961) Ectotrophic mycorrhizal fungi and growth of trees on *Calluna* heathlands. *IUFRO 13th Congress, Vienna*.

Handley, W. R. C. (1963) Mycorrhizal associations and *Calluna* heathland afforestation. *For. Comm. Bull.*, **36**, HMSO, London.

Krause, H. H.(1986) Ericaceous vegetation as a site factor in jack pine growth of New Brunswick plantations. *IUFRO Conference*, Fredericton, N. B.

Mallik, A. U. (1987) Allelopathic potential of *Kalmia angustifolia* to black spruce (*Picea mariana*). *For. Ecol. Manage.*, **20**, 43–51.

Mallik, A. U. and Newton, P. F. (1988) Inhibition of black spruce seedling growth by forest floor substrates of central Newfoundland. *For. Ecol. Manage.*, **23**, 272–83.

Mallik, A. U. and Roberts, B. A. Natural regeneration of red pine on burned and unburned sites in Newfoundland. *For. Ecol. Manage.* (Submitted).

Meades, W. J. (1983a) The origin and successional status of anthropogenic dwarf shrub heath in Newfoundland. *Adv. Space Res.*, **2**, 97–101.

Meades, M. J. (1983b) Heathlands, in *Biogeography and Ecology of Newfoundland* (ed. G. R. South), Dr W. Junk Publishers, The Hague.

Page, G. (1970) The development of *Kalmia angustifolia* on a black spruce cutover in central Newfoundland. *Nfld. For. Res. Centre, St. John's, Int. Rep. N-27*, 7pp.

Peterson, E. B. (1965) Inhibition of black spruce primary roots by a water soluble substance in *Kalmia angustifolia*. *For. Sci.*, **11**, 473–9.

Rice, E. L. (1984) *Allelopathy*. Academic Press, New York.

Richardson, J. (1974a) A comparison of two direct seedling techniques on scarified and unscarified ground. *Environ. Canada, Can. For. Serv., Info. Rep. N-X-107*, 23pp.

Richardson, J. (1974b) Natural regeneration after disturbance in the forests of Labrador. *Environ. Canada, Can. For. Serv., Info. Rep. N-X-113*, 34pp.

Richardson, J. (1975) Establishment of a study to compare different means of reforesting sites invaded by *Kalmia*. *Nfld. For. Res. Centre*, St John's, Nfld. File Rep., Study 8-8, 12pp.

Richardson, J. (1979) A comparison of methods of reforesting sites invaded by *Kalmia angustifolia*, using black spruce. *Environ. Canada, Can. For. Serv., Info. Rep. N-X-174*.

Richardson, J. (1981) Black spruce research by the Canadian Forestry Service in Newfoundland. *Environ. Canada, Can. For. Serv., Info. Rep. N-X-206*.

Richardson, J. and Peter Hall, J. (1973) Natural regeneration after disturbance in the forests of central Newfoundland. *Can. Dept. Environ., Can. For. Serv., Info. Rep. N-X-86*, 63pp.

Richardson, J. and Peter Hall, J. (1973b) Natural regeneration after disturbance in the forests of eastern Newfoundland. *Can. Dept Environ., Can. For. Serv., Info. Rep. N-X-90*, 46pp.

Robinson, R. K. (1971) Importance of soil toxicity in relation to the stability of plant communities, in *Scientific Management of Plant and Animal Communities of Conservation* (eds E. A. G. Duffey, and A. S. Watt), Brit. Ecol. Soc. Symp., No. 11, Blackwell, Oxford and Edinburgh, pp. 105–13.

Robinson, R. K. (1972) The production of roots of *Calluna vulgaris* of a factor inhibitory to growth of some mycorrizal fungi. *J. Ecol.*, **60**, 219–24.

Ryan, A. G. (1978) *Native Trees and Shrubs of Newfoundland and Labrador*. Parks Division, Dept of Tourism, Government of Nfld and Labrador, St John's, 116pp.

Scoggan, H. J. (1979) *Flora of Canada*. Natl. Mus. Nat. Sci., Ottawa, pp. 1–4.

Sokal, R. R. and Rohlf, F. J. (1981) *Biometry*, 2nd edn. Freeman, San Francisco, CA.

Stewart, R. E. (1975) Allelopathic potential of western bracken. *Ecology*, **1**, 161–9.

Thompson, I. D. and Mallik, A. U. (1990) Effects of moose browsing and allelopathy of forest floor substrates as factors for growth inhibition of balsam fir seedlings in central Newfoundland. *Can. J. For. Res.* **19**(4), 524–6.

van Nostrand, R. S. (1971) Strip cutting black spruce in central Newfoundland to induce regeneration. *Can. For. Serv. Pub. No. 1294*, p. 21.

Vincent, A. B. (1965) Black spruce, a review of its silvics, ecology and silviculture. *Can. Dept For. Publ. 1100*, 49pp.

Wall, R. E. (1977) Ericaceous ground cover on cutover sites in southwestern Nova Scotia. *Can. Dept Fish. Environ., Can. For. Serv., Info. Rep. N-X-71*, 55pp.

CHAPTER 20

Allelopathy in Quebec forestry – case studies in natural and managed ecosystems

R. Jobidon

20.1 INTRODUCTION

Allelopathy is currently gaining importance in Quebec as a factor of ecological significance. However, there are few reports on the phenomenon. Most of the work was done in forestry and surprisingly little in agriculture. Research in allelopathy in Quebec was initiated in 1974 as part of a project on the ecology of balsam fir (*Abies balsamea* (L.) Mill.) and balsam poplar (*Populus balsamifera* L.) at the Faculty of Forestry and Geodesy of Laval University (Quebec city). Then, the research was orientated on (1) the adverse allelopathic effect of forest weed species on regeneration and (2) the allelopathic control of competing species in forest plantations.

20.2 STUDIES ON BALSAM FIR AND BALSAM POPLAR

The allelopathic activity of balsam fir, a climax species, and balsam poplar, a transitional species, was studied to evaluate the restriction of nitrification beneath the canopy of these species. Laboratory experiments were done to determine the allelopathic effect of foliar leachates, leaf extracts, and bud extracts of the studied species on nitrification, using a soil percolation apparatus (Thibault *et al.*, 1982). Balsam fir foliar leachates proved to be more inhibiting on nitrification than those of balsam poplar. Five percent aqueous extracts of balsam fir needles and dormant buds of balsam poplar completely prevented oxidation of ammonium. The results suggest that both species, balsam fir and balsam poplar, inhibit nitrification and that the inhibition increases with succession (Thibault *et al.*, 1982). Inhibition of nitrification by allelopathic substances in forest ecosystems was studied by Lodhi (1978, 1979) and Lodhi

Allelopathy: Basic and applied aspects Edited by S. J. H. Rizvi and V. Rizvi
Published in 1992 by Chapman & Hall, London ISBN 0 412 39400 6

TABLE 20.1 *Morphological modifications observed on* Alnus crispa *var.* mollis *radicle after treatment with different plant parts of* Populus balsamifera *(after Jobidon and Thibault, 1981)*

	Morphological modifications (% of plants affected per treatment)			
Treatments (% wt/v)	Total inhibition of root hair development	Partial inhibition of root hair development	Total meristematic necrosis	Partial meristematic necrosis
Leaf litter extracts:				
2	50	50	62	33
1	14	43	14	40
0.5	0	0	0	0
0.1	0	0	0	0
Bud extracts:				
2	80	20	100	0
1	62	38	77	23
0.5	0	73	13	73
0.1	0	0	0	0
Leaf leachates:				
2	83	8	8	16
1	62	27	0	18
0.5	0	0	0	0
0.1	0	0	0	0

and Killingbeck (1980). They noted that low nitrate concentrations in climax areas are due to inhibition of nitrification by climax plants. As part of a study in grassland communities, Rice (1984) confirms that successional stages exert an allopathic influence on nitrification processes and thus on conservation of nitrogen and energy.

The allelopathic activity of balsam poplar was specifically studied in relation to succession patterns. Degraded sites in Quebec are commonly invaded first by alder species, a nitrogen-fixing species, which are then replaced by poplar species, including balsam poplar. A growth reduction of alders is frequently observed near a newly established balsam poplar stand. In Jobidon and Thibault (1981), the allelopathic effect of fresh leaf leachates, leaf litter and bud extracts of balsam poplar on green alder (*Alnus crispa* var. *mollis* Fern.) seed germination was tested. The results show that leachates and extracts significantly inhibit seed germination, as well as hypocotyl and radicle growth. Moreover, severe morphological modifications were observed (Table 20.1), such as impaired root hair development and meristematic necrosis of the radicle. The inhibition of

TABLE 20.2 *Effects of different plant parts of* Populus balsamifera *on nodulation of* A. crispa *var.* mollis *seedlings (±SE) (after Jobidon and Thibault, 1982)*[a]

Treatments (% wt/v)	Mean nodule no.	% of inhibition
Leaf leachates:		
2	22.90 ± 1.57a	57.87
1	26.30 ± 1.76a	51.61
0.5	29.20 ± 2.23ab	46.27
0.1	32.95 ± 2.04b	39.37
Control	54.35 ± 2.66c	0.00
Bud extracts:		
2	22.90 ± 2.41a	48.31
1	23.40 ± 2.51ab	47.18
0.5	26.70 ± 2.59ab	39.73
0.1	31.10 ± 2.98b	29.80
Control	44.30 ± 2.47c	0.00
Leaf litter extracts:		
2	22.40 ± 2.10a	46.79
1	25.60 ± 2.10a	39.19
0.5	28.50 ± 4.73a	32.30
0.1	27.80 ± 2.19a	33.97
Control	42.10 ± 4.14b	0.00

[a] Within each treatment, means followed by the same letter are not different at 0.05 level, Duncan's new multiple range test.

root hair development on green alder radicle may have direct effects on its infection by the actinomycete *Frankia* spp. Lalonde (1979), Lalonde and Quispel (1977), and Callahan *et al.* (1979) have shown that the endophyte penetrates the root hair to initiate nodule formation. Thus, the inhibition of root hair formation resulting from allelopathic activity deserves more attention in relation to success of nodulation and nitrogen fixation. Weston *et al.* (1987) showed that phytotoxic compounds from quackgrass may indirectly influence snapbeans–*Rhizobium* symbiosis, by reducing root growth and root hair formation.

In Jobidon and Thibault (1982), the allelopathic effect of fresh leaf leachates and bud extracts on nodulation, nitrogenase activity (acetylene reduction), and growth of nodulated and unnodulated green alder seedlings was studied. The purpose of the study was to test if symbiotically associated species have the same response pattern to allelopathic treatments as unassociated ones. Initiation of nodulation was achieved by inoculating green alder seedlings with three-month-old colonies of *Frankia*

TABLE 20.3 *Effects of different plant parts of* Populus balsamifera *on nitrogenase activity (mole C_2H_4 per 3 h per root dry weight × 10^{-11}) (±SE). (After Jobidon and Thibault, 1982)*[a]

Treatments (% wt/v)	Mole C_2H_4 per 3 h per root dry weight × 10^{-11}	% of inhibition
Bud extracts:		
2	4.36 ± 0.59a	61.72
1	4.12 ± 0.49a	63.88
0.5	5.26 ± 0.26ab	53.89
0.1	6.54 ± 0.63b	42.67
Control	11.41 ± 0.99c	0.00
Leaf litter extracts:		
2	4.25 ± 0.96a	62.74
1	6.19 ± 0.78a	45.74
0.5	4.14 ± 0.54a	63.71
0.1	5.20 ± 0.55a	54.42
Control	11.41 ± 0.99b	0.00

[a] Within each treatment, means followed by the same letter are not different at 0.05 level, Duncan's new multiple range test.

spp., strain ACN1[AG]. Water extracts of different plant parts of balsam poplar significantly inhibited height growth, root elongation, dry weight increment of shoots and roots, foliar nitrogen content, nodule formation and nitrogenase activity of green alder seedlings after a two-month-growth period under controlled conditions. Balsam poplar leaf leachates (1%, wt/v) significantly inhibited height growth (45% of inhibition, as compared with control), root elongation (19%), dry weight increment of shoots (55%) and roots (37%), and foliar nitrogen content (11%) of nodulated green alder seedlings. The average number of nodules per green alder seedling treated with any of the three extracts used, was only 51% that of control seedlings (Table 20.2). There was a 62% decrease in acetylene reduction (nitrogen fixation) by green alder seedlings treated with bud and leaf litter extracts (2%, wt/v) of balsam poplar (Table 20.3). Unnodulated green alder seedlings were generally more affected by the treatments than nodulated seedlings. These results suggest that an allelopathic mechanism might be involved in the succession process from green alder to balsam poplar stands. Additional evidence is put forward by Perradin *et al.* (1983), who worked on the effect of plant phenolics (some phenolics tested were involved in the allelopathic activity of balsam poplar) on *in vitro* growth of six *Frankia* strains isolated from *Alnus* and *Elaeagnus*. The three cinnamic acids (*o*-coumaric, *p*-coumaric and ferulic acids) were highly effective in limiting the growth of *Frankia* colonies, increasing the ramification of hyphae and decreasing the number and size of sporangia produced. The two benzoic acids (benzoic and *p*-hydroxybenzoic acids) did not significantly influence the total growth of *Frankia* colonies but were extremely effective in stimulating the production of spherical septate vesicles by two *Frankia* strains (ACN1^{AG} and AGN1g).

20.3 ALLELOPATHIC EFFECTS OF FOREST WEED SPECIES ON REGENERATION

The problem of regeneration failure on commercial forest lands in Quebec is of great importance. Unfortunately, the research effort devoted towards a complete understanding of the problem, including allelopathy, is still limited. As noted by Sutton (1985), harvesting and regeneration operations are generally separated in traditional Canadian forestry practices. This method has led to tremendous weed problems. Furthermore, a detrimental effect of competing species is often observed not only on natural regeneration but also on planted tree seedlings. Obviously, such detrimental effect could originate from a competition for water, nutrients and light between competing species and tree seedlings. However, research has shown (Horsley, 1977a,b; Rice, 1984; Fisher, 1987) that allelopathy could, in many instances, play a major role and thus explain, at least partially, the harmful effect of many competing species on tree seedlings. Many of the most common competing species found on Quebec forest lands are believed to have such a detrimental allelopathic effect (Table 20.4).

The biological role of lichen substances was recently reviewed by Lawrey (1986). He pointed out the antimicrobial and antiherbivore characteristics, as well as the allelopathic activities of lichen substances. Of great interest in Quebec forestry are the allelopathic activities of lichen substances on tree seedlings. The Quebec boreal forest communities are characterized by open stands of black spruce (*Picea mariana*) associated with lichen species. The potential adverse effect of fruticose soil lichens was studied by Arsenault (1978) in a *Cladino stellaris–Piceetum marianae betuletosum glandulosae* association located in 'Les Grands Jardins à Caribou' (mercator 19TCC 595 795). Arsenault (1978) tested several lichen extracts of *Cladina stellaris*, *C. mitis*, *C. rangiferina* and *Stereocaulon paschale* on seed germination and early growth of black spruce under both sterile and non-sterile conditions. In most cases, the allelopathic lichen extracts did not impair black spruce seed germination, but inhibited initial shoot growth. The results obtained by Arsenault on the effect of lichen extracts confirmed those previously obtained by Gagnon (1966) working on the effect of *Lecidae granulosa* (Enrh.) on black spruce seed germination in the Gaspesia region of

TABLE 20.4 *Most common species of competing vegetation on commercial forest lands in Quebec*

Botanical name	Common name
Acer pennsylvanicum	Striped maple
A. rubrum	Red maple
A. spicatum	Mountain maple
Alnus rugosa	Speckled alder
Aster spp.	Aster
Betula papyrifera	Paper birch
Carex spp.	Misc. sedges
Corylus cornuta	Beaked hazel
Epilobium angustifolium	Fireweed
Graminae spp.	Misc. grasses
Kalmia angustifolia	Lambkill
Populus spp.	Aspen
Prunus pennsylvanica	Pin cherry
Pteridium aquilinum	Bracken fern
Rubus spp.	Raspberry
Salix spp.	Willows
Solidago spp.	Goldenrod
Vaccinium spp.	Blueberries

Quebec. Moreover, Arsenault (1978) demonstrated that lichen extracts strongly inhibit growth of various fungi, including black spruce mycorrhizal fungi: *Paxillus involutus* being the most inhibited, followed by *Cenococcum graniforme* > *Euscoboletinus glandulosus* > *Tricholoma focale* > *Boletinus capipes* = *Suillus granulatus* > *Mycelium radicis-atrovirens*. The results obtained by Arsenault (1978) are similar to those obtained by Brown and Mikola (1974) and Fisher (1979), working with some of the same lichen species. Thus, it appears that some fruticose soil lichens affect the survival and growth of some forest tree seedlings, including black spruce, by producing and releasing secondary substances that limit growth of mycorrhizal fungi and formation of mycorrhizal root tips. The inhibition of a mycorrhizal symbiosis may eventually prove to be of much greater significance than direct inhibition of seed germination.

Among the major competing species encountered on Quebec forest lands (Table 20.4), red raspberry (*Rubus idaeus* L.) is certainly the most common. This species strongly competes for light, water and nutrients with naturally regenerated tree seedlings as well as with planted tree seedlings. Norby and Kozlowski (1980) first reported the potential allelopathic activity of red raspberry. Water extracts of red raspberry foliage affected radicle elongation and growth of red pine (*Pinus resinosa* Ait.) seedlings, but had no effect on seed germination. Côté and Thibault

(1988) studied the influence of red raspberry foliar leachates on seven species of ectomycorrhizal fungi able to colonize black spruce root system. They demonstrated that leachates of red raspberry significantly inhibited *Paxillus involutus*, *Laccaria proxima*, *L. bicolor*, *Telephora terrestris* and *Cortinarius pseudomapus* on agar plates, whereas *Hebeloma cylindrosporum* and *Cenococcum geophilum* were stimulated. The authors suggested that forest tree nurseries should produce seedlings associated with the proper fungal symbiont, in order to minimize the adverse potential allelopathic effect from red raspberry.

Research on the allelopathic effect of forest weed species should include testing the effect of allelochemicals on ectomycorrhizal fungi and on formation of mycorrhizal root tips on tree seedlings. Such an interaction could play a major role in the regeneration failure of forest lands. It could also partially explain the temporary dominance of deciduous pioneer species (most species forming endomycorrhizae) on coniferous species (species forming ectomycorrhizae) following a major forest disturbance such as clearcutting.

20.4 ALLELOPATHIC CONTROL OF WEED SPECIES IN FOREST PLANTATIONS

Foresters have long viewed allelopathy as a problem rather than an opportunity. Most of the research describes the deleterious effect of one plant on another, as part of a global understanding of forest ecosystems. Most of the past research was devoted to the problem of forest regeneration failure (e.g. Fisher, 1987). In agricultural sciences, a lot of research effort was recently devoted to directly exploiting allelopathy as a weed management strategy (Putnam and Tang, 1986). Such research includes selection of allelopathic types in germplasm collection of crops, and utilization of allelopathic rotational crops or companion plants in annual or perennial cropping systems. Through the study of allelopathy, industrial research now focuses on the development of natural products to be used as pesticides. New agricultural and forest chemicals may be produced or synthesized, based on the structure and characteristics of allelochemicals. Unfortunately, very few research projects are orientated towards the exploitation of allelopathy for forest weed control.

Under the auspices of the Quebec Forest Research and Development Service, a research project was initiated in 1984 to study the use of allelopathy as a weed control treatment in forestry. The strategy is first to develop a methodology allelopathically to prevent the establishment of competing species after forest harvesting operations, and secondly to develop a methodology allelopathically to release conifer plantations. The objective is to use allelopathy as a preventive as well as a curative method for vegetation control in forest plantations. Most of the research accomplished so far was orientated towards the preventive aspect of the study. The work done on the curative method is only preliminary and will not be discussed in this chapter.

Eastern Quebec is notably characterized by abandoned agricultural fields which are now under a reafforestation programme. Following site preparation, old-field well-established weed species strongly compete with young tree seedlings for light, water and nutrients. Tree seedlings may possibly also be affected by the allelopathic potential of weed species. A few years following site preparation, chemical herbicides are often used to release conifer plantations. A promising alternative to such use of chemicals could be found in allelopathy. Leaf residues of various conifer species could efficiently prevent the re-establishment of old-field weed species following site preparation and thus help tree seedling growth without use of chemical herbicides (Jobidon, unpublished results). Under laboratory conditions, Jobidon (1986) studied the allelopathic potential of fresh leaf and leaf litter extracts of balsam fir, black spruce, Norway pine (*Pinus resinosa* Ait.), jack pine (*Pinus divaricata* (Ait.) Dumont) and white cedar (*Thuja occidentalis* L.) on seed germination and growth of four selected old-field weed species: timothy (*Phleum pratense* L.), Kentucky bluegrass (*Poa pratensis* L.), couchgrass (*Agropyron repens* (L.) Beauv.) and fireweed (*Epilobium angustifolium* L.). Seed germination of the tested weed species was significantly inhibited by the extract solutions used. Seed germination of fireweed and couchgrass ranged from 26 to 57% and 21 to 55% of control respectively, with the use of fresh leaf leachates and leaf litter extracts of the coniferous species. Height growth and root elongation of couchgrass, Kentucky bluegrass and timothy were variously inhibited by the extract solutions used. Black spruce fresh leaf leachates were the most inhibitory to height growth and root elongation of weed species under study. It is important to note that such experiments are only partial analyses on the use of allelopathy as a biological weed control. Long-term research is needed to make sure that tree seedlings would not be negatively affected by the treatments, and that the inhibition of some weed species would not favour the establishment of any other opportunistic weed species after a few years. Therefore, the duration of the allelopathic effect is of great importance in relation to tree seedling growth, free of competition (or interference), during the first few years.

As part of the investigation on an allelopathic preventive method to weed species invasion, we also worked on deforested lands, following the same research pattern used for abandoned fields. The allelopathic activity of barley, oat and wheat straw mulches is well documented, at least in agricultural sciences (Rice, 1984). In a series of laboratory and field experiments, we tested the allelopathic potential of barley, oat and wheat straw mulches, as well as a mixture of the three kinds of straw, to prevent the establishment of competing species, mainly red raspberry,

Fig. 20.1 Aerial view of an experimental 5 ha site in autumn 1985, showing barley, oat and wheat straws placed on the furrows. The three straw mulches restricted the establishment of red raspberry and associated competing species. Planted black spruce seedlings growth and vigour were not affected by the treatments.

after clearcutting (complete removal of all trees) and site preparation in eastern Quebec (Jobidon *et al.*, 1989a, b).

Three sites, measuring 5 ha each, were prepared in 1985 and the straws placed in the centre of the created furrows (Fig. 20.1). The use of *Populus* wood shavings as a control mulch was used to separate physical and chemical effects of the residues (Putnam and DeFrank, 1983). Black spruce seedlings were planted during spring 1986. Phytosociological surveys were carried out during summers 1986 and 1987 with special attention to red raspberry (Jobidon *et al.*, 1989a).

Under laboratory conditions, cold water extracts of barley, oat and wheat straws strongly inhibited red raspberry propagule growth. Shoot dry weight was 10, 44 and 68% of the control for oat, wheat and barley straw extracts respectively. Under field conditions, red raspberry seed germination and growth were inhibited due to the treatments. Red raspberry plants from treated plots had a significantly lower foliar N-content (mean N-content 1, 48%) as compared with control plants (mean N-content 2, 23%). There were no significant differences between treated and control red raspberry plants for the other nutrients analysed (P, K,

TABLE 20.5 *Mean number and height growth (cm) of red raspberry seedlings per 1 m² subplot for the 1986 and 1987 growing seasons (after Jobidon* et al., *1989a)*

		1986		1987	
Treatment	Site	Mean number	Height growth	Mean number	Height growth
Barley	P1	0.72	7.0	2.85	18.9
	P2	0.28	5.9	2.70	17.1
	P3	0.76	5.5	2.95	17.1
	Average[a]	0.59a	6.1a	2.83a	17.7a
Oat	P1	0.48	7.5	2.35	15.7
	P2	0.20	7.2	1.85	15.3
	P3	0.76	5.8	3.61	18.7
	Average[a]	0.48a	6.8a	2.61a	16.6a
Wheat	P1	0.16	10.5	2.40	14.0
	P2	0.16	9.0	2.60	13.1
	P3	0.68	6.6	4.10	18.4
	Average[a]	0.33a	7.6a	3.03a	15.1a
Mixture	P1	0.12	4.3	1.85	16.1
	P2	0.24	5.2	2.00	19.4
	P3	0.28	5.7	3.08	16.1
	Average[a]	0.21a	5.2a	2.31a	17.2a
Control mulch	P1	6.08	14.6	5.85	48.3
	P2	2.92	15.3	3.33	53.2
	P3	6.36	27.8	7.00	42.6
	Average[a]	5.12b	19.6b	5.39b	48.0b
Control	P1	5.68	27.5	8.10	49.6
	P2	3.85	22.5	6.78	53.4
	P3	6.44	27.5	7.68	54.3
	Average[a]	5.30b	26.1c	7.52c	52.4b

Each value represents mean of 40 1 m² subplots survey.
[a] Average values in the same vertical column followed by the same letter do not differ ($p < 0.05$), according to Duncan's new multiple range test.

Ca, Mg, Mn and Fe). The mean number and height growth of red raspberry shoots was significantly lower on treated plots, as compared with both control plots (Table 20.5). The reduction in red raspberry establishment on the treated plots from the three sites did not favour establishment of any other opportunistic species. After two growing seasons, the mean cover percent for the treated plots was only 20%, and reached 61% for the control plots. The results obtained could not be

Fig. 20.2 After clearcutting and site preparation, barley, oat and wheat straws were placed to cover the soil surface. After two growing seasons, far less competing vegetation had invaded the treated plots (B), as compared with the control plots (A).

attributed to physical effect from the straws, nor to any differences in soil characteristics between treated and control plots, including soil texture, pH and nutritional status (Jobidon *et al.*, 1989a). Figure 20.2 illustrates the results obtained from a preliminary experiment. Barley, oat or wheat straws covered the soil surface. After two growing seasons, competing vegetation was far less present on treated plots than on the control plot.

In light of these results, the hypothesis is that the straw mulches used exerted an inhibitory activity on nitrification processes, which in turn impaired the establishment of red raspberry, a notorious nitrophilous species (Jobidon *et al.*, 1989b). It is a known factor that nitrification greatly increases after a major disturbance, such as clearcutting (Romell, 1935; Likens *et al.*, 1970; Vitousek, 1981). The rapid increase in nitrification could be attributed to: (1) variations of abiotic factors; (2) reduction in plant uptake of nitrogen; (3) increase of organic matter decomposition and nitrogen mineralization; and (4) the removal of sources of potentially inhibitory biochemicals. The results obtained over two consecutive years confirmed our hypothesis. The use of barley, oat and wheat straw mulches significantly reduced NO_3–N soil production and NH_4^+–oxidizers. The five phenolic acids, *p*-coumaric, syringic, vanillic, ferulic and

TABLE 20.6 *Mean shoot increment and basal diameter of black spruce seedlings, at the first and second growing season after straw placement (after Jobidon* et al., *1989b)*

Site	Treatment[a]	Shoot increment 1986 (cm)	Shoot increment 1987 (cm)	Basal diameter 1987 (mm)
P1	BA	9.7d	17.1d	6.40f
	OA	8.1c	14.7c	4.37abc
	WH	8.6cd	15.4c	5.21cde
	CM	4.3a	10.8ab	3.57a
	C	7.5bc	12.7b	4.44bc
P2	BA	7.5bc	16.7d	6.35f
	OA	8.4cd	15.4c	5.56de
	WH	8.2cd	16.4c	6.66f
	CM	5.2a	10.7a	4.39bc
	C	7.6bc	12.6b	4.50bc
P3	BA	7.9bc	16.2c	6.03ef
	OA	8.3cd	15.2cd	4.70bcd
	WH	8.1c	15.2cd	5.81ef
	CM	6.5b	11.9ab	4.15ab
	C	8.3cd	12.6b	4.35bc

Means followed by the same letter do not differ ($p < 0.05$), according to Duncan's new multiple range test.
[a] BA, barley straw; OA, oat straw; WH, wheat straw; CM, control mulch; C, control.

p-hydroxybenzoic acids, responsible for toxicity of the straws, significantly inhibited NH_4^+–oxidizers at concentrations ranging from 10^{-3} to 10^{-6} M. Furthermore, we periodically measured the nitrate reductase activity (NRA) in red raspberry leaves, as an indication of NO_3–N availability in soil. The NRA in red raspberry foliage confirmed the results obtained for NO_3–N content of soil. The results suggest that the inhibition of NO_3–N production under straw mulching restricts the establishment of red raspberry seedlings (Jobidon *et al.*, 1989b).

The third part of the study was devoted to the potentially harmful effect of straw mulches on planted black spruce seedlings (Jobidon *et al.*, 1989b). The three straw treatments enhanced shoot and stem diameter growth of black spruce seedlings (Table 20.6). Treated black spruce seedlings showed a significantly higher foliar N-content than control seedlings. The enhancement of black spruce seedlings growth and foliar N-content could be the result of a reduction in competing vegetation, the result of an inhibition of nitrification from the straws, or both. On one

TABLE 20.7 *Mean nutrient content in black spruce needles (after Jobidon* et al., *1989c)*

Treatment	N	P	K	Ca	Mg	Mn	Fe
Barley straw	1.89ns	3811c	6356bc	8486ns	1815ns	188c	342ns
Oat straw	1.87ns	4805a	6089c	8824ns	1935ns	213b	371ns
Wheat straw	1.93ns	4843a	6967a	8962ns	1881ns	218b	298ns
Straw mixture	1.85ns	4742ab	6569b	8725ns	1840ns	198bc	310ns
Control	1.83	4394b	6674ab	9176	1952	301a	333

Note: Nitrogen is measured as a percentage. All other nutrients are measured in ppm. Each value represents the mean of three repetitions (30 analyses). Values followed by the same letter do not differ ($p < 0.05$) according to Duncan's multiple range test; ns, not significant ($p < 0.05$).

out of three study sites, barley and oat straws significantly depressed black spruce foliar Mn-content. Under greenhouse conditions, the three kinds of straws produced the same detrimental effect on Mn uptake or translocation to the shoot (Jobidon *et al.*, 1989c). In both cases, however, the Mn levels obtained were clearly not at a deficiency level. Table 20.7 provides foliar mineral contents of black spruce seedlings under greenhouse conditions, after mulching treatments. Under greenhouse conditions, the straws do not produce any beneficial effect on spruce seedling growth, whereas a growth increase was noticed under field conditions. These results strongly suggest that the reduction of competing vegetation, due to the allelopathic effect of straws, is a factor which contributes significantly to the increase of seedling growth under field conditions.

In light of these experiments, it is obvious that use of barley, oat and wheat straw mulches could efficiently prevent the establishment of forest weed species, especially red raspberry, as well as have a significantly positive effect on growth and nitrogen nutrition of black spruce seedlings. The use of straw mulches to prevent the establishment of competing species after forest harvesting operations could become a valuable silvicultural tool, where use of chemical herbicides are restricted or in areas of intensively managed plantations. From an ecological point of view, it is of interest to note that nitrogen nutrition of a pioneer species, red raspberry, was significantly depressed by the treatments whereas nitrogen nutrition of black spruce, a climax species, was significantly enhanced. The only plausible explanation for the opposite response seems to be that red raspberry has adapted well to rapidly released N in the form of NO_3^-. In contrast it appears that black spruce has evolved as an efficient user of NH_4^+. This is of importance for conservation of nitrogen within the site, especially in areas where nitrogen nutrition is critical.

20.5 CONCLUSIONS AND PROSPECTS

We are presently conducting experiments to determine how different site preparation techniques could be combined with allelopathic straw mulch treatments in order to prevent the establishment of competing species, including red raspberry and fireweed. The measurements include the establishment of the vegetation over time, and the response of black spruce seedlings to the treatments. The results of these studies should help us determine how allelopathic vegetation control could be part of site preparation. Such a method could, in turn, restrict the need for chemical herbicides. The use of allelopathic straw mulches will soon help Quebec private forest land owners to prevent the establishment of competing species after clearcutting. In Quebec, these forest lands, which are relatively small in size, are located near agricultural farms and in many instances are already under intensive management. All these characteristics make those lands a good choice for the implementation of an allelopathical–silvicultural management system to control competing species during the establishment phase of the plantation. We are also conducting experiments on vegetation control in plantations with deciduous species, such as butternut (*Juglans cinerea* L.), red oak (*Quercus rubra* L.), yellow birch (*Betula alleghaniensis* Britton) and white ash (*Fraxinus americana* L.). The purpose of these experiments is to compare the effectiveness of mechanical, chemical and allelopathical vegetation control methods, in relation to tree seedling growth, development and physiological response of tree seedlings.

When competing species cover the site, chemical herbicides are often used to release conifer plantations. It would be of great interest to develop allelopathic (biological) control of such dense cover, and thus develop a curative method for vegetation control in conifer plantations. The urgent need for forest protection technologies that are environmentally acceptable (biorational) is increasingly obvious. In this respect, governments and industries need to increase substantially the research effort devoted towards the study of a curative strategy to control competing species in forest plantations, and thus significantly decrease or limit the use of chemical herbicides. Microbially produced phytotoxins (marasmin, c.f. Rice, 1984) offer many advantages over chemical herbicides, and interest in such products is growing. Among these, bialaphos (L-2-amino-4-[(hydroxy)(methyl)phosphinoyl]-butyrl-L-alanyl-L-alanine) is currently marketed as a herbicide in Japan. It is produced by *Streptomyces viridochromogenes* with non-selective properties, killing both dicot and monocot species. As part of our research effort to use biorational technologies in forestry, we initiated a research project on the use of bialaphos to control competing species in Quebec coniferous forest plantations. The results obtained so far indicate that bialaphos is highly effective to release conifer plantations (Jobidon, 1991). Bialaphos, and other microbially produced phytotoxins, may be part, in the near future, of the silvicultural tools devoted to forest protection.

REFERENCES

Arsenault, P. (1978) Effets allélopathiques causés par les lichens fucticuleux terricoles sur *Picea mariana*. MS thesis, Université Laval, Québec (Canada).

Brown, R. T. and Mikola, P. (1974) The influence of fruticose soil lichens upon the mycorrhizae and seedling growth of forest trees. *Acta For. Fenn.*, **141**, 1–22.

Callahan, D., Newcomb, W., Torrey, J. G. and Peterson, R. L. (1979) Root hair infection in actinomycete-induced root nodule initiation in *Casuarina, Myrica* and *Comptonia*. *Bot. Gaz.*, **140** (suppl.): S1–S9.

Côté, J. F. and Thibault, J. R. (1988) Allelopathic potential of raspberry foliar leachates on growth of black spruce associated ectomycorrhizal fungi. *Amer. J. Bot.*, **75**, 966–70.

Fisher, R. F. (1979) Possible allelopathic effects of Reindeer-moss (*Cladonia*) on jack pine and white spruce. *Forest Sci.*, **25**, 256–60.

Fisher, R. F. (1987) Allelopathy: a potential cause of forest regeneration failure, in *Allelochemicals: Role in Agriculture and Forestry* (ed. G. R. Waller), ACS Symp. Ser. 330, Amer. Chem. Soc., Washington, DC, pp. 176–84.

Gagnon, J. D. (1966) Le lichen *Lecidea granulosa* constitue un milieu favorable à la germination de l'épinette noire. *Naturaliste Can.*, **93**, 89–98.

Horsley, S. B. (1977a) Allelopathic inhibition of black cherry by fern, grass, goldenrod, and aster. *Can. J. For. Res.*, **7**, 205–16.

Horsley, S. B. (1977b) Allelopathic inhibition of black cherry. II. Inhibition by woodland grass, ferns, and club moss. *Can. J. For. Res.*, **7**, 515–19.

Jobidon, R. (1986) Allelopathic potential of coniferous species to old-field weeds in eastern Quebec. *Forest Sci.*, **32**, 112–18.

Jobidon, R. (1991) Potential use of bialaphos, a microbially produced phytotoxin, to control red raspberry in forest plantations and its effect on black spruce. *Can. J. For. Res.*, **21**, 489–97.

Jobidon, R. and Thibault, J. R. (1981) Allelopathic effects of balsam poplar on green alder germination. *Bull. Torrey Bot. Club*, **108**, 413–18.

Jobidon, R. and Thibault, J. R. (1982) Allelopathic growth inhibition of nodulated and unnodulated *Alnus crispa* seedlings by *Populus balsamifera*. *Amer. J. Bot.*, **69**, 1213–23.

Jobidon, R., Thibault, J. R. and Fortin, J. A. (1989a) Phytotoxic effect of barley, oat, and wheat straw mulches in eastern Quebec forest plantations. I. Effects on red raspberry (*Rubus idaeus* L.) *For. Ecol. Manage.*, **29**, 277–94.

Jobidon, R., Thibault J. R. and Fortin, J. A. (1989b) Phytotoxic effect of barley, oat, and wheat straw mulches in Eastern Quebec forest plantations. II. Effects on nitrification and black spruce (*Picea mariana* (Mill.) BSP) seedling growth. *For. Ecol. Manage.*, **29**, 295–310.

Jobidon, R., Thibault, J. R. and Fortin, J. A. (1989c) Effect of straw residues on black spruce seedling growth and mineral nutrition, under greenhouse conditions. *Can. J. For. Res.*, **19**, 1291–3.

Lalonde, M. (1979) Immunological and ultrastructural demonstration of nodulation of the European *Alnus glutinosa* (L.) Gaertn. host plant by an actinomycetal isolate from the North American *Comptonia peregrina* (L.) Coult. root nodule. *Bot. Gaz.*, **140** (suppl.): S35–S43.

Lalonde, M. and Quispel, A. (1977) Ultrastructural and immunological demonstration of the nodulation of the European *Alnus glutinosa* (L.) Gaertn. host plant by the North American *Alnus crispa* var. *mollis* Fern. root nodule endophyte. *Can. J. Microbiol.* , **23**, 1529–47.

Lawrey, J. D. (1986) Biological role of lichen substances. *The Bryologist*, **89**, 111–22.

Likens, G. E., Bormann, F. H., Johnson, N. M. *et al.* (1970) Effects of forest cutting and herbicide treatment on nutrient budgets in the Hubbard Brook watershed-ecosystem. *Ecol. Monogr.*, **40**, 23–47.

Lodhi, M. A. K. (1978) Comparative inhibition of nitrifiers and nitrification in a forest community as a result of the allelopathic nature of various tree species. *Amer. J. Bot.* , **65**, 1135–7.

Lodhi, M. A. K. (1979) Inhibition of nitrifying bacteria, nitrification and mineralization in spoil soils as related to their successional stages. *Bull. Torrey Bot. Club*, **106**, 284–9.

Lodhi, M. A. K. and Killingbeck. K. T. (1980) Allelopathic inhibition of nitrification and nitrifying bacteria in a ponderosa pine (*Pinus ponderosa* Dougl.) community. *Amer. J. Bot.*, **67**, 1423–9.

Norby, R. J. and Kozlowski, T. T. (1980) Allelopathic potential of ground cover species on *Pinus resinosa* seedlings. *Plant and Soil*, **57**, 363–74.

Perradin, Y., Mottet, M. J. and Lalonde, M. (1983) Influence of phenolics on *in vitro* growth of *Frankia* strains. *Can. J. Bot.*, **61**, 2807–14.

Putnam, A. R. and DeFrank, J. (1983) Use of phytotoxic plant residues for selective weed control. *Crop Prot.*, **2**, 173–81.

Putnam, A. R. and Tang, C. S. (1986) Allelopathy: state of the science, in *The Science of Allelopathy*. (eds A. R. Putnam and C. S. Tang), Wiley-Interscience, New York, pp. 1–19.

Rice, E. L. (1984) *Allelopathy*, 2nd edn, Academic Press, Orlando, Florida.

Romell, L. G. (1935) Ecological problems of the humus layer in the forest. *Cornell Univ. Agric. Exp. Stn. Mem.*, **170**, p. 28p.

Sutton, R. F. (1985) Vegetation management in Canadian forestry. Great Lakes Forest Research Centre, Can. For. Serv., Sault Ste-Marie, Ontario. Inf. Rep., O-X-369. p. 34p.

Thibault, J. R., Fortin, J. A. and Smirnoff, W. A. (1982) *In vitro* allelopathic inhibition of nitrification by balsam poplar and balsam fir. *Amer. J. Bot.*, **69**, 676–9.

Vitousek, P. (1981) Clear-cutting and the nitrogen cycle, in *Terrestrial Nitrogen Cycles. Processes, Ecosystem Strategies and Management Impacts* (eds F. E. Clark and T. Rosswall), *Ecol. Bull.* (Stockholm) **33**, 641–2.

Weston, L. A., Burke, B. A. and Putnam, A. R. (1987) Isolation, characterization and activity of phytotoxic compounds from quackgrass (*Agropyron repens* (L.) Beauv.). *J. Chem. Ecol.*, **13**, 403–21.

CHAPTER 21

Pigeon pea and velvet bean allelopathy

P. Hepperly, H. Aguilar-Erazo, R. Perez, M. Diaz and C. Reyes

21.1 INTRODUCTION

Pigeon pea (*Cajanus cajan* L. Millsp.) and velvet bean (*Mucuna deeringiana* Bort. Merr.) are large seeded tropical legumes of Indian and southeast Asian origin, respectively (Duke, 1981). Pigeon pea was introduced into the Caribbean by African slaves and fresh pigeon peas are a popular vegetable. They are stewed and served with rice or tuber crops. Velvet bean was introduced into the new world as a forage and cover crop.

Velvet bean can be used as a soil improvement crop. King *et al.* (1965) in Australia reported 331 kg N ha^{-1} in harvested velvet bean green matter within one year. In Honduras, the high price and unavailability of fertilizer nitrogen greatly limits the principal food crop, maize (*Zea mays* L.). In humid areas of Honduras, farmers have developed a maize–velvetbean intercropping system which greatly reduces fertilizer and weed control costs and results in stable high yields of maize without rotating fields from this *staple* (Nuñez, 1982).

In this system, maize is established and weeded twice in the first 40 days of growth. At 40 days, velvet bean seeds are planted at 6 m intervals throughout the field. At harvest, velvet bean covers the ground, greatly reducing weed growth and soil erosion. When ready for the following maize crop, farmers chop the green cover with machetes and allow three weeks for its decomposition. Seed is hand planted by jab planters into the soil covered by partially decayed velvet bean residues. In the second cropping cycle, fields are weeded once only at 40 days after planting when velvet bean is planted. In the third cycle and those thereafter, weeds are reduced to such low levels that many farmers do not weed at all, relying instead only on velvet bean for weed control.

Allelopathy: Basic and applied aspects Edited by S. J. H. Rizvi and V. Rizvi
Published in 1992 by Chapman & Hall, London ISBN 0 412 39400 6

Farmers in the Caribbean have recognized the weed suppression of pigeon pea (Gooding, 1962). Considering that weed suppressing legumes eliminate faster growing plants, it is likely that allelopathy is involved. Besides weed control, allelopathy may be related to poor crop development. Under continuous culture, pigeon pea yield decline is a serious worldwide problem. Autotoxicity may be relevant to this phenomenon.

21.2 WEED CONTROL

To determine the effect of pigeon pea leaf litter mulch on weed development, five square frames were randomly tossed over a tilled ultisol at the College of Agriculture Experimental Farm in Mayaguez, Puerto Rico (Hepperly and Diaz, 1983). Weeds were allowed one month of growth and the framed areas and adjacent equal areas were harvested and weeds counted and weighed. After one month, the mean number of weeds per frame was 40 with pigeon pea litter and 322 for the control areas. Pigeon pea litter (24 g) significantly ($p = 0.01$) controlled weeds compared with the control (105 g). Effect of the pigeon pea litter was selective. After two months, pigweed (*Amaranthus dubius* Mart.) was increased threefold in the pigeon pea litter treatment compared with that of the control. Wild poinsettia (*Euphorbia heterophylla* (L.) Small) was not significantly changed by pigeon pea leaf litter treatment; whereas, grass populations were reduced with pigeon pea litter to approximately one-seventh of those in the control. Pigeon pea litter gave excellent weed control for at least two months.

Field populations of weeds in pigeon pea were monitored in the field in the southern coast of Puerto Rico at the Fortuna Research Station in a mollisol. In these trials, grasses were the dominant weed until 37 days after planting. Maximum grass yield occurred at 23 days after harvest and production diminished to about one-sixth of the maximum by 49 days. Spider flower (*Cleome gynandra* L.) dominated from 37 to 60 days after planting, reaching a maximum yield of about one half of the maximum grass yield at 42 days after planting. Throughout the experiment, pigweed increased steadily and became the dominant weed at 60 days after planting, and thereafter. These field studies support the inhibition and compatible plant reactions found in the Mayaguez experiment in a divergent soil environment.

To determine the effects of velvet bean on weed populations, a time course study was conducted in Mayaguez, Puerto Rico, using four replicates of 25 square metres each (Aguilar-Erazo, 1984) (Table 21.1). To obtain weed analyses, 0.5 by 0.5 m square frames were randomly thrown into the experimental replicates and weeds were harvested and identified. Sixteen velvet bean plants were established in each replicate. Weed populations were determined at 1 and 4 months after planting

TABLE 21.1 *Changes of weed populations under velvet bean on an ultisolo in Mayaguez, Puerto Rico*[a]

Scientific name	Population 30 DAP[b]	Population 120 DAP
Cyperus rotundus L.	28	31
Dichromena ciliata Vahl	22	36
Cynodon dactylon (L.) Pers.	4	4
Digitaria sanguinalis (L.) Scop.	24	4
Eleusine indica (L.) Gaertn.	18	4
Panicum maximum Jacq.	0	10
Paspalum fimbriatum H. B. K.	8	2
Setaria geniculata (Lam.) Beauv.	20	20
Commelina diffusa Burm. f	25	32
Ditremexa occidentalis (L.) Britton and Rose	4	0
Bidens pilosa L.	9	0
Senecio confusa (D.C.) J. Britten	6	0
Ipomoea tiliacea (Willd.) Choisy	2	1
Chamaesyce hirta (L.) Millsp.	16	0
Euphorbia heterophylla L.	22	8
Phyllanthus niruri L.	7	4
Mimosa pudica L.	7	4
Centrosema pubescens Benth.	1	0
Indigofera tinctoria L.	3	0
Macroptilium lathyroides (L.) Urban.	9	4
Vigna vexillata (L.) Rich	1	0
Serjania polyphylla Poir. ex Steud.	3	3

[a] Each mean obtained from three samples (0.5 × 0.5 m frames) taken randomly from four (25 m^2) replicates.
[b] DAP = days after planting velvet bean.

velvet bean. As in the pigeon pea studies, weed populations declined over time with velvet bean. As with pigeon pea, weed control was selective but the spectrum of control was decidedly different. Broadleaf weeds were most suppressed by velvet leaf. Over the course of the experiment broadleaf populations declined by 78%. Less losses were noted for grasses, which declined 58%, and cyperaceous weeds increased by 34%.

21.3 EFFECTS ON CROPS

While suppression and stimulation of weeds may be of great importance, the direct effect of pigeon pea and velvet bean crop plants is crucial to crop management for interplanting and rotation. The inhibiting effect of pigeon pea leaf litter on other legumes and pigeon pea itself was marked (Hepperly and Diaz, 1983). Reduction of weed plants may occur from competition as well as allelopathy and residues may also reduce plant growth by serving as substrates for pathogens. Field beans and pigeon peas planted in fields with undecomposed pigeon pea litter show increased development of southern blight caused by *Sclerotium rolfsii* which grows from the pigeon pea residue.

Pigeon pea leaves harboured soil-borne *Pythium*, *Fusarium* and *Rhizoctonia* populations which are known pathogens in a wide range of plants. To eliminate pathogenic effects of microorganisms, leaf extracts were made using concentrated potassium hydroxide. These were diluted with a monobasic phosphate buffer giving a final solution of pH 6.2 with nutrient levels within ranges used in ordinary nutrient solutions. Three grams of extracted leaf litter applied per metre of germination surface had a significant negative effect on germination and root elongation in pigeon peas. This showed that the detrimental effects of pigeon pea leaf residue were mostly chemical and not related to physical effects or its value as a pathogen base and food source. Significant reductions of germination and root elongation of soybean (*Glycine max* (L.) Merr.), lab-lab bean (*Lablab purpureus* (L.) Sweet) and field beans (*Phaseolus vulgaris* L.) were also found in the laboratory with leaf residue at concentrations of $+4\,g\,M^{-2}$. Besides legumes, maize (*Zea mays* L.), tomato (*Lycopersicon esculentum* Mill.) and sorghum (*Sorghum bicolor* (L.) Moench.) were notably reduced in germination, root and shoot growth in *in vitro* tests. Winter squash (*Curcurbita moschata* (Duchesne) Poir.) showed no negative reaction to pigeon pea leaf litter but showed root growth to increase. Watermelon (*Citrullus vulgaris* Schrad) and cucumber (*Cucumis sativus* L.) also showed little detrimental effect from the pigeon pea leaf litter.

Crude sap of whole plants of *M. deeringiana* was expressed and diluted with sterile distilled water to give 0, 25, 37.5 and 50% concentrations. As in *in vitro* pigeon pea tests, moistened cellulose pads were used for germination and seedling growth tests (Aguilar-Erazo, 1984). Twenty-five to 75% inhibition of soybean roots and shoots were found at the test concentrations; whereas, germination was affected by less than 10%. As in pigeon pea tests, seedling parameters were more sensitive indices of toxicity than germination *per se*. In general, more germination losses were noted in the pigeon pea testing than with velvet bean. In sorghum,velvet bean inhibitions ranged from 50 to 85% and germination losses exceeded 15% at the highest concentration. In pigeon pea, velvet bean extract at 50% completely eliminated shoot development. These tests were also repeated in greenhouse pot trials with similar results.

Seedling growth abnormalities were common in the velvet bean treatments. These included: production of multiple shoots, swollen shoots and abnormal root proliferations. These reactions were reminiscent of those caused by plant growth regulators and hormones and were not observed in pigeon pea work. *In vitro* tests correlated well with *in vivo* techniques but obtained information more efficiently and economically. The elucidation of the growth-regulating activity of velvet bean chemicals should be a fertile area of future research.

21.4 LOCALIZATION AND NATURE OF ALLELOCHEMICALS

Pigeon pea leaf, stem and root extracts were prepared and tested using a tomato seed and seedling assay on moist cellulose pads. Germination inhibition with leaf extracts ranged from 80 to 5% compared with 80 to 70% for stems and 75 to 65% for root extracts and 85% in the non-treated control (Table 21.2) Radicle and epicotyl growth reductions were similiar to decreases in germination rates. Leaves are the main source of allelopathic chemicals in pigeon pea.

Yellow resinous microscopic secretions were found on pigeon pea

TABLE 21.2 *Influence of pigeon pea leaf, stem and root extracts on germination and growth of tomato seedlings* in vitro[a]

Pigeon pea extract	Aqueous extract (%)	Germination (%)	Growth (mm)	
			Radicle	Epicotyl
Leaf	5	80	47.4	17.5
	25	60	26.0	14.2
	50	30	4.3	2.5
	100	5	3.0	0.0
Stem	5	80	41.8	15.8
	25	75	33.8	23.0
	50	85	32.8	24.5
	100	70	31.9	30.1
Root	5	75	46.7	23.0
	25	65	34.5	23.0
	50	70	28.7	25.4
	100	75	26.8	18.1
None (control)	0	85	54.0	19.3

[a] Means are based on four seedlots of 25 tomato seeds, each germinated on moist cellulose pads at 27°C and 95% RH. Germination readings were taken at 10 days after planting and seedling growth measurements at 14 days.

TABLE 21.3 *The density of glandular leaf secretions on the upper and lower surfaces of leaves of 18 pigeon pea lines grown in Mayaguez, Puerto Rico*[a]

	Secretion density (no. cm^{-2})	
Pigeon pea line	Upper surface	Lower surface
Blanco	267	587
Kaki	237	663
Pinto	197	663
2B-Bushy	377	920
59	263	467
98	173	517
147	293	657
BDN-1	380	860
HY3C	297	673
PPE-45-2-4-EB	117	557
ICPL-289	237	637
ICPL-319	200	683
ICPL-7035	250	717
TCF-644	163	503
75452	140	660
76004	90	603
78341	130	477
78343	203	677
mean[b]	223	640

[a] Each mean is based on the leaf secretion counts of three random areas of 1 cm^2 on both the upper and lower leaf surfaces of three different leaflets.
[b] Highly significant difference between the upper and lower surface (mean based on paired *t*-test).

leaves. Similar secretions at much reduced frequency were found on stems and pods but not roots. Secretion density averaged 640 cm^{-2} on lower leaf surfaces and 223 cm^{-2} on the upper surfaces (Table 21.3). In 18 pigeon pea lines from India and Puerto Rico, little variation in the concentrations of these secretions was noted.

Pigeon peas under stress or pathogen attack fold their leaves, exposing the lower surface. This reaction appears to be a defensive one which exposes both the heaviest pubescence and the greatest density of secretions. In this way, both physical and chemical protection from the outside environment may be optimized.

To test the role of glandular secretions in allelopathy, it was necessary to separate these from the rest of the plant's constitutents. Seven solvents were employed to test the solubility of pigeon pea leaf secretions.

TABLE 21.4 *Tomato germination and growth as influenced by extracts from pigeon pea leaves using solvents with increasing ability to extract glandular secretions*[a]

Extract solvent	Germination (%)	Growth (mm)	
		Radicle	Epicotyl
Water	90	49.0	27.2
Hexane	90	49.2	19.6
Butanol	75	44.3	0.0
Benzene	30	3.6	0.0
Chloroform	10	2.0	0.0
Methanol	5	1.0	0.0
Acetone	0	0.0	0.0

[a] Means are based on two samples of 25 seeds for each extract. Extracts were applied to filter paper discs (15 cm) and solvents were completely evaporated before testing. Discs were moistened in sterile distilled water for the seed assay.

Secretions were very soluble in acetone, methanol and chloroform; slightly soluble in benzene and butanol; and hardly soluble in water and hexane. Solubility was determined by submerging leaflets in test solvents for 2 min and microscopically analysing erosion of secretions. In tomato seedling bioassay, acetone, methanol and chloroform extracts of leaves gave 0–10% germination, benzene and butanol 30–75% and hexane and water 90% (Table 21.4). Extracts were applied to filter paper discs and solvents were completely evaporated before bioassay. The leaf residues were also tested for their toxicity. Efficient solvents of leaf secretion were effective in detoxifying leaf residues, intermediate solvents gave partial detoxification, and non-solvents did not change leaf toxicity compared with non-extracted leaves. Besides tomato, sorghum was also tested with similiar growth inhibitions.

Pigeon pea leaf extracts were separated with silica gel thin layer chromatography using a benzene–methanol (90:10) solvent system. The separations revealed eight fluorescent spots. A single light-blue fluorescent compound (Rf = 0.3) inhibited tomatoes in seedling bioassay. Using silica gel column chromatography (benzene–methanol elution), 20 fractions were collected with the 11th fraction showing toxicity in bioassays of tomato and sorghum. This fraction, when analysed in thin layer chromatography, showed a single bule spot (Rf = 0.3). Analysis of the toxic fraction by nuclear magnetic resonance showed a profusion of peaks, suggesting a chemical mixture which prevented positive identification of any single compound. Nevertheless, many of peaks were consistent with the presence of a terpenoid compound.

TABLE 21.5 *The effect of pumping air over volatiles obtained by distillation of macerated stems and leaves of velvet bean on the growth of soybean, pigeon pea and sorghum*[a]

Test species	Plant organ	Growth response (mm) Treatments[b]	
		Control	Volatiles
Pigeon pea: *Cajanus cajan*	Shoots	6.8	4.1
	Roots	10.3	4.2
Soybean: *Glycine max*	Shoots	16.4	12.8
	Roots	13.6	10.1
Sorghum: *Sorghum bicolor*	Shoots	9.0	5.1
	Roots	12.8	4.2

[a] Means based on two replicates of 25 plants each.
[b] All comparisons between the controls and volatiles were statistically significant at $p = 0.01$.

In velvet bean, we were interested in the possible involvement of volatile chemicals in the allelopathic effects of this species. Under desert environments, volatile inhibitors were shown as primary sources of allelopathic interactions (Muller *et al.*, 1964). Volatiles of velvet bean were tested using *in vitro* and *in vivo* systems. To obtain volatiles for *in vitro* studies, 2 kg of macerated fresh leaves and stems of velvet bean were suspended in 3 litres of distilled water. In a distillation apparatus the macerated suspension was heated to 130°C for 1 h and 250 ml of distillate with the concentrated volatile component was harvested and sealed for later use. Using a pump ($0.5\,kg\,cm^{-2}$), air was circulated over the concentrated volatiles to secondary flasks where pigeon pea, soybean or sorghum were seeded. The same system using a water blank for the treatment flask served as the non-treated control. Inhibition of root and shoot growth in these experiments ranged from 20 to 25% in soybean and from 40 to 60% in sorghum and pigeon pea (Table 21.5). Residues after extraction of volatiles also reduced root and shoot growth of test species, except for an increase in sorghum root growth (Tables 21.5 and 21.6). Negative growth responses varied from 35 to 64% compared with non-treated control.

In *in vivo* studies, air was constantly circulated from potted velvet plants enclosed in an inflated plastic bag with lead tubing inflating other plastic bags enclosing germination trays of the test species. The set in which air circulated directly to the bags without passing over velvet bean plants served as controls. Velvet bean treatment reduced the root and

TABLE 21.6 *Non-volatile residues from distillation of velvet bean stem and leaf macerate and their effect on the growth of pigeon pea, soybean and sorghum seedlings*[a]

		Growth (mm)[b]	
Test species	Plant organ	Control	Residue
Cajanus cajan	Shoot	7.8	2.8
	Root	26.8	16.3
Glycine max	Shoot	4.7	3.0
	Root	6.9	4.2
Sorghum bicolor	Shoot	37.3	14.8
	Root	38.8	44.7

[a] Means based on four replications per treatment (50 seeds/replicate) using a randomized complete block design. Seedling were grown on cellulose pads at 28°C and 95% RH for 8 days. Residues were applied at 500 g m^{-2}.
[b] All differences between treated and controls were statistically significant ($p = 0.05$).

TABLE 21.7 *The effect of volatiles from intact live velvet bean plants on the growth of pigeon pea, soybean and sorghum*[a]

		Growth (mm)	
Plant species	Organ	Control	Volatiles
Cajanus cajan (L.) Millsp.	Shoots	8.8	6.2
	Roots	5.3	6.9
Glycine max (L.) Merr.	Shoots	5.2	3.7
	Roots	14.3	8.6
Sorghum bicolor Moench.	Shoots	3.9	4.1
	Roots	17.0	13.9

[a] Means based on two replicates of 25 plants each.

shoot growth of soybean by 29 and 40%, respectively, compared with the control (Table 21.7). Sorghum showed an 18% reduction in root growth and no effect on shoot growth. Pigeon pea was perhaps the most disturbed with a 30% reduction of shoot growth and 31% increase in root growth. In these experiments, there were many examples of multiple shoots, swollen stems and other growth abnormalities similar to those observed in other experiments with velvet bean extracts.

Unlike pigeon pea, where the allelopathic source was localized in leaf tissue, velvet bean allelochemicals do not appear to be restricted to foliar tissues. The inhibitors of velvet bean appear in both volatile and non-volatile portions of the extracted plant and probably are more complex than in the case of pigeon pea. The growth-regulating properties add an interesting characteristic to the velvet bean allelochemicals.

Mucuna species are known as the greatest natural source of L-dopa, which gives symptomatic relief of Parkinson's disease (Daxenbichler *et al.*, 1971). Rehr and co-workers (1973) document the inhibition of insects on velvet bean seeds due to L-dopa content. Upon wounding, velvet plants rapidly produce red dopachrome pigments which quickly polymerize to black phytomelanin. Pathways involving L-dopa may be of major importance in the defensive reactions of the velvet bean.

Mears (1971) studied the bicyclic tryptophan alkaloids finding nicotine, prurinine, prurinidine, and two yet to be named compounds in velvet bean. Di-hydroxyphenylalanine is reported to cause vomiting and diarrhoea in pigs fed on velvet beans (Duke, 1981). Further study of velvet bean could well lead to important discovery of growth-regulating compounds with useful properties in agriculture, as growth alterations were consistently found in our tests.

21.5 COMPARATIVE ALLELOPATHY

Using tomato and soybean as bioassay plants, foliar residues of maize, Johnson grass (*Sorghum halapense* (L.) Pers.), pigeon pea, velvet bean, *Lantana camara* L., a verbenaceous weed, were tested. Less than 10 and 20% root length reductions were found with maize and Johnson grass, respectively (Table 21.8). Twenty to 50% reductions were for pigeon pea

TABLE 21.8 *Comparative toxicity of leaf residues of pigeon pea, velvet bean, maize, Johnson grass and* Lantana camara *determined by inhibition of tomato and soybean seedling growth in* in vitro *bioassays*[a]

	Root growth (mm)		Shoot growth (mm)	
Plant residue[b]	Tomato	Soybean	Tomato	Soybean
None (control)	28.9	64.2	36.8	83.4
Maize (*Zea mays*)	26.4	61.5	35.2	28.2
Johnson grass (*S. halapense*)	20.8	60.2	31.0	26.6
Pigeon pea (*Cajanus cajan*)	23.9	36.6	23.8	10.5
Velvet bean (*M. deeringianum*)	20.4	20.3	12.7	12.7
Lantana camara	11.0	24.1	9.7	5.7

[a] Each mean based on two replicates of 50 seeds each.
[b] Plant residue consisted of 50 g dried foliage per square metre.

and velvet bean, and *L. camara* reduced growth by over 50% in all cases. Perhaps the more efficient and aggressive growth of tropical C_4 grasses may make them less dependent on chemical defences than legumes and other plants lacking their efficient physiology. Persistent slow-growing species that are not excluded under intense competition, are probably excellent candidates for identifying allelopathic interactions.

21.6 TOXICITY MANAGEMENT

Soil type can affect chemical toxicity by adsorption of compounds, making them less available for absorption by plant roots (Sanchez, 1976). Clay (Marshall, 1975) and organic matter (Alexander, 1977) are major determinants of the absorptive capacity of soil, and soil microflora presence and activity. Soil microorganisms can be major contributors to degradation of toxic organic compounds. In our studies, pigeon pea leaf toxicity was greatest in sand. Kaolinitic clay, organic soil and a mixture of sand and the other components reduced observed toxicity (Table 21.9).

Burgos-Leon (1976) reported autotoxicity from sorghum to be severe in sandy soils and unobservable in loamic soils. Dried leaves in dry soil did not show reduction in toxicity over 8 weeks incubation in our studies. In moist warm mollisol soil from the major pigeon pea zone in Puerto Rico, pigeon pea leaves maintained an observable toxicity for approximately 7 weeks (Table 21.10). Under natural conditions, residues can remain on the surface for extended periods without significant precipitation and

TABLE 21.9 *Effect of soil type on pigeon pea leaf toxicity as measured in the tomato seedling bioassay*[a]

Soil type	Pigeon pea leaf residue[b]	Germination %	Root length (mm)
Organic (decomposed filter mud)	No	85	39
	Yes	72	27
Sand	No	82	28
	Yes	59	19
Kaolinite clay	No	62	24
	Yes	54	18
Organic:sand:clay (1:1:1) mix	No	79	25
	Yes	72	24

[a] Each mean based on three replications of 25 tomato seeds.
[b] Residues were applied at the rate of 50 g m^{-2}.

TABLE 21.10 *Pigeon pea germination and growth inhibition from pigeon pea leaf residue in progressive stages of decomposition*[a]

Decomposition time (weeks)	Percent inhibition[b] Germination (%)	Root growth	Shoot growth
1	15	85	50
2	35	18	18
3	21	26	15
4	37	11	23
5	17	14	14
6	18	22	−11
7	−5	9	−4
8	−7	−3	−7

[a] Means based on two (50 seed) replicates. At each date three treatments were prepared – no leaf residue, 150 g pigeon pea leaf residue per square metre non-incubated, and decomposing leaf residues under moist conditions at 27°C.
[b] Percentage inhibition = [(Without residue − Decomposed residue) ÷ Without] × 100. Undecomposed residues after over eight trials showed 25% inhibition of germination, 43% inhibition of radicle length and 29% inhibition of shoot growth.

their degradation would be much reduced compared with the optimum conditions that were provided in this study. Early incorporation of pigeon pea residues and proper incubation should greatly reduce crop damage through allelopathy. Supplemental irrigation to stimulate residue decomposition might also be warranted in some cases.

21.7 CONCLUSIONS AND PROSPECTS

In the tropical world, a large portion of the rapidly growing population are peasant farmers with increasingly meagre and degraded land resources. Cereals and grain legumes are major portions of the diet and legume culture is the most important method of maintaining soil fertility. Besides soil fertility, weed and pest controls are major costs to farmers. Legumes are notable producers of a great variety of toxic chemicals which form useful defences against a variety of organisms (weeds, pests and diseases) and environmental stresses which hamper their development. Considering the need for intensifying tropical food production and conserving shrinking resources, greater emphasis in legume use as food and in soil conservation and pest control are urgently needed. Considerable basic and applied research on these crops and their integration in agricultural systems is also warranted.

Allelopathic potentials of tropical legumes should be thoroughly studied to make the best use of these plants for soil improvement and in pest control. As shown in these studies, legumes can be effective in specific weed control, as the pigeon pea has been used experimentally in Guatemala to control, successfully, purple nutsedge (*Cyperus rotundus* L.), the most damaging of all weeds world wide (D. Dardon, personal communication). Pigeon pea shows compatible reactions with plants such as winter squash and pigweed which may be managed to advantage in rotation with grain amaranth or curcurbits as well as other crops. Velvet bean shows good ability to control broad leaf weeds and appears to be an excellent intercrop (compatible) with maize. Recent studies by the University of Puerto Rico Nematology group headed by Dr N. Acosta has shown that velvet bean has excellent potential for controlling root-knot nematodes in tomatoes.

Allelopathy depends on the accumulation of plant chemicals which is highly dependent on soil and environmental factors, especially those governing their rates of production and degradation. Allelopathic studies need to consider residue management more thoroughly. This could include the evaluation of the microflora involved in degradation and the factors which affect the development and kinetics of this phenomenon. Many studies have focused on the interspecific tolerance or resistance to allelopathic chemicals but few studies have focused on the intraspecific variation, for example, autotoxicity reactions. Yield declines, which occur in long-term crops with autotoxicity, might be avoided if greater resistance to allelochemicals within the crop species could be identified and/or varieties without allelopathic potential are detected and used.

Because of its complexity, allelopathy research should continue to emphasize multi- and interdisciplinary approaches. Greater emphasis should be given to economic development of discoveries from allelopathic research and publicizing these contributions. Novel pest control arising from basic work in allelopathy may be one of the most promising areas for economic impacts from this research area.

REFERENCES

Aguilar-Erazo, H. (1984) Evaluation of velvet bean allelopathy. MSc Thesis (in Spanish), Univ. of Puerto Rico, Dept. Crop Prot., Mayaguez.

Alexander, M. (1977) Organic matter decomposition, in *Soil Microbiology*, 2nd edn, Wiley, New York, pp. 128–47.

Burgos-Leon, W. (1976) Phytoxicity of sorghum residue in soils of the western African savannah. PhD Thesis, Nancy Univ., France.

Daxenbichler, M. E., Van Etten, C. H., Hallinan, E. A. *et al.* (1971) Seeds as sources of dopa. *J. Med. Chem.*, **14**, 463–5.

Duke, J. A. (1981) *Cajanus cajan* and *Mucuna* spp., in *Handbook of Legumes of World Economic Importance*, Plenum Press, New York, pp. 33–7 and 170–3.

Gooding, H. J. (1962) The agronomic aspects of pigeon peas. *Field Crop Abstr.*, **15**, 1–15.

Hepperly, P. R. and Diaz, M. (1983) The allelopathic potential of pigeon peas in Puerto Rico, *J. Agric. Univ. P.R.*, **67**, 453–63.

King, N. J., Mungomery, R. W. and Hughes, C. G. (1965) *Manual of Cane Growing*, Angus and Robertson, Sydney.

Marshall, K. C. (1975) Clay mineralogy in relation to survival of soil bacteria. *Ann. Rev. Phytopath.*, **13**, 357–74.

Mears, J. A. (1971) Alkaloids in the Leguminosae, in *Chemotaxonomy of the Leguminosae*, Academic Press, New York.

Muller, C. H., Muller, W. H. and Haines, B. L. (1964) Volatile growth inhibitors produced by *Salvia* species. *Bull. Torrey Bot. Club*, **91**, 327–30.

Nuñez, M. A. (1982) Alternative weed control in maize for small farmers on the humid northern coastal plain of Honduras (in Spanish), *Proc. PCCMCA, San Jose, Costa Rica*.

Rehr, S. S., Janzen, D. H. and Feeny, P. P. (1973) L-dopa in legume seeds: a chemical barrier to insect attack. *Science*, **181**, 81–3.

Sanchez, P. A. (1976) Clay mineralogy and ion exchange processes and soil organic matter, in *Properties and Management of Soils in the Tropics*, Wiley, New York, pp. 135–83.

CHAPTER 22

Allelopathy in forest and agroecosystems in the Himalayan region

N. P. Melkania

22.1 INTRODUCTION

A great variety of metabolic chemicals potentially involved in plant–plant chemical interactions are released from plants primarily through leaching from aboveground parts, thus, play a significant role in plant interactions on a day-to-day basis (Tukey, 1970). The plant leachates have an effect upon the soil structure, texture, aeration, permeability and exchange. Roots, especially those of large plants, extract substances from deep and distant areas of the soil from where they are translocated into the plant system, later to be returned to the soil by leaching and litter fall. Forest litter has long been recognized as a possible deterrent to tree seedling establishment (Koroleff, 1954), and differences in germination and growth of plant species beneath trees of various species are well known (Telfer 1972). Thus, trees build a characteristic profile of their own by this method; however, the magnitude of materials supplied to the soil layers depends upon the rate of leaching, litter fall and chemical disintegration of the compounds in the environment.

It is often asserted uncritically that dominant plant species exert influence on the floor conditions and understorey vegetation through indirect means as shade level, moisture status of soil and nutrient availability; however, in many circumstances allelopathy may determine directly the habitat characteristics and also may often interact with the other factors (Muller, 1974; Gleissman and Muller, 1978; Rice, 1984; Chou, 1986). The wide variations in quality and quantity of metabolites that are released and can accumulate beneath a plant, finally determine the development of species and community underneath the canopy. Leachates also influence the number and behaviour of soil microorganisms that affect the soil-forming processes, soil fertility and susceptibility, and immunity of plant species to pests (Gigon and Ryser, 1986).

Allelopathy: Basic and applied aspects Edited by S. J. H. Rizvi and V. Rizvi
Published in 1992 by Chapman & Hall, London ISBN 0 412 39400 6

Allelopathy is increasingly gaining acceptance in India as a factor having ecological significance in plant dominance, patterning of vegetation (Melkania, 1983), succession, crop productivity and agroforestry systems (Melkania, 1987a). During 1979, I initiated research on allelopathy and its possible significance in tree–herb associations in the subtropical and temperate Himalayan plant species. This chapter describes the role of allelopathy in forest- and agroecosystems in the Himalayan region.

22.2 ALLELOPATHIC MECHANISM OF PATTERNING OF UNDERSTOREY VEGETATION IN A FOREST ECOSYSTEM

I observed over a period of several years that herbaceous plants exhibit characteristic pattern under different tree species and also around some herbaceous and half-shrub species. In the case of the former (tree elements) this is evident even under lopped trees (poor canopy forests) that allow nearly full sunlight to penetrate entirely to the trunk during part of the day. The condition persisted even under no or negligible biotic interference. Consequently, I started to determine what sort of interference on the part of the woody species and certain nonwoody species is responsible for understorey patterns and, subsequently, regeneration. The prima facie evidence expressing allelopathic behaviour of the plant species studied, is given in Table 22.1.

22.2.1 Chemical interactions between canopy species and herbaceous flora

I selected predominant canopy species, viz. *Pinus roxburghii*, *Cedrus deodara*, *Quercus leucotrichophora*, *Q. floribunda*, *Myrica esculenta* and *Rhododendron arboreum* for the study. Individuals of these tree species were selected at different topographic and altitudinal situations in natural forests. The herbaceous vegetation under these species was analysed qualitatively and quantitatively by means of a permanent quadrat of 1 × 1 m for species composition, importance values and aboveground biomass. The role of habitat conditions, viz. light intensity and soil characteristics, if any, was also analysed. The allelopathic mechanism was investigated using leachates of plant parts and soil (around the trunk) through bioassay. The leachates were prepared by extracting 50 g of plant material with 1.0 litre of distilled water (wt/v ratio) for 48 h. The drip water was simulated by passing water sprays (1.0 l) on the fresh leaf materials (100 g) by a nozzle for 2 min. This was referred to as dilution A. The stock solution was further diluted by mixing 10 ml of dilution A into 100 ml of distilled water. This was referred to as dilution B. In each case, the control was run with distilled water. The effect of leachates was studied by examining their influence on seed germination of two phytometer species, viz. *Lepidium virginicum* and *Lolium perenne*. The

TABLE 22.1 *Allelopathic phenomena in Himalayan species (based on Melkania, 1983, 1984, 1987a, b; Melkania* et al., *1982; Melkania, unpublished studies)*

Allelopathic species	Test species affected		Allelochems present	Allelopathic effects	'Prima facie' evidence
	Inhibition	Stimulation			
Forest tree species					
Abies pindrow	*Lepidium virginicum* *Lolium perenne*	–	Hydrophilic metabolite(s) ?	Inhibition of seed germination, seedling growth and yield	Species composition within 'circle' effect of trees is different from outside, creation of bare areas around trees at some places
Cedrus deodara	*Lepidium virginicum* *Lolium perenne*	–	Terpenes, terpenoids, hydrophilic metabolite(s) ?	Inhibition of seed germination	Suppression of growth and gradual exclusion from the zone around the tree
Myrica esculenta	*Lepidium virginicum* *Lolium perenne*	–	Hydrophilic metabolite(s) ?	Inhibition of seed germination	Suppression of growth under canopy
Pinus roxburghii	*Lepidium virginicum* *Lolium perenne*	–	Terpenes, terpenoids	Delayed and reduced seed germination	Suppression of growth and yield beneath the tree canopy

TABLE 22.1 *Cont'd*

Allelopathic species	Test species affected		Allelochems present	Allelopathic effects	'Prima facie' evidence
	Inhibition	Stimulation			
Quercus floribunda	*Lepidium virginicum* *Lolium perenne*	–	Tannins, hydrophilic substance(s) ?	Reduced and delayed seed germination	Suppression of growth and regeneration by seedlings
Q. leucotrichophora	*Lepidium virginicum* *Lolium perenne*	–	Tannins, hydrophilic metabolite(s) ?	Reduced and delayed seed germination	Suppression of growth and regeneration of seedlings
Rhododendron arboreum	*Lepidium virginicum* *Lolium perenne*	–	Hydrophilic metabolite(s) ?	Reduced and delayed seed germination	Species composition within the zone of influence of the tree is different from outside areas
Farmland tree species					
Bauhinia purpurea	*Brassica campestris* *Echinochloa frumentacea* *Eleusine coracana* *Fagopyrum esculentum* *Lepidium sativum* *Raphanus sativus* *Setaria italica*	*Glycine soja*	Tannins, hydrophilic metabolite(s) ?	Reduced seed germination, seedling growth and yield	Suppression of growth under canopy

B. retusa	*B. campestris* *E. frumentacea* *E. coracana* *F. esculentum* *L. sativum* *R. sativus* *S. italica*	*G. soja*	Tannins, hydrophilic metabolite(s) ?	Reduced seed germination, seedling growth and yield	Suppression of growth beneath canopy
Celtis australis	*B. campestris* *E. frumentacea* *E. coracana* *F. esculentum* *G. soja* *L. sativum* *R. sativus* *S. italica*	–	Phenolic acids ?	Reduced seed germination, seedling growth and yield	Suppression of growth beneath canopy
Grewia optiva	*B. campestris* *E. frumentacea* *E. coracana* *F. esculentum* *L. sativum* *R. sativus* *S. italica*	*G. soja*	Phenolic acids ?	Reduced germination, seedling growth and yield	Suppression of growth under canopy
Juglans regia	*B. campestris* *E. frumentacea* *E. coracana* *F. esculentum* *G. soja* *L. sativum* *R. sativus* *S. italica*	–	Hydrophilic metabolite(s) ?	Reduced germination, seedling growth and yield	Suppression of growth under canopy

TABLE 22.1 *Cont'd*

Allelopathic species	Test species affected: Inhibition	Test species affected: Stimulation	Allelochems present	Allelopathic effects	'Prima facie' evidence
Melia azedarach	*B. campestris* *E. frumentacea* *E. coracana* *F. esculentum* *G. soja* *L. sativum* *R. sativus* *S. italica*	–	Hydrophilic metabolite(s) ?	Reduced germination, seedling growth and yield	Suppression of growth under canopy in case of dense planting
Half-shrub species					
Artemisia vulgaris	*Chrysopogon fulvus,* *Lepidium virginicum* *Lolium perenne*	–	Terpenes	Reduced germination and growth	Zonation around the bush and encroachment on new areas
Herb species					
Salvia lantana	*Alysicarpus rugosus* *Chrysopogon fulvus* *Lepidium virginicum* *Lolium perenne*	–	Volatiles, terpenes ?	Reduced germination, seedling growth and yield	Zonation around the plant and rapid encroachment in new areas. The effect is more pronounced under biotic stresses, viz., over-grazing and burning

TABLE 22.2 *Species richness in understorey of some woody species (based on Melkania, 1983)*[a]

Ground flora category	Woody species[a]						Total ground flora across woody species
	Pr	Cd	Ql	Of	Me	Ra	
Bryophytes							
Musci	2	4	4	5	4	6	8
Jungermannials	0	1	0	1	2	4	4
Hepaticae	4	5	4	0	4	5	8
Total	6	10	8	6	10	15	20
Pteridophytes							
Non-filiaceous	1	1	1	1	1	1	1
Filiaceous	4	14	10	6	8	17	27
Total	5	15	11	7	9	18	28
Gymnosperms (seedlings)	1	3	1	2	0	0	3
Angiosperms							
Leguminous	14	6	7	3	5	3	24
Non-leguminous	48	62	70	61	61	73	160
Total dicots	62	68	77	64	66	76	184
Graminoids	19	16	16	13	19	16	51
Non-graminoids	10	11	13	14	12	10	29
Total monocots	29	27	29	27	31	26	80

[a] Pr = *Pinus roxburghii*; Cd = *Cedrus deodara*; Ql = *Quercus leucotrichophora*; Qf = *Q. floribunda*; Me = *Myrica esculenta*; Ra = *Rhododendron arboreum*.

pH and electrical conductivity of the extracts were measured immediately after filtration.

The canopy tree species influenced the floor conditions appreciably. Except for *Quercus floribunda*, soil beneath all other tree species was slightly acidic in reaction. The greater amount of organic carbon coupled with greater accumulation of litter, low temperature and light intensity in understorey spaces resulted in a higher amount of gravitational soil moisture, compared with the outside areas of the forest floor. The woodland floors under broadleaf tree species, compared with conifers, included more shrubs. The species richness in understorey differ markedly among canopy species (Table 22.2). The number of ground flora species exclusively associated with a tree species, was in the order: *Quercus floribunda* (39) > *Cedrus deodara* (34) > *Pinus roxburghii* (29) >

TABLE 22.3 *Relative performance of common ground flora species (biomass per unit plant; grams per plant) (based on Melkania, 1983)*

Ground flora species	Woody species[b]					
	Pr	Cd	Ql	Qf	Me	Ra
Phanerogams						
Berberis asiatica (seedling)	1.5	0.4	0.6	0.7	0.7	0.6
Bidens biternata	1.1	0.5	0.4	0.4	0.1	<0.1
Commelina bengalensis	0.7	0.3	0.3	0.5	0.2	0.1
Crataegus crenulata (seedling)	0.6	1.2	1.3	0.8	1.5	1.2
Galium asperifolium	0.5	0.4	0.3	0.4	0.5	0.2
Geranium nepalense	0.8	0.7	0.4	0.5	0.5	0.5
Myrsine africana (seedling)	0.4	0.5	1.2	0.7	1.2	1.2
Oxalis corniculata	0.4	0.6	0.2	0.4	0.3	0.1
Valariana wallichii	0.5	0.6	0.2	0.4	0.3	0.3
Viola serpens	<0.1	<0.1	<0.1	0.6	0.5	<0.1
Cryptogams						
Fern species	1.6	1.8	0.2	0.6	1.3	2.5
Selaginella sp.[a]	0.002	0.001	0.002	0.001	0.001	0.003

[a] The values are grams per square centimetre of plant cover.
[b] For abbreviations, see Table 22.2.

Rhododendron arboreum (25) > *Quercus leucotrichophora* (20) > *Myrica esculenta* (11). The high numbers indicated the selective influence of the canopy species on the ground flora. Twelve understorey species were common under all the canopy species but the differences in their biomass values with respect to tree species were statistically significant ($p < 0.05$) (Table 22.3).

No definite relation between the characteristics of the extracts and the degree of inhibition was noticed, as, in some cases, the extracts having higher conductivity expressed lower inhibition compared with the extracts of relatively lower concentration. All the canopy species expressed allelopathic potential and inhibited rate and percent seed germination of the test species. The allelopathic potential of the water-soluble extracts

TABLE 22.4 *Inhibitory potential in leachates of various plant parts and soil of allelopathic species (based on Melkania, 1983; Melkania and Singh, 1987; Melkania* et al., *1982; Melkania, unpublished studies)*

Allelopathic species	Orders of inhibition by leachates
	Lepidium virginicum
Abies pindrow	YDL > LL > OML >DW
Artemisia vulgaris	YDL > OML > SE
Cedrus deodara	LL > YDL > OML > S > WL > DW
Myrica esculenta	LL > WL > S > DW > OML > YDL
Pinus roxburghii	LL > S > WL > YDL > OML > DW
Quercus floribunda	LL > S > WL > YDL > OML > DW
Q. leucotrichophora	LL > WL > S > YDL > OML > DW
Rhododendron arboreum	S > LL > YDL > WL > OML > DW
Salvia lantana	L_1 > S > R > DW > L_2
	Lolium perenne
Abies pindrow	OML > YDL > DW > LL
Artemisia vulgaris	YDL > OML > S
Cedrus deodara	YDL > WL > S > LL > OML > DW
Myrica esculenta	LL > WL > OML > DW > S > YDL
Pinus roxburghii	S > WL > DW > LL > YDL
Quercus floribunda	LL > YDL > S > WL > OML > DW
Q. leucotrichophora	S > WL > LL > OML > YDL > DW
Rhododendron arboreum	S > WL > LL > YDL > OML > DW
	Lepidum sativum, Glycine soja, Setaria italica
Bauhinia purpurea, B. retusa, Celtis australis, Grewia optiva, Juglans regia, Melia azedarach	L_1 > WL > LL

DW, drip water; YDL, young developing leaf; OML, old mature leaf; LL, leaf litter; WL, wood litter; L_1, leaf (young plus old); L_2, litter (leaf plus wood); S, soil; R, Root.

from soil and plant parts of the canopy species, appeared to differ considerably with the phytometer species. Drip water in dilution A from *Myrica esculenta* and *Quercus floribunda* were most inhibitory to the germination of both the test species, compared with the same from other tree species. In dilution B, drip water from *Rhododendron arboreum* stimulated seed germination in *Lepidium virginicum*. Similarly, drip water in dilution B from *Quercus leucotrichophora* and *Rhododendron arboreum* stimulated germination in *Lolium perenne*. Among leaf extracts, young leaf material was more inhibitory to seed germination of both the test species, compared with old leaf material. The percent inhibition across the tree species increased in the order: leaf litter > soil > wood litter > young leaf > old leaf > drip water for *Lepidium virginium* and soil > wood litter > leaf litter > young leaf > old leaf > drip water for *Lolium perenne* (Melkania and Singh, 1987).

The allelopathic potential of plant parts and soil of the plant species studied at dilution A is presented in Table 22.4. Among plant parts, litter was more inhibitory to seed germination than leaf materials, soil and drip water. The relatively lower inhibitory effect of old leaf, compared with young leaf, conformed to the inverse relationship between allelopathic potential and phenophases reported by Koeppe *et al.* (1970). This could be attributed to degradation of allelochemicals into some less toxic form (molecular changes) or reduction in their concentration owing to dilution, or both (Bokhari, 1978). The greater tolerance exhibited by *Lolium perenne* to extracts, compared with *Lepidium virginicum*, indicated that the allelopathic effect may be selective. The existence of exclusive species and variations in relative performance of common understorey species and group composition are evidently related to the selective allelopathic potential of the canopy species. However, it was not possible to segregate the effects of leachates from those of light intensity and litter biomass. *Abies pindrow* leachates also expressed inhibition of seedling growth and dry matter yield of the test species.

22.2.2 Chemical interactions among herbaceous species

Many species of *Artemisia* and *Salvia* are known to be allelopathic for associated herbaceous flora (Muller and Muller, 1964; Muller, 1965, 1966; Muller *et al.*, 1964; Qasem and Abu-Irmaileh, 1985). While in the natural stands of *Artemisia vulgaris* there is a definite zone of reduced growth of associated herbaceous elements, the rapid encroachment of *Salvia lantana* over ground flora species presented an interesting problem in relation to the vegetational patterns. Initial studies on the analysis of site characteristics around the *A. vulgaris* and *S. lantana* and away from them revealed that the differences in soil organic matter and water content were insignificant. Therefore, investigations on the leaf leachates were undertaken to investigate whether the competitive ability of *S. lantana* and the suppressing nature of *A. vulgaris* were due to any allelopathic potential.

The extracts of plant materials and soil around these were prepared in the same manner as described for canopy species. In the case of *S. lantana*, the effect of volatile emanations was also studied. The effect of decaying leaf litter was studied using a pot culture technique. The percent seed germination was recorded after six days. The plant population was later maintained to five individuals per pot for an additional two weeks to examine the effect of decaying leaf litter on seedling growth and phytomass.

The pH and concentration have no marked inhibitory effect on the seed germination. The extracts of *A. vulgaris* and *S. lantana* adversely affected the rate and percentage of seed germination of the test species. The inhibitory effect of young developing leaf extract was greater than that of old mature leaf extract. The inhibition was low at lower concentrations. Seasonal variation in allelopathic potential revealed that the leaf material had greater phytotoxic effects in summer followed by the rainy season, while the winter material had no phytotoxicity. Perhaps, in summer, the chemical substances accumulate in higher amounts because of the lack of rainfall and leaching loss; while in winter, possibly due to the low temperature, the production of these substances was reduced. Soil extracts had also shown toxic effects.

The volatile emanations from young *Salvia* leaves were also inhibitory to seed germination. Decaying leaf material also inhibited seed germination, seedling growth and phytomass of *Lepidium virginicum* (Table 22.5). Although, the inhibitory effect may be selective for the test species, it is concluded that the rapid encroachment of *S. lantana* and the suppressive nature of *A. vulgaris* may be due to their phytotoxic ability.

22.3 ALLELOPATHY IN FARMLAND WOODY SPECIES

In the Himalayan region, many tree species, viz. *Bauhemeria rugulosa*, *Bauhinia purpurea*, *B. retusa*, *Celtis australis*, *Grewia optiva*, etc., are

TABLE 22.5 *Effect of decaying* Salvia lantana *leaf material on seed germination, growth and biomass of* Lepidium virginicum *seedlings*

Treatment	Germination (%)	Seedling growth (cm)	Biomass (mg)
Distilled water	54.66	4.74	62.33
Decaying leaf material	27.66[a]	4.44[a]	40.00[a]

[a] $p < 0.05$.

cultivated on terrace risers in an agroecosystem for tree foliage. The suppression in growth of agronomic crops is evident underneath these tree species. Since these woody species receive more nutrients and water from the lower stratum, compared with annual crops, and their branches are extensively lopped for top feeds and fuel, it appears that physical factors play a negligible role in the reduced growth of agronomic crops. Therefore, leaf leachates of *Bauhinia purpurea*, *B. retusa*, *Celtis australis*, *Juglans regia*, *Grewia optiva* and *Melia azedarach* were screened for possible phytotoxic effects. The leachates of fresh leaves and leaf litter were obtained by emersing 200 leaves of each woody species in 1.0 litre of distilled water for 24 h as recommended by Tukey and Mecklenburg (1964). In the case of *Melia azedarach*, 200 leaflets were used. The wood litter extract was prepared by soaking 50 g of wood litter to 1.0 litre of distilled water for 48 h. The influence of leachates was studied using a bioassay test (Guenzi *et al.*, 1967). The agronomic crops grown widely under these tree species, viz. *Brassica compestris*, *Echinochloa frumentacea*, *Eleusine coracana*, *Fagopyrum esculentum*, *Glysine soja*, *Lepidium sativum*, *Raphanus sativus* and *Setaria italica* were used as phytometer species (Melkania, 1984). The effect of decaying leaf litter was studied through a pot culture technique. In addition, two exotic species, viz. *Eucalyptus tereticornis* and *Populus deltoides*, introduced as agroforestry tree species in the foothill zone (*Tarai regime*), were also examined for their allelopathic potential. The extracts of the leaf litter and soil around the tree species were prepared in 1:10 wt/v ratio. The further dilution was prepared by adding 100 ml of solution A to an equal amount of distilled water. The rain drip was simulated by passing 1.0 litre of distilled water on 100 g of fresh leaf material for 10 min through a sprayer. *Triticum aestivum* and *Cicer arietinum* were used as test species.

The water-soluble substances in the leachates, except for a few cases, influenced significantly ($p < 0.05$) the seed germination of the test species (Table 22.6). The inhibitory effect varied among test species. Maximum inhibition was recorded for *Setaria italica, Lepidium sativum* and *Glysine soja*. Germination of *Echinochloa frumentacea* and *Eleusine coracana* was least inhibited. The degree of toxicity of the leachates was in the sequence: *Celtis australis* > *Juglans regia* > *Grewia optiva* > *Bauhinia retusa* > *B. purpurea* > *Melia azedarach*. The decaying leaf litter influenced negatively the seedling growth and phytomass production in the cases of *Setaria italica*, *Lepidium sativum* and *Glysine soja*. Among plant parts, the inhibition was recorded as leaf > leaf litter > wood litter.

Among exotic species, *Populus deltoides* has an inhibitory effect on the radicle/rootlet growth and seed germination of both the test species, while *Eucalyptus tereticornis* had no marked effect on seed germination of *Triticum aestivum* but had greater inhibitory potential than *Populus deltoides* on *Cicer arietinum*. In both the woody species, the leaf litter leachate had maximum inhibitory potential followed by soil leachates and

TABLE 22.6 *Effect of farmland tree species leaf leachates on germination of agronomic crops (based on Melkania, 1984, 1987a)*

Agronomic crops	Control (distilled water)	Tree species					
		Br	Bp	Ca	Go	Jr	Ma
Vegetables							
Lepidium sativum	92.0	64.7[a]	68.5[a]	72.0[a]	70.7[a]	68.0[a]	84.0[a]
Raphanus sativus	99.3	96.0[a]	96.2[a]	85.0[a]	68.0[a]	67.3[a]	95.3[a]
Cereals							
Echinochloa frumentacea	100.0	78.0[a]	82.0[a]	94.7[a]	91.3[a]	93.3[a]	100.00
Eleusine coracana	92.7	86.0[a]	87.5[a]	78.7[a]	88.7	86.0	92.7
Setaria italica	95.7	84.7[a]	86.5[a]	47.3[a]	50.7[a]	90.0	88.0
Pseudo-cereal							
Fagopyrum esculentum	94.0	85.4[a]	85.4[a]	86.7	66.0[a]	52.0[a]	82.7[a]
Pulse							
Glycine soja	85.3	95.0[a]	92.6[a]	41.3[a]	89.0	79.0[a]	40.0[a]
Oil seed							
Brassica campestris	100.0	95.3[a]	96.2[a]	86.0[a]	86.0[a]	71.3[a]	96.0[a]

[a] Significantly different from control at 0.05 level.
Br, *Bauhinia retusa*; Bp, *B. purpurea*; Ca, *Celtis australis*; Go, *Grewia optiva*; Jr, *Juglans regia*; Ma, *Melia azedarach*.

drip water. The decaying leaf litter of *Populus deltoides* significantly reduced the seed germination and biomass in *Triticum aestivum*, while *Eucalyptus tereticornis* leaf litter had no inhibitory influence on either test species.

Stickney and Hoy (1881) suggested that the failure of most herbs to grow under *Juglans nigra* was due to toxins produced by the tree species. Similar results were also recorded by Lodhi and Rice (1971) while working on *Celtis laevigata*. In India, less attention has been paid to the crop–tree allelopathy, but there are numerous reports in the literature indicating that the germination and biomass of herbaceous plants may decrease when grown under the influence of woody elements (Saxena and Singh, 1978; Singh and Bawa, 1982; Melkania, 1987a). My studies concluded that possibly the phytotoxic effects of the leachates of farmland tree species may be the cause of suppressed growth of agronomic crops beneath them. Such effects of woody species on the agronomic crops must be considered and investigated before planning agroforestry programmes.

22.4 ALLELOPATHY IN RANGELANDS

The Himalayan rangelands experience overgrazing throughout the year, thus have shown signs of deterioration in the form of decline in species richness (particularly the palatable species), low productivity, etc. Many herbaceous weedy species are encroaching rapidly on such degraded sites. These are: *Amaranthus spinosus*, *Anaphalis cordata*, *Argemone mexicana*, *Avena fatua* (in submarginal lands), *Chenopodium botrys*, *Echinochloa crusgalli* (in wet places), *Osmunda regalis*, *Polygonum polystachys*, *Rottboellia exaltata*, *Rumex hastatus*, *R. nepalensis*, *Tagetes minor* and *Verbascum thapsus*. Preliminary studies with the leaf leachates indicated inhibitory potential in these species. However, detailed allelopathic analyses on (i) the effect of leachates of different plant parts and soil on different grassland species, (ii) the effect of dilution and seasonality on allelopathic potential, and (iii) identification of the allelochemicals are in progress.

22.5 ALLELOPATHY IN RELATION TO ENVIRONMENTAL FACTORS

The allelopathic mechanism is always related to other environmental factors and it is very difficult to isolate it from the environmental complex (Muller, 1974) and competition (Wilson and Rice, 1968). Several cases of allelopathic influences in relation to environmental stresses have been recorded (Koeppe *et al.*, 1970). Many allelochemicals produced by plants are governed by such environmental factors as temperature, light intensity, water potential of environment, nutrients, soil microorganisms, etc. (Koeppe *et al.*, 1976). Besides leaching, the chemicals are also released through volatilization, decomposition of litter and root exudation (Muller, 1974; Chou, 1983; Rice, 1984). The terpenoids are released first under drought conditions by volatilization, followed by the release of water-borne phenolics and alkaloids through leaching, and finally the scopoletin and hydroquinone into the soil (rhizosphere) through root exudation (Chou, 1986).

In my studies on canopy tree species of forest ecosystems, I observed that overstorey vegetation aspect and altitude act in an interactive manner. While the overstorey exerts influence both through indirect changes in physical conditions as soil moisture, nutrients and shade level, and directly by means of allelopathy; the altitude and other aspects strictly determine the microclimatic characteristics (e.g. temperature, light, rainfall, and physicochemical and biological determinants of soil) of the site. For farmland woody species and other herbaceous species, allelopathy alone was identified as the factor determining vegetational pattern around them and for harmful consequences on growth and biomass of the associated flora.

In Himalayan forests, since the amount of shade, litter thickness and allelopathic potential of the canopy act in an interactive manner, therefore, it may be concluded that the overstorey vegetation (tree stratum) has an overriding influence on the floor conditions, patterning of herbaceous vegetation and regeneration (Melkania, 1983). Litter perhaps exerts its influence on the forest floor dynamics both physically and chemically. Physically, it may provide a too-thick cushion for seeds and deter the seedling establishment and growth, while chemically, it may act through the production of allelochemicals. The accumulation of phytotoxic metabolites in woodland floors corresponds to the amount of litter produced and its chemical nature and rate of decomposition.

22.6 CONSTRAINTS IN ALLELOPATHIC RESEARCH

In India, most of the allelopathic investigations are based on bioassay studies (Datta and Chatterjee, 1980; Melkania, 1987a) and very little successful attempts could be made to analyse the allelochemicals and to conform allelopathy in field conditions. In fact, under natural conditions, the fate of allelochemicals depends on several interrelated factors including chemical disintegration of the allelochemicals. Therefore, *in vitro* results cannot be perfectly true under field conditions. Further, only a limited number of species have been screened against the allelopathic potential of overstorey species and other herbs. Recall that the allelopathy is selective, the toxic potential of a species cannot be generalized for all affected associated species. Studies on the relative allelopathic potential of the plant parts and soil, the effect of seasonality and phenophases, etc., on the toxic potential are meagre even under laboratory conditions. Also, it is not evident in many studies whether the allelochemicals are really produced by the plant species or are due to chemical transformation or microbial activities.

In my studies, I attempted to identify the allelopathic potential of the canopy species both in laboratory and under field conditions. However, attempts could not be made to isolate allelopathy in relation to microbial activities and other physical factors. Initial analysis of allelochemicals revealed the presence of terpenes, terpenoids and phenolic acids in the selected species (Table 22.1). Further work on the identification of allelochemicals is in progress.

22.7 CONCLUSIONS AND PROSPECTS

The major constraints developed due to allelopathy in the field are gradual elimination of associated species, decline in seed germination, growth and production potential, soil sickness and finally the creation of bare areas (Rice, 1984). The regeneration of the area is important to ecological consideration. Many monoculture systems often lead to a soil sickness problem which is presumed to be due to the unbalance of soil microbes, accumulation of plant toxins in soil, mineral deficiency or abnormal soil pH and salt concentration, which leads to the failure of understorey production. These constraints will lead to the failure of integrated perennial–annual cultivation systems as agroforestry, silvipasture, etc., and even the integrated afforestation programmes. Thus, analysing allelopathic potential of the canopy species before including them in such integrated programmes, is a prerequisite.

To eliminate phytotoxic effects, experiments on screening of a large number of plant species need to be made to identify the agronomic crops or wild species that could overcome much of the intolerance of allelopathic potential of overstorey species and also to identify the tree species with least allelopathic activity (Melkania, 1987a). In agroecosystems, crop rotation is the most suitable control measure to eliminate the cause of the problem – allelopathy. In forest ecosystems and silvipastures, where the understorey cannot be feasibly replaced annually, use of absorbents (Putnam and Duke, 1978) and the introduction of companion crops will be the most practical approach for ecological conservation of the area and also for sustained yield in the long run.

ACKNOWLEDGEMENTS

I am highly indebted to Professor, Dr J. S. Singh, Nainital (now at BHU, Varanasi), for his guidance and suggestions. The studies cited herein were conducted at Kumaun University, Nainital (Almora Campus), Vivekananda Laboratory for Hill Agriculture, Almora; and G. B. Pant University of Agriculture and Technology, Pantnagar. My wife, Dr Uma Melkania, offered her constructive criticism and unrestricted help during studies. Mr A. Y. Joseph, Bhopal, legibly typed the manuscript. My thanks are due to them.

REFERENCES

Bokhari, U. G. (1978) Allelopathy among prairie grass and its possible ecological significance. *Ann. Bot.*, **42**, 127–36.

Chou, C. H. (1983) Allelopathy in agroecosystems in Taiwan, in *Allelochemicals and Pheromones* (eds C. H. Chou and G. R. Waller), Academia Sinica, Monograph Ser. No. 9, Acad Sinica, Taipei, ROC., pp. 27–64.

Chou, C. H. (1986) The role of allelopathy in subtropical agroecosystems in Taiwan, in *The Science of Allelopathy* (eds A. R. Putnam and C. S. Tang), Wiley, New York, pp. 57–73.

Datta, S. C. and Chatterjee, A. K. (1980) Pollution by plants. *Proc. Symp. Environmental Pollution and Toxicology*, Today and Tomorrow's Printers and Publishers, New Delhi, pp. 195–214.

Gigon, A. and Ryser, P. (1986) Positive interactions between plant species. I. Definition and examples from grassland ecosystems. *Veroff. Geobot. Inst. ETH, Stiftung Rubel, Zurich*, **87**, 372–87.

Gleissman, S. R. and Muller, C. H. (1978) The allelopathic mechanisms of dominance in bracken (*Pteridium aquilinum*) in southern California. *J. Chem. Ecol.*, **4**, 337–62.

Guenzi, W. D., McCalla, T. M. and Norstadt (1967) Presence of phytotoxic substances in wheat, oat, corn and sorghum residues. *Agron. J.*, **59**, 163–5.

Koeppe, D. E., Rohrbaugh, L. M., Rice, E. L. and Wender, S. H. (1970) Tissue age and caffeoylquinic acid concentration in sunflower. *Phytochemistry*, **9**, 297–301.

Koeppe, D. E., Southwick, L. M. and Bittell, J. E. (1976). The relationship of tissue chlorogenic acid concentrations and leaching of phenolics from sunflowers grown under varying phosphate nutrient conditions. *Can. J. Bot.*, **54**, 593–9.

Koroleff, A. (1954) Leaf litter as a killer. *J. For.*, **52**, 178–82.

Lodhi, M. A. K. and Rice, E. L. (1971) Allelopathic effects of *Celtis laevigata*. *Bull. Torrey Bot. Club*, **98**, 83–9.

Melkania, N. P. (1983) Influence of certain selected tree species on ground flora. PhD Thesis, Kumaun Univ. Nainital.

Melkania, N. P. (1984) Influence of leaf leachates of certain woody species on agricultural crops. *Indian J. Ecol.*, **11**, 82–6.

Melkania, N. P. (1987a) Allelopathy and its significance on production of agroforestry plant associations. *Proc. Workshop on Agroforestry for Rural Needs*, Vol. 1, ISTS, Solan, India, pp. 221–4.

Melkania, N. P. (1987b) Analysis of the allelopathic potential of selected non-forest tree species on crops in Himalaya, India. *Proc XIV Int. Bot. Congr., Berlin, Germany*.

Melkania, N. P. and Singh, J. S. (1987) Allelopathy in Himalayan forest species. *Proc. IX Int. Symp. on Tropical Ecology and Int. Conf. on Rehabilitation of Disturbed Ecosystems: A Global Issue*, Banaras Hindu Univ., Varanasi, India.

Melkania, N. P., Singh, J. S. and Bisht, K. K. S. (1982) Allelopathic potential of *Artemisia vulgaris* L. and *Pinus roxburghii* Sargent: a bioassay study. *Proc. Indian Nat. Sci. Acad.*, **B48**, 685–8.

Muller, C. H. (1965) Inhibitory terpenes volatilised from *Salvia* shrubs. *Bull. Torrey Bot. Club*, **92**, 38–45.

Muller, C. H. (1966) The role of chemical inhibition (allelopathy) in vegetational composition. *Bull. Torrey Bot. Club*, **93**, 332–51.

Muller, C. H. (1974) Allelopathy in the environmental complex, in *Handbook of Vegetation Science, Part VI, Vegetation and Environment* (eds B. R. Strain, and W. D. Billings), W. Junk, The Hague, pp. 37–85.

Muller, C. H., Muller, W. H. and Haines, B. L. (1964) Volatile growth inhibitors produced by shrubs. *Science*, **143**, 471–3.

Muller, W. H. and Muller, C. H. (1964) Volatile growth inhibitors produced by *Salvia* species. *Bull. Torrey Bot. Club*, **91**, 327–30.

Putnam, A. R. and Duke, W. B. (1978) Allelopathy in agroecosystems. *Ann. Rev. Phytopathol.*, **16**, 431–51.

Qasem, J. R. and Abu-Irmaileh, B. E. (1985) Allelopathic effect of *Salvia syriaca* L. (Syrian sage) in wheat. *Weed Res.*, **25**, 47–52.

Rice, E. L. (1984) *Allelopathy*, Academic Press, New York.

Saxena, S. and Singh, J. S. (1978) Influence of leaf leachates from *Eucalyptus globulus* Labill. and *Aesculus indica* Colebr. on the growth of *Vigna radiata* (L.) Wilczek and *Lolium perenne* L. *Indian J. Ecol.*, **5**, 148–58.

Singh, R. and Bawa, R. (1982) Effect of leaf leachates from *Eucalyptus globulus* Labill. and *Aesculus indica* Colebr. on seed germination of *Glaucium flavum* Crantz. *Indian J. Ecol.*, **9**, 21–8.

Stickney, J. S. and Hoy, P. R. (1881) Toxic action of black walnut. *Trans. Wis. State Hort. Sci.*, **11**, 166–7.

Telfer, E. S. (1972) Understorey biomass in five forest types in southwestern Nova Scotia. *Can. J. Bot.*, **50**, 1263–7.

Tukey, H. B. Jr (1970) The leaching of substances from plants. *Ann. Rev. Pl. Physiol.*, **21**, 305–24.

Tukey, H. B. Jr and Mecklenburg, R. A. (1964) Leaching of metabolites from foliage and subsequent reabsorption and redistribution of the leachates in plants. *Amer. J. Bot.*, **51**, 737–43.

Wilson, R. E. and Rice, E. L. (1968) Allelopathy as expressed by *Helianthus annuus* and its role in old field succession. *Bull. Torrey Bot. Club*, **95**, 432–48.

CHAPTER 23

Neem allelochemicals and insect control

O. Koul

23.1 INTRODUCTION

Most living plants have some degree of resistance to natural enemies due to the presence of secondary plant compounds (allelochemicals). It is well known that plants are used as food sources by many insects and plants, therefore, have evolved many barriers against herbivores (Janzen, 1981) including competitive production of biomass, morphological features (e.g. spines) and secondary plant compounds (the allelochemicals) which act as protective agents against herbivorous insects. While monophagous insects are usually limited to plants whose constitution is apparently not acceptable to other phytophagous species because they contain unpalatable and/or toxic substances, adapted herbivores, have evolved the ability to cope with these plant toxins and avoid deleterious effects.

The role of plant allelochemicals, in plant–insect interactions, has received considerable attention in recent years. Insects have evolved various mechanisms for manipulating these allelochemicals, yet a few of these phytochemicals have provided leads to new pesticides. Knowledge of the efficacy of these chemicals, particularly as allomones (which give adaptive advantage to the producing organisms) comes from ethnobotanical folklore of many nations, i.e. the traditional use of plant material. Plant products have been, and continue to be, sold as drugs or insecticides in many parts of the world. An example *par excellence* among such plants in the ancient pharmacopea of India is the versatile Indian neem tree, *Azadirachta indica* A. Juss (Meliaceae), which has received substantial attention during the last decade particularly in terms of its potential for insect pest control. Almost every part of this tree has been shown to possess some property of use to man (Dymock *et al.*, 1885; Mangunath, 1948; Koul *et al.*, 1989).

The neem tree is native to the arid regions of the Indian subcontinent and has been subsequently introduced into the arid zones of Africa during the last century. It is presently grown in many Asian countries and the

Allelopathy: Basic and applied aspects Edited by S. J. H. Rizvi and V. Rizvi
Published in 1992 by Chapman & Hall, London ISBN 0 412 39400 6

tropical areas of new world. This avenue tree has been extensively grown and cultivated in the Old World tropics and in recent years has become the choice species for cultivation and propagation in the New World tropics (Pliske, 1984). Several comprehensive reviews of the distribution, propagation, pharmacology and toxicology, chemistry and efficacy of crude extracts of neem products have been published recently (Jacobson, 1988; Siddiqi *et al.*, 1988; Koul *et al.*, 1989). The subject of this chapter, therefore, is limited to a compilation of data on the effects of various isolated allelochemicals from *A. indica* on insect behaviour, growth, development and physiology.

23.2 NEEM AS A PESTICIDE

It has been an age-old practice in rural India to mix dried neem leaves with stored grain or to place them among warm clothes to repel insects. However, scientific reports in this regard date back to the 1930s (Pruthi, 1937), and since then the information on the performance of various neem products against different pests has continued to flood the scientific literature. Most of the information refers to the insecticidal, antifeedant, growth inhibitory, oviposition deterring, antihormonal and antifertility activities against a broad spectrum of insects. Antifungal, antiviral and nematicidal properties of this versatile plant have also been recently reviewed (Parmar, 1987). These biological activities have been demonstrated for the oil expelled from seeds and leaves, leaf and seed extracts, neem cake, fruit extracts and various isolated compounds. These materials are efficacious against a wide range of insects that are pests on vegetables, tree fruits, ornamental crops and stored grains, as well as household pests of medical and hygienic importance (Jacobson, 1988).

Comparatively little information on the fungicidal activity of neem products is available (Khan *et al.*, 1974; Pant *et al.*, 1986; Gunasekaran *et al.*, 1986; Jeyarajan *et al.*, 1987). Recently it was reported that 10% neem oil diluted from an EC formulation completely inhibited *Aspergillus niger*, *Fusarium moniliforme*, *Macrophormina phaseolina* and *Drechlera rostrata in vitro*. Some studies at least in part attribute control of plant pathogens to the sulphur compunds present in the oil (Singh *et al.*, 1980, 1984; Radwanski and Wickens, 1981).

Use of neem oil for the control of insect-transmitted viral diseases in plants is another noteworthy property of neem (Simons, 1981). The application of neem oil and neem cake, for instance, has been shown to reduce rice seedling infection by rice tungro bacilliform and spherical virus particles (Saxena *et al.*, 1985).

Aqueous extracts of leaf, flower, fruit, bark, root and gum have also been shown to be toxic to plant nematodes (Siddiqi and Alam, 1985a, b). Similarly neem oil and neem cake have been found to be highly useful in practical nematode control (Mishra and Gaur, 1984; Vijayalakshmi *et al.*, 1984).

In summary, results of studies using neem preparations as pesticides show that alcholic, aqueous or enriched neem fractions have excellent potential for pest control, particularly in developing countries where neem is readily available. This would help to remove the use of synthetic pesticides, the indiscriminate use of which has resulted in widespread environmental hazards, by providing another alternative control material. It is important to note that estimated costs of neem application to repel insects could be as low as one-tenth the cost of, for instance, malathion (Redknap, 1981). Accordingly, the use of neem would be both environmentally and economically desirable.

Efforts to commercialize neem preparations are currently underway. Vikwood Ltd (Wisconsin, USA) has gained Environmental Protection Agency registration for a pesticide formulation of neem seed extract Margosan-O-Concentrate (W. R. Grace and Co., Washington, DC, currently owns the rights to the product), which has been recommended for use against ornamental plants. Neemark, a product of West Coast Herbochem Ltd, India, is a well-known insect antifeedant and repellent bioderived neem product. Safer Ltd, a Canadian company, is also developing neem-based insecticidal products.

The main incentive for using these products is mostly based on environmental safety. Neem preparations are highly biodegradable, and most of the activity is lost within two weeks. It has been suggested that if neem extracts are to be used in field applications, special formulations for protection from sunlight will be necessary for the extension of the potency (Stokes and Redfern, 1982). This property would likely make neem-based pesticides much safer than the big three synthetic groups of chlorinated hydrocarbons, organophosphates and carbamates.

23.3 CHEMICAL NATURE

The Meliaceae, to which the neem tree belongs, is a large, chemically diverse family very rich in limonoid-type terpenoid compounds. *A. indica*, one of the most important species of this family, contains many such compounds. Pioneering work on the isolation and identification of neem constituents was initiated in India in 1942 and has continued in various parts of the world. Renewable plant parts have received the most attention (leaves and seeds), although the bark, heartwood and fruits have also been examined for their chemistry. Today we have many fascinating and structurally complex substances on record, representing a number of chemical classes.

The first two crystalline compounds isolated from neem oil were nimbin and nimbinin (Fig. 23.1) along with an amorphous bitter substance called

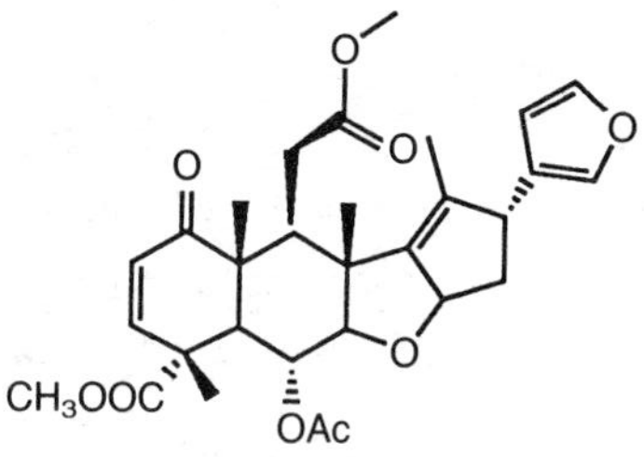

O

CH_3OOC

OAc

O

O

Nimbin

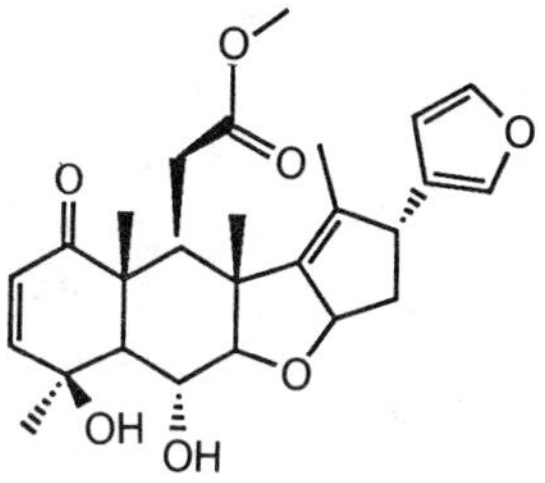

O

O

O

OH

OH

O

Nimbandiol

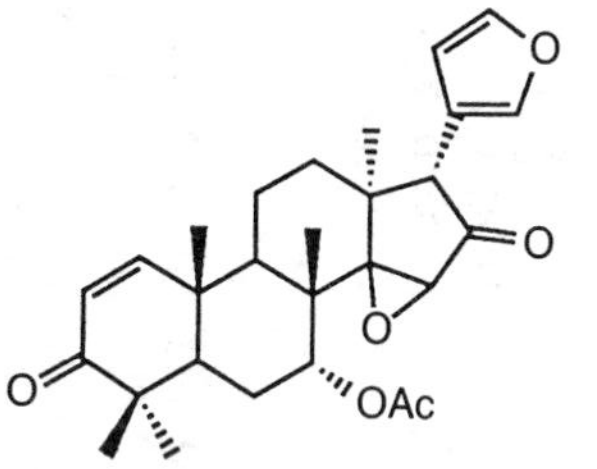

O

O

O

OAc

O

Nimbinin

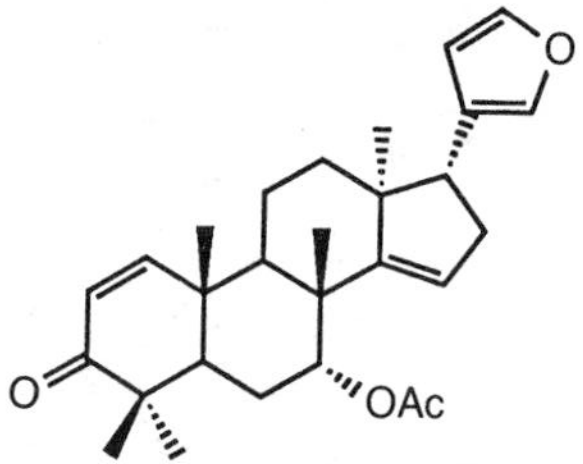

O

OAc

O

Azadirone

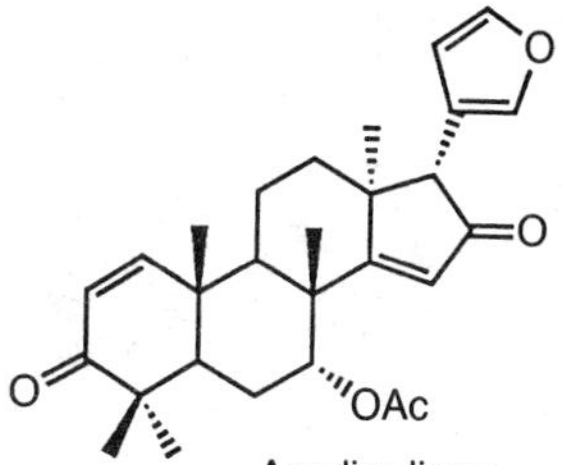

O

OAc

O

O

Azadiradione

O

OAc

O

O

O

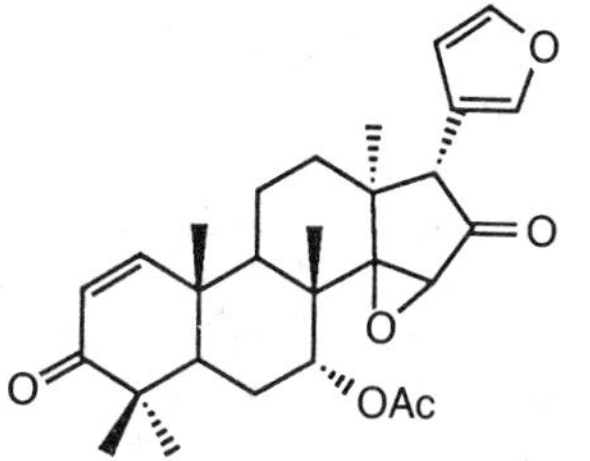

14-Epoxyazadiradione

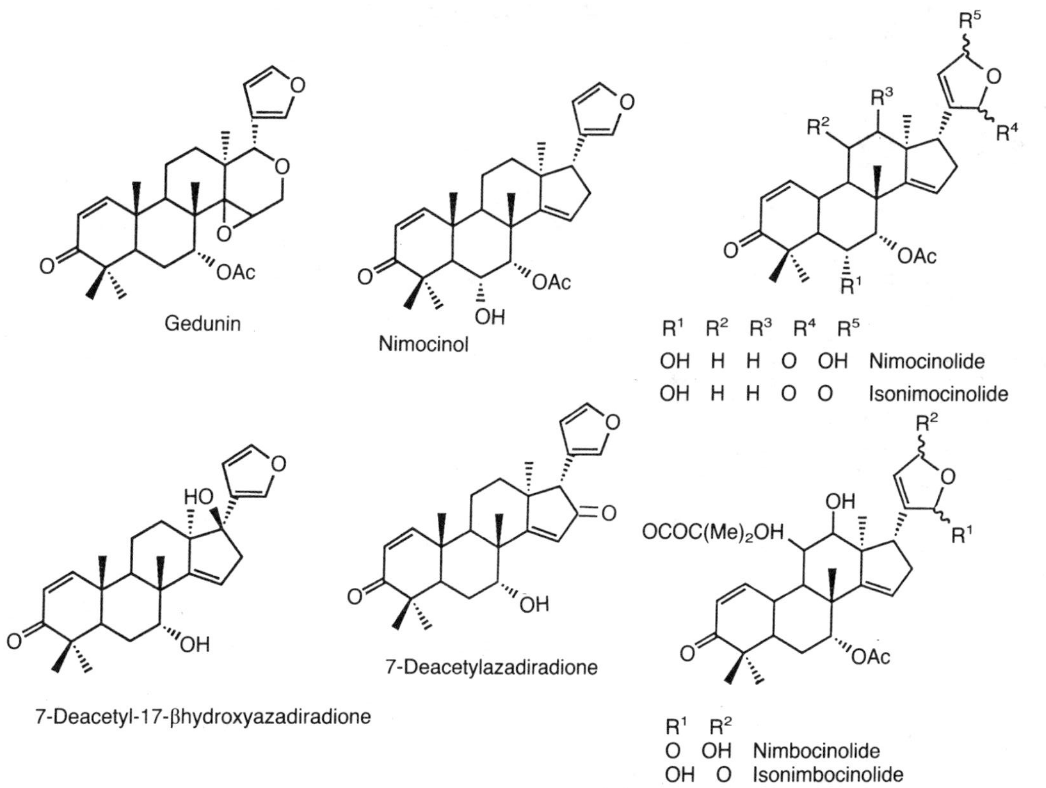

Gedunin

Nimocinol

R^1	R^2	R^3	R^4	R^5	
OH	H	H	O	OH	Nimocinolide
OH	H	H	O	O	Isonimocinolide

7-Deacetyl-17-βhydroxyazadiradione

7-Deacetylazadiradione

OCOC(Me)$_2$OH

R^1	R^2	
O	OH	Nimbocinolide
OH	O	Isonimbocinolide

Fig. 23.1.

nimbidin (Narayanan *et al.*, 1962, 1967). Since then nearly a hundred more constituents have been isolated from different parts of the tree and their structures elucidated. These include protolimonoids, limonoids or tetranortriterpenoids, pentanortriterpenoids, hexanortriterpenoids and nortriterpenoidal compounds.

Protolimonoids include C-30 tetracyclic triterpenes (the euphol/tirucallol derivatives) and the limonoids, which are considered to be Δ^7-euphane or Δ^7-tirucallane derivatives due to the oxygen function at C-7, and are degraded C-26 triterpenoids. However, to describe these compounds individually would be beyond the scope of this chapter, and therefore, I will deal only with those limonoids which are biologically active against insects. Detailed descriptions of limonoid chemistry is available in several comprehensive reviews (Kraus, 1983; Siddiqi *et al.*, 1988; Jacobson, 1988).

Compounds other than nortriterpenoids isolated from different parts of the neem tree include hydrocarbons, fatty acids, diterpenoids, sterols, phenols, flavonoids and glycosides. Sugicol and nimbiol are tricyclic diterpenes isolated from trunk bark (Sengupta *et al.*, 1960). Nimboflavone isolated from leaves is the first example of an isoprenyl flavenone isolated from this tree (Garg and Bhakuni, 1984). Other interesting compounds are nimbochalcin (a dihydrochalcone), nimbocelin (a substituted aromatic ester), quercetin, various alkanes, tri- and tetrasulphides from neem leaf oil, and a series of volatile sulphides from crushed neem seeds (Balandrin *et al.*, 1988). Recently an antineoplastic drug has been isolated from neem bark (Shimizu *et al.*, 1985) and scopoletin has also been isolated from neem leaves, the first example of a coumarin from this species (Siddiqi *et al.*, 1988).

23.4 DEFENCE CHEMICALS AND INSECT INTERACTIONS

Neem derivatives have been found to be effective against more than 120 species of insect pests (Jacobson, 1986). Although most of these studies are based on crude neem seed extracts or other chemical fractions, there are reports of the bioactivity of individual isolates as well. Limonoid compounds contained primarily in the seeds and fruits are apparently the first line in the defensive chemistry of the neem tree. To my knowledge this tree is the only example of a plant which contains such a wide range of triterpenoids, their derivatives, and other highly oxidized natural products. These active ingredients represent a complex array of compounds which cause diverse behavioural and physiological effects on insects such as repellency, feeding and oviposition deterrency, reproduction and growth inhibition, and other physiological disorders. However, most of these compounds have only been shown to deter feeding.

Among the various active compounds which will be described in this section, the most potent to date remains azadirachtin – the ring C-seco tetranortriterpenoid for which substantial bioefficacy data are available and on the basis of which some commercial formulations are being developed.

23.4.1 Azadirachtins

The first indication of this complex triterpenoid from the neem tree was recorded when Butterworth and Morgan (1968) reported the isolation of a substance that inhibited feeding in desert locusts; they named it azadirachtin (Fig. 23.2). Subsequently Zanno *et al.* (1975) gave the first complete structure for this compound. Azadirachtin is comparatively difficult to isolate and several techniques have been developed for its preparation from neem seed kernels using adsorption chromatography and reverse phase high-pressure liquid chromatography (HPLC) (Uebel *et al.*, 1979; Warthen *et al.*, 1984; Yamasaki *et al.*, 1986; Schroeder and Nakanishi, 1987). In recent years extensive chemical investigations resulted in a reassignment of the structure of this compound (Turner *et al.*, 1987; Bilton *et al.*, 1987; Kraus *et al.*, 1987a) which now unequivocally gives the basis for the structural elucidation of other isomeric azadirachtins called azadirachtins A to G (Rembold, 1988) (Fig. 23.2). Of these, azadirachtin A remains the prominent growth-inhibiting neem compound and its structure is that isolated as azadirachtin (Fig. 23.2) (originally isolated by Zanno *et al.*, 1975).

Azadirachtin A (hereafter referred to simply as azadirachtin) is now well known to have growth-inhibitory effects on species of most if not all economically important insect orders (Butterworth and Morgan, 1971; Ruscoe, 1972; Steets, 1975, 1976; Steets and Schmutterer, 1975; Qadri and Narsaiah, 1978; Warthen, 1979; Meisner *et al.*, 1981; Koul, 1984, 1985; Ladd *et al.*, 1984; Rembold, 1988). What is obvious from many studies is that azadirachtin is biologically active against many pests of vegetables and fruit trees, but these studies are restricted to laboratory bioassays, largely because it is too expensive to extract and purify sufficient amounts of this compound for field evaluations and its synthesis is not economically feasible at this time. However, the best results in controlling insect pests are obtained with seed extracts containing the highest amounts of azadirachtin (Ermel *et al.*, 1987). In various experiments these azadirachtin-rich extracts have been shown to control pests of cabbage, cauliflower, Chinese kale and other crucifers, cucurbits, beans, cowpea and other legumes, okra, tomato, egg plant, onions, solanum, potato and sweet potato, citrus and some forest trees (Schmutterer and Hellpap, 1988). Pests which are affected by these neem materials include diamond back moth, *Plutella xylostella* (Adhikary, 1985; Kirsch, 1987); Colorado potato beetle, *Leptinotarsa decemlineata* (Lange and Feuerhake, 1984); cabbage webworm, *Crocidolomia binotalis* (Fagoonee and Lange,

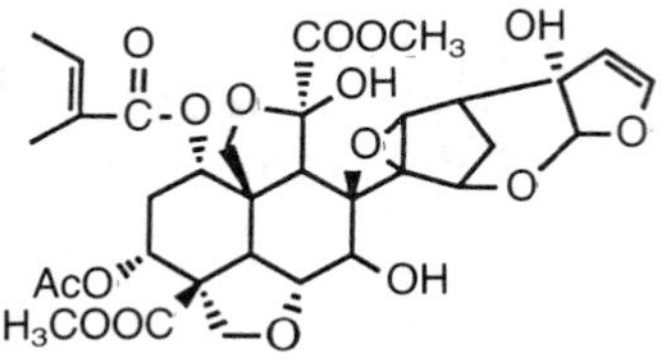

Azadirachtin

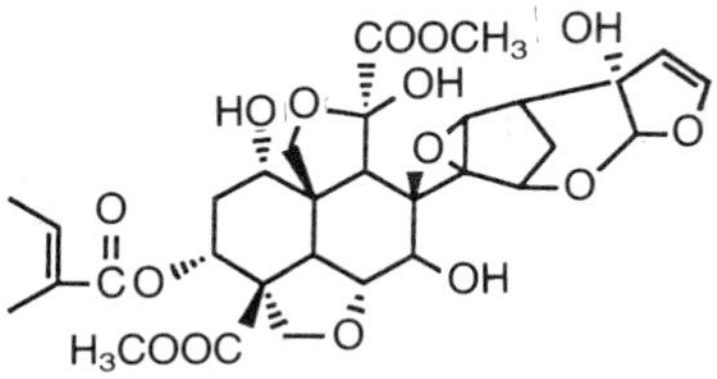

Azadirachtin B

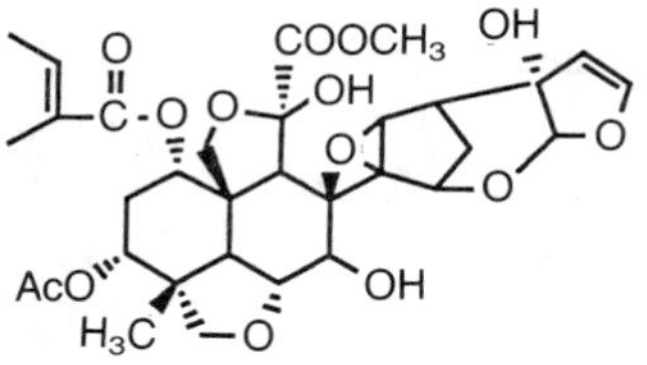

Azadirachtin D

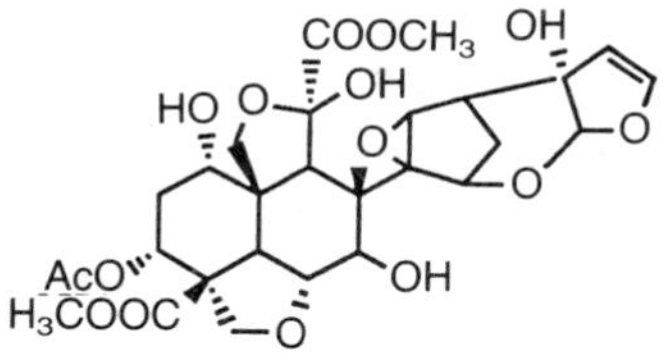

Azadirachtin E

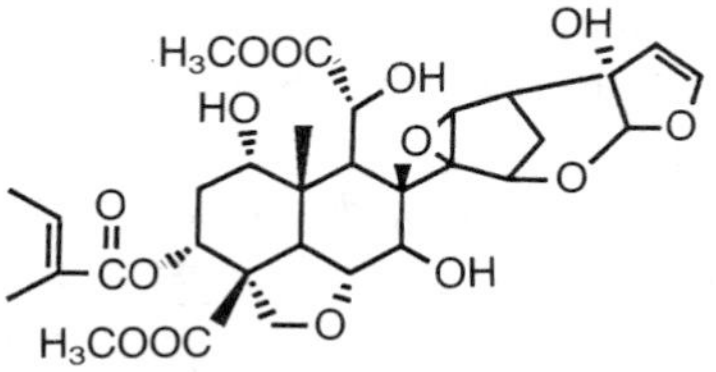

Azadirachtin F

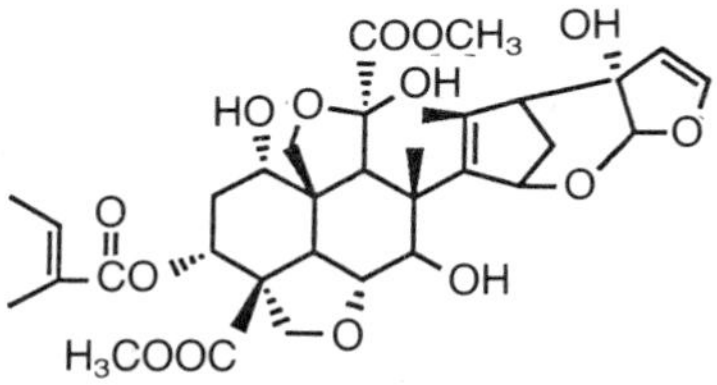

Azadirachtin G

Fig. 23.2.

1981); cabbage looper, *Tricoplusia ni* (Reed and Reed, 1986); cabbage-worm, *Pieris canidia* (Kirsch, 1987); flea beetle, *Phyllotreta downsei* (AbdulKareem, 1981); woolybear, *Amsacta moorei* (Saxena, 1982), tobacco whitefly, *Bemisia tabaci* (Dreyer, 1984; Coudriet *et al.*, 1985), African melon ladybird, *Henosepilachna elaterii* (Redknap, 1981) and the brown plant hopper, *Nilaparvata lugens* (Saxena, 1989). All of these studies indicate that these neem preparations are efficacious against a variety of insects belonging to diverse insect orders and activity is apparently due to the most active component azadirachtin.

However, pure azadirachtin at 0.01 and 0.1% concentrations prevented 98% of foliar damage by the striped cucumber beetle, *Acalymma vittatum* and the spotted cucumber beetle, *Diabrotica undecimpunctata* (Reed *et al.*, 1982). *Epilachna varivestis*, the common Mexican bean beetle, has been found to be quite susceptible to azadirachtin (Steets and Schmutterer, 1975; Schmutterer and Rembold, 1980; Ascher and Gsell, 1981). Toxicity of azadirachtin against the larval, prepupal and early pupal stages of the Japanese beetle, *Popillia japonica*, has been extensively studied in the laboratory, with appreciable efficacy by topical application at 1–4 μg per insect (Ladd *et al.*, 1984). Recently azadirachtin has also been tested against the vegetable leaf miner, *Liriomyza sativae* and caused high mortality in larvae and pupae (Webb *et al.*, 1984). This compound has recently been shown to inhibit the growth, feeding and survival of the variegated cutworm, *Peridroma saucia*, with an EC_{50} and LC_{50} of 0.36 and 2.7 ppm in diet respectively (Champagne *et al.*, 1989). In contrast, no antifeedant activity was observed by these workers against nymphs of the migratory grasshopper *Melanoplus sanguinipes*. However, moult inhibition in this species was reported with an oral ED_{50} of 11.3 $\mu g\,g^{-1}$ insect fresh weight (Champagne *et al.*, 1989). Results from trials against insect pests of trees indicate that neem seed extracts containing azadirachtin as the active ingredient successfully controlled the infestations of the gypsy moth, *Lymantria dispar*, at 0.02–0.5% concentrations (Meisner and Ascher, 1986); sawfly, *Pristophora abietina*, where 50% larval population was reduced six days after treatment (Schmutterer, 1985); birch leaf miner, *Fenusa pusilla*, at 1.0% level (Larew *et al.*, 1987); and orange striped oakworm, *Anisota senatoria*, at 0.2% concentration (Schultz and Coffelt, 1987) when applied as a spray.

Compared with agriculture pests much less is known about the effects of azadirachtin on insects of medical and veterinary importance, on household pests or on beneficial insects. Azadirachtin at low concentrations of 0.1 ppm has been shown to repel adult house flies for three days (Warthen *et al.*, 1978). Gaaboub and Hayes (1984a, b) exposed third instar larvae of face fly, *Musca autumnalis*, to filter paper discs treated with 5 ml solution of azadirachtin in acetone at concentrations of 10 pg to 100 $\mu g\,ml^{-1}$ which inhibited the development of adults, induced deleterious effects in pupae and reduced fecundity and egg hatch. Similarly, biological effects of azadirachtin on blowfly, *Calliphora vicina*, larvae have revealed several dose- and stage-dependent effects such as delayed pupation, reduced pupal weight and inhibition of adult emergence (Bidmon *et al.*, 1987). The ovicidal, insect growth regulatory and repellent effects of azadirachtin against blood-sucking flies like stable flies and mosquitoes have also been observed (Gill and Lewis, 1971; Zebitz, 1986; Ludlum and Sieber, 1988). The azadirachtin based neem formulation, Margosan-O (3000 ppm azadirachtin) has also been tested against mosquito larvae. In these studies an LC_{50} of 8.8 and 17.0 ppm AI against fifth instar larvae of house mosquito, *Culex pipiens fatigans*, and yellow fever mosquito, *Aedes aegypti*, respectively, has been recorded (Koul, 1988). Garcia and Rembold (1984) investigated the influence of azadirachtin administered in a blood meal on the feediing and ecdysis of fourth instar *Rhodnius prolixus* nymphs. The antifeedant ED_{50} was 25 μg ml^{-1} which was much higher than the ecdysis inhibiting ED_{50} of 40 μg ml^{-1}. Similarly, the LD_{50} of azadirachtin injected into the blood stream of last instar male and female cockroach nymphs, *Periplaneta americana*, was determined by Qadri and Narsaiah (1978) to be 1.5 μg g^{-1}. It also induced reduction of haemocytes by 20–25%.

Although pure azadirachtin has not been evaluated against ants and termites, azadirachtin-rich fractions have been shown to affect the fecundity of ants (Schmidt and Pesel, 1987) and induce toxicity in termites. However, Butterworth and Morgan (1971) showed that azadirachtin had no effect on the feeding or the survival of the Mediterranean wood termite, *Reticulitermes santonensis*. Interesting results have been obtained from 10 mg kg^{-1} azadirachtin diet treatments against newly hatched crickets, *Acheta domesticus* (Warthen and Uebel, 1981). The average weight of the crickets was inversely proportional to the concentration of azadirachtin in the diet. When first instar nymphs were fed for six weeks on this diet and then transferred to untreated diet they recovered quickly from the strong antifeedant effect.

The effect of azadirachtin on stored grain pests has not been well studied, although neem seed extracts and deoiled cakes have been reported to protect stored grains (Ambika and Mohandas, 1982; Saxena *et al.*, 1988). Mukherjee and Ramachandran (1989) have shown that azadirachtin incorporated in diets reduced growth of red flour beetle, *Tribolium castaneum*, at concentrations >1 ppm. However, no antifeedant effect was observed in these insects. Topical application of 1, 2 and 5 μg on eggs and larvae did not induce any adverse effects but emergence of normal adults was reduced when applied to pupae less than six hours old. Azadirachtin incorporated in wheat flour has also been shown to be hormetic to red flour beetles at doses <1 ppm and adversely affected development and growth at 3 ppm (Ramachandran *et al.*, 1988). In contrast, in another recent study azadirachtin has been reported as an active antifeedant against adults of grain weevil, *Sitophilus granarius*, and confused flour beetle, *T. confusum*, and also the larvae of khapra beetle, *Trogoderma granarium*, and confused flour beetle (Nawrot *et al.*, 1987).

Among the beneficial insects azadirachtin has been tested against silkworm, *Bombyx mori* (Koul *et al.*, 1987) and causes growth inhibition in these insects when administered by injection at a dose of 2 μg g^{-1} body weight. Studies with honey bees, *Apis mellifera*, have been conducted using neem extracts and various neem fractions, but have failed to show any antifeedant effects in these insects (Rembold *et al.*, 1982). However, larval abnormalities could be seen which caused larval mortality. Survival of third instar larvae was greatly reduced by doses of 0.25 and 0.5 μg per larva azadirachtin.

It is thus evident that azadirachtin possesses antifeedant, growth-inhibitory and endocrine-disrupting effects in a variety of insect species. It seems that azadirachtin deters feeding at much lower concentrations in hemimetabolus insects than in holometabolus insects but this generalization is weakened by extensive variation which exists in both groups.

The question still remaining to be answered is: what is the growth-disrupting mode of action of this compound and how does it influence endocrine events? Several recent studies have attempted to track changes in the endocrine system induced by azadirachtin, particularly in terms of delay or inhibition of moulting, disturbance of the moulting process and disruption of reproduction. The salient feature which has emerged from these studies is the decrease in the titres of morphogenetic hormones resulting from azadirachtin treatment (Sieber and Rembold, 1983; Schluter *et al.*, 1985; Mordue *et al.*, 1986; Dorn *et al.*, 1986; Koul *et al.*, 1987; Fritzsche and Cleffmann, 1987). It is not clear whether this effect is a direct or indirect one, however, a direct effect of azadirachtin on ecdysone synthesis and prothoracico tropic hormone (PTTH) has been ruled out in the silkworm, *Bombyx mori* (Koul *et al.*, 1987). Following administration of azadirachtin in *Locusta migratoria* turnover of L-(^{35}S) cysteine-labelled neurosecretory material is poor in the corpus cardiacum (Subramanyam *et al.*, 1989). Electrophoresis and autoradiography of neurosecretory proteins reveal that no change occurs quantitatively in the corpora cardiaca (Subramanyam *et al.*, 1989). It has been concluded that the release of neurohormones from the corpus cardiacum is probably inhibited which in turn is responsible for the disturbance of endocrine events by azadirachtin (Rembold *et al.*, 1989). On the whole, further studies are required to pinpoint the actual mode of action of this phytochemical.

Recently several techniques have been described to isolate various so-called isomers of azadirachtin (Rembold, 1988). Azadirachtin B (Fig. 23.2) separated by preparative HPLC on a SiO_2–RP_8 column with methanol/water (43:57) as isocratic solvent (Rembold *et al.*, 1987) has been shown to inhibit larval growth of the Mexican bean beetle, *Epilachna varivestis*. Its structure has been simultaneously confirmed in other laboratories (Kubo *et al.*, 1984; Klenk *et al.*, 1986). Similarly, azadirachtin C has been characterized as partially having a *trans*-decalin ring substituted as in azadirachtin. Azadirachtin D (Fig. 23.2) differs from azadirachtin by reduction of the ester group in position 4 to a methyl group. Azadirachtin E (Fig. 23.2) is detigloylazadirachtin, while azadirachtin F is lacking either linkage in position 19 and opened by formation of a C-19 methyl group. Azadirachtin G is reported to be an isomer of azadirachtin B with a double bond instead of the epoxide ring and with a hydroxy group in position 17. All these compounds have been tested in the *Epilachna* bioassay and 50% metamorphosis inhibition concentration (MC_{50}) has been recorded (Rembold, 1988).

Azadirachtin	A	B	C	D	E	F	G
MC_{50} (ppm)	1.66	1.30	12.69	1.57	0.57	1.15	7.69

However, these are the only evaluations regarding the bioactivity of these compounds, and further investigation is required before any conclusions regarding structure–activity relations can be drawn.

23.4.2 Azadirones

Azadirones are simple limonoids closely related to the tetracyclic triterpenoids. The azadirone, azadiradione, 14-epoxyazadiradione and gedunin (Fig. 23.1) are isolated from the seed oil of neem and possess antifeedant activity against Mexican bean beetles, *Epilachna varivestis* (Lavie *et al.*, 1971; Schwinger *et al.*, 1984). These were the first benzoyl allelochemic derivatives of the neem tree tested against this insect at concentrations of 0.001–1.0%. Azadiradione is apparently the most active, followed by gedunin, 14-epoxyazadiradione and finally the azadirone which remains least active among these compounds (Kraus *et al.*, 1987b). Recently another azadirone, 7-deacetyl-17β-hydroxyazadiradione (Fig. 23.1) has been isolated (Lee *et al.*, 1988) from the seeds of *A. indica* having C-7 and C-17 hydroxyl groups. This compound has been shown to inhibit the growth of first instar larvae of tobacco budworm, *Heliothis virescens* (EC_{50} = 240 ppm; 105–550 fiducial limit). In these studies other compounds like azadiradione and 7-deacetylazadiradione (Fig. 23.1; also called nimbocinol, Siddiqi *et al.*, 1986d) were less active exhibiting EC_{50} of 560 and 1600 ppm, respectively (Lee *et al.*, 1988). However, the establishment of any structure–activity relationships from these studies is impossible until information regarding various modified structures and their bioactivity data is available. Recently nimocinol (Fig. 23.1) which is an oxygenated derivative of azadirone (Siddiqi *et al.*, 1984), has been isolated from different parts of the neem tree. Preliminary tests of this compound and its mother fraction have shown that they possess insect growth regulatory properties (Naqvi, 1987) when tested on house flies.

23.4.3 Nimo- and nimbocinolides

Nimocinolide and isonimocinolide (Fig. 23.1) are characteristic tetranortriterpenoid-γ-hydroxybutenolides obtained from fresh neem leaves (Siddiqi *et al.*, 1986a) and are the first γ-hydroxybutenolides from neem with an apoeuphane/apotirucallane intact carbocyclic skeleton. They possess insect growth-inhibitory properties and affect the fecundity of house flies at a dosage of 100–500 ppm. They also caused mutagenesis in the yellow fever mosquito (*Aedes aegypti*), producing larval-pupal intermediates (LC_{50} nimocinolide = 0.625 ppm; isonimocinolide = 0.74 ppm).

Nimbocinolide and isonimbocinolide (Fig. 23.1) are obtained from the same source and bear a 2-methyl-2-hydroxypropionate function at C-11 (Siddiqi *et al.*, 1986b); only the mother fraction of these have been shown to disrupt metamorphosis of *Aedes aegypti* and to produce larval–pupal intermediates (Naqvi, 1987), which apparently shows that these two latter compounds may be responsible for this activity. However, this remains speculative until the pure compounds are evaluated for their bioactivity.

23.4.4 Salannins

Salannin (Fig. 23.3) isolated from neem oil has been shown to possess strong antifeedant activity against the Mexican bean beetle, *Epilachna varivestis*, at 0.05% concentration (Kraus and Cramer, 1981). Salannin is a C-seco tetranortriterpenoid with saturated A-ring and oxygen function at C-1 and C-3 (Schwinger *et al.*, 1984). When tested against striped cucumber beetles, *Acalymma vittatum*, and spotted cucumber beetles, *Diabrotica undecimpunctata*, in laboratory and greenhouse trials on leaf discs, salannin has been shown to deter feeding at concentrations of 0.01 and 0.1% (Reed *et al.*, 1982).

Against Egyptian cotton leaf worm, *Spodoptera littoralis*, and bollworm, *Earias insulana*, feeding deterrence occurred at 0.005–0.01% levels (Meisner *et al.*, 1981). Salannin added to sugar solutions at 0.5, 0.25 and 0.1% completely prevented the settling of house flies on the diet (Warthen *et al.*, 1978). Similarly, salannin has been tested against the yellow fever mosquito, *Aedes aegypti*, by the US Department of Agriculture, Beltsville, but found to be less effective than azadirachtin (Jacobson, 1981). A fresh 15 min air-dried treatment of 1 mg cm^{-2} during 1 min exposure on human subjects resulted in 25 bites (27 bites at 0.5 mg cm^{-2} treatment) in comparison to azadirachtin where only four bites were seen during 1 min exposure at 0.2 ng cm^{-2}.

In a recent study 14 derivatives of salannin have been prepared and bioassayed for antifeedant activity against Colorado potato beetle, *L. decemlineata*, to derive possible structure–activity relationships (Yamasaki and Klocke, 1989). Changes in antifeedant activity have been attributed to chemical modifications at four points of the salannin molecule, i.e. hydrogenation of the furan ring, modification of the tigloyl group, saponification of the methyl ester and replacement of the acetoxy group (Yamasaki and Klocke, 1989). Although hydrogenation of the furan ring and the tigloyl group increased the activity against Colorado beetles, *L. decemlineata*, the hydrogenated compound did not affect the fall army worm, *Spodoptera frugiperda*, to a similar extent. However, the authors believe that the effectiveness of salannin and its derivatives as antifeedants for various insects is difficult to predict and should be determined empirically (Yamasaki and Klocke, 1989).

Some other derivatives/related compounds of salannin such as 3-deacetylsalannin, salannol, 3-desacetylsalannin (Fig. 23.3) and nimbandiol

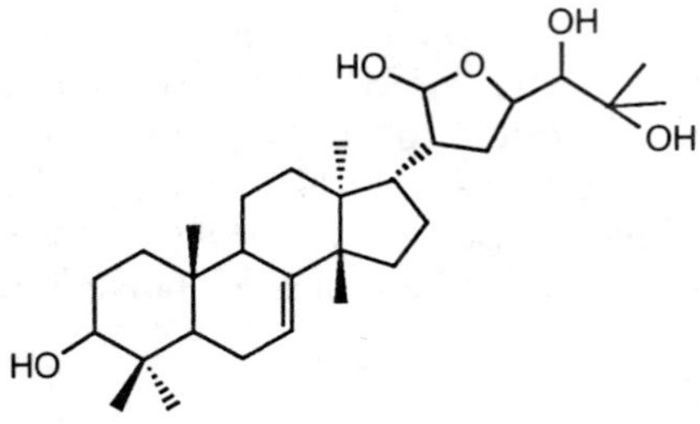

Meliantriol

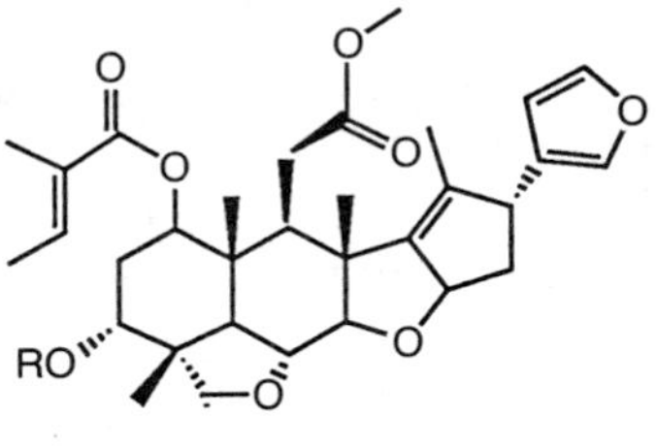

R = Ac Salannin
R = H 3-Deacetylsalannin

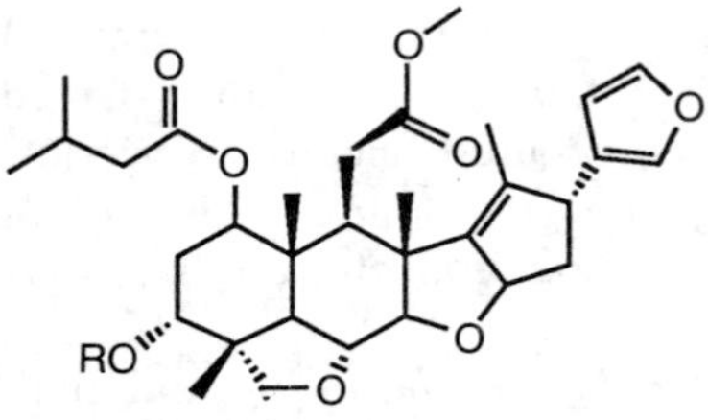

R = H Salannol
R = Ac Salannolacetate

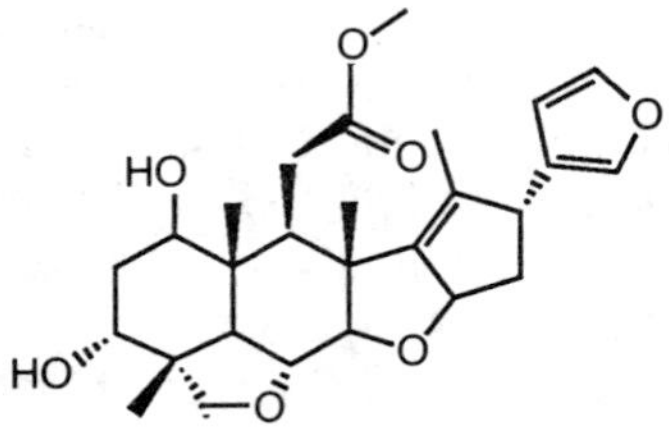

3-Desacetylsalannin

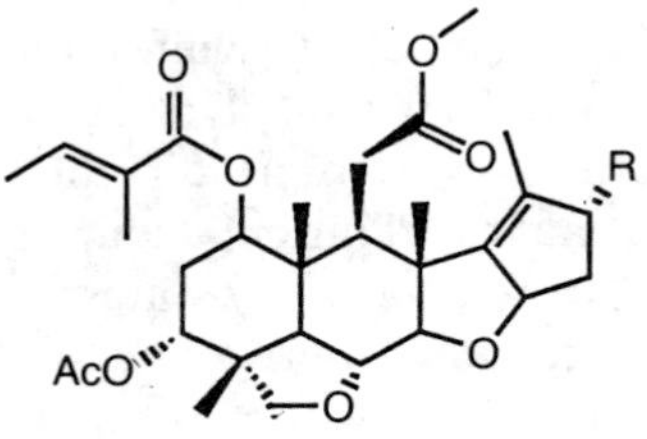

Salannolactame-(21)

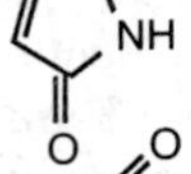

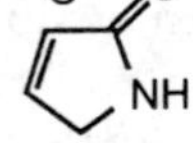

Sallanolactame-(23)

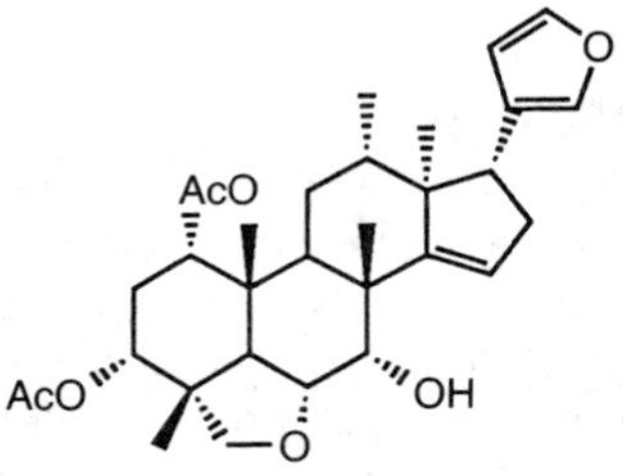

1,3-Diacetylvilasinine

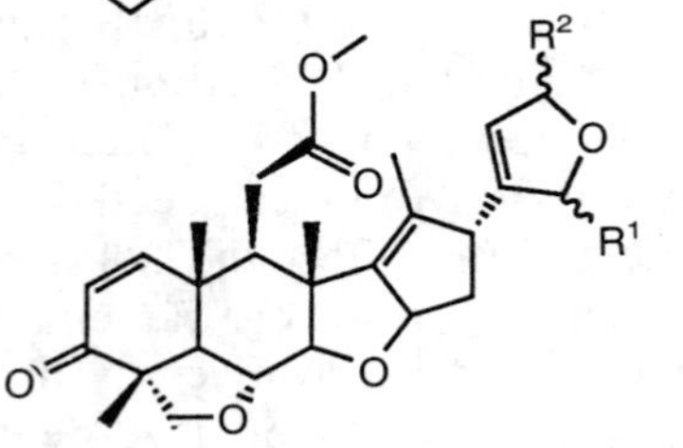

R^1	R^2	
O	OH	Margosinolide
OH	O	Isomargosinolide

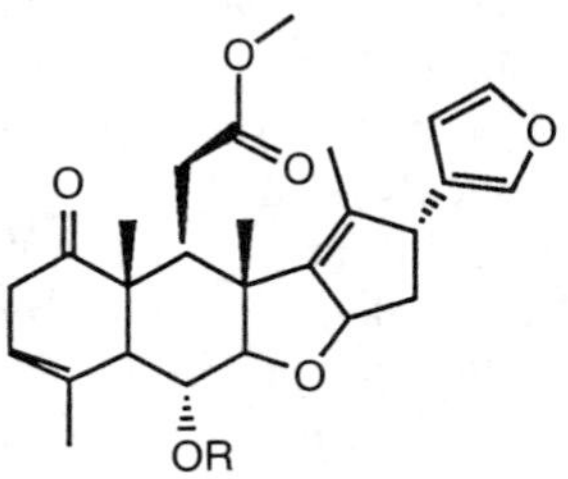

R = Ac Nimbinene
R = H 6-Deacetylnimbinene

Fig. 23.3.

(Fig. 23.1) also deter feeding of the Mexican bean beetles, *Epilachna varivestis* (Schwinger *et al.*, 1984) at 0.01–0.05% levels. Activity has been assigned to these compounds due to additional oxidation at C-28, the oxygen bridges C-28/C-6 and C-7/C-15 and C-12 ester groups. The activities of salannin, 3-deacetylsalannin and salannol have been shown to be similar to azadirachtin at least against the Mexican bean beetles (Kraus *et al.*, 1987b).

23.4.5 Vilasinines

Vilasinines are limonoids with an intact ring system, initially isolated by Pachapurkar *et al.* (1974), stereochemistry was confirmed by Kraus and Cramer (1981) who also isolated vilasinine-1, 3-diacetate (Fig. 23.3) from neem oil having C-1 and C-3 acetoxy functions instead of the OH-group of vilasinine. This 1, 3-diacetylvilasinine was obtained as an active compound by partitioning between petroleum ether and methanol: water and was found to deter feeding of Mexican bean beetles, *E. varivestis*. No quantitative data for its efficacy have been recorded.

23.4.6 Nimbinene and 6-deacetylnimbinene

Nimbinene and 6-deacetylnimbinene (Fig. 23.3) are new pentanortriterpenoids in which the double bond and carbonyl group are not conjugated, and C-4 methyls have been lost along with four carbons of the side chain. They are constituents of leaves and bark of the neem tree (Kraus and Cramer, 1981) and active antifeedants in *Epilachna* bioassays. Nimbinene has been found to be slightly less active than deacetylnimbinene in the concentration range of 0.001–0.1% (Schwinger *et al.*, 1984).

23.4.7 Margosinolides

Margosinolide and isomargosinolide (Fig. 23.3) are ring C-seco tetranortriterpenoid-γ-hydroxybutenolides isolated from fresh neem twigs (Siddiqi *et al.*, 1986c). These are the only constituents so far reported from the twigs with a unique feature of ether linkage between C-6 and C-28 with a 1-en-3-one ring A. These compounds are of biological significance because the growth-inhibitory activity of ring C-seco limonoids is enhanced by the presence of an α-β-unsaturated ketone system in ring A (Kubo *et al.*, 1986). However, no detailed activity data for these compounds are available at present.

23.4.8 Meliantriol

Meliantriol (Fig. 23.3) is a C-30 tetracyclic protolimonoid from neem oil and the fresh fruits of the closely related species *Melia azedarach*. The structure of meliantriol is based on a butyrospermol skeleton and is apparently the only active protolimonoid from the neem tree which has caused 100% feeding deterrence against desert locusts in a contact test (Lavie *et al.*, 1967). To my knowledge no further detailed investigation of this allelochemical has been conducted.

23.4.9 Alkanes

Recently an active fraction called NP-2 was separated from the dried leaves of neem and found to be a mixture of saturated long-chain alkanes of $CH_3-(CH_2)_{32}-CH_{23}$, $CH_3-(CH_2)_{16}-CH_3$, $CH_3-(CH_2)_{24}-CH_3$ and $CH_3-(CH_2)_{17}-CH_3$ configuration (Chavan, 1984). This fraction has been found to produce 100% mortality of mosquito larvae, *Culex pipiens fatigans*. Effects against house flies, bedbugs and cockroaches were observed at 2% formulation concentration which induced 82, 60 and 35% mortality respectively. This activity has been attributed to alkanes present in the fraction.

23.4.10 Sulphur-containing compounds

Ripening neem fruits and pressed neem seed oil give off a strong alliaceous odour due to the sulphur-containing compounds present in the material (Nadkarni and Nadkarni, 1954; Jacobson, 1986). A number of studies are available showing the presence of these compounds in the neem tree (Nadkarni and Nadkarni, 1954; Dey and Mair, 1973; Radwanski, 1977a,b; Sinniah and Baskaran, 1981; Radwanski and Wickens, 1981; Balandrin *et al.*, 1988), and have described the isolation and characterization of these compounds. Although tri- and tetrasulphides of C-3, C-5, C-6 and C-9 units have been only tested against fungi (e.g. *Trichophyton mentagrophytes*) at $125\,\mu g\,ml^{-1}$ level to show antifungal action (Pant *et al.*, 1986), only diallyl disulfide and di-*n*-propyldisulfide have been tested against third instar larvae of yellow fever mosquito, *Aedes aegypti*, in aqueous suspension and third instar larvae of corn earworm, *Heliothis zea*, and tobacco budworm, *H. virescens*, in feeding trials. While diallyl disulphide was active at 6 ppm (LC_{50}) against the mosquitoes, the latter compound was less active (LC_{50} = 66 ppm). LC_{50} values when tested on *H. zea* were 280 and 980 ppm for the two compounds respectively. Against tobacco budworm, *H. virescens*, these values are slightly higher (340 and 1000 ppm respectively; Balandrin *et al.*, 1988). The presence of these volatile compounds and their activity may explain, at least in part, the insect-repellent activity of neem leaves and seeds (Nadkarni and Nadkarni, 1954; Irvine, 1961; Dey and Mair, 1973; Kunkel, 1978).

23.5 CONCLUSIONS AND PROSPECTS

The neem tree, with such a wide range of allelochemicals – particularly the triterpenoids, derived compounds and highly oxidized products – remains a versatile plant with diverse biosynthetic pathways and substantial bioactivities against insects. It is, however, surprising that no biosynthetic studies have been made, particularly in view of the importance of these compounds. It seems an opportune time to study this subject which will apparently enlighten the biochemical mechanisms and allow us to understand the production of these compounds. Perhaps the production of these compounds could be controlled and will benefit from biotechnology research.

On comparing the activities of the neem compounds, azadirachtin remains the only compound with the potential to deter feeding and inhibit growth, reproduction and endocrine events. However, the complexity of active principles, particularly the azadirachtins, precludes the possibility of synthesis on a large scale. Therefore, the use of simple formulations based on preparations from seeds needs to be popularized for use. Some progress has been made in this regard such as the development of Margosan 'O'™ (Larson, 1988), the efforts of Safer Ltd in Canada and the availability of neem-based formulations like nimbosol, neemark, neemol, biosol, etc., in India (Ketkar, 1989; personal communication). Future goals, however, include the need to establish neem plantations, obtain the best neem seed variety with highest azadirachtin content, improve extraction procedures to obtain the highest possible yield at the lowest possible costs, develop stable formulations, and possibly find synergists to improve the efficacy of such formulations.

ACKNOWLEDGEMENTS

I thank Murray B. Isman, Michael Smirle, Jan Nawrot and Tom Lowery for their critical comments on the manuscript. Thanks are also due to Don Champagne for his help in producing the figures.

REFERENCES

AbdulKareem, A. (1981) Neem as an antifeedant for certain phytophagous insects and a bruchid on pulses. *Proc. Ist Int. Neem Conf. RottachEgern*, GTZ Press, Germany, pp. 223–49.

Adhikary, S. (1985) Results of field trials to control the diamond back moth, *Plutella xylostella* L. by application of crude methanolic extract and aqueous suspension of seed kernels and leaves of neem, *Azadirachta indica* A. Juss, in Togo. *Z. Angew. Ent.*, **100**, 27–33.

Ambika, D. and Mohandas, N. (1982) Relative efficacy of some antifeedants and deterrents against insect pests of stored paddy. *Entomon.*, **7**, 261–4.

Ascher, K. R. S. and Gsell, R. (1981) The effect of neem seed kernel extract on *Epilachna varivestis* Mus. larvae. *J. Plant Dis. Protection*, **88**, 764–7.

Balandrin, M. F., Lee, S. M. and Klocke, J. A. (1988) Biologically active volatile organosulfur compounds from seeds of the neem tree, *Azadirachta indica* (Meliaceae). *J. Agric. Food Chem.*, **36**, 1048–54.

Bidmon, H.-J., Kauser, G., Mobus, P. and Koolman, J. (1987) Action of azadirachtin on blowfly larvae and pupae. *Proc. 3rd Int. Neem Conf., Nairobi, Kenya*, GTZ Press, Germany, pp. 253–71.

Bilton, J. N., Broughton, H. B., Jones, P. S. *et al.* (1987) An X-ray crystallographic, mass spectroscopic and NMR study of the limonoid insect antifeedant azadirachtin and related derivatives. *Tetrahedron*, **43**, 2805–15.

Butterworth, J. H. and Morgan, E. D. (1971) Investigation of the locust feeding inhibition of the seeds of the neem tree *Azadirachta indica. J. Insect Physiol.*, **17**, 969–77.

Butterworth, J. H. and Morgan E. D. (1968) Isolation of a substance that suppresses feeding in locusts. *Chem. Commun.*, 23–4.

Chavan, S. R. (1984) Chemistry of alkanes separated from leaves of *Azadirachta indica* and their larvicidal/insecticidal activity against mosquitoes. *Proc. 2nd Int. Neem Conf., Rauischholzhausen*, GTZ Press, Germany, pp. 59–66.

Champagne, D. E., Isman, M. B. and Towers, G. H. N. (1989) Insecticidal activity of phytochemicals and extracts of meliaceae, in *Insecticides of Plant Origin* (eds J. T. Arnason, B. J. R. Philogene, and P. Morand), ACS Symp. Ser. 387, Amer. Chem. Soc., Washington, DC, pp. 95–109.

Coudriet, D. L., Prabhaker, N. and Meyerdirk, D. E. (1985) Sweet potato white fly (Homoptera: Aleyroididae): effects of neem seed extract on oviposition and immature stages. *Environ. Entomol.*, **14**, 776–9.

Dey, K. L. and Mair, W. (1973) *The Indigenous Drugs of India*, 2nd edn, Pama Primlane, Chronica Botanica, New Delhi, India, pp. 186–7.

Dorn, A., Rademacher, J. M. and Sehn, E. (1986) Effects of azadirachtin on the moulting cycle, endocrine system and ovaries in last instar larvae of the milkweed bug *Oncopeltus fasciatus. J. Insect Physiol.*, **32**, 231–8.

Dreyer, M. (1984) Effects of aqueous neem extracts and neem oil on the main pests of *Cucurbita pepo* in Togo. *Proc. 2nd Int. Neem Conf., Rauischholzhausen*, GTZ Press, Germany, pp. 435–44.

Dymock, W. (1885) *The Vegetable Materia Medica of Western India*, Trubner, London.

Ermel, K., Pahlich, E. and Schmutterer, H. (1987) Azadirachtin content of neem kernels from different geographical locations and its dependence on temperature, relative humidity and light. *Proc. 3rd Int. Neem Conf., Nairobi, Kenya*, GTZ Press, Germany, pp. 171–84.

Fagoonee, I. and Lange, G. (1981) Noxious effects of neem extracts on *Crocidolomia binotalis. Phytoparasitica*, **9**, 111–18.

Fritzsche, U. and Cleffmann, G. (1987) The insecticide azadirachtin reduces predominantly cellular RNA in *Tetrahymena. Naturwiss.*, **74**, 191–2.

Gaaboub, I. A. and Hayes, D. K. (1984a) Biological activity of azadirachtin component of the neem tree, inhibiting moulting in the face fly, *Musca autumnalis* DeGeer (Diptera: Muscidae). *Environ. Entomol.*, **13**, 803–12.

Gaaboub, I. A. and Hayes, D. K. (1984b) Effect of larval treatment with

azadirachtin, a moulting inhibitory compound of the neem tree, on reproductive capacity of the face fly, *Musca autumnalis* DeGeer (Diptera: Muscidae). *Environ. Entomol.*, **13**, 1639–43.
Garcia, E. S. and Rembold, H. (1984) Effects of azadirachtin on ecdysis of *Rhodnius prolixus. J. Insect Physiol.*, **30**, 939–41.
Garg, H. S. and Bhakuni, D. S. (1984) An isoprenylated flavanone from leaves of *Azadirachta indica. Phytochemistry*, **23**, 2115–18.
Gill, J. S. and Lewis, C. T. (1971) Systemic action of an insect feeding deterrent. *Nature*, **232**, 402–3.
Gunasekaran, M., Ramadoss, N., Ramiajh, M. *et al.* (1986) Role of neem cake in the control of Thanjavur wilt of coconut. *Indian Coconut J.*, **17**, 7–12.
Irvine, F. R. (1961) *Woody Plants of Ghana*, Oxford University, London, p. 512.
Jacobson, M. (1981) Neem research in the US Department of Agriculture: chemical, biological and cultural aspects. *Proc. Ist Int. Neem Conf., Rottach-Egern*, pp. 33–42.
Jacobson, M. (1986) The neem tree: natural resistance par excellence, in *Natural Resistance of Plants to Pests, Role of Allelochemicals* (eds M. B. Green and P. A. Hedin), ACS Symp. Ser. 296, Amer. Chem. Soc., Washington, DC, pp. 220–32.
Jacobson, M. (1988) *Focus on Phytochemical Pesticides*, Vol. 1. *The Neem Tree*, CRC Press, Florida.
Janzen, D. H. (1981) In *Physiological Ecology* (eds C. R. Townsend and P. Calow), Blackwell, Oxford, pp. 145–64.
Jeyarajan, R., Doraiswamy, S., Bhaskaran, P. and Jayaraj, S. (1987) Effect of neem and other plant products in the management of plant diseases in India. *Proc. 3rd Int. Neem Conf., Nairobi, Kenya*, GTZ Press, Germany, pp. 635–44.
Khan, M. M., Khan, M. and Saxsena, S. K. (1974) Rhizosphere fungi and nematodes of egg plant as influenced by oil cake amendments. *Indian Phytopath.*, **27**, 480–4.
Kirsch, K. (1987) Studies on the efficacy of neem extracts in controlling major pests in tobacco and cabbage. *Proc. 3rd Int. Neem Conf., Nairobi, Kenya*, GTZ Press, Germany, pp. 495–515.
Klenk, A., Bokel, M. and Kraus, W. (1986) 3-Tigloylazadirachtol (tigloyl-2 methylcrotonyl), an insect growth regulating constituent of *Azadirachta indica. Chem. Commun.*, 523–4.
Koul, O. (1984) Azadirachtin I. Interaction with the development of red cotton bugs. *Entomol. Exp. Appl.*, **36**, 85–8.
Koul, O. (1985) Azadirachtin interaction with the development of *Spodoptera litura* F. *Ind. J. Exp. Biol.*, **23**, 160–3.
Koul, O. (1988) Mosquito larvicidal effects of neem formulation Margosan-*O*-concentrate. *Neem Newsletter*, **5**, 45–7.
Koul, O., Amanai, K. and Ohtaki, T. (1987) Effect of azadirachtin on the endocrine events of *Bombyx mori. J. Insect Physiol.*, **33**, 103–8.
Koul, O., Isman, M. B. and Ketkar, C. M. (1989) Properties and uses of Indian neem tree, *Azadirachta indica* A. Juss. *Can. J. Bot.*, **68**, 1–11.
Kraus, W. (1983) Biologically active compounds from Meliaceae. Chemistry and biotechnology of biologically active natural products. *2nd Int. Congr., Budapest*, pp. 331–45.
Kraus, W., Bokel, M., Bruhn, A. *et al.* (1987a) Structure determination by NMR

of azadirachtin and related compounds from *Azadirachta indica* A. Juss (Meliaceae). *Tetrahedron*, **43**, 2817–30.

Kraus, W., Baumann, S., Bokel, M. *et al.* (1987b) Control of insect feeding and development by constituents of *Melia azedarach* and *Azadirachta indica*. Proc. 3rd Int. Neem Conf., *Nairobi*, *Kenya*, GTZ Press, Germany, pp. 111–25.

Kraus, W. and Cramer, R. (1981) New tetranortriterpenoids with insect antifeedant activity from neem oil. *Liebigs Ann. Chem.*, 181–9.

Kubo, I., Matsumoto, T., Matsumoto, A, and Shoolery, J. N. (1984) Structure of deacetylazadirachtinol. Application of 2D $^{1}H-^{1}H$ and $^{1}H-^{13}C$ shift correlation spectroscopy. *Tett. Lett.*, **25**, 4729–32.

Kubo, I., Matsumoto, T. and Klocke, J. A. (1986) New ecdysis inhibitory limonoid deacetylazadirachtinol isolated from *Azadirachta indica* (Meliaceae) oil. *Tetrahedron*, **42**, 489–96.

Kunkel, G. (1978) *Flowering Trees in Subtropical Gardens*, Dr W. Junk, The Hague.

Ladd, T. L., Warthen, J. D. Jr and Klein, M. G. (1984) Japanese beetle (Coleoptera: Scarabaeidae): the effects of azadirachtin on the growth and development of the immature forms. *J. Econ. Entomol.*, **77**, 903–5.

Lange, W. and Feuerhake, K. (1984) Improved activity of enriched neem seed extracts with synergists (piperonyl butoxide) under laboratory conditions. *Z. Angew. Ent.*, **98**, 368–75.

Larew, H. G., Knodel, J. J. and Marion, D. F. (1987) Use of foliar applied neem (*Azadirachta indica* A. Juss) seed extract for the control of birch leaf miner, *Fenusa pusilla* (Hymenoptera: Tenthredinidae). *J. Environ. Hort.*, **5**, 17–19.

Larson, R. O. (1988) The commercialization of neem, in *Focus on Phytochemical Pesticides*, Vol. I. *The Neem Tree* (ed. M. Jacobson), CRC Press, pp. 156–68.

Lavie, D., Jain, M. K. and Shpan-Gabrielith, S. R. (1967) A locust phagodeterrent from two *Melia* species. *Chem Commun.*, 910–11.

Lavie, D., Levy, E. C. and Jain, M. K. (1971) Limonoids of biogenetic interest from *Melia azadirachta* L. *Tetrahedron*, **27**, 3927–39.

Lee, S. M., Olsen, J. I., Schweizer, M. P. and Klocke, J. A. (1988) 7-Deacetyl-17-hydroxyazadiradione, a new limonoid insect growth inhibitor from *Azadirachta indica*. *Phytochemistry*, **27**, 2773–5.

Ludlum, C. T. and Sieber, K. P. (1988) Effects of azadirachtin on oogenesis in *Aedes aegypti*. *Physiol. Entomol.*, **13**, 177–84.

Mangunath, B. L. (1948) *The Wealth of India*, Vol. I. The Council of Scientific and Industrial Research, New Delhi.

Meisner, J. and Ascher, K. R. S. (1986) Extracts of neem (*Azadirachta indica*) and their effectiveness as pesticides for different insects. *Phytoparasitica*, **14**, 171–80.

Meisner, J., Ascher, K. R. S., Aly, R. and Warthen, J. D. Jr (1981) Response of *Spodoptera littoralis* (Boisd) and *Earias insulana* (Boisd) larvae to azadirachtin and salannin. *Phytoparasitica*, **9**, 27–32.

Mishra, S. D. and Gaur, H. S. (1984) Control of nematodes infesting mung with nematicidal seed treatment and field applications of dasanit, aldicarb and neem cake. *Natl. Symp on Soil Pest and Soil Organisms, BHU, India*, p. 71.

Mordue, A. J., Evans, K. A. and Charlet, M. (1986) Azadirachtin, ecdysteroids and ecdysis in *Locusta migratoria. Comp. Biochem. Physiol.*, **85C**, 297–301.

Mukherjee, S. N. and Ramachandran, R. (1989) Effects of azadirachtin on the feeding, growth and development of *Tribolium castaneum* (Herbst.) (Col: tenebrionidae). *J. Appl. Ent.*, **107**, 145–9.

Nadkarni, K. M. and Nadkarni, A. K. (1954) *Indian Materia Medica*, 3rd edn, Popular Book Depot, Bombay, India, Vol. I, pp. 776–84.

Naqvi, S. N. H. (1987) Biological evaluation of fresh neem extracts and some neem components with reference to abnormalities and esterase activity in insects. *Proc. 3rd Int. Neem Conf. Nairobi, Kenya*, GTZ Press, Germany, pp. 315–30.

Narayanan, C. R., Pachapurkar, R. V. and Sawant, B. M. (1967) Nimbinin: A new tetranortriterpenoid. *Tett. Lett.*, 3563–5.

Narayanan, C. R., Pradhan, S. K., Pachapurkar, R. V. and Narasimhan, N. S. (1962) The molecular formula of nimbin. *Chemistry and Industry*, 1283.

Nawrot, J., Harmatha, J. and Bloszyk, E. (1987) Secondary plant metabolites with antifeeding activity and their effects on some stored grain insects. *Proc. 4th Int. Work Conf. Stored Product Protection, Tel Aviv, Israel* (eds E. Donahaye and S. Navarro), pp. 591–7.

Pachapurkar, R. V., Kornule, P. M. and Narayanan, C. R. (1974) A new hexacyclic tetranortriterpenoid. *Chem. Lett.*, 357–8.

Pant, N., Garg, H. S., Madhusudanan, K. P. and Bhakuni, D. S. (1986) Sulfurous compounds from *Azadirachta indica* leaves. *Filoterapia*, **57**, 302–4.

Parmar, B. S. (1987) An overview of neem research and use in India during the years 1983–1986. *Proc. 3rd Int Neem Conf., Nairobi, Kenya*, GTZ Press, Germany, pp. 55–80.

Pliske, T. E. (1984) The establishment of neem plantation in the American tropics. *Proc. 2nd Int. Neem Conf., Rauischholzhausen*, GTZ Press, Germany, pp. 521–6.

Pruthi, H. S. (1937) *Report of Imperial Entomologists 1935–1936*. Scientific Report of the Agricultural Research Institute, New Delhi, India.

Qadri, S. S. H. and Narsaiah, J. (1978) Effect of azadirachtin on the moulting process of last instar nymphs of *Periplaneta americana* (L). *Ind. J. Exp. Biol.*, **16**, 1141–3.

Radwanski, S. A. (1977a) Neem tree 1. Commercial potential, characteristics and distribution. *World Crops and Livestock*, **29**, 62–3, 65–6.

Radwanski, S. A. (1977b) Neem tree 2. Uses and potential uses. *World Crops and Livestock*, **29**, 111–13.

Radwanski, S. A. and Wickens, G. E. (1981) Vegetative fallows and potential value of the neem tree (*Azadirachta indica*) in the tropics. *Econ Bot.*, **35**, 398–414.

Ramachandran R., Mukherjee, S. N. and Sharma, R. N. (1988) Hermetic effects of azadirachtin on *Tribolium castaneum* (Herbst.) (Coleoptera: Tenebrionidae). *Ind. J. Exp. Biol.*, **26**, 913–14.

Redknap, R. S. (1981) The use of crushed neem berries in the control of some insect pests of Gambia. *Proc. 1st Int. Neem Conf., Rottach-Egern*, GTZ Press, Germany, pp. 205–14.

Reed, D. K. and Reed, G. L. (1986) Control of vegetable insects with neem seed extracts. *Proc. Indian Acad. Sci.*, **94**, 335.

Reed, D. K., Warthen, J. D. Jr, Uebel, E. C. and Reed, G. L. (1982) Effects of two triterpenoids from neem on feeding by cucumber beetles (Coleoptera; Chrysomelidae). *J. Econ. Entomol.*, **75**, 1109–13.

Rembold, H. (1988) Isomeric azadirachtins and their mode of action, in *Focus on Phytochemical Pesticides*, Vol. I. *The Neem Tree* (ed. M. Jacobson), CRC Press, Florida, pp. 48–67.

Rembold, H, Forster, H. and Czoppelt, C. H. (1987) Structure and biological activity of azadirachtin A and B. *Proc. 3rd Int. Neem Conf., Nairobi, Kenya*, GTZ Press, Germany, pp. 149–60.

Rembold, H., Muller, T. and Subramanyam, B. (1989) Tissue specific incorporation of azadirachtin in the malpighian tubules of *Locusta migratoria* *Z. Naturforsch.*, **43C**, 903–7.

Rembold, H., Sharma, G. K., Czoppelt, C. H. and Schmutterer, H. (1982) Azadirachtin a potent insect growth regulator of plant origin. *Z. Angew. Ent.*, **93**, 12–17.

Ruscoe, C. N. E. (1972) Growth disruption effects of an insect antifeedant. *Nature New Biology*, **236**, 159–60.

Saxena, R. C. (1989) Insecticides from neem, in *Insecticides of Plant Origin* (eds J. T. Arnason, B. J. R. Philogene and P. Morand), ACS Symp Ser. 387, Amer. Chem. Soc., Washington, DC, pp. 110–35.

Saxena, R. C. (1982) Note on the use of neem kernel for the protection of dew gram against *Amsacta moorei* Butler. *Ind. J. Agric. Sci.*, **52**, 51–3.

Saxena, R. C., Jilani, G. and AbdulKareem, A. (1988) Effects of neem on stored grain insects, in *Focus on Phytochemical Pesticides*, Vol. I. *The Neem Tree* (ed. M. Jacobson), CRC Press, Florida, pp. 98–111.

Saxena, R. C., Khan, Z. R. and Bajet, N. B. (1985) Neem seed derivatives for preventing rice tungro virus transmission by the green leafhopper, *Nephotettix virescens* (Distant). *Phil. Phytopath.*, **21**, 88–102.

Schluter, U., Bidmon, H. J. and Grewe, S. (1985) Azadirachtin affects growth and endocrine events in larvae of the tobacco hornworm *Manduca sexta*. *J. Insect Physiol.*, **31**, 773–7.

Schmidt, G. H. and Pesel, E. (1987) Studies of the sterilizing effect of neem extracts in ants. *Proc. 3rd Int. Neem Conf., Nairobi, Kenya*, GTZ Press, Germany, pp. 361–73.

Schmutterer, H. (1985) Which insect pests can be controlled by application of neem seed kernel extracts under field conditions? *Z. Angew. Ent.*, **100**, 468–75.

Schmutterer, H. and Hellpap, C. (1988) Effects of neem on pests of vegetables and fruit trees, in *Focus on Phytochemical Pesticides*, Vol. I. *The Neem Tree* (ed. M. Jacobson), CRC Press, Florida, pp. 70–86.

Schmutterer, H. and Rembold, H. (1980) Zur wirkung einiger reinfraktionen aus samen von *Azadirachta indica* auf frassaktivitat und metamorphose von *Epilachna varivestis* (Coleoptera: Coccinellidae). *Z. Angew. Ent.*, **89**, 179–88.

Schroeder, D. and Nakanishi, K. (1987) A simplified isolation procedure of azadirachtin. *J. Nat. Prod.*, **50**, 241–4.

Schultz, P. B. and Coffelt, M. A. (1987) Orange striped oakworm control on willow oak, 1985. *Insect. Acar. Tests*, **11**, 415.

Schwinger, M., Ehhammer, B. and Kraus, W. (1984) Methodology of the *Epilachna varivestis* bioassay of antifeedants demonsrated with some compounds from *Azadirachta indica* and *Melia azedarach. Proc. 2nd Int. Neem Conf., Rauischholzhausen*, GTZ Press, Germany, pp. 181–98.

Sengupta, P., Chowdhury, S. N. and Khastgir, H. N. (1960) Terpenoids and related compounds, I. Constituents of the trunk bark of *Melia azadirachta* L. and the structure of the ketophenol nimbiol. *Tetrahedron*, **10**, 45–54.

Shimizu, M., Sudo T. and Nomura, T. (1985) Patentschrift (Switz.) CH 650404 (1985), cf. *Chem. Abstr.*, **103**, 109926.

Siddiqi, M. A. and Alam, M. M. (1985a) Evaluation of nematicidal properties of different parts of Margosa and Persian lilac. *Neem Newsl.*, **2**, 1–4

Siddiqi, M. A. and Alam, M. M. (1985b) Further studies on the nematode toxicity of Margosa and Persian lilac. *Neem Newsl.*, **2**, 43–7.

Siddiqi, S., Faizi, S., Mahmood, T. and Siddiqi, B. S. (1986a) Two new insect growth regulatory meliacines from *Azadirachta indica* A. Juss (Meliaceae). *J. Chem. Soc. Perkin Trans.*, **I**, 1021–5.

Siddiqi, S., Faizi, S., Mahmood, T. and Siddiqi, B. S. (1986b) Isolation of new tetranortriterpenoid from *Azadirachta indica* A. Juss (Meliaceae). *Heterocycles*, **24**, 1319–24.

Siddiqi, S., Faizi, S., Mahmood, T. and Siddiqi, B. S. (1986c) Margosinolide and isomargosinolide two new tetranortriterpenoids from *Azadirachta indica* A. Juss (Meliaceae). *Tetrahedron*, **42**, 4849–56.

Siddiqi, S., Faizi, S. and Siddiqi, B. S. (1986d) Chemical constituents of *Azadirachta indica* A. Juss (Meliaceae) VII. *Z. Naturforsch.*, **41B**, 922–6.

Siddiqi, S., Siddiqi, B. S., Faizi, S. and Mahmood, T. (1988) Tetratricyclic triterpenoids and their derivatives from *Azadirachta indica. J. Nat. Prod.*, **51**, 30–43.

Siddiqi, S., Siddiqi, B. S., Faizi, S. and Mahmood, T. (1984) Isolation of a tetranortriterpenoid from *Azadirachta indica. Phytochemistry*, **23**, 2899–901.

Sieber, K.-P. and Rembold, H. (1983) The effects of azadirachtin on the endocrine control of moulting in *Locusta migratoria. J. Insect Physiol.*, **29**, 523–7.

Simons, J. N. (1981) Innovation methods of control of transmitted viral diseases, in *Vector of Disease Agents: Interaction with Plants, Animals and Men* (eds J. K. McKelvey Jr, B. F. Eldnge and K. Maramorsch), Praeger, New York, pp. 169–78.

Singh, U. P., Singh, H. B. and Chauhan, V. B. (1984) Effects of some fungicides, plant extracts and oils on inoculum density of different nodal leaves of pea infected by *Erysiphe polygoni. Z. Pflanzenk. Pflanzensch.*, **91**, 20–6.

Singh, U. P., Singh, H. B. and Singh, R. B. (1980) The fungicidal effect of neem (*Azadirachta indica*) extracts on some soil borne pathogens of gram (*Cicer arietinum*). *Mycologia*, **72**, 1077–93.

Sinniah, D. and Baskaran, G. (1981) Margosa oil poisoning as a cause of Rey's syndrome. *Lancet*, **i**, 487–9.

Steets, V. R. (1975) Die wirkung von rohextrakten aus den meliaceen *Azadirachta indica* und *Melia azedarach* auf verschiedene insektenarten. *Z. Angew. Ent.*, **77**, 306–12.

Steets, V. R. (1976) Zur wirkung eines gereinigten extraktes aus fruchten von *Azadirachta indica* A. Juss auf *Leptinotarsa decemlineata* Say (Coleoptera; Chrysomelidae). *Z. Angew. Ent.*, **82**, 169–76.

Steets, V. R. and Schmutterer, H. (1975) The effect of azadirachtin on the longevity and reproduction of *Epilachna varivestis* Muls. (Coleoptera: Coccinellidae). *Z. Pflkrankh. Pfl. Schutz.*, **82**, 176–9.

Stokes, J. B. and Redfern, R. E. (1982) Effects of sunlight on azadirachtin antifeeding potency. *J. Environ. Sci. Health*, **17A**, 57–65.

Subrahmanyam, B., Muller, T. and Rembold, H. (1989) Inhibition of turnover of neurosecretion by azadirachtin in *Locusta migratoria. J. Insect Physiol.*, **35**, 493–500.

Turner, C. J., Temperta, M. S., Taylor, R. B. *et al.* (1987) An NMR spectroscopic study of azadirachtin and its trimethyl ether. *Tetrahedron*, **43**, 2789–803.

Uebel, E. C., Warthen, J. D. Jr and Jacobson, M. (1979) Preparative reversed phase liquid chromatographic isolation of azadirachtin from neem kernels. *J. Liq. Chromatogr.*, **2**, 875–82.

Vijayalakshmi, K., Gaur, H. S. and Goswami, B. K. (1985) Neem for the control of plant parasitic nematodes. *Neem Newsl.*, **2**, 35–42.

Warthen, J. D. Jr (1979) *Azadirachta indica*: A source of insect feeding inhibitors and growth regulators. *USDA, Sci. and Educ. Adm. Agric. Reviews and Manuals, Northeastern Series* **4**, 1–21.

Warthen, J. D. Jr, Stokes, J. B. and Jacobson, M. (1984) Estimation of azadirachtin content in neem extracts and formulations. *J. Liq. Chromatogr.*, **7**, 591–8.

Warthen, J. D. Jr and Uebel, E. C. (1981) Effect of azadirachtin on house crickets, *Acheta domesticus. Proc. Ist Int. Neem Conf., Rottach-Egern*, GTZ Press, Germany, pp. 137–48.

Warthen, J. D. Jr, Uebel, E. C., Dutky, S. R. *et al.* (1978) Adult house fly feeding deterrent from neem seeds, *USDA, SEA, Agric. Res. Results, Series ARR-NE-2*, pp. 1–11.

Webb, R. E., Larew, H. G., Weiber, A. M. *et al.* (1984) Systemic activity of neem seed extract and purified azadirachtin against *Liriomyza* leaf miners. *Proc. 4th Ann. Industry Conf. on Leaf Miners, Sarasota, Florida*, pp. 118–27.

Yamasaki, R. B. and Klocke, J. A. (1989) Structure–bioactivity relationships of salannin as an antifeedant against the Colorado potato beetle (*Leptinotarsa decemlineata*). *J. Agric. Food Chem.*, **37**, 1118–24.

Yamasaki, R. B., Klocke, J. A., Lee, S. M. *et al.* (1986) Isolation and purification of azadirachtin from neem (*Azadirachta indica*) seeds using flash chromatography and high performance liquid chromatography. *J. Chromatogr.*, **356**, 220–6.

Zanno, P. R., Miura, I., Nakanishi, K. and Elder, D. L. (1975) Structure of the insect phagorepellent azadirachtin. Application of PRFTCWD carbon-13-nuclear magnetic resonance. *J. Amer. Chem. Soc.*, **97**, 1975–7.

Zebitz, C. P. W. (1986) Effect of three different neem seed kernel extracts and azadirachtin on larvae of different mosquito species. *Z. Angew. Ent.*, **102**, 455–63.

CHAPTER 24

Allelopathy in the management of root-knot nematodes

A. Hasan

24.1 INTRODUCTION

Based on the worldwide survey carried out by Sasser and Freckman (1987), the root-knot nematode (*Meloidogyne*) was ranked as number one out of the ten most important phytoparasitic nematode genera (*Meloidogyne* and *Pratylenchus*, the lesion nematode; *Ditylenchus*, the stem and bulb nematode; *Globodera* and *Heterodera*, the cyst nematodes; *Tylenchulus*, the citrus nematode; *Xiphinema*, the dagger nematode; *Radopholus*, the burrowing nematode; *Rotylenchulus*, the reniform nematode; and *Helicotylenchus*, the spiral nematode) with wide geographical distribution (Sasser, 1989), phytophagous food habit and infecting over 2200 plant species. It consists of more than 63 species. Out of them, *M. incognita*, *M. javanica*, *M. arenaria* and *M. hapla* are the most commonly occurring and devastating species in various agricultural crops. The first three species are mainly distributed in the tropical and the subtropical zones (between 35° S and 35° N latitudes) while *M. hapla* is found in the cold climate areas of the world (Sasser, 1977).

The average annual losses (both direct and indirect) to various life-sustaining and economically important crops caused by various phytoparasitic nematodes account for as much as 12.3 per cent of the total losses (Sasser, 1989), of which 5 per cent losses are attributed to the root-knot nematodes alone (Taylor and Sasser, 1978) though a reliable estimate is still lacking.

Owing to the cost, availability, mode of application, phytotoxicity, environmental and health hazards associated with the use of pesticides (Fassuliotis, 1985; Thomason, 1987; Thomason and Caswell, 1987), other approaches to nematode management, including allelopathy, appear to be potential alternatives (Minton, 1977; Rich, 1986).

Plant parts contain allelochemicals which are released into the environment as a result of volatilization, root exudation, leaching and decomposition of plant residues (Rice, 1984, Putnam and Tang, 1986). There is

Allelopathy: Basic and applied aspects Edited by S. J. H. Rizvi and V. Rizvi
Published in 1992 by Chapman & Hall, London ISBN 0 412 39400 6

a fairly good possibility that these allelochemicals may in turn regulate the population of nematodes leading to a better nematode management. During the past few decades the possible role of root exudates and decomposition products of plants and their residues in suppression of the root-knot nematodes have been under investigation. Some of the important approaches related to these aspects are described here and the possibility of their exploitation for nematode management is discussed.

24.2 APPROACHES TO NEMATODE MANAGEMENT

24.2.1 Antagonistic plants

Certain plant species have been found adversely to affect abnoxious root-knot nematodes by preventing the juveniles from reaching the root surface and/or killing them in the rhizosphere, thus regulating their population under natural conditions. Roots of actively growing plants secrete a variety of allelochemicals in the soil (Fisher, 1979; Rice, 1984). Some of these have been found to suppress various nematodes in soil. Griffin and Waite (1971) reported that when eggmass of *M. hapla* was placed on a water agar plate equidistant from germinating susceptible and resistant alfalfa seeds, approximately 75% of the hatched juveniles were attracted to the susceptible and only 25% to the resistant seedlings. Excised roots from bitter cucumber attracted significantly fewer *M. incognita* juveniles than did roots from non-bitter isogenic lines (Haynes and Jones, 1976). These effects were attributed to the repellent action of the exudates emanating from the resistant plants. Tanda *et al.* (1988) demonstrated in axenic tissue culture that exudates originating from sesame (*Sesamum* sp.) inhibited root penetration and development of *M. incognita* juveniles in okra (*Abelmoschus esculentus*) and subsequent gall formation on the roots.

In the pot culture tests the soil population of *M. incognita* and gall development on the roots were found suppressed around cabbage (*Brassica oleracea capitata*), cauliflower (*Brassica oleracea botrytis*), egg-plant (*Solanum melongena*) and tomato (*Lycopersicon esculentum*) when intercultured with margosa, sesame (*Sesamum orientale*) or zinia (*Zennia elegans*) (Atwal and Manger, 1969; Tiyagi *et al.*, 1986a; Siddiqi and Saxena, 1987). The leaf extracts of *Calotropis procera*, *Datura stramonium* and *Xanthium strumarium* were very effective in reducing juvenile penetration of eggplant roots and subsequent gall development under greenhouse conditions (Nandal and Bhatti, 1987). The soil and root population of root-knot nematodes and subsequent gall development were found to be reduced under field conditions when susceptible crops were intercultured with caster bean (*Ricinus communis*), mustard (*Brassica campestris*), rocket salad (*Lactuca* sp.) or sesame (Atwal and

Manger, 1969; Prasad and Dasgupta, 1969; Hackney and Dickerson, 1975; Alam *et al.*, 1976a). There was considerably less severe galling by *M. incognita* on potato roots intercropped with onion (*Allium* sp.) or corn (*Zea mays*) than on potatoes cultured alone (Midmore, 1983; Raymundo, 1983). In the Ivory Coast when cassava (*Manihot esculenta*) was grown simultaneously along with other crops, the damage caused by root-knot nematode was less severe than when it was mixed grown with eggplant (*Solanum melongena*) or okra (Luc, 1968).

Among angiospermic plants tested for their allelopathic potential against various phytoparasitic nematodes including the root-knot nematodes, marigolds (*Tagetes* spp.) of the family compositae have been studied most extensively (Suatmadji, 1969; Winoto, 1969; Gommers, 1973, 1981; Gommers and Bakker, 1988). Alam *et al.* (1975) showed that 63% of *M. incognita* juveniles were killed within 72 hours when treated with root exudates obtained from 1-month-old marigold (*T. erecta*) seedlings. There was also an appreciable reduction in the soil population and gall development induced by root-knot nematodes in various susceptible crops, both under greenhouse (Medhane *et al.*, 1985; Siddiqi and Alam, 1987a; Perwez *et al.*, 1988) and field conditions when such crops were intercultured with marigolds (Hackney and Dickerson, 1975; Alam *et al.*, 1977b; Rickard and Durpee, 1978; Choudhury, 1981; Baghel and Gupta, 1986).

24.2.2 Resistant plants

The mechanism of plant resistance against root-knot nematodes has not been understood fully. However, its nature has been studied to a great extent (Giebel, 1982; Kaplan and Davis, 1987). The resistance of plants against nematodes has been generally attributed to the facts that nematodes either fail to penetrate the roots of plants or if they enter, fail to develop to the adult stage. Even if some of them developed to the egg-laying stage, hatching was likely to be impeded.

(a) Penetration and development inhibition

Sasser (1954) in an intensive study of the host–parasite relationship of root-knot nematode found that resistance was due to lack of penetration as well as to the factors (chemicals) preventing development of the nematode after penetration. Juveniles of *M. hapla* would not penetrate oat (*Avena sativa*) or rye (*Secale cereale*) roots, while juveniles of *M. incognita* entered the roots of peanut (*Arachis hypogoea*) and strawberry (*Fragaria* sp.), though in reduced numbers. Resistant sweet potato cultivars were reported to be penetrated by a fewer number of juveniles of *M. incognita* than the susceptible cultivars (Jatala and Russell, 1972).

But the hypothesis that resistance was based on lack of penetration became more perplexing with the publication of the results of Goplen and

Stanford (1959) that fewer *M. hapla* juveniles entered the roots of resistant than the susceptible alfalfa plants, but juveniles of *M. javanica* entered equally in both the resistant and susceptible cultivars. In other studies of a similar nature employing other susceptible (Ranger) and resistant (#298) alfalfa cultivars, Griffin and Elgin (1977) reported that *M. hapla* juveniles entered the roots of these cultivars in equal numbers. Against this, line M-4 offered resistance to penetration because very few juveniles were recorded within the roots throughout the period of observation. They also found that the number of *M. hapla* juveniles in the resistant line (#298) declined after 8 days of inoculation, and even less than 5% of the juveniles were observed within the roots of the resistant line as against 37% in the susceptible cultivar (Ranger). Similar observations that juveniles left the roots of a resistant plant within a few days of penetration have also been made earlier (Reynolds *et al.*, 1970). Thus, it could be presumed that even when juveniles penetrate the resistant crop plants, they leave the roots after a week or so. Therefore, it appears quite pertinent that resistance to penetration is an exception rather than the rule.

Steiner (1941) observed that a large number of juveniles of root-knot nematode entered the roots of 40 horticultural varieties of *Tagetes patula* but, with few exceptions, all usually failed to develop to adult egg-laying stage. Similarly, the development of root-knot nematode juveniles was impeded at various life stages in certain resistant plants including grasses, *Brachiaria decumbens*, *Crotalaria striata*, *C. retusa*, *C. spectabilis*, *Panicum maximum* cv. Guine and marigolds (Minton, 1962; Daulton and Curtis, 1964; Good *et al.*, 1965; Brito and Ferraz, 1987). Not only was the development found to be arrested, but even a distinct shift in the sex ratio in favour of males was observed in *M. incognita acrita* on resistant *Cucumis* spp. (Fassuliotis, 1970). The most common response in a compatible host–nematode (*Meloidogyne* spp.) interaction is inducement of giant cells in the vascular parenchyma of the host plant by the invading juveniles on which they feed and develop to adult (egg-laying) stage. These giant cells may fail to develop wholly or partially on resistant plants which adversely affect the growth, development and reproduction of these nematodes (McClure *et al.*, 1974a, b; Kaplan and Keen, 1980; Bleue-Zacheo *et al.*, 1982; Sosa-Moss *et al.*, 1983; Singh *et al.*, 1985). However, the most common feature exhibited by such plants is the hypersensitivity which results in instantaneous death of the host cells surrounding the nematode head and thereby leading to death of the nematode (Bird, 1962; Khana and Nirula, 1964; Smart and Brodie, 1977). Pi and Rohde (1967) observed higher concentrations of chlorogenic acid, a phenolic compound, in the resistant (Nemared) compared with moderately resistant (Hawaii 7153) or susceptible (B-5) tomato cultivars. Keeping in view several other similar reports it may be suggested that the impaired development leading to death of the nematodes may be a consequence of the presence of various allelochemicals such as alkaloids, phenolics, etc., in the resistant plant or phytoalexins produced in response to nematode invasion (Milne *et al.*, 1965; Pi and Rohde, 1967; Singh and Choudhury, 1973; Hung and Rohde, 1973; Veech, 1981).

(b) Hatching inhibition

Hatching is an important biological phenomenon in the life cycle of root-knot nematodes where all the eggs are deposited outside the female body in a gelatinous matrix (eggmass) attached to the root surface. While studying the effect of root leachates on *in vitro* hatching, Ahmad and Khan (1964) observed that it was suppressed by the leachates of some non-host plants. Later, similar observations were also made by various workers (Yadav, 1970; Alam *et al.*, 1975, 1977b; Jain and Hasan, 1984; Patel *et al.*, 1985a; Tiyagi *et al.*, 1985, 1986b; Bano *et al.*, 1986a, b; Goswami and Vijayalakshmi 1986a, b; Jain *et al.*, 1986). The excised roots of sesame were also shown to inhibit the hatching of *M. incognita* in tissue culture studies (Tanda *et al.*, 1988). Grasses, such as *Brachiaria decumbens* and *Panicum maximum* cv. Guine, were also shown to suppress the hatching in *M. javanica* under greenhouse conditions (Brito and Ferraz, 1987). Similar inhibition in hatching of this nematode was reported to be caused in soils amended with oil-cakes (Khan, 1969, 1976; Singh *et al.*, 1980), which was considered due to the presence of a certain water-soluble antinematode compound (Khan *et al.*, 1974b).

24.2.3 Crop rotation

The host–parasite relationship can be measured in terms of parasite growth, host damage or the parasite growth coupled with host damage. The designation, non-host or poor-host, relates to the nematode multiplication on plants when it is less than 1 ($P_f/P_i < 1$, where P_f is the final population and P_i the initial (Jones, 1956). Thus, if root-knot nematode susceptible crops were spaced at intervals of 3–5 years, a good control could be achieved (DiMuro, 1975), provided that the intervening periods are adjusted with non-host/poor-host plants.

Despite the fact that root-knot nematodes multiply on a wide range of crops and weeds, certain non-host/poor-host plants have been identified and found suitable for temporal cropping sequences.

The inclusion of *Daucus carota*, *Zea mays*, *Crotalaria usaramoensis*, *Glycine javanica*, *Allium porrum*, *Allium cepa*, *Arachis hypogea*, *Oryza sativa*, *Sorghum* sp., *Fragaria* sp., *Tagetes* spp., *Crotalaria juncea*, *Ipomoea batatas* and *Triticum aestivum* in rotation sequence with nematode susceptible crops has been shown to suppress the root-knot nematode populations in the fields (Cortado and Davide, 1968; Widjaja and Windrich, 1974; Muro, 1975; Netscher and Taylor, 1976; Smit and Bos, 1976; Sivapalan, 1981; Sharma and Scolari, 1984). It has been reported that, in the Ivory Coast, heavily reduced commercial production of tomato due to root-knot nematodes could be improved by growing rice for two years (Merny, 1976). Also, in Burma the rotation sequence of Kohlrabi (*Brassica oleracea caulorapa*)–tomato–bitter gourd (*Momordica charantia*)–sweet potato–tomato, reduced the root-knot damage quite appreciably (Myint, 1981). For rice-based cropping systems in the Philippines, sequences such as rice–mungbean, corn–cabbage–rice, rice–tobacco–rice, rice–cotton–rice and rice–watermelon–rice have been found to be quite effective in combating the root-knot nematode menace (Davide and Zorilla, 1983). Johnson (1985), after conducting an exhaustive field trial to determine the suitability of various non-host/poor-host plants for various cropping systems, reported that sweetcorn–soybean–wheat–soybean–spinach, turnip–peanut–snapbean, turnip–peanut–turnip, turnip–peanut–cucumber–turnip–soybean and turnip–fieldcorn–southern pea sequences were quite encouraging in controlling *M. incognita* infestation. Out of these rotations, sweetcorn–soybean–wheat–soybean–spinach, turnip–peanut–snapbean and turnip–peanut–turnip were found to be the most effective. The Pangola grass (*Digitaria decumbens*) has also been shown to reduce the populations of *M. incognita*, *M. javanica*, *M. arenaria* and *M. hapla* below detectable level (Winchester and Hayslip, 1960; Overman, 1961; Haroon and Smart, 1983).

24.2.4 Organic additives

Organic matter is an essential soil component which not only improves the soil conditions but also suppresses various plant diseases. The importance of organic additives in soil in controlling root-knot and other nematodes was realized and some review articles have been published during the past few years (Muller and Gooch, 1982; Castillo, 1985; Rodriguez-Kabana, 1986).

Nematode mortality, to a varying extent, was recorded under laboratory conditions when juveniles of root-knot nematodes were exposed for different periods and various concentrations of water/organic solvent extracts of different parts of plants (Goswami and Vijayalakshmi, 1986a; Mahmood *et al.*, 1982; Nandal and Bhatti, 1983; Jain and Hasan, 1984; Devakumar *et al.*, 1985; Maqbool *et al.*, 1987; Osman and Viglierchio, 1988; Stephan and Al-Askari, 1989).

In greenhouse test, the simple dipping of root in leaf extracts of margosa and Persian lilac caused significant inhibition in the penetration of *M. incognita* juveniles and in subsequent galling of the roots of tomato cv. Pusa Ruby and eggplant cv. Pusa Purple Long. The root-galling gradually decreased with an increase in the concentration of the extracts and the dip duration (Siddiqi and Alam, 1988). Similar findings were also reported earlier. However, when the seeds of eggplant, greengram, okra and tomato were coated with either water-soluble fractions of various parts of margosa and Persian lilac or with oil-cake powder/oil of margosa, and seedlings raised from these seeds were inoculated with nematodes, there was a pronounced suppressive effects on juvenile penetration and subsequent gall development.

24.2.5 Soil amendments

The C:N ratio of various nitrogenous organic additives plays a significant role in nematode suppression. It was shown that where the C:N ratio was low, suppression of nematode number and root gall formation was higher in the soil (Kirmani *et al.*, 1975; Khan, 1976; Mian and Rodriguez-Kabana, 1982a, b; Rodriguez-Kabana, 1986). Additives with low C:N ratios were also proved to be phytotoxic (Khan, 1976; Rodriguez-Kabana, 1986). Chitin amendments have also given good control of root-knot nematodes (Mankau and Das, 1969) but have proved phytotoxic due to their narrow C:N ratio (Mian *et al.*, 1982; Culbreath *et al.*, 1986). The phytotoxic effect could be eliminated or reduced substantially by raising the C:N ratio and that was achieved by supplementing it with cellulosic materials rich in carbon (Culbreath *et al.*, 1985). Amendment of soil with sawdust as the carbon source and oil-cake as the nitrogen source significantly reduced the population of root-knot nematode in soil and gall development on the roots with least phytotoxic effects (Khan, 1976).

The suppressive effects of soil organic amendments in the form of oil-cake, green manuring, compost, chitin, sawdust and others have been studied on the control of root-knot nematodes. The results are summarized in Table 24.1. All the additives were found to suppress the nematodes to a varying extent. In certain cases plant growth was also reported to have improved. Oil-cakes, specially *Mahua* cake, sawdust at higher dosages and chitin were all found to be phytotoxic (Khan, 1969, 1976; Rodriguez-Kabana, 1986). The effectiveness of the additives varied greatly owing to variations in the experimental conditions. The dosages employed by different workers were noticeably different (oil-cakes, 1–105 t ha^{-1}; green and dry plant manuring, 10–200 kg ha^{-1}; compost and sawdust, 2–100 t ha^{-1}; chitin, 2–80 t ha^{-1}). The population of *M. incognita* and of other parasitic nematodes on tomato var. LaBonita, Nemared, Anahu and 67B 1169 was reported to be suppressed by the application of castor, *Mahua*, margosa, mustard, peanut cakes, compost and inorganic fertilizers (Ismail *et al.*, 1976). They further reported that varieties themselves influenced the nematode populations supporting the allelopathic mechanism. Alam *et al.* (1977a, c) also indicated that several crops reacted to nematode infection differently in comparison to other crops and pointed out that this difference might be attributed to the presence of allelochemicals.

TABLE 24.1 *Organic soil amendments implicated to possess allelopathic potentials against root-knot nematodes,* Meloidogyne spp.

Nature of Amendment (1)	Experimental condition (2)	Nematode species (3)	Test plant (4)	Remark (5)	Reference (6)
Oil-cakes					
Castor	Field	*M. javanica*	Okra	Galling on roots was reduced	Singh (1969)
Castor, *Mahua*, margosa, mustard and peanut	Field	*M. incognita*	Spinach	Soil population of the nematode was suppressed and root galling reduced	Alam and Khan (1974)
		M. incognita	Several vegetables	Soil population of the nematode was suppressed, root galling reduced and plant growth improved; *Mahua* cake proved to be phytotoxic	Khan (1969, 1976)
Castor, *Mahua*, margosa and peanut	Field	*M. incognita*	Eggplant and tomato	Soil populations of the nematode and pathogenic root infecting fungi were suppressed, saprozoic nematodes and fungi increased but *Trichoderma lignorum* fungus remained unaffected	Khan *et al*. (1969 1973, 1974b)
Castor, karange, linseed, *Mahua*, mustard and peanut	Field	*M. javanica* and *Meloidogyne* spp.	Okra and tomato	Soil population of the nematode was suppressed, root galling reduced and yield increased	Singh (1965); Singh and Sitara-maiah (1969, 1973)

Mustard and peanut	Field	*M. incognita*	Pepper	Root galling was suppressed	Trivedi *et al.* (1978)
Karange	Field	*Meloidogyne* spp.	Okra and tomato	Soil population of the nematodes was suppressed and root galling reduced	Singh (1965)
Mahua	Field	*M. incognita*	Cowpea	Soil population of the nematode was suppressed	Jain and Hasan (1986)
Margosa and Persian lilac	Field	*M. incognita*	Tomato	Soil population of the nematode was suppressed and root galling reduced	Verma (1986)
Mahua	Field	*M. incognita*	Betelvine	Nematode population and root galling were suppressed, number of leaves on treated plants increased	Jagdale *et al.* (1985)
	Field	*M. incognita*	Potato	Tuber infestation and soil population of the nematode was reduced significantly	Sharma and Raj (1987)
Mustard	Greenhouse	*M. incognita*	Jute	Nematode population was suppressed and galling reduced; amendment proved to be phytotoxic	Bora and Phukan (1983)
Mustard	Field	*M. incognita* and *M. javanica*	Okra	Nematode population was suppressed and root galling reduced	Patel *et al.* (1985b)
Castor, *Mahua*, margosa, mustard, peanut and sesame	Field	*M. incognita*	Tomato	Soil population of the nematode were suppressed and galling reduced	Gowda and Setty (1973), Khan *et al.* (1973)
Castor	Field	*M. javanica* and *Meloidogyne* spp.	Sugarbeet and tomato	Soil populations of the nematodes were suppressed and galling reduced	Lear (1959)

TABLE 24.1 *Cont'd*

Nature of Amendment (1)	Experimental condition (2)	Nematode species (3)	Test plant (4)	Remark (5)	Reference (6)
Margosa and mustard	Field	*Meloidogyne* spp.	Tomato	Low incidence of root-knot in soil amended with neem cake	Hameed (1970)
Karange and margosa	Field	*M. incognita*	Okra	Low incidence of root-knot; neem cake to be highly efficacious	Desai *et al.* (1979)
Cotton seed and peanut	Greenhouse	*M. arenaria*	Squash	Amendment at the rate of 0.4% or more practically eliminated the galls but proved to be phytotoxic	Mian and Rodriguez-Kabana (1982a)
Green-manuring					
Chopped leaves of various plants	Greenhouse	*M. incognita*	Eggplant	Nematode population was suppressed in all the treatments but the greatest reduction occurred in *Calotropis procera*	Haseeb *et al.* (1978)
Chopped shoots of latex bearing plants	Greenhouse	*M. incognita*	Cabbage, cauliflower, eggplant and tomato	Soil population of the nematode was suppressed and root galling reduced; the greatest reduction occurred in *Ficus elastica*; plant growth improved in treated pots	Siddiqi *et al.* (1987)
Chopped shoots of weeds	Greenhouse	*M. incognita*	Eggplant	Nematode population was suppressed in all the treatments but *Solanum xanthocarpum* was highly toxic; plant growth improved significantly	Alam (1986)

Chopped plant parts of marigold	Greenhouse	*M. incognita*	Eggplant and tomato	Flower part was more toxic to the nematode than leaf or shoot, plant growth improved in treated pots	Siddiqi and Alam (1987b)
Neem leaves	Greenhouse	*M. javanica*	Tomato	Galling on roots reduced and plant growth improved	Jain and Bhatti (1988)
Chopped floral parts and leaves of various plants	Greenhouse	*M. incognita*	Eggplant	The greatest reduction in nematode population occurred in pots treated with *Mentha viridis* leaves or *Iresine herbastii* flower	Haseeb and Alam (1984), Haseeb *et al.* (1984)
Chopped leaves of margosa, Persian lilac, sesbania, crotalaria	Field	*M. javanica*	Okra and Tomato	Soil population of the nematode was suppressed, root galling reduced and plant growth increased. Residual inhibitory effects on the nematode was also observed in the succeeding crop	Singh and Sitaramaiah (1967, 1969)
Miscellaneous					
FYM and Pressmud	Field	*Meloidogyne* spp.	Okra	Root galling was reduced	Alam *et al.* (1979)
Mushroom spent up compost, sawdust and dried leaf powder of Persian lilac	Field	*M. incognita*	Tomato	Root galling was significantly reduced	Verma (1986)
Chicken manure, sawdust and decaffeinated tea waste	Greenhouse	*M. incognita*	Jute	Soil population of the nematode was suppressed and root galling and egg masses production reduced	Bora and Phukan (1983)

TABLE 24.1 *Cont'd*

Nature of Amendment (1)	Experimental condition (2)	Nematode species (3)	Test plant (4)	Remark (5)	Reference (6)
Chicken manure	Field	*M. incognita*	Several plants species	Galling on the roots was reduced	Badra *et al.* (1979)
			Tomato	Galling on the roots was reduced	Derrico and Maio (1980)
	Greenhouse	*M. arenaria*	Squash	Galling was adversely affected when applied at the rate of 1% or above and was nonphytotoxic	Mian and Rodriguez-Kabana (1982a)
Sawdust	Greenhouse and field	*M. javanica*	Okra and tomato	Incidence of root-knot was reduced and plant growth improved in treated plots. Residual inhibitory effects of the nematode was also observed in the succeeding crop	Singh and Sitaramaiah (1967)
	Field	*M. javanica*	Okra	Intensity of root-knot was reduced	Singh (1969)
	Field	*M. incognita*	Tomato	Root galling was significantly reduced	Verma (1986)
			Betelvine	Soil population of the nematode was suppressed and galling reduced; number of leaves increased on treated plants	Jagdale *et al.* (1985)
			Potato	Tuber infestation and soil population of the nematode decreased	Sharma and Raj (1987)
Dried azolla and leaf powder of *Clerodendron enermi*	Field and Greenhouse	*M. incognita* and *M. javanica*	Okra	Root galling was greatly reduced and plant growth increased	Patel *et al.* (1985a, b)

Alfalfa, lespedeza and oat hays residue	Field	*M. incognita*	Tomato	Incidence of root-knot was decreased	Johnson *et al.* (1967)
Dry powder of crotalaria, Kentucky bluegrass and marigold	Field	*M. hapla*	Sugarbeet	Incidence of root-knot was adversely affected	Yuhara (1971)
Dried powder of various plants	Greenhouse	*M. incognita*	Muskmelon	Incidence of root-knot was affected to a varying degree but the highest reduction in galls occurred in marigold and xanthium	Sharma *et al.* (1985)
Chitin	Field	*M. incognita*	Tomato	Incidence of root-knot was adversely affected	Mankau and Das (1969)
	Greenhouse	*M. arenaria*	Squash	Root galling was almost eliminated but amendment proved to be phytotoxic which was nullified by supplementing with hemi-cellulosic wastes	Mian *et al.* (1982); Culbreath *et al.* (1985, 1986)
Various plant byproducts	Greenhouse	*M. arenaria*	Squash	The nematicidal efficacy of all the 15 amendments was inversely related to their C:N ratios; materials with narrow C:N ratios proved to be phytotoxic too; spent coffee (*Coffea arabica*) grinds, Crotalaria (*Crotalaria spectabilis*), Kudzu (*Pueraria lobata*) or Ramie (*Boehmeria nivea*) hays applied at the rate of 1% were most effective in reducing the root galls	Mian and Rodriguez-Kabana (1982b)
FYM	Field	*M. javanica*	Eggplant	Cow and sheep manures applied in furrows at the rate of 1 kg m^{-2} suppressed the nematode population and increased the yield	Stephan *et al.* (1989)

24.2.6 Purified chemicals

The nature of allelochemicals has been reviewed in recent years (Rice, 1984; Thompson, 1985; Waller, 1987; Gommers and Bakker, 1988). However, it is apparent that only a few allelochemicals have been isolated, elucidated and characterized, particularly against root-knot nematodes. Some examples of these chemicals are given below.

(a) Terthienyl

The earliest record that marigolds possessed resistance factors which suppressed the root-knot nematodes in soil and on plant (Steiner, 1941) got a sound footing when Uhlenbroek and Bijloo (1958, 1959) isolated a nematicidal compound α-terthienyl (2,2′–5′,2″-terthienyl) and a related compound 5(3-buten-1-ynyl)-2,2′-bithienyl. They exhibited strong nematicidal properties under laboratory conditions and most of the nematodes were rapidly killed even at concentrations as low as 1 ppm (Daulton and Curtis, 1963). However, when added to soil infested with *Meloidogyne* sp., even at 200 ppm, there was no suppression in the nematode population (Uhlenbroek and Bijloo, 1960; Handele, 1972).

(b) Cucurbitacin

The cucurbitacins are triterpenoids that accumulate in bitter cucumber genotypes carrying a dominant allele at the *Bi* locus for bitterness (DaCosta and Jones, 1971). It was observed that such genotypes significantly repelled more juveniles of *M. incognita* from infecting them than did the non-bitter genotype (*bibi*), and the cucurbitacins were implicated with such repellent actions (Haynes and Jones, 1976).

(c) Asparaguic acid

It has been shown that the population of *Trichodorus christiei* around tomato declined more rapidly when intercultured with *Asparagus officinalis* than in fallow soil or in a field containing tomato only. Similar observations were also made for *M. incognita acrita* on *A. officinalis* var. Mary Washington (Crittenden, 1952). The phenomenon was ascribed to a glycoside, secreted in the rhizosphere of this plant, having anticholinesterase activity. This glycoside was later identified as asparaguic acid (1,2-dithiolane-4-carboxylic acid) by Takagusi *et al.* (1975, 1977) and was also found to be toxic to juveniles of *M. hapla*, *Globodera rostochiensis*, *Pratylenchus penetrans* and *P. curvitatis*.

(d) Alkaloids

Margosa root exudates were shown to be highly toxic to the juveniles of *M. incognita* and the toxicity was ascribed to nimbidin and thionemone alkaloids (Alam *et al.*, 1975). These alkaloids have shown similar toxicity (*in vitro*) to various phytoparasitic nematodes (Khan *et al.*, 1974a, b).

The root-knot nematode in peaches was controlled when *Crotalaria spectabilis* was interplanted for a period of two years. It was attributed to the failure of juveniles to infect the peachroots (McBeth and Taylor, 1944). Later it was found that the roots of *C. spectabilis* secrete a toxic principle which adversely affects mobility in juveniles of *M. incognita* (Ochse and Brewton, 1954). This toxic principle was isolated and characterized as monocrotaline, a pyrolizidine ester (Fassuliotis and Skukas, 1969).

(e) Phenolics
Pyrocatechol is a naturally occurring phenolic compound which accumulates in high concentration in the roots of *Eragrostis curvula* and has been shown to possess nematicidal properties against *M. javanica*, *M. hapla*, *M. arenaria thamesi* and *M. incognita acrita* (Scheffer *et al.*, 1962).

(f) Isothiocyanate
Mahler *et al.* (1986) reported that when high and low glucosinate lines of swede rape (*Brassica* sp.) were grown in fields, the root-knot nematode and the fungus pathogen *Aphanomyces enteiches* in peas were significantly inhibited. Preliminary studies indicated that lines even with moderate contents of glucosinolates produced levels of isothiocyanate equivalent to those currently recommended for soil fumigation.

Available data are strongly in favour of the fact that the inhibition of nematode penetration, development and hatching by plant exudates, leachates, decomposition products of plants residues and purified allelochemicals could be used as a tool to control the root-knot nematodes. However, a comprehensive study is a prerequisite for any such venture.

24.3 MECHANISM OF NEMATODE CONTROL

The basic principles involved in the control of nematodes have been studied (Cook and Baker, 1983), but much has to be learned about the mode/mechanism of nematode control. During the past few decades various approaches have been adopted to explain this mechanism, however, any discussion here would be restricted to those aspects involving allelochemicals. Studies have clearly shown that allelochemicals may either affect nematodes directly or may exert their effects through indirect means.

Studies on the mechanism of action of α-terthienyl is probably one of the best examples of investigation in this line. Gommers (1972) showed that terthienyl was strongly photoactivated in near UV light when served as a sensitizer for singlet oxygen (Bakker *et al.*, 1979). Singlet oxygen rapidly oxidizes the amino acids, histidine, methionine, tryptophan and proteins containing these amino acids (Nilsson *et al.*, 1972; Matheson *et al.*, 1975). This results in aerobic inactivation of neurosensory cholinesterase and glucose-6-phosphate dehydrogenase enzymes in nematode (Bakker *et al.*, 1979). But under field conditions it appears to be unrealistic as the quantum of light required to activate terthienyl exuded by marigolds might not reach the roots although plants do pipe light downwards in soil (Gommers and Bakker, 1988). Thus, there might be some mechanism involved other than the photoactivation of α-terthienyl. It is suggested that 1,2-dioxetenes, which remain chemically active in the dark, give rise to carbonyl fragments upon cleavage, an electronically excited triplet state, which generates singlet oxygen that oxidizes the nematode enzymes. Or, the alternative view being forwarded implicates the involvement of indole-3-aldehyde – an electronically excited triplet state – which readily transfers the energy in singlet oxygen and inactivates the nematode enzyme, synthesized due to peroxidase catalysed oxidation of indole-3-acetic acid (Gommers and Bakker, 1988).

The formation of H_2S from fatty acids (present in additives) under anaerobic conditions in flooded soils, and of nitrite/nitrate and NH_3 during the decomposition of nitrogenous organic additives through the activities of proteolytic and deaminating urease enzymes, were suggested to be the cause of the decline of root-knot and other nematodes (Khan *et al.*, 1974a; Mian and Rodriguez-Kabana, 1982a; Mian *et al.*, 1982; Heubner *et al.*, 1983).

The nematoxic chemicals released in soil from organic additives might also inhibit the nematode population development indirectly through host plants. It was reported that root-knot development and soil population of the nematode was more inhibited on susceptible plants grown in soil amended with oil-cake than on those plants grown in unamended soils (Alam *et al.*, 1976b, 1977a, 1978; Sitaramaiah and Singh, 1978). These workers postulated that plants grown in amended soils acquired resistance to root-knot nematodes as a higher concentration of phenolic compounds was estimated in those plants than in plants grown in unamended soils. This assumption appears to be quite plausible as it was shown by Kononova (1966) and Hurst and Burges (1967) that phenols were readily absorbed by roots from the soil. A more conclusive evidence was provided by Alam *et al.* (1980) who had shown that when seedlings of chilli (*Capsicum annuum*) cv. NP-46A, eggplant cv. Pusa Purple Long and tomato cv. Marglobe, susceptible to *M. incognita*, were raised in the potted soils amended with different oil-cakes and then transplanted to autoclaved soil or to acid leached sand, juvenile penetration decreased and fewer galls were formed on the roots compared with seedlings raised in untreated soils.

In addition to the allelochemicals produced by higher plants, allelochemicals of microbial origin have also been implicated in the control of nematodes, and some studies have been made to explain their mode of action. It has been found that toxicity of culture filtrate of

Aspergillus niger, a ubiquitous fungus in soil, was due to oxalic acid secreted by the fungus. The culture filtrates of certain soil-borne fungi, when tested *in vitro*, proved toxic to nematodes (Alam *et al.*, 1973; Kirmani *et al.*, 1978). The embryonic obliteration, alterations in the eggshell make-up and reduced hatching in *M. incognita* and *Globodera pallida* were ascribed to the activities of exopathic diffusible toxic metabolites originating from *Paecilomyces lilacinus* and *Penicillium anatolicum* (Jatala *et al.*, 1985; O'Hara, 1985; Jatala, 1986).

The metabolic product of *Streptomyces overmitilis* has been shown to possess potent, broad spectrum nematicidal activities and is designated as avermictins (Burg *et al.*, 1979; Miller *et al.*, 1979). The movement, invasion and the post-infectional development of *M. incognita* in cucumber roots were adversely affected when the juveniles were exposed to avermectins under soil-free conditions (Stretton *et al.*, 1987).

Thus a wide range of chemicals belonging to various chemical groups – viz. tannins, phenols, fatty acids, organic acids, nitrites, nitrates, ammonia, hydrogen sulphide, etc. – are known to be released into the soils as a result of the decomposition of organic additives. These chemicals in turn have been found to be effective against nematodes (Ahmad *et al.*, 1972; Singh and Sitaramaiah, 1973; Khan *et al.*, 1974a; Mian and Rodriguez-Kabana, 1982b). Most of these chemicals have been also implicated in at least one aspect of allelopathy (Rice, 1984; Chou and Waller, 1989). Though, it is beyond doubt that allelochemicals play a very important role in regulating the population of nematodes, how it actually happens has yet to be seen in detail.

24.4 CONCLUSIONS AND PROSPECTS

Leachates and exudates from several allelopathic plants or their residues have been found to be useful in controlling root-knot nematodes. It is suggested that the use of such plants in inter-cropping or in a rotational sequence should be exploited to control these nematodes. Besides, organic additives could also be a good alternative for the control of nematodes. In these approaches the involvement of allelochemicals cannot be ruled out. Thus, it is imperative that the allelopathic phenomenon should be extensively researched for possible exploitation in the control and management of nematodes.

In view of the health and environmental hazards posed by the pesticides in present use, it is essential to look for alternative methods. Allelochemicals, owing to their biodegradable nature, are supposed to be environmentally clean and therefore are strong competitors as commercial pesticides. Thus, there is a pressing need to screen plants, including microorganisms, for their allelopathic (anti-nematode) properties. Although attempts have been made to screen some members of the family compositae (Gommers, 1973), the available information on the vast vegetational resources is very scanty. Many workers have suggested the mixed cropping of marigolds and margosa as a potent method of controlling nematode as these crops are known to exude allelochemicals through their roots. Thus, allelochemicals have the necessary potential, and if efficacious allelochemicals were discovered, their commercial development could be possible, as has already been done in the development of pesticides of botanical origin (Putnam and Tang, 1986; Waller, 1987). The genetic manipulation technique could also be employed in achieving this goal.

The organic amendments have proved beyond doubt their utility as nematode suppressants, but the dosages employed by different workers have varied greatly (1 to 200 t ha^{-1}). Such a high range of dosages is indicative of the fact that experimental conditions might have been quite variable. Therefore, optimization of the dose is required. Organic additives not only suppress the nematodes, they also act as an organic manure which improves the soil conditions for the current crops and for the crops of the following season.

Thus, besides well-planned experiments to optimize the dosages of organic amendments, isolation and identification of allelochemicals prior to and after microbial degradation, the effects of these allelochemicals on nematodes and host plants, their mode and mechanism of action and their fate in the environment are some of the aspects that need to be studied extensively. An assessment of any residues of allelochemicals in crop plants, and of any possible effects they may have on the consumers of these plants, would also be worth contemplating.

REFERENCES

Ahmad, A. and Khan, A. M. (1964) Factors influencing larval hatching in the root-knot nematodes, *Meloidogyne incognita* (Kofoid & White) Chitwood, 1949. II. Effects of root leachates and certain chemicals. *Indian Phytopathol.*, **17**, 102–9.

Ahmad, R., Khan, A. M. and Saxena, S. K. (1972) Changes resulting from amending the soil with oil-cakes and analysis of oil-cakes (Abstr.). *Proc. 59th Sess., Indian Sci. Cong. Calcutta*, Part III, p. 164.

Alam, M. M. (1986) Possible use of weeds as soil amendment for the management of root-knot and stunt nematodes attacking eggplant. *Agricultural Wastes*, **16**, 97–102.

Alam, M. M., Ahmad, M. and Khan, A. M. (1980) Effect of organic amendments on the growth and chemical composition of tomato, eggplant and chilli and their susceptibility to attack by *Meloidogyne incognita*. *Plant* and *Soil*, **57**, 231–6.

Alam, M. M., Ali, Q. G., Masood, A. and Khan, A. M. (1976a) Studies on the chemical changes induced by infection of root-knot nematode (*Meloidogyne incognita*) in tomato and eggplant, and the stunt nematode (*Tylenchorhynchus brassicae*) in cabbage and cauliflower roots. *Indian J. Exp. Biol.*, **14**, 517–18.

Alam, M. M. and Khan, A. M. (1974) Control of phytonematodes with oil-cakes amendments in spinach fields. *Indian J. Nematol.*, **4**, 239–40.

Alam, M. M., Khan, M. W. and Saxena, S. K. (1973) Inhibitory effect of culture filtrates of some rhizosphere fungi of okra on the mortality and larval hatch of certain plant parasitic nematodes. *Indian J. Nematol.*, **3**, 94–8.

Alam, M. M., Khan, A. M. and Saxena, S. K. (1977a) Effect of bone meal on the population of nematodes infesting certain crops. *Geobios*, **4**, 196–8.

Alam, M. M., Khan, A. M. and Saxena, S. K. (1978) Mechanism of control of plant parasitic nematodes as a result of the application of organic amendments to the soil. IV. Role of formaldehyde and acetone. *Indian J. Nematol.*, **8**, 172–4.

Alam, M. M., Khan, A. M. and Saxena, S. K. (1979) Mechanism of control of plant parasitic nematodes as a result of the application of organic amendments to the soil. V. Role of phenolic compounds. *Indian J. Nematol.*, **9**, 136–42.

Alam, M. M., Kirmani, M. R. and Khan, A. M. (1976b) Studies on the role of root exudates for nematode control by the interculture of mustard and rocket salad with wheat and barley. *Fert. Technol.*, **13**, 289–92.

Alam, M. M., Masood, A. and Husain, S. I. (1975) Effect of margosa and marigold root exudates on mortality and larval hatch of certain nematodes. *Indian J. Exp. Biol.*, **13**, 412–14.

Alam, M. M., Saxena, S. K. and Khan, A. M. (1977b) Influence of interculture of marigold and margosa with some vegetable crops on plant growth and nematode population. *Acta Botanica Indica*, **5**, 33–9.

Alam, M. M., Siddiqi, S. A. and Khan, A. M. (1977c) Mechanism of control of plant parasitic nematodes as a result of the application of organic amendments to the soil. III. Role of phenols and amino acids in host roots. *Indian J. Nematol.*, **7**, 27–31.

Atwal, A. S. and Manger, A. (1969) Repellent action of root exudates of *Sesamum orientale* against the root-knot nematode, *Meloidogyne incognita* (Heteroderidae: Nematoda). *Indian J. Entomol.*, **31**, 286.

Badra, T., Saleh, M. A. and Oteifa, B. A. (1979) Nematicidal activity and composition of some organic fertilizers and amendments. *Rev. Nematol.*, **2**, 29–36.

Baghel, P. P. S. and Gupta, D. C. (1986) Effect of intercropping on root-knot nematode (*Meloidogyne javanica*) infesting grapevine (var. *Perlette*). *Indian J. Nematol.*, **16**, 283–4.

Bakker, J., Gommers, F. J., Nieuwenhuis, I. and Wynberg, H. (1979) Photoactivation of the nematicidal compound α-terthienyl from roots of marigolds (*Tagetes* species). A possible singlet oxygen role. *J. Biol. Chem.*, **254**, 1841–4.

Bano, M., Anver, S., Tiyagi, S. A. and Alam, M. M. (1986b) Evaluation of nematicidal properties of some members of the family compositae. *Int. Nematol. Network Newsl.* **3**, 10.

Bano, M., Tiyagi, S. A., Anver, S. and Alam, M. M. (1986a) Antinemic action of some plants of the family compositae. *Indian J. Bio. Res.*, **2**, 311–17.

Bird, A. F. (1962) The inducement of giant cells by *Meloidogyne javanica*.

Nematologica, **8**, 1–10.

Bleue-Zacheo, T., Zacheo, G., Melillo, M. T. and Lamberti, F. (1982) Ultrastructural aspects of the hypersensitive reaction in tomato root cells resistant to *Meloidogyne incognita*. *Nematol. Medit.*, **10**, 81–90.

Bora, B. C. and Phukan, P. N. (1983) Organic amendments for the control of root-knot in jute. *J. Res. Assam Agric. Univ.*, **4**, 50–4.

Brito, J. A. De and Ferraz, S. (1987) Antagonismo de *Brachiaria decumbens* e *Panicum maximum* cv. Guine a *Meloidogyne javanica*. *Nematol. Brasil.*, **11**, 270–85.

Burg, R. W., Miller, B. M., Baker, E. E. *et al.* (1979) Avermectins, a new family of potent antihelminthic agents: producing organisms and fermentations. *Antimicrob. Agents Chemother.*, **15**, 361–7.

Castillo, M. B. (1985) Some studies on the use of organic soil amendments for nematode control. *Philipp. Agric.*, **68**, 76–93.

Chou, C. H. and Waller, G. R. (eds) (1989) *Phytochemical Ecology: Allelochemicals, Mycotoxins, and Insect Pheromones and Allomones*, Academia Sinica Monograph Ser. No. 9, Acad. Sinica, Taipei, ROC.

Choudhury, B. C. (1981) Root-knot nematode problem on various crop plants in Bangladesh, in *Proc. 3rd IPM Res. Plann. Conf. on Root-Knot Nematodes, Meloidogyne* spp. Region VI, Jakarta, Indonesia, pp. 142–5.

Cook, R. J. and Baker, K. F. (1983) *The Nature and Practice of Biological Control of Plant Pathogens*, Amer. Phytopathol. Soc., St Paul, USA.

Cortado, R. V. and Davide, R. G. (1968) Survey, identification and pathogenicity test of nematodes associated with tobacco in the Cagayan Valley and Ilocas Region. *Philipp. Agric.*, **51**, 779–801.

Crittenden, H. W. (1952) Resistance of asparagus to *Meloidogyne incognita* var. *acrita*. *Phytopathology*, **42**, 6 (Abstr.).

Culbreath, A. K., Rodriguez-Kabana, R. and Morgan-Jones, G. (1985) The use of hemicellulosic waste matter for reduction of phytotoxic effects of chitin and control of root-knot nematodes. *Nematropica*, **15**, 49–75.

Culbreath, A. K., Rodriguez-Kabana, R. and Morgan-Jones, G. (1986) Chitin and *Paecilomyces lilacinus* for control of *Meloidogyne arenaria*. *Nematropica*, **16**, 153–66.

DaCosta, C. P. and Jones, C. M. (1971) Cucumber beetle resistance and mite susceptibility controlled by the bitter gene in *Cucumis sativus* L. *Science*, **172**, 1145–6.

Daulton, R. A. C. and Curtis, R. F. (1964) The effect of *Tagetes* spp. on *Meloidogyne javanica* in Southern Rhodesia. *Nematologica*, **9**, 357–62.

Davide, R. G. and Zorilla, A. (1983) *National Crop Protection Center (NCPC) Leaflet Sr. No. 2*, Los Banos, Laguna, Philippines.

Derrico, F. P. and Maio, F. D. (1980) Effect of some organic materials on root-knot nematodes on tomato in field preliminary experiments. *Nematol. Medit.*, **8**, 107–11.

Desai, M. V., Shah, H. M., Pillai, S. N. and Patel, S. A. (1979) Oil-cakes in control of root-knot nematodes. *Tobacco Res.*, **5**, 105–8.

Devakumar, C., Goswami, B. K. and Mukherjee, S. K. (1985) Nematicidal principles from neem (*Azadirachta indica* A. Juss). Part I. Screening of neem kernel fractions against *Meloidogyne incognita*. *Indian J. Nematol.*, **15**, 121–4.

DiMuro, A. (1975) Effecto di varie piante sui nematodi galligeni del tabacco in rotazioni biennali. *Ann. Ist Sper. Tabacco*, **2**, 93–106.

Fassuliotis, G. (1970) Resistance of *Cucumis* spp. to the root-knot nematode, *Meloidogyne incognita acrita*. *J. Nematol.*, **2**, 174–8.

Fassuliotis, G. (1985) The role of nematologist in the development of resistant cultivars, in *An Advanced Treatise on Meloidogyne*, Vol. I. *Biology and Control* (eds J. N. Sasser and C. C. Carter), A cooperative publication of the Dept. Plant Pathol. and the US Agency for Intern. Dev., North Carolina State University Press, Raleigh, USA, pp. 233–40.

Fassuliotis, G. and Skukas, G. P. (1969) The effect of pyrolizidine alkaloid ester and plants containing pyrolizidine alkaloid ester on *Meloidogyne incognita acrita*. *J. Nematol*, **1**, 287–8.

Fisher, R. F. (1979) Allelopathy, in *Plant Disease* (eds J. G. Horsfall and E. C. Cowling), Vol. IV, Academic Press, New York, pp. 313–30.

Giebel, J. (1982) Mechanism of resistance to plant nematodes. *Ann. Rev. Phytopathol.*, **20**, 257–79.

Gommers, F. J. (1972) Increase of the nematicidal activity of α-terthienyl and related compounds by light. *Nematologica*, **18**, 458–62.

Gommers, F. J. (1973) *Nematicidal Principles in Compositae*, Dissertation, Agricultural Univ. Wageningen, The Netherlands.

Gommers, F. J. (1981) Biochemical interactions between nematodes and plants and their relevance to control. *Helminthol. Abstr. Sr. B. Plant Nematol.*, **50**, 9–24.

Gommers, F. J. and Bakker, J. (1988) Physiological diseases induced by plant responses or products, in *Diseases of Nematodes* (eds G. O. Poinar Jr and H. B. Janson), Vol. II, CRC Press, Florida, pp. 3–22.

Good, J. M., Minton, N. A. and Jaworski, C. A. (1965) Relative susceptibility of selected cover crops and coastal bermudagrass to plant nematodes. *Phytopathology*, **55**, 1026–30.

Goplen, B. P. and Stanford, E. H. (1959) Studies on the nature of resistance in alfalfa to two species of root-knot nematodes. *Agronomy J.*, **51**, 486–8.

Goswami, B. K. and Vijayalakshmi, K. (1986a) Nematicidal properties of some indigenous plant materials against root-knot nematode *Meloidogyne incognita* on tomato. *Indian J. Nematol.*, **16**, 65–8.

Goswami, B. K. and Vijayalakshmi, K. (1986b) Effect of some indigenous plant materials and oil-cake amended soil on the growth of tomato and root-knot nematode population. *Annals Agric. Res.*, **7**, 363–6.

Gowda, D. N. and Setty, K. G. H. (1973) Studies on comparative efficacy of various organic amendments on control of root-knot of tomato. *Mysore J. Agric. Sci.*, **7**, 419–23.

Griffin, G. D. and Elgin Jr, J. H. (1977) Penetration and development of *Meloidogyne hapla* in resistant and susceptible alfalfa under differing temperatures. *J. Nematol.*, **9**, 51–6.

Griffin, G. D. and Waite, W. W. (1971) Attraction of *Ditylenchus dipsaci* and *Meloidogyne hapla* by resistant and susceptible alfalfa seedlings. *J. Nematol.*, **3**, 215–19.

Hackney, R. W. and Dickerson, O. J. (1975) Marigold, castor bean and chrysanthemum as controls of *Meloidogyne incognita* and *Pratylenchus alleni*. *J. Nematol.*, **7**, 84–90.

Hameed, S. F. (1970) Note on the effect of some organic additives on the incidence of root-knot nematodes in tomato (*Lycopersicon esculentum*).

Indian J. Agric. Sci., **40**, 207–10.

Handele, M. J. (1972) Synthese spectra en nematicide activiteit van 1,2-dithienylethenen en 1-fenyl-2-thienylethenen. Dissertation, State Univ., Utrecht, The Netherlands.

Haroon, S. and Smart Jr, G. C. (1983) Effect of Pangola Digitgrass on *Meloidogyne arenaria, M. javanica* and *M. hapla. J. Nematol.*, **15**, 649–50.

Haseeb, A., Alam, M. M., Khan, A. M. and Saxena, S. K. (1978) Nematode population as influenced by soil amendments. *Geobios*, **5**, 152–5.

Haseeb, A. and Alam, M. M. (1984) Use of chopped floral plant parts in suppressing population of plant parasitic nematodes. *Indian J. Plant Pathol.*, **2**, 194–5.

Haseeb, A., Siddiqi, M. A. and Alam, M. M. (1984) Toxicity of latex bearing plants to phytonematodes, in *Environmental and Biotic-Interaction* (eds A. K. Dattagupta, and R. P. Maleyvar), *Proc. IV All India Symp. Environ. Biol.*, pp. 67–71.

Haynes, R. L. and Jones, C. M. (1976) Effects of the *Bi* locus in cucumber on reproduction, attraction and response of the plant to infection by the southern root-knot nematode. *J. Amer. Hortic. Sci.*, **101**, 422–4.

Heubner, R. A., Rodriguez-Kabana, R. and Patterson, R. M. (1983) Hemicellulosic waste and urea for control of plant-parasitic nematodes: effect of soil enzyme activities. *Nematropica*, **13**, 37–54.

Hung, Ch. L. and Rohde, R. A. (1973). Phenol accumulation related to resistance in tomato to infection by root-knot and lesion nematodes. *J. Nematol.*, **5**, 253–8.

Hurst, H. M. and Burges, N. A. (1967) Lignin and humic acids, in *Soil Biochemistry* (eds A. D. McLaren and G. H. Peterson), Marcel Dekker, New York, pp. 260–86.

Ismail, W., Kirmani, M. R., Alam, M. M. and Khan, A. M. (1976) Control of phytophagous nematodes with oil-cake amendments around the roots of different varieties of tomato in field. *Proc. Nat. Acad. Sci., India*, **46B**, 165–8.

Jagdale, G. B., Pawar, A. B. and Darekar, K. S. (1985) Effect of organic amendments on root-knot nematodes infecting betelvine. *Int. Nematol. Network Newsl.*, **2**, 7–10.

Jain, R. K. and Bhatti, D. S. (1988) Effect of degradation of neem leaves on incidence of root-knot nematode in tomato. *Int. Nematol. Network Newsl.*, **5**, 7–9.

Jain, R. K. and Hasan, N. (1984) Toxicity of koo-babool (*Leucaena leucocephala* L.) extracts to *Meloidogyne incognita* and *Helicotylenchus dihystera. Indian J. Nematol.*, **14**, 179–81.

Jain, R. K. and Hasan, N. (1986) Efficacy of 'neem' cake on fodder production, photosynthetic pigments associated with oats and its residual effect on cowpea. *Indian J. Nematol.*, **16**, 98–100.

Jain, U., Datta, S., Trivedi, P. C. and Tiagi, B. (1986) Allelochemic effects of some plants on hatching in *Meloidogyne incognita. Indian J. Nematol.*, **16**, 215–76.

Jatala, P. (1986) Biological control of plant-parasitic nematodes. *Ann. Rev. Phytopathol.*, **24**, 453–89.

Jatala, P. and Russell, C. C. (1972) Nature of sweet potato resistance to *Meloidogyne incognita* and the effects of temperature on parasitism.

J. Nematol., **4**, 1–7.

Jatala, P., Franco, J., Gonzalez, A. and O'Hara, C. M. (1985) Hatching stimulation and inhibition of *Globodera pallida* eggs by enzymatic and exopathic toxic compounds of some biocontrol fungi. *J. Nematol.*, **17**, 501.

Johnson, A. W. (1985) Specific crop rotation effects combined with cultural practices and nematicides, in *An Advanced Treatise on Meloidogyne Vol. I. Biology and Control* (eds J. N. Sasser and C. C. Carter), A cooperative publication of the Dept Plant Pathol. and the US Agency for Int. Dev., North Carolina State Univ. Press, North Carolina, USA, pp. 283–301.

Johnson, L. F., Chambers, A. Y. and Reed, H. E. (1967) Reduction of root-knot of tomatoes with crop residues amendment in field experiments. *Plant Dis. Reptr.*, **51**, 219–22.

Jones, F. G. W. (1956) Soil populations of beet celworm (*Heterodera schachtii* Schm.) in relation to cropping. II. Microplot and field plot results. *Ann. Appl. Biol.*, **44**, 25–56.

Kaplan, D. T. and Davis, E. L. (1987) Mechanisms of plant incompatibility with nematodes, in *Vistas on Nematology: A Commemoration of the 25th Anniversary of the Society of Nematologists* (eds J. A. Veech and D. W. Dickson), Society of Nematologists, Inc, Hyattsville, Maryland, USA, pp. 267–76.

Kaplan, D. T. and Keen, N. T. (1980) Mechanisms conferring plant incompatibility to nematodes. *Rev. Nematol.*, **3**, 123–34.

Khan, A. M. (1969) *Studies on Plant Parasitic Nematodes Associated with Vegetable Crops in Uttar Pradesh*. Final Tech. Report, PL-480 Grant No. FG-In-225, Project No. A7-CR-65, Bot. Dept, Aligarh Muslim Univ., Aligarh, India.

Khan, A. M. (1976) *Control of Diseases Caused by Nematodes by the Application of Oil-Cake Manures*, Final Tech. Report, PL-480 Grant No. FG-In-426, Project No. A7-CR-400, Bot. Dept, Aligarh Muslim Univ., Aligarh, India.

Khan, A. M., Alam, M. M. and Saxena, S. K. (1974a) Mechanism of the control of the plant parasitic nematodes as a result of the application of the oil-cakes to the soil. *Indian J. Nematol.*, **4**, 93–6.

Khan, A. M., Khan, M. W. and Saxena, S. K. (1969) Effect of organic amendments on the population of nematodes and fungi in the rhizosphere of eggplant (*Solanum melongena*). *All India Nematol. Symp.*, IARI, New Delhi, pp. 67–8.

Khan, M. W., Khan, A. M. and Saxena, S. K. (1973) Influence of certain oil-cake amendments on nematodes and fungi in tomato field. *Acta Botanica Indica*, **1**, 49–51.

Khan, M. W., Khan, A. M. and Saxena, S. K. (1974b) Rhizosphere fungi and nematodes of eggplant as influenced by oil-cake amendments. *Indian Phytopath.*, **27**, 480–4.

Khana, M. L. and Nirula, K. K. (1964) Breeding potatoes for resistance to root-knot nematode. *Curr. Sci.*, **33**, 314.

Kirmani, M. R., Alam, M. M., Khan, A. M. and Saxena, S. K. (1975) Effect of different carbon:nitrogen ratios on the population of nematodes and fungi and plant growth of cabbage. *Indian J. Mycol. Plant Pathol.*, **5**, 22.

Kirmani, M. R., Saxena, S. K. and Khan, A. M. (1978) Growth and development of root-knot on eggplant as influenced by some fungi. *Indian J. Nematol.*, **8**, 153–5.

Kononova, M. M. (1966) *Soil Organic Matter its Nature, its Role in Soil Formation and in Soil Fertility*, Pergamon Press, New York.

Lear, B. (1959) Application of castor pomace and cropping of castor beans to soil to reduce nematode populations. *Plant Dis. Reptr.*, **43**, 459–60.

Luc, M. (1968) Nematological problems in the former French African tropical territories and Madagascar, in *Tropical Nematology* (eds G. C. Smart and V. G. Perry), Univ. Florida Press, Gainesville, pp. 93–112.

Mahler, K. A., Auld, D. L. and Davis, J. B. (1986) Potential of high glucosinolate lines for possible biological suppression of soil borne plant pathogens, in *Proc. Crucifer Genetics Workshop*, III, Univ. Guelph, Ontario, Canada.

Mahmood, I., Saxena, S. K. and Zakiuddin (1982) Effect of certain plant extracts on the mortality of *Rotylenchulus reniformis* Linford & Oliveira, and *Meloidogyne incognita* (Kofoid & White) Chitwood. *Bangladesh J. Bot.*, **11**, 154–7.

Mankau, R. and Das, S. (1969) The influence of chitin amendments on *Meloidogyne incognita*. *J. Nematol.*, **1**, 15–16.

Maqbool, M. A., Hashmi, S. and Ghaffar, A. (1987) Effect of latex extracts from *Euphorbia caducifolia* and *Calotropis procera* on root-knot nematode *Meloidogyne incognita* infesting tomato and eggplant. *Pakistan J. Nematol.*, **5**, 43–7.

Matheson, I. B. C., Etheridge, R. D., Kratowich, N. R. and Lee, J. (1975) The quenching of singlet oxygen by amino acids and proteins. *Phytochem. Photobiol.*, **21**, 165–71.

McBeth, C. W. and Taylor, A. L. (1944) Immune and resistant cover crops valuable in root-knot infested peach orchards. *Proc. Amer. Soc. Hort. Sci.*, **45**, 158–66.

McClure, M. A., Ellis, K. C. and Nigh, E. L. (1974a) Resistance of cotton to the root-knot nematode *Meloidogyne incognita*. *J. Nematol.*, **6**, 17–20.

McClure, M. A., Ellis, K. C. and Nigh, E. L. (1974b) Post infection development and histopathology of *Meloidogyne incognita* in resistant cotton. *J. Nematol.*, **6**, 21–6.

Medhane, M. S., Jagdale, G. B., Pawar, A. P. and Darekar, K. S. (1985) Effect of *Tagetes erecta* on root-knot nematodes infecting betelvine. *Int. Nematol. Network Newsl.*, **2**, 11–12.

Merny, G. (1976), Root-knot problems in Ivory Coast, in *Proc. IPM Res. Plann. Conf. on Root-knot Nematodes, Meloidogyne* spp., IITA, Ibadan, Nigeria, pp. 60–5.

Mian, I. H. and Rodriguez-Kabana, R. (1982a) Soil amendments with oil-cakes and chicken litter for control of *Meloidogyne arenaria*. *Nematropica*, **12**, 205–20.

Mian, I. H. and Rodriguez-Kabana, R. (1982b) Survey of nematicidal properties of some organic materials available in Alabama as amendments to soil for control of *Meloidogyne arenaria*. *Nematropica*, **12**, 235–46.

Mian, I. H., Godoy, G., Shelby, R. A. *et al.* (1982) Chitin amendments for control of *Meloidogyne arenaria* in infested soil. *Nematropica*, **12**, 71–84.

Midmore, D. (1983) Shading and its influence in intercropping system with a potato base, in *Training Course on Agronomy for the Potato in the Hot Tropics*, Int. Potato Centre, Lima, Peru, p. 8.

Miller, T. W., Charet, L., Cole, D. J. *et al.* (1979) Avermictins, a new family of potent antihelminthic agents: Isolation and chromatographic properties. *Antimicrob. Agents Chemother.*, **15**, 368–71.

Milne, D. L., Bashoft, D. N. and Buchan, P. W. (1965) The nature of resistance of *Nicotiana rependa* to the root-knot nematode *Meloidogyne javanica. S. Afr. J. Agric. Sci.*, **8**, 557–67.

Minton, N. A. (1962) Factors influencing resistance of cotton to root-knot nematodes (*Meloidogyne* spp.). *Phytopathology*, **52**, 272–9.

Minton, N. A. (1977) Plant nematode relationships of an allelopathic nature, in *Report on the Res. Plann. Conf. on the Role of Secondary Compounds in Plant Interactions (Allelopathy)*, Mississippi State Univ., State College, Mississippi, pp. 17–32.

Muller, R. and Gooch, P. S. (1982) Organic amendments in nematode control – an examination of the literature. *Nematropica*, **12**, 319–26.

Muro, A. D. (1975) Root-knot nematode (*Meloidogyne incognita*) control by two- and three-year crop rotation. *Ann. 1st Sper. Tab. (Rome)*, **2**, 89–91.

Myint, Y. Y. (1981) Country report on root-knot nematode in Burma, in *Proc. 3rd IPM Res. & Plann. Conf. on Root-knot Nematodes, Meloidogyne* spp. *Region* VI, Jakarta, Indonesia, pp. 163–70.

Nandal, S. N. and Bhatti, D. S. (1983) Preliminary screening of some weeds, shrubs for their nematicidal activity against *Meloidogyne javanica. Indian J. Nematol.*, **13**, 123–7.

Nandal, S. N. and Bhatti, D. S. (1987) Effect of some weed and shrub extracts on penetration and gall formation by *Meloidogyne javanica* on brinjal. *Nematol. Medit.*, **15**, 159–62.

Netscher, C. and Taylor, D. P. (1976) *Meloidogyne* research at O.R.S.T.O.M., Dakar, in *Proc. IPM Res. & Plann. Conf. on Root-knot Nematodes, Meloidogyne* spp. IITA, Ibadan, Nigeria, pp. 66–79.

Nilsson, R., Merkel, P. B. and Kearns, D. R. (1972) Unambiguous evidence for the participation of singlet oxygen in photodynamic oxidation of aminoacids. *Photochem. Photobiol.*, **16**, 117–24.

Ochse, J. J. and Brewton, W. S. (1954) Preliminary report on *Crotalaria* versus nematodes. *Proc. Florida State Hort. Soc.*, **67**, 218–19.

O'Hara, C. M. (1985) Effectos de compuestos enzimaticos y exopaticos difusibles de hongos en el control biologico de nematodos. MS Thesis, University of Peru, Cayetano Heredia, Fac. Cienc. Filosof, Lima, Peru.

Osman, A. A. and Viglierchio, D. R. (1988) Efficacy of biologically active agents as non traditional nematicides for *Meloidogyne javanica. Rev. Nematol.*, **11**, 93–8.

Overman, A. J. (1961) Nematodes associated with Pangola grass pastures. *Proc. Flo. State Hort. Soc.*, **74**, 201–4.

Patel, H. R., Thakar, N. A. and Patel, C. C. (1985a) Inhibitory effect of *Cleodendron enermi* on root-knot at Okra. *Madras Agric. J.*, **72**, 470–2.

Patel, H. R., Thakar, N. A. and Patel, C. C. (1985b) Azolla and mustard cake against nematicides for root-knot nematode management in Okra. *Madras Agric. J.*, **72**, 593–4.

Perwez, M. S., Rahman, M. F. and Haider, S. R. (1988) Effect of *Tagetes erecta* on *Meloidogyne javanica* infecting lettuce. *Int. Nematol. Network Newsl.*, **5**, 18–19.

Pi. C. L. and Rohde, R. A. (1967) Phenolic compounds, and host reaction in tomato to injury caused by root-knot and lesion nematodes. *Phytopathol.*, **57**, 344 (Abstr.).

Prasad, S. K. and Dasgupta, D. R. (1969) A preliminary field trial on the effect of *Sesamum* intercropping on the population development of plant parasitic nematodes. *Proc. All India Nematol. Symp.*, IARI, New Delhi, p. 6.

Putnam, A. R. and Tang, C. S. (1986) State of the science, in *The Science of Allelopathy* (eds A. R. Putnam and C. S. Tang), Wiley, New York, pp. 1–19.

Raymundo, S. A. (1983) Major pests of potatoes in hot humid climates and their control. *Training Course on Agronomy for Potato in the Hot Tropics*, International Potato Centre, Lima, Peru, p. 7.

Reynolds, H. W., Carter, W. W. and O'Banon, J. H. (1970) Symptomless resistance of alfalfa to *Meloidogyne incognita acrita*. *J. Nematol.*, **2**, 131–4.

Rice, E. L. (1984) *Allelopathy*, Academic Press, Orlando, Florida.

Rich, J. R. (1986) New trends in managing plant parasitic nematodes – Introductory remarks. *J. Nematol.*, **18**, 121–2.

Rickard, D. A. and Durpee Jr, A. W. (1978) The effectiveness of ten kinds of marigolds and five other treatments for the control of four *Meloidogyne* spp. *J. Nematol.*, **10**, 296–7.

Rodriguez-Kabana, R. (1986) Organic and inorganic amendments to soil as nematode suppressants. *J. Nematol.*, **18**, 129–35.

Rodriguez-Kabana, R., Godoy, G., Morgan-Jones, G. and Shelby, R. A. (1983) The determination of soil chitinase activity. Conditions for assay and ecological studies. *Plant and Soil*, **75**, 95–106.

Rohde, R. A. (1960) Acetylcholinesterase in plant parasitic nematodes and anticholinesterase from *Asparagus*. *Proc. Helminthol. Soc. Washington*, **27**, 121–3.

Rohde, R. A. and Jenkins, W. R. (1958) Basis for resistance of *Asparagus officinalis* var. *altilis* L to the stubby root nematode *Trichodorus christiei* Allen, 1957. *Bull. Maryland Agric. Exp. Stn*, **A-97**, 19.

Rovira, A. D. (1965) Plant root exudates and their influence upon soil microorganisms, in *Ecology of Soil-Borne Plant Pathogens – Prelude to Biological Control* (eds K. F. Baker and W. C. Snyder), Univ. Calif. Press, Berkeley, pp. 179–84.

Salem, F. M. and Osman, G. Y. (1988) Effectiveness of *Tagetes* natural exudates on *Meloidogyne javanica* (Chitwood) nematode. *Anz. Schad. Pflanz. Umwz.*, **61**, 17–19.

Sasser, J. N. (1954) Identification and host–parasite relationships of certain root-knot nematodes (*Meloidogyne* spp.). *Bull.* Maryland *Agric. Exp. Stn*, **A-97**, p. 19.

Sasser, J. N. (1977) Worldwide dissemination and importance of the root-knot nematodes (*Meloidogyne* spp.), *J. Nematol.*, **9**, 26–9.

Sasser, J. N. (1989) *Plant Parasitic Nematodes: The Farmer's Hidden Enemy*, A cooperative publication of the Dept Plant Pathol. and the Consortium for International Crop Protection, p. 115.

Sasser, J. N. and Freckman, D. W. (1987) World perspective on nematology: the role of society, in *Vistas on Nematology: A Commemoration of the 25th Anniversary of the Society of Nematologists* (eds J. A. Veech and D. W.

Dickson), Society of Nematologists Inc., Hyattsville, Maryland, pp. 7–14.

Scheffer, F., Kickuth, R. and Visser, J. H. (1962) Die Wirzelausscheidungen von *Eragrostis curvula* (Schrad.) Nees und ihr einfluss auf wurzel-knoten-Nematoden. *Z. Pflanzen Bodenk.*, **98**, 114–20.

Sharma, C., Trivedi, P. C. and Tiagi, B. (1985) Effect of green manuring on populations of *Meloidogyne incognita* on muskmelon. *In. Nematol. Network Newsl.*, **2**, 7–9.

Sharma, R. D. and Scolari, D. D. G. (1984) Efficacy of green manure and crop rotation in the control of nematodes under savannah conditions. *Nematol. Brasil.*, **8**, 193–218.

Sharma, R. K. and Raj, D. (1987) Effect of nematicides and organic amendments on root-knot nematodes infecting potato. *Int. Nematol. Network Newsl.*, **4**, 8–10.

Siddiqi, M. A. and Alam, M. M. (1987a) Control of plant parasitic nematodes by intercropping with *Tagetes minuta*. *Nematol. Medit.*, **15**, 205–11.

Siddiqi, M. A. and Alam, M. M. (1987b) Utilization of marigold plant wastes for the control of plant parasitic nematodes. *Biological Wastes,* **21**, 221–9.

Siddiqi, M. A. and Alam, M. M. (1988) Control of root-knot and reniform nematodes by bare root dip in leaf extracts of margosa and Persian lilac. *Z. Pflanzen Pflanzensch.*, **95**, 138–42.

Siddiqi, M. A. and Saxena, S. K. (1987), Effect of interculture of margosa and Persian lilac with tomato and eggplant on root-knot and reniform nematodes. *Int. Nematol. Network Newsl.*, **4**, 5–8.

Siddiqi, M. A., Haseeb, A. and Alam, M. M. (1987) Evaluation of nematicidal properties in some latex bearing plants. *Indian J. Nematol.*, **17**, 99–102.

Singh, B. (1969) Observations on the effect of organic soil amendments and fertilizers on incidence of root-knot and yield of okra in nematode infested soil. *Proc. All India Nematol. Symp.*, IARI, New Delhi, pp. 68–9.

Singh, R. S. (1965) Control of root-knot of tomato with organic soil amendments. *F.A.O. Plant Prot. Bull.*, **13**, 35–7.

Singh, B. and Choudhury, B. (1973) The chemical characteristics of tomato cultivars resistant to root-knot nematodes (*Meloidogyne* spp.). *Nematologica*, **19**, 443–8.

Singh, D. B., Reddy, P. P. and Shyamasundar, J. (1985) Histological, histopathological and histochemical investigations on root-knot nematode resistant and susceptible lines of cowpea. *Nematol. Medit.*, **13**, 213–19.

Singh, I., Sharma, S. K. and Singh, P. K. (1980) Efficacy of oil-cakes extract on the hatching of root-knot nematodes, *Meloidogyne incognita*. *Indian J. Mycol. Plant Pathol.*, **9**, 115–16.

Singh, R. S. and Sitaramaiah, K. (1967) Effect of decomposing greenleaves, sawdust and urea on the incidence of root-knot of okra and tomato. *Indian Phytopathol.*, **20**, 349–55.

Singh, R. S. and Sitaramaiah, K. (1969) Control of root-knot through organic and inorganic amendments of soil. I. Effect of oil-cakes and sawdust. *Proc. All India Nematol. Symp.*, IARI, New Delhi, pp. 63–4.

Singh, R. S. and Sitaramaiah, K. (1973) *Control of Plant Parasitic Nematodes with Organic Amendments of Soil.* Res. Bull., G. B. Pant University of Agriculture and Technology, Pantnagar, India.

Sitaramaiah, K. and Singh, R. S. (1978) Effect of organic amendments on phenolic content of soil and plant and response of *Meloidogyne javanica* and its host to related compounds. *Plant* and *Soil*, **50**, 671–9.

Smart, G. C. and Brodie, B. B. (1977) Reaction of Katahdin breeding line M-905-1-1 potato to *Meloidogyne incognita*. *Nematropica*, **7**, 7 (Abstr.).

Smit, J. J. and Bos, W. S. (1976) Root-knot in northern Nigeria, in *Proc. IPM Res. Plann. Conf. on Root-knot Nematodes, Meloidogyne* spp., IITA, Ibadan, Nigeria, pp. 41–3.

Sosa-Moss, C., Barker, K. R. and Daykin, M. E. (1983) Histopathology of selected cultivars of tobacco infected with *Meloidogyne* species. *J. Nematol.*, **15**, 392–7.

Stephan, Z. A. and Al-Askari, A. A. (1989) Effect of *Haplophyllum tuberculatum* plant extract on root-knot nematode. *Int. Nematol. Network Newsl.*, **6**, 31–2.

Stephan, Z. A., Michbas, A. H. and Shakir, I. (1989) Effect of organic amendments, nematicides and solar heating on root-knot nematodes infecting eggplant. *Int. Nematol. Network Newsl.*, **6**, 34–5.

Steiner, G. (1941) Nematodes parasitic on and associated with roots of marigolds (*Tagetes* hybrids). *Proc. Biol. Soc. Washington*, **54**, 31–4.

Stretton, A. O. W., Campbell, W. C. and Babu, J. R. (1987) Biological activity and mode of action of avermictins, in *Vistas on Nematology: A Commemoration of 25th Anniversary of the Society of Nematologists* (eds J. A. Veech and D. W. Dickson), Soc. Nematol., Inc., Hyattsville, Maryland, pp. 136–46.

Suatmadji, R. W. (1969) *Studies on the Effect of* Tagetes *Species on Plant Parasitic Nematodes*, Dissertation, Agricultural Univ., Wageningen.

Takagusi, M., Toda, H., Takagusi, Y. *et al.* (1977) The relation between structure of asparaguic acid-related compounds and their nematicidal activity. *Res. Bull. Hokkaido National Agric. Exp. Stn*, No. 118, 105–11.

Takagusi, M., Yachida, Y., Anetai, M. *et al.* (1975) Identification of asparaguic acid as a nematicide occuring naturally in the roots of *Asparagus*. *Chemical Lett.*, **43**, 44.

Tanda, A. S., Atwal, A. S. and Bajaj, Y. P. S. (1988) Antagonism of sesame to the root-knot nematode (*Meloidogyne incognita*) on okra in tissue culture. *Nematologica*, **34**, 78–87.

Taylor, A. L. and Sasser, J. N. (1978) *Biology, Identification and Control of Root-knot Nematodes (Meloidogyne species)*. Coop. Pub. Dept Plant Path., North Carolina State University and US Agency Int. Dev., Raleigh, NC.

Tiyagi, S. A., Mukhtar, J. and Alam, M. M. (1985) Preliminary studies on the nematicidal nature of two plants of the family compositae. *Int. Nematol. Network Newsl.*, **2**, 19–21.

Tiyagi, S. A., Anver, S., Bano, M. and Siddiqi, M. A. (1986a) Feasibility of growing zinia as a mix crop along with tomato for control of root-knot and reniform nematodes. *Int. Nematol. Network Newsl.*, **3**, 6–7.

Tiyagi, S. A., Siddiqi, M. A. and Alam, M. M. (1986b) Toxicity of an insect repellent plant to plant-parasitic nematodes. *Int. Nematol. Network Newsl.*, **3**, 16–17.

Thomason, I. J. (1987) Challenges facing nematology, in *Vistas on Nematology: A Commemoration of 25th Anniversary of the Society of Nematologists* (eds J. A. Veech and D. W. Dickson), Soc. Nematol., Inc., Hyattsville, Maryland, pp. 469–76.

Thomason, I. J. and Caswell, E. P. (1987) Principles of nematode control, in *Principles and Practice of Nematode Control in Crops* (eds R. H. Brown and B. R. Kerry), Academic Press, Australia, pp. 87–130.

Thompson, A. C. (ed.) (1985) *The Chemistry of Allelopathy Biochemical Interactions Among Plants*, ACS Symp. Ser. 268, Amer. Chem. Soc., Washington, DC.

Trivedi, P. C., Bhatnagar, A. and Tiagi, B. (1978) Control of nematodes on *Capsicum annuum* by application of oil-cakes. *Indian Phytopathol.*, **31**, 75–6.

Uhlenbroek, J. H. and Bijloo, J. D. (1958) Investigations on nematicides. I. Isolation and structure of nematicidal principle occurring in *Tagetes* roots. *Recueil Trav. Chim. des Pays-Bas*, **77**, 1004–8.

Uhlenbroek, J. H. and Bijloo, J. D. (1959) Investigations on nematicides. II. Structure of a second nematicidal principle isolated from *Tagetes* roots. *Recueil Trav. Chim. des Pays-Bas*, **78**, 382–90.

Uhlenbroek, J. H. and Bijloo, J. D. (1960) Investigations on nematicides. III. Polythienyls and related compounds. *Recueil Trav. Chim. des Pays-Bas*, **79**, 1181–96.

Veech, J. A. (1981) Plant resistance to nematodes, in *Plant Parasitic Nematodes* (eds B. M. Zuckerman and R. A. Rohde), Vol. III, Academic Press, New York, pp. 377–403.

Verma, R. R. (1986) Efficacy of organic amendments against *Meloidogyne incognita* infesting tomato. *Indian J. Nematol.*, **16**, 105–6.

Waller, G. R. (ed.) (1987) *Allelochemicals: Role in Agriculture and Forestry*, ACS Symp. Ser. 330. Amer. Chem. Soc., Washington, DC.

Widjaja, W. and Windrich, R. (1974) Studies on nematological problems in horticulture, in *Agricultural Cooperation Project, Indonesia–The Netherlands*, Res. Reports, Section II, Tech. Contribution Bogar, pp. 393–402.

Winchester, J. A. and Hayslip, N. C. (1960) The effect of land management practices on the root-knot nematode *Meloidogyne incognita acrita* in South Florida. *Proc. Fl. State Hort. Soc.*, **73**, 100–4.

Winoto, R. S. (1969) *Studies on the Effect of Tagetes Species on Plant Parasitic Nematodes*, Veenman and Zonen, Wageningen.

Yadav, B. S. (1970) Tests for the nematicidal properties of some used plants on *Meloidogyne incognita*. *Proc. Indian Sci. Congr. Assoc.*, **57**, 551.

Yuhara, I. (1971) Effect of soil treatment with organic powder on the population of *Meloidogyne hapla* attacking sugarbeet. *Bull. Sugarbeet Res.*, **13**(suppl.), 201–5.

CHAPTER 25

Exploitation of allelochemicals in improving crop productivity

S. J. H. Rizvi and V. Rizvi

25.1 INTRODUCTION

During the recent past, the importance of allelopathy in nature as well as in agroecosystems has attracted worldwide attention. In the beginning, allelopathic studies were limited to the effect of one plant on the other, without much knowledge about the chemicals responsible for such effects. Later, active involvement of scientists belonging to various disciplines, made allelopathy a multidisciplinary subject. Consequently, several allelochemicals which were at the helm of the entire allelopathic affair were isolated, identified and their effects were studied not only on germination, growth or metabolism but also at the molecular level. These developments slowly transformed allelopathic research from basic to applied, and ideas were floated to exploit allelopathy/allelochemicals in various ways in agriculture and forestry (Rice, 1984; Thompson, 1985; Putnam and Tang, 1986; Waller, 1987). We started working with allelochemicals during 1976 and remained concerned mainly with exploring various aspects of allelopathy that may lead to increased crop production. In this chapter we shall discuss some of our work on the potential of allelochemicals in pest control, crop rotation and also discuss the implications of allelopathy in agroforestry, a relatively younger aspect of allelopathic research.

25.2 POTENTIAL OF ALLELOCHEMICALS IN PEST CONTROL

Plants synthesize allelochemicals which possess a wide range of biological activities leading to diverse types of interactions with plants and microorganisms. These interactions include largely a negative effect on germination, growth, development, distribution and behaviour of other organisms. Such interactions may thus lead to recognition and utilization (i.e. synthesis, chemically directed selection, gene insertion) of new

Allelopathy: Basic and applied aspects Edited by S. J. H. Rizvi and V. Rizvi
Published in 1992 by Chapman & Hall, London ISBN 0 412 39400 6

molecules designed to attack a particular plant or organism from a plant protection viewpoint. Attempts to develop pest-resistant lines of cultivars by employing analyses to determine whether correlations between the allelochemicals and resistance are statistically significant, has only recently been made on a substantive scale. Studies have shown that allelochemicals often impart plant resistance to insects, nematodes, pathogens and, following their release into the environment, regulate distribution, growth and development of plants including weeds (Kubo and Nakanishi, 1977; Putnam and DeFrank, 1983; Rice, 1984; Einhellig, 1984; Hedin, 1986; Wink, 1987; Cooper-Driver and Le Quesne, 1987).

A consideration of the above facts suggests an immediate utility for the management of pests. Several groups are exploring these possibilities through the identification of cultivars and accessions with high allelopathic potential, to be exploited through crop rotation, mix-cropping, intercropping and mulching. Alternatively, allelochemicals responsible for inducing resistance or inhibiting growth and development of other plants and organisms can be isolated, identified and used as natural pesticides. Both the above approaches have several advantages compared with other approaches to pest control, chiefly because of environmental and health considerations. We would, however, in this chapter, deal with our work related to the potential use of allelochemicals in pest control.

25.2.1 Allelochemicals as herbicides

To isolate and identify allelochemicals with herbicidal activity, several plants were screened for their inhibitory activity against the test weed *Amaranthus spinosus*. For this, water-soluble residues of crude ethanolic extracts from leaves and seeds were first assayed. Results showed that out of 120 samples screened, the extract from seeds of *Coffea arabica* exhibited strong inhibitory activity (Rizvi *et al.*, 1980a). Therefore, its active principle was isolated through chromatographic techniques and was identified as 1,3,7-trimethylxanthine (1,3,7-T, an alkaloid) by spectroscopic studies (Rizvi *et al.*, 1981). 1,3,7-T was shown completely to inhibit germination of the test weed at a minimum inhibitory concentration (MIC) of 1200 ppm. Similar treatment of seeds of *Vigna mungo*, a crop in which the weed is a problem, however, showed no effect. The alkaloid also inhibited seven other noxious weeds at various concentrations (Table 25.1).

25.2.1 Possible mode of action of 1,3,7-T

In order to study the selective mode of action of the active principle, certain parameters were studied in *A. spinosus* and *V. mungo*.

TABLE 25.1 *Weedicidal spectrum of 1,3,7-trimethylxanthine isolated from seeds of* Coffea arabica *(after Rizvi* et al., *1981)*

	% Inhibition of seed germination					
	Concentration (ppm)					
Weeds tested	1200	1600	2000	2500	5000	10 000
Amaranthus spinosus	100	100	100	100	100	100
Avena fatua	3.2	20.1	40.1	58.0	88.0	100
Echinochloa colonum	42.6	81.3	100	100	100	100
Echinochloa crusgalli	52.2	73.0	91.2	100	100	100
Lathyrus aphaca	–	–	–	0.0	31.5	100
Lathyrus sp.	–	–	–	3.0	40.0	100
Vicia sativa	–	–	2.0	12.5	42.0	100
Vicia hirsuta	–	–	4.0	15.0	100	100

(a) Studies with *A. spinosus*

The germination of seed largely depends on enzymatic hydrolysis of stored products like protein and starch in cotyledons. The seeds of *A. spinosus* contain about 3.0% protein and 8.0% total carbohydrate. As the carbohydrate content of the seed is higher than the protein, its hydrolysis and solubilization during seed germination will be crucial. The effect of 1,3,7-T on germination could, therefore, result from an effect on the activity of starch hydrolysing enzyme, amylase. Seeds of *A. spinosus* treated with 1000 ppm of 1,3,7-T showed a marked decline in amylase activity during germination. The extent of enzyme inhibition varied with the duration of germination, and was found to be 30% (maximum) after

TABLE 25.2 *Effect of 1000 ppm of 1,3,7-trimethylxanthine on amylase activity of seedlings of* Amaranthus spinosus *at different age (after Rizvi* et al., *1987)*

Age (hours)	% Inhibition
18	1.0
27	8.0
36	10.0
45	20.0
54	24.0
63	30.0
72	28.0

TABLE 25.3 *Amylase activity in cell free extracts obtained from germinating seeds of* Amaranthus spinosus *treated with different concentrations of 1,3,7-trimethylxanthine (after Rizvi* et al., *1987)*

Concentration (ppm)	Enzyme activity (mg maltose $h^{-1} g^{-1}$ f.w.)
0	256.5 ± 1.5
300	257.3 ± 2.0
600	250.2 ± 1.8
900	253.6 ± 1.3
1200	255.1 ± 1.7
1500	258.7 ± 1.2

63 hours (Table 25.2). On the other hand, treatment of seeds of *V. mungo* with 1200 ppm of 1,3,7-T showed no effect on amylase activity.

The observed inhibition of amylase activity could largely be due to: (i) inhibited catalytic property of the enzyme; (ii) inhibited synthesis of the enzyme; or (iii) both of these causes. The effect on catalytic property can be determined from studies on such kinetic parameters as substrate saturation and the Km value of the enzyme. The Km value (0.45%) as derived from double reciprocal plot, however, remained unaffected, showing no effect on the catalytic property of the enzyme (Rizvi *et al.*, 1987). *In vitro* treatment of the enzyme, extracted from untreated germinating seeds, similarly had no effect at any of the concentrations (300, 600, 900, 1200 and 1500 ppm) used (Table 25.3), nor was any effect evident when the incubation time was varied (Table 25.4).

TABLE 25.4 *Amylase activity in cell free extracts obtained from untreated germinating seeds of* Amaranthus spinosus *when incubated for different periods in the presence of 1000 ppm of 1,3,7-trimethylxanthine (after Rizvi* et al., *1987)*

Incubation period (minutes)	Enzyme activity (mg maltose $h^{-1} g^{-1}$ f.w.)
0	251.2 ± 1.4
10	250.8 ± 1.8
20	254.5 ± 1.3
30	248.3 ± 2.2
40	250.4 ± 2.1
50	252.3 ± 1.9
60	249.5 ± 2.0

TABLE 25.5 *Amylase activity in germinating seeds of* Amaranthus spinosus *treated with 1,3,7-trimethylxanthine (1000 ppm) in the presence of different concentrations of GA_3 (after Rizvi* et al.*, 1987)*

Concentration (μM)	% Inhibition
0	30.0 ± 2.0
2	31.2 ± 1.7
5	30.6 ± 1.6
10	31.3 ± 2.0
15	29.8 ± 1.9
20	30.2 ± 1.5

It, therefore, is clear from the above results that 1,3,7-T most likely affected the synthesis of amylase. The synthesis of this enzyme during germination is regulated by gibberellic acid (GA_3). Thus, one of the possible causes of reduced synthesis might be a reduced production of GA_3. In this context, therefore, experiments with GA_3 treatment were performed. While all the concentrations (0.5, 1.0, 1.5, 3.0, 3.5 μM) of GA_3 (in the absence of 1,3,7-T) caused an appropriate increase in amylase activity, it could not overcome the inhibitory effect of 1,3,7-T even at concentrations much higher than the saturating one. When 1,3,7-T (1000 ppm) treated seeds were simultaneously treated with various concentrations (2.0, 5.0, 10.0, 15.0, 20.0 μM) of GA_3, amylase activity decreased by nearly 30% at all the concentrations (Table 25.5).

(b) Studies with *V. mungo*

The selectivity of 1,3,7-T was established on the basis of its differential effects on seed germination of the test weed and of *V. mungo*. Its possible utility as a selective herbicide, however, would depend on how it affects the crop plant in its later life. Thus, its effects on growth, nitrate reductase activity and some aspects of nitrogen metabolism were examined (because *V. mungo* is a protein rich seed) at MIC for the test weed.

(c) Biochemical studies

A direct correlation between germination and amylase activity indicated that the germination might at least be partly inhibited through the inhibition of amylase activity. Such an inhibition of the enzyme activity could be expected to occur in more than one way. GA_3 is known to regulate *de novo* amylase production during seed germination (Chrispeels and Varner, 1976). However, the fact that the loss in enzyme activity could not be effectively overcome by the addition of GA_3, rules out the

TABLE 25.6 *Water-soluble sugar and starch content in the seedlings of* Amaranthus spinosus *treated with 1000 ppm of 1,3,7-trimethylxanthine (after Rizvi* et al., *1987)*

Hours after germination	Sugar content ($mg\,g^{-1}$ d.w.)		Starch content ($mg\,g^{-1}$ d.w.)	
	Control	Treated	Control	Treated
18	15.10 ± 0.7	14.80 ± 1.1	66.04 ± 0.5	67.24 ± 0.6
27	17.10 ± 0.3	16.31 ± 0.8	65.12 ± 0.6	66.60 ± 0.5
36	21.78 ± 0.8	20.65 ± 0.5	60.01 ± 1.0	63.52 ± 1.0
45	23.50 ± 0.9	21.48 ± 0.3	50.51 ± 1.1	54.52 ± 1.0
54	29.24 ± 0.8	24.00 ± 0.7	35.10 ± 0.6	41.02 ± 0.3
63	33.82 ± 1.2	27.33 ± 0.3	27.23 ± 0.9	34.10 ± 0.8
72	35.50 ± 0.6	29.50 ± 0.8	24.12 ± 0.8	29.20 ± 0.2

TABLE 25.7 *Ethanol-soluble and ethanol-insoluble nitrogen in the seedlings of* Amaranthus spinosus *treated with 1000 ppm of 1,3,7-trimethylxanthine (after Rizvi* et al., *1987)*

Hours after germination	Soluble nitrogen ($mg\,g^{-1}$ d.w.)		Insoluble nitrogen ($mg\,g^{-1}$ d.w.)	
	Control	Treated	Control	Treated
18	0.71 ± 0.010	0.70 ± 0.001	4.80 ± 0.011	4.87 ± 0.002
27	0.73 ± 0.001	0.71 ± 0.010	4.63 ± 0.010	4.73 ± 0.007
36	0.75 ± 0.010	0.72 ± 0.010	4.44 ± 0.021	4.56 ± 0.008
45	0.92 ± 0.003	0.87 ± 0.080	3.40 ± 0.040	3.62 ± 0.001
54	1.22 ± 0.006	1.10 ± 0.010	2.31 ± 0.022	2.60 ± 0.010
63	1.40 ± 0.010	1.25 ± 0.090	1.60 ± 0.031	1.85 ± 0.013
72	1.50 ± 0.020	1.30 ± 0.040	1.40 ± 0.021	1.65 ± 0.012

possibility of the effect being GA_3 mediated. Alternatively, the alkaloid could affect either the catalytic properties of the enzyme or its biosynthesis, or both. No variation was observed in such catalytic parameters as substrate saturation and apparent Km value in the enzyme extracts obtained from the germinating seeds. Thus, by the process of elimination, it might be concluded that the enzyme inhibition, most likely, was caused by an effect on its biosynthesis.

Consequently, as revealed by a general biochemical analyses, the starch/sugar ratio increased in the seedlings grown from treated seeds (Table 25.6). The biochemical analysis also showed a higher ratio for insoluble/soluble nitrogen (Table 25.7). Possibly, this might have been caused through an adverse effect on the activity or level of proteases. The retarded germination, besides being mainly a consequence of the reduced amylase level, could also result, at least partly, due to a retarded synthesis of proteases. Here it is relevant to note that nucleic acid metabolism and protein synthesis studied in bacteria and animals are known to be variously affected by different alkaloids (Grollman, 1967; Mahler and Baylor, 1967; Perlman and Penman, 1970; Robinson, 1974; Jimnez *et al.*, 1975). The reports that 1,3,7-T binds to a part of an operon and inhibits nucleic acid synthesis (Papas *et al.*, 1973; Robinson, 1974) is of more direct relevance to the present study.

The utility of 1,3,7-T as a weedkiller would depend on its general non-toxic effects on crop plants; such studies as its effects on NR activity and nitrogen content are therefore important. Nitrate reductase is important for nitrogen metabolism as its activity has been shown to be directly correlated with the yield of crops (Jonson *et al.*, 1976). The observed lack of toxicity with respect to NR activity is thus of considerable significance (Rizvi *et al.*, 1987). The fact that the compound has no effect on various nitrogen fractions (Table 25.8) firmly established its (at the concentration used) non-toxicity with respect to *V. mungo*. The non-toxicity of 1,3,7-T to *V. mungo* was in agreement with our earlier studies (Rizvi *et al.*, 1980a, b, d; Rizvi and Rizvi, 1984).

Currently, we are engaged in isolation and identification of promising natural herbicides from *Cassia sericea* and lemongrass. Apart from the isolation of allelochemicals, we are also screening several identified allelochemicals for their herbicidal activities. The results obtained with some of these are summarized in Table 25.9. Several other groups are also engaged in such research (Putnam and Tang, 1986; Hedin, 1986; Waller, 1987) and we look forward to an interesting breakthrough in this direction.

It would not be out of place to mention here that even if 1,3,7-T or other similar products may not be used as natural herbicides, work on such compounds would provide novel biomolecules and fresher insights into phytotoxicity. Alterations to the molecular structure of such compounds could then be initiated to modify or increase their toxicity, selectivity and other desired characteristics. Otherwise to discover as well more cost-effective, efficacious, selective and environmentally safe herbicides, new strategies have to be formulated because of the increasing diminishing returns with traditional herbicide syntheses and screening methodologies.

25.2.2 Allelochemicals as fungicides

Many plants produce chemicals either prior to or after infection by certain pathogens which render the plants resistant against diseases (Rice, 1984). Evenari (1949) pointed out that the seeds of *Brassica oleracea*

TABLE 25.8 *Ethanol-soluble nitrogen, ethanol-insoluble nitrogen and protein content in the primary leaves of plants of* Vigna mungo *raised from seeds treated with 1200 ppm of 1,3,7-trimethylxanthine and of those from untreated seeds (after Rizvi* et al., *1987)*

	Soluble nitrogen (mg g^{-1} d.w.)		Insoluble nitrogen (mg g^{-1} d.w.)		Protein (mg g^{-1} d.w.)	
Days after germination	Control	Treated	Control	Treated	Control	Treated
6	0.58 ± 0.009	0.59 ± 0.010	2.50 ± 0.20	2.56 ± 0.08	15.63 ± 0.30	16.00 ± 0.23
10	1.52 ± 0.018	1.54 ± 0.009	6.40 ± 0.09	6.31 ± 0.07	40.00 ± 0.70	39.44 ± 0.51
14	2.72 ± 0.023	2.69 ± 0.101	13.20 ± 0.07	13.32 ± 0.06	83.06 ± 0.50	83.25 ± 0.62
18	3.92 ± 0.110	3.84 ± 0.110	13.76 ± 0.10	13.43 ± 0.03	86.00 ± 0.40	83.94 ± 0.42

TABLE 25.9 *Germination, radicle and plumule length as affected by various concentrations of allelochemicals in* Amaranthus spinosus

Allelochemical	Concentration (mM)	Germination (%)	Radicle length (cm)	Plumule length (cm)
Control	–	75 ± 2	2.1 ± 0.33	2.0 ± 0.11
Citral	1.0	25 ± 2	2.3 ± 0.13	1.9 ± 0.22
	2.0	17 ± 1	0.5 ± 0.08	1.5 ± 0.11
	3.0	0	0	0
Citronellol	1.0	71 ± 1	2.6 ± 0.08	2.2 ± 0.09
	2.0	58 ± 2	1.8 ± 0.11	1.8 ± 0.08
	3.0	47 ± 3	1.6 ± 0.09	1.9 ± 0.10
3,5-Dihydroxybenzoic acid	1.0	75 ± 2	2.28 ± 0.10	2.41 ± 0.10
	2.0	79 ± 2	2.0 ± 0.08	3.2 ± 0.13
	3.0	79 ± 3	2.3 ± 0.13	3.0 ± 0.14
2,4-Dihydroxybenzoic acid	1.0	75 ± 2	3.2 ± 0.11	2.6 ± 0.09
	2.0	60 ± 3	1.71 ± 0.13	2.9 ± 0.13
	3.0	55 ± 2	2.0 ± 0.09	2.1 ± 0.14
Geraniol	1.0	74 ± 1	1.95 ± 0.09	1.18 ± 0.05
	2.0	76 ± 2	0.8 ± 0.12	1.1 ± 0.11
	3.0	71 ± 1	0	1.2 ± 0.12
Geranylacetate	1.0	75 ± 2	2.9 ± 0.11	2.6 ± 0.13
	2.0	77 ± 1	3.2 ± 0.13	2.8 ± 0.12
	3.0	74 ± 1	3.1 ± 0.14	2.5 ± 0.15
Terpenene	1.0	62 ± 3	2.2 ± 0.11	2.6 ± 0.16
	2.0	20 ± 2	1.7 ± 0.21	2.2 ± 0.13
	3.0	9 ± 2	1.5 ± 0.16	2.1 ± 0.14

TABLE 25.10 *Antifungal properties of 1,3,7-trimethylxanthine (after Rizvi* et al., *1980c)*

Properties	Mycelial inhibition (%)
Lethal dose (ppm)	
500	52.0
1000	83.0
1500	100
Effect of temperature (°C)	
40	100
60	100
80	100
100	100
120	69.8
140	52.1
Effect of autoclaving (at 1 bar for 15 min)	45.2
Fungicidal/fungistatic at MIC	
Treated	100
Reinoculated	100

contain a microbial inhibitor belonging to mustard oil and that the resistance of crucifers to clubroot disease caused by *Plasmodiophora brassicae* is attributed to such oils. One of the best-known examples of the protective role of allelochemicals is that of protocatechuic acid, and catechol in onion, against infection of *Colletotrichum circinans* (Farkas and Kiraly, 1962). These water-soluble phenolics diffuse out from the dead cell layers of the scales and inhibit spore germination and/or hyphal penetration of the pathogen (Rice, 1984). It is interesting to note that many of the same allelochemicals that have been implicated in other phases of allelopathy are important in protecting plants against diseases. If allelochemicals are able to induce disease resistance in plants, exploration of their fungitoxic activities deserves exploitation. However, this aspect of allelopathy, as compared with others, has remained largely neglected. Therefore, fungitoxicity of chemicals, implicated in other aspects of allelopathy, was tested against a number of plant pathogenic fungi. The antifungal activity of 1,3,7-T was evaluated *in vitro* against *Helminthosporium maydis*, a parasitic pathogen of maize. It inhibited the mycelial growth of *H. maydis* at a MIC of 1500 ppm and proved to be fungicidal in nature. The antifungal properties of 1,3,7-T are summarized

TABLE 25.11 *Fungitoxic spectrum of 1,3,7-trimethylxanthine (after Rizvi* et al., *1980c)*

	Mycelial inhibition (%)		
Test fungi	1000	2000	4000
Alternaria solani	51.8	91.0	100
A. tenuis	82.0	100	100
Aspergillus niger	52.5	92.0	100
A. flavus	20.2	67.5	100
A. wentii	82.5	100	100
Cladosporium herbarum	35.5	72.6	100
C. sphaerospermum	31.2	65.2	100
Helminthosporium oryzae	89.3	100	100
H. tercicum	90.5	100	100
Syncephalastrum racemosum	40.0	78.2	100

TABLE 25.12 *Fungitoxic spectrum of citronellol and geraniol (at MIC)*

	Concentration (ppm)	
Test fungi	Citronellol	Geraniol
Aspergillus niger	500	300
Claviceps purpurea	400	400
Cladosporium sphaerospermum	400	350
Helminthosporium oryzae	950	600
H. tercicum	400	400
H. maydis	450	600
Pyracularia oryzae	400	350

in Table 25.10. The compound also showed a broad range of toxicity and was found to be potent against all the fungi tested (Table 25.11). Similarly, other allelochemicals have also been found to be potent against other plant pathogenic fungi (Rizvi and Rizvi, 1987; Rizvi *et al.*, 1988). Interestingly, many of the allelochemicals tested have shown a broad fungitoxic spectrum (Table 25.12). These results strongly support the idea of exploitation of allelochemicals as natural fungicides. A comprehensive study to this effect may yield some useful results.

25.2.3 Allelochemicals as multipurpose pesticides in integrated pest management

An integrated pest management system (IPMS) depends upon integrating various approaches to pest control in such a way that the benefits of the different possible combinations are maximized while the disadvantages are minimized. Despite many attempts to develop IPMS, control by synthetic chemicals is still the dominant technique. Increasing global concern about the indiscriminate use, overdependence and hazards of synthetic pesticides has prompted exploration of natural plant products, since these could be expected to be comparatively safe (Fawcett and Spencer, 1970; Mathur *et al.*, 1982; Rizvi and Rizvi, 1987; Rizvi *et al.*, 1987). Any IPMS requires simultaneous use of different chemicals and there are indications that although some of the synthetic pesticides may be harmless when used singly, they may become poisonous after interacting with other pesticides (Samersov and Prishchepa, 1978; Ramakrishna and Ramachandran, 1978; Verma *et al.*, 1984). The cost of applying many pesticides in any IPMS is also of concern to plant protectionists. Accordingly, we propose that the number of chemicals used in any IPMS should be reduced to minimize the chances of synergistic toxicity as well as to cut costs. Surprisingly, while much emphasis has been given to the integration of various strategies of pest control, little effort appears to have been made to develop multipurpose pesticides capable of controlling more than one agricultural pest.

The fact that plants can affect the vegetation and microflora of their habitat by producing allelochemicals (Rice, 1984) prompted us to explore the possibility of using allelochemicals as multipurpose pesticides.

Various allelochemicals were tested for their herbicidal and fungicidal activities in terms of the inhibition of germination and growth of the test weed (*A. spinosus*) and inhibition of mycelial growth of the test fungus (*Alternaria solani*), respectively. The non-target toxicity of the selected allelochemicals was established on the basis of its effect on tomato plants (both the weed and the fungus are common pests of tomato).

Among the allelochemicals tested, citral, citronellol and geraniol showed considerable herbicidal activity. However, geraniol was the most effective: it completely inhibited germination, radicle and plumule growth of *A. spinosus* at a concentration of 3 mM. At 2.8 mM it also completely inhibited the mycelial growth of *A. solani* (Rizvi *et al.*, 1988). The probit analysis revealed a LD_{90} of 1.65 and 2.23 mM for *A. solani* and *A. spinosus*, respectively (Table 25.13). Thus geraniol proved to be the most potent of the allelochemicals tested.

The possible utility of geraniol in plant protection, however, depends on how it affects the host crop. Hence, its effects on seed germination and growth of tomato were studied. None of the parameters considered was adversely influenced, thus geraniol had the necessary selectivity (Table 25.14).

The terpenes are well known for their allelopathic properties (Muller and del Moral, 1966; Duke *et al.*, 1988). However, little or no attempt has been made to exploit their allelopathic potential in plant protection. Total inhibition of seed germination of *A. spinosus* and that of mycelial

TABLE 25.13 *LD_{90} of allelochemicals for* A. spinosus *and* A. solani *(after Rizvi* et al., *1988)*

	LD_{90} (mM)	
Allelochemical	*A. spinosus*	*A. solani*
Citral	5.01 ± 0.12	2.34 ± 0.00
Citronellol	5.01 ± 0.16	2.63 ± 0.08
Geraniol	2.23 ± 0.08	1.65 ± 0.09

TABLE 25.14 *Effect of geraniol on seed germination, seedling growth and total biomass production of* Lycopersicon esculentum

Treatment (mM)	Germination (%)	Radicle length (cm)	Plumule length (cm)	Biomass production (mg)
0.0	73.66 ± 2.65	3.08 ± 0.83	2.93 ± 0.03	9.2 ± 0.012
2.5	74.33 ± 1.98	3.05 ± 0.53	3.08 ± 0.18	10.2 ± 1.110

growth of *A. solani* by geraniol at very low concentration, together with its non-toxicity for the crop plant, is very promising. The non-toxicity of geraniol for the crop plant, in this instance, concurs with earlier reports on plant products that have been stated to be safer owing to their readily biodegradable nature (Beye, 1978).

Citral, citronellol and geraniol are structurally very closely related. However, they significantly differ in their pest-controlling properties. A comparison of the molecular structures revealed that geraniol differs from citral only in having an alcoholic group instead of an aldehydic one, and from citronellol in possessing an extra double bond at the second carbon position. This leads to a tentative conclusion that the strong pesticidal activity of geraniol may be attributed to both of these structural differences.

25.2.4 Natural vs synthetic pesticides

A critical appraisal of the relative merits of the natural and synthetic pesticides would not be out of place here. Understandably, not much work has been done to compare the natural and synthetic pesticides. However, extrapolation of the available observations made with respect to pesticides and natural compounds can be useful in assessing their relative merits. During the past few decades the extensive use of pesticides (synthetic) has been a cause of concern, for most of them have been shown to be hazardous due to long persistence, non-target toxicity, teratogenic, carcinogenic, pollutive, mutagenic and residual toxicity (Epstein *et al.*, 1967; Matsunaka and Kwatsuka, 1975; Wild, 1975; Paul and Vadlamudi, 1976; Beniwal and Chaubey, 1976; Javorska, 1978; Raza and Bano, 1980; Smith and Stratton, 1986; Mowbray, 1986; Westcott *et al.*, 1987; Duke, 1988; Duke *et al.*, 1987; Ali *et al.*, 1988; Lydon and Duke, 1989). By contrast the botanicals (allelochemicals), with a few exceptions, have been shown to be less toxic and environmentally safer owing to their biodegradable nature. Further, the application of synthetic chemicals involves a 'source to sink' system, i.e. a system with little or no scope for recovery of the used pesticides. This results in depletion of resources. On the other hand, use of natural products should lead to conservation of resources, for they are likely to be recycled through nature. Plants, thus, may constitute a never-ending reservoir of such biochemicals and with their potential biodegradability would nullify the problems raised by synthetic chemicals.

25.3 ALLELOCHEMICALS IN CROP ROTATION

25.3.1 The concept

In modern agriculture, mixed cropping or crop rotation has been identified as a tool to increase the total yield of an agricultural field. Lockhart and Wiseman (1970) have defined that 'a rotation is a cropping system in which two or more crops are grown in a fixed sequence'. A number of factors, like maintenance of soil fertility, soil structure, plant nutrient, choice of suitable crops, etc., are considered while developing a crop rotation system. However, the literature is not very rich in reports on consideration of allelopathy as a factor in developing a rotation system.

Allelochemicals may get assimilated in soil either by direct exudation from living plant tissues or by decomposition of plant residues. Experimental evidence is substantially in favour of the fact that allelochemicals may either be inhibitory (autotoxic/phytotoxic) or stimulatory to the following crops (Rice, 1984; Waller, 1987). Börner (1959) reported that German nurserymen encountered problems in replanting apple trees in soils which were earlier used for apple seedling cultivation even for 1–2 years. After a series of experiments, several allelochemicals – viz. phlorozin, phloretin, phloroglucinol, *p*-hydroxyhydrocinnamic acid and *p*-hydroxybenzoic acid – were isolated from the roots of apple. It was concluded that the apple replant problem may be primarily due to the allelopathic effects of these compounds. Similar problems were also observed while replanting peach trees following the removal of old orchards. Patrick *et al.* (1964), after long experimentation, concluded that an allelochemical, amygdalin, produced by the peach roots, appears to be the main factor in peach replant problem. In annual crops also cases of autotoxicity are well known. Kimbler (1973) found autotoxicity in wheat and reported that wheat straw residues depressed the yield of subsequent wheat crop. Similar observations were made by Leon (1976), Bhowmik and Doll (1982) and Waller *et al.* (1983, 1987) in the case of sorghum, maize, coffee and wheat, respectively. The phytotoxicity of allelochemicals is also well documented. The phytotoxic effects of *Brassica campestris* on *Vigna radiata* (Kuo *et al.*, 1981), *Polygonum aviculare* on *Cynodon dactylon* (L.) (Alsaadawi and Rice, 1982; Alsadawi *et al.*, 1983), *Sorghum halepense* (L.) on *Amaranthus retroflexus* L. (Mikulas, 1984), *Glycine max* on *Triticum aestivum* (L.) (Huber and Abney, 1986) and of *Chrysanthemum morifolium* on *Coleus blumen* (Kil and Lee, 1987) are some of the common examples.

Unfortunately, negligible work has been done on the beneficial aspects of allelopathy. Several legumes have been shown to increase maize growth (Dzyubenno and Petrenko, 1971). Rakhteenko *et al.* (1973) have seen that exudates of pea roots stimulate photosynthesis and absorption of P^{32} in barley. Increase in growth or yield of rice, corn, tomato, cauliflower and cucumber by a supplement of *Medicago sativa* leaves have been demonstrated by Ries *et al.* (1978) and Ries and Houtz (1983). Hydrophobic root exudate of sorghum has been found active in stimulation of witch-weed parasite (Netzly *et al.*, 1988).

The above examples clearly demonstrate the importance of harmful (autotoxic/phytotoxic) and beneficial aspects of allelopathy. Therefore, experimentation is desirable to evaluate the allelopathic compatibility of crop plants while formulating any rotational sequence of crops.

25.3.2 Experimental support

To test this view, we selected a tobacco–maize cropping system. In the northern part of India chewing tobacco is normally grown in rotation with maize. However, sometimes rice is also planted after tobacco. Every harvested crop of chewing tobacco adds approximately 5.0 kg nicotine ha^{-1}. This appears to be a fairly high quantity which can significantly affect the next crop of tobacco, maize or rice. Therefore, the effect of nicotine on these crops was studied.

Treating tobacco seeds with 2.5 and 5.0 mM of nicotine did not exert any adverse effect on radicle growth but 2.5, 5.0, 7.5 and 10.0 mM of nicotine adversely affected the germination. The maximum inhibition was caused by 10 mM which was about 80, 90 and 36% for germination, germination relative index (GRI) and radicle length, respectively (Table 25.15). In rice, germination, GRI, radicle length and seedling vigour (SV) were adversely affected by all concentrations of nicotine used, but there was no significant effect on the length of the plumule (Table 25.15). In

TABLE 25.15 *Effect of different concentrations of nicotine on* Nicotiana tabacum, Oryza sativa *and* Zea mays *(after Mishra, 1986, and Rizvi* et al.*, 1989a)*

Test plant	Treatment (mM)	Germination (%)	GRI[a]	Radicle length (cm)	Plumule length (cm)	Seedling vigour
N. tabacum	0.0	68.6	136.6	0.53	–	36.4
	2.5	54.0	95.3	0.96	–	51.8
	5.0	42.6	73.6	0.60	–	25.5
	7.5	20.0	28.0	0.48	–	9.6
	10.5	14.0	15.0	0.34	–	4.7
SE	–	3.08	8.81	0.04	–	1.20
CD at 5%	–	9.71	27.76	0.12	–	4.31
O. sativa	0.0	84.7	155.7	5.0	1.8	422.5
	2.5	80.7	144.0	5.0	2.1	442.8
	5.0	75.3	130.3	4.0	2.2	298.3
	7.5	73.3	114.3	3.0	1.8	217.0
	10.0	45.3	60.7	0.1	2.4	41.8
SE	–	1.90	3.50	0.27	0.18	1.70
CD at 5%	–	6.01	11.04	0.76	0.53	5.88
Z. mays	0.0	95.5	27.7	3.4	1.5	323.0
	2.5	92.1	26.3	4.9	1.9	450.8
	5.0	100.0	29.7	5.8	2.0	580.0
	7.5	95.3	27.0	6.7	2.8	630.5
	10.5	73.3	20.0	4.5	1.9	329.8
SE	–	3.29	0.73	0.13	0.11	2.11
CD at 5%	–	10.39	2.30	0.37	0.37	5.51

[a] GRI = Germination relative index.

maize, however, all the above parameters were stimulated by nicotine and a maximum induction was caused by 7.5 mM (Table 25.15). On the basis of these results, 5.0 and 7.5 mM concentrations were selected for further studies. Plants grown from treated and untreated seeds showed a similar growth pattern, whereas the height of the plants, specific leaf weight and chlorophyll content were significantly increased in treated plants (Tables 25.16 and 25.17). The number of leaves per plant was equal in both the treated and control plants except during the initial phase when it was higher in the treated plants (Table 25.18). In the above studies, nicotine showed a selective nature in its effects on tobacco, rice and maize. Several other workers have also reported the selective behaviour of various allelochemicals or plant products (Talwar *et al.*, 1983; Rizvi and Rizvi, 1987; Rizvi *et al.*, 1987; Rizvi *et al.*, 1989b). Interestingly, a similarity was found in the results of laboratory experiments with nicotine and in field observations. The analyses of data

TABLE 25.16 *Mean height of plants (cm) raised from seeds treated with different concentrations of nicotine and those grown from untreated seeds of* Zea mays *(var. Diara composite) (after Rizvi* et al., *1989a)*

Treatment (mM)	Age (week)								
	1	2	3	4	5	6	7	8	Mean
0.0	4.4	6.0	9.3	17.2	25.6	41.3	81.4	92.3	34.3
5.0	5.6	6.5	10.1	15.3	23.7	38.7	72.7	96.5	29.0
7.5	6.0	6.7	10.6	20.6	28.3	46.8	100.3	109.3	38.8
Mean	5.28	6.19	9.83	17.53	24.03	37.37	76.77	95.27	

		CD at 5%
SE (nicotine)	Mean = 1.46	4.10
SE (week)	Mean = 2.39	6.72
SE (interaction)	Mean = 4.16	11.64

TABLE 25.17 *Specific leaf weight and total chlorophyll content of leaves of plants raised from seeds treated with different concentrations of nicotine and those grown from untreated seeds of* Zea mays *(after Rizvi* et al., *1989a)*

Treatment (mM)	Specific leaf weight	Total chlorophyll (mg g^{-1})
0.0	4.7	3.5
5.0	5.5[a]	3.7[a]
7.5	5.4[a]	4.3[a]

[a] Significant at 5% level.

TABLE 25.18 *Total number of leaves per plant raised from seeds treated with different concentrations of nicotine and those grown from untreated seeds of* Zea mays *(var. Diara composite) (after Rizvi* et al., *1989a)*

Treatment (mM)	Age (weeks)								
	1	2	3	4	5	6	7	8	Mean
0.0	3	5	7	8	10	12	12	13	9
5.0	3	5	7	9	10	12	13	14	9
7.5	4	6	7	8	10	12	13	14	9

		CD at 5%
SE (nicotine)	Mean = 0.16	–
SE (week)	Mean = 0.26	0.73
SE (interaction)	Mean = 0.45	–

obtained from long-term experiments on tobacco–maize rotation always showed a higher yield of maize grown after tobacco. Thus, the selective allelopathic nature of nicotine may play an important role in crop rotation involving tobacco. But careful considerations of the concentration and its persistence in soil under the action of microorganisms, would be a prerequisite before reaching any definite conclusion. Nevertheless, the peach replant and citrus decline problems due to an accumulation of allelochemicals, amygdaline and homovanillic acid respectively, in soil (see Rice, 1984), emphasize the potentialities of these findings for exploitation in a crop rotation system.

25.4 ALLELOCHEMICALS IN AGROFORESTRY

The term 'agroforestry' is used for a suitable land management system which increases the yield of a particular piece of land; it combines the production of crops (including tree crops) and forest plants and/or animals, simultaneously or sequentially on the same unit of land and applies management and cultural practices that are compatible with the practices of the local population (King and Chandler, 1978). The agroforestry system has been designed to fulfil the economical, social and cultural needs of the local population, keeping an eye on the ecological balance. Trees which are recommended for an agroforestry system ought to have some specific characters, e.g. fast growth rate, good forage, food and wood value and morphology, which should permit the penetration of light to the ground, etc.

It has been assumed that viability of the tree–herb association is governed by physical factors, viz. status of light, moisture, nutrients, etc. Nevertheless, recent research has provided convincing evidence that, apart from the physical factors, the naturally occurring phenomenon of allelopathy may also play a crucial role in determining the success of tree–crop/fodder plant association. One of the aspects of agroforestry is mixed cropping of trees and crops. Thus, the possibilities are fair that allelochemicals produced by the intercropped trees will affect the companion crops.

Since the idea of agroforestry was floated, a lot of work has been done (Baker, 1966; Singh and Bawa, 1982; Huxley, 1983; Mathur *et al.*, 1984; Chingape, 1985; Brewbaker, 1989). However, the possible allelopathic interaction among crop plant and trees has not been given the attention it deserves.

Among the trees recommended for plantation under an agroforestry programme, multipurpose, nitrogen-fixing trees are seen as ideal as they contribute to the overall productivity and help to stabilize and enrich the soil. *Leucaena leucocephala* is among some of the most popular trees belonging to above class. As it is multipurpose, *Leucaena* is known as 'miracle tree' in many regions. It has been designated as a perfect tree because the farmer can feed its products to livestock, harvest it for fuel wood or charcoal and use it for house building and can also spread it as green manure in fields or on plants to enrich the soil (Brewbaker, 1989).

25.4.1 Mimosine allelopathy – a case study

In spite of all the above uses, large-scale plantation of *L. leucocephala* is threatened by the presence of a non-protein amino acid mimosine (α-amino-3-hydroxy-4-oxo-1-pyridine propanoic acid) in its leaves and seeds. The air-dried leaves of various species of *Leucaena* have been found to contain 2.5–5.75% mimosine (Kuo *et al.*, 1983). This is a sufficient amount to affect the germination and growth of plants.

Kuo *et al.* (1983) found that only a few understorey species grow under the *Leucaena* canopy but a substantial number of *Leucaena* seedlings were able to grow. Experiments with the light, moisture content, texture, pH and nutrient status of the soil ruled out the possibility that competition for these factors was the major cause for the phenomenon of the exclusion of vegetation below the *Leucaena* canopy. After a series of experiments, they concluded that mimosine might be playing a major role through its allelopathic effect. They further demonstrated that mimosine causes significant inhibition of the radicle growth of lettuce and rice at 10 and 20 ppm, respectively. Seed germination and radicle growth of *Acacia confusa* and *Casuarina glauca* were significantly inhibited by mimosine at 50 ppm. The radicle growth of seedlings of *Alnus formosana*, *Liquidambar formosana* and *Pinus taiwanensis* were also significantly inhibited.

Aqueous extracts of air-dried leaves of six *Leucaena* varieties also showed strong phytotoxicity for the test plants. Among them, K-29 was the most toxic, followed by K-28, S_1, K-72, K-18 and K-67. The radicle growth of *Lactuca sativa*, *Lolium multiflorum*, *A. confusa*, *C. glauca*, *A. formosana* and *L. formosana* were significantly suppressed by a 1% aqueous extract of K-28 leaves but those of *P. taiwanensis* and *Miscanthus floridulus* were not inhibited. Seed germination and radicle growth of *L. leucocephala* itself were not retarded by either mimosine or the leaf extracts. The bioassay results had a correlation with those of field experiments showing that seedlings of *Leucaena* grow normally underneath mature plants of the species (Kuo *et al.*, 1983).

Thus, the available findings indicate that the allelopathic interactions in tree–crop association have more bearing on crop production under an integrated land use system, i.e. agroforestry, than on agriculture alone. However, most of these inferences have been drawn by practising agroforestry, but the research efforts from scientific means are limited.

L. leucocephala, due to its several merits, is considered an ideal tree for agroforestry (Brewbaker, 1989). Consequently, large-scale plantations have been done in various countries including India. In fact, in certain

TABLE 25.19 *Effect of 1.0 mM mimosine on per cent increase/decrease of germination, radicle and plumule length of various crop plants (after Rizvi* et al., *1990)*

Test plant	Germination (%)	Radicle length (cm)	Plumule length (cm)
Abelmoschus esculentus	+13	−42[a]	−38[a]
Brassica campestris	−42[a]	–	+12
Oryza sativa	0	−6	0
Phaseolus aureus	−3	−86[a]	−85[a]
Vigna mungo	−14	−83[a]	−47[a]

[a] Significant at 5% level.

instances *Leucaena* plantations have increased the total productivity of a particular field when grown in association with crop plants. However, the presence of mimosine has been a persistent cause of concern. There are reports that mimosine causes hair loss, induces cataract disease in domestic animals and affects their growth (ter-Meulen and El-Harith, 1985), but there are only a few reports regarding the allelopathic effects of mimosine on crop plants.

(a) Inhibition of seed germination and seedling growth of various crops by mimosine

To study the possible allelopathic effects of mimosine on crop plants, representative plants belonging to pulse, vegetable, cereal and oilseed crops were selected. The effect of an 1.0 mM aqueous solution of mimosine was studied employing a seed germination bioassay (SGB), as described by Rizvi *et al.* (1981). The introduction of mimosine adversely affected all the parameters studied in almost all the test crops except seed germination in *Abelmoschus esculentus*, plumule length in *Brassica campestris* and germination and plumule length in *Oryza sativa*, where either there was no effect or only a slight induction was observed; however, the promotion was not significant (Table 25.19). On the basis of preliminary screening, *Vigna mungo* (L.) Hepper was selected for the detailed study as it is widely used as one of the pulses in the Indian subcontinent.

(b) Detailed studies of the effect of mimosine on *V. mungo* (L.) Hepper

When the seeds of *V. mungo* were treated with different concentrations of mimosine, all the concentrations except 5.0 mM were found inhibitory to seed germination. The inhibition increased with an increase in the concentration. Similarly, the radicle and plumule length, weight of embryonic axis and seedling vigour were also adversely affected (Table 25.20). Besides these, the morphology of root and root hairs was changed and the tip of the roots turned brownish. The dry weight of the radicle

TABLE 25.20 *Effect of different concentrations of mimosine on per cent inhibition of germination, radicle and plumule length and fresh weight of embryonic axis of* Vigna mungo, *at 72 hours age (after Rizvi* et al., *1990)*

Concentration (mM)	Germination	Radicle length	Plumule length	Embryonic axis (f.w.)	Seedling vigour[a]
0.5	2	28[b]	9	6	23
1.0	4	76[b]	47[b]	42[b]	75[b]
1.5	14[b]	83[b]	88[b]	77[b]	86[b]
2.0	18[b]	83[b]	100[b]	89[b]	87[b]
3.0	23[b]	86[b]	100[b]	87[b]	90[b]
4.0	38[b]	86[b]	100[b]	89[b]	92[b]

[a] Seedling vigour (SV) = % Germination × Radicle length.
[b] Significant at 5% level.

and plumule of the seedlings grown from the seeds treated with 1.0 mM mimosine were also significantly decreased. The dry weight of the radicle was worst affected as the inhibition was found to be as high as 79% (Table 25.21). However, the dry weight of the cotyledons of the seedlings treated with mimosine was always higher in comparison to the cotyledons of untreated seedlings. This indicated that mimosine treatment may possibly affect the mobilization of the stored food from cotyledons to the embryonic axis. This possibility was substantiated by the fact that the food mobilization efficiency of the treated seedlings was found to be 48% less in comparison to the untreated seedlings (Table 25.21).

The above findings clearly demonstrated that mimosine strongly inhibited the seed germination and early seedling growth of *V. mungo*. In order to understand its mode of action it was felt necessary to discover how mimosine inhibits these processes. For this, some analyses were performed and it was found that the level of protein in the cotyledons of the treated seedlings was always higher in comparison to the seedlings grown from untreated seeds. On the other hand, the level of total amino acids was always lower in the embryonic axes of the seedlings grown from treated seeds. Possibly, this might have been caused through an adverse effect on the activity/level of proteases, the protein hydrolysing enzymes. In addition to the alteration of the hydrolysis of protein, seeds treated with 1.0 mM mimosine also showed a marked decline in amylase activity during germination. Substantiating the results of reduced amylase activity, starch level of the treated seedlings was higher than that of the control while the level of sugar in the treated seedlings was always lower (Table 25.22). A direct correlation between all the parameters studied clearly indicated that germination and seedling growth were inhibited through the inhibition of amylase and solubilization of the stored protein in the cotyledons.

Here it would be relevant to note that, though studied in animals and

TABLE 25.21 *Cotyledon, radicle and plumule dry weight and food mobilization efficiency of* Vigna mungo *as affected by 1.0 mM of mimosine, at 72 hours of age (after Rizvi* et al., *1990)*

Parameter studied	% Inhibition/induction
Cotyledon dry weight	+66
Radicle dry weight	−79
Plumule dry weight	−35
Food mobilization efficiency[a]	−48

[a] Food mobilization efficiency (ME)

$$ME = 100 \times \frac{\text{Increase in dry weight of axis}}{\text{Decrease in dry weight of cotyledons}}$$

TABLE 25.22 *Effect of 1.0 mM mimosine on per cent increase/decrease of amylase activity and various biochemical constituents of cotyledon and embryonic axis of seedlings of* Vigna mungo, *grown from seeds treated with mimosine (after Rizvi* et al., *1990)*

	Age (hours)		
Parameter studied	24	48	72
Protein (C)[a]	+21	+29	+25
Amino acid (EA)[b]	−15	−22	−19
Total starch (C)	+19	+25	+20
Soluble sugars (C)	−17	−22	−17
Amylase activity (C)	−25	−31	−22

[a] C = estimated in cotyledons.
[b] EA = estimated in embryonic axis.

bacteria, mimosine inhibits the enzymes tyrosine decarboxylase from animal tissue, aspartate glutamate transaminase of pig heart and affects the syntheses of DNA, RNA and protein in *Paramoecium tetraurellia* (see Bell, 1972).

Further, many allelochemicals in general have also been found to affect various enzymes and metabolic processes in higher plants (Rice, 1984; Einhellig, 1984; Thompson, 1985; Waller, 1987, Rizvi *et al.*, 1987, 1989b). The retarded germination/growth, besides being a consequence of the reduced amylase/protease level, could also result, at least partly, through an effect on the hormonal balance in the plant, for many allelochemicals have been found adversely to affect the level of growth hormones in plants (Green and Corcoran, 1975; Zimmerman *et al.*, 1977; Rice, 1984).

The results presented here thus clearly demonstrate that the allelopathic compatibility of trees with food crops should be carefully studied before being recommended for an agroforestry programme. Based on the above findings, it can be concluded that, in any agroforestry programme, consideration of allelopathic interactions could prove to be beneficial in more ways than one; for example:

(i) the best suited companion crop/variety could be selected;
(ii) production could be increased either by minimizing losses due to negative allelopathic effects or by exploiting the positive allelopathic interactions among the trees and the companion crops; and
(iii) any reduction in loss or increase in production due to allelopathic interactions could be made without increasing the farm input.

Although comprehensive *in vivo* trials are needed to extrapolate the laboratory results to the field conditions, the results presented here establish the fact that such studies can be of great help in exploiting the naturally occurring phenomenon of allelopathy in improving total production of agroforestry systems.

25.4.2 Mimosine and animal nutrition

Apart from the allelopathic effects of mimosine and other tree allelochemicals on crop plants, their presence should also be viewed seriously with respect to their use in animal nutrition. It is a well-established fact that *Leucaena* forage is a good source of nutrient, roughage and protein. The crude protein, which is normally over 25% of the dry matter, has a high nutritional quality and its amino acid contents are comparable to those contained in soybeans. Also, *Leucaena* leaf meal is a good source of β-carotene, vitamin K, calcium and phosphorus (see ter-Meulen and El-Harith, 1985). In recent decades *Leucaena* green meal has been incorporated in the feeds of several farm animals which has often led to adverse effects on the performance of livestock. ter-Meulen and El-Harith (1985) have summarized several such results which include hair loss, retarded growth, fertility problems development of goitre and general adverse effects on health which may eventually lead to mortality.

Mimosine is known to inhibit a number of biochemical reactions. It acts as a tyrosine antagonist and competes with tyrosine to inhibit the activity of tyrosinase. Supplementary tyrosine added to the diet of rats has been reported to compensate the growth inhibition caused by mimosine. Many other reports, however, have indicated that dietary supplementation with tyrosine was not successful in preventing the adverse effects of mimosine in experimental and farm animals (ter-Meulen and El-Harith, 1985). Thus, the mechanism of mimosine toxicity is complex and multifactorial. Though not studied with respect to humans, adverse effects similar to those observed in other mammals may be expected.

In agroforestry, *Leucaena* and crop plants are grown together. Thus, crop plants are exposed to mimosine released from living/dead tissue of *Leucaena*. This mimosine is bound to enter the food chain through food/forage crops and affect their consumers. It has often been suggested that, in the soil, mimosine is degraded into simpler and less toxic chemicals leading to the elimination of toxic effects. However, detailed studies of the degradation of mimosine have presented an entirely different story. It has been found that mimosine is degraded to 3,4-dihydroxypyridine (DHP), which is a potent goitrogen (Bray, 1986) equally as toxic as mimosine itself. Several partly successful methods to reduce the deleterious effects of mimosine are on record. These include the development of new hybrids of *L. leucocephala* with lower mimosine content, the chemical treatment of *Leucaena* meal, and the washing or soaking of leaves in water. Besides these methods, ter-Meulen and El-Harith (1985) suggest that the maximum mimosine-tolerance capacity of various domestic animals, in terms of grams per kilogram of weight, would also be worth following. Recent developments have also shown that ruminants which lack the appropriate stomach enzyme to digest mimosine do it with the help of certain bacteria, now commercially available in Australia (Brewbaker, 1989). Thus, the problem of mimosine toxicity from an animal nutrition viewpoint is coming closer to the solution; but the study of allelopathic effects of mimosine on crop plants is in its infancy and needs a thorough investigation.

25.5 CONCLUSIONS AND PROSPECTS

The prospects of work on the various topics addressed in this chapter have been discussed in each section. It could be said that the results presented here clearly demonstrate the potential of allelochemicals for pest control, integrated pest management, crop rotation and agroforestry systems. It is a well-established fact that most of the plant products, blessed with some biological activity, are better than the currently used commercial pesticides owing to both their biodegradable nature and to health and environmental considerations. Being plant products, most of the allelochemicals are also in an advantageous position and, consequently, are strong contenders against synthetic pesticides. Work in this line is in progress at several laboratories and we should soon expect an array of new generation natural pesticides in the arsenal of plant protectionists. Also, the novel concept of developing multipurpose allelochemical pesticides and their potential for integrated pest management is worth developing for further exploitation.

Recent investigations on the possible role of allelochemicals in crop rotation have revealed that allelochemicals produced by a predecessor crop may either favour a preceeding crop or adversely affect it. Thus, production could be enhanced simply by avoiding inhibitory effects or by exploiting favourable interactions. Both of these can be achieved through a selection of crop plants based on their allelopathic compatibility. Such considerations would definitely lead to increased agricultural production without increasing farm inputs.

The introduction of tree species in agroforestry systems, without evaluating their allelopathic effects on crop/fodder plant, should be a cause for concern. It has been well demonstrated that, besides having deleterious effects on crops, tree allelochemicals enter the food chain and may also affect humans and livestock; therefore, a closer investigation of this problem remains unavoidable.

Thus research in these areas needs to be further strengthened. It is very encouraging that an increasing number of groups are conducting imaginative experiments to reveal the mysteries of allelopathy. Research conducted so far has begun to yield fruit any many such researches have found commercial application in several directions. However, to ensure a rich harvest from allelopathic research, it is essential to bring together the talents of several scientific disciplines like botany, ecology, plant chemistry, biochemistry, physiology, agronomy, soil science, plant breeding, genetics, microbiology, pestology, forestry and biotechnology. In conclusion it could be said that the almost untapped phenomenon of allelopathy has much to offer for the welfare of mankind, and Rice (1985) has rightly pointed out that: 'The horizons for future researches in allelopathy are unlimited and applications of the knowledge gained will be phenomenal.'

REFERENCES

Ali, A., Nigg, H. N., Stamper, J. H. *et al.* (1988) Diflubenzuron application to citrus and its impact on invertebrates in an adjacent pond. *Bull. Environ. Contam. Toxicol.*, **41**, 781–90.

Alsaadawi, S. I. and Rice, E. L. (1982) Allelopathic effects of *Polygonum aviculare* L. I. Vegetational patterning. *J. Chem. Ecol.*, **8**, 993–1009.

Alsaadawi, S. I., Rice, E. L. and Karns, T. K. B. (1983) Allelopathic effects of *Polygonum aviculare* L. III. Isolation, characterization, and biological activities of phytotoxins other than phenols. *J. Chem. Ecol.*, **9**, 761–74.

Baker, H. G. (1966) Volatile growth inhibitors produced by *Eucalyptus globulus*. *Madrono S. Francisco*, **18**, 207–10.

Bell, E. A. (1972) Toxic amino acids in the leguminosae, in *Phytochemical Ecology* (ed. J. B . Harborne.), Academic Press, London and New York, pp. 163–77.

Beniwal, S. P. S. and Chaubey, S. N. (1976) Thiram, a fungicide, effective against *Xanthomonas citri*. *Pesticides*, **10**, 31–6.

Beye, F. (1978) Insecticides from the vegetable kingdom. *Plant Res. Dev.*, **7**, 13–31.

Bhowmik, P. C. and Doll, J. D. (1982) Corn and soybean response to allelopathic effects of weed and crop residues. *Agron. J.*, **74**, 601–6.

Borner, H. (1959) The apple replant problem. I. The excretion of phlorizin from apple root residues. *Contrib. Boyce Thompson Inst.*, **20**, 39–56.

Bray, R. A. (1986) Leucaena in northern Australia – a review. *Forest Ecol. Manage.*, **16**, 345–54.

Brewbaker, J. L. (1989) Can there be such a thing as a perfect tree? *Agroforestry Today*, **1**, 4–7.

Chingape, T. M. (1985) Early growth of *Eucalyptus camaldulensis* under agroforestry conditions at Matiga, Morogora, Tanzania. *Forest Ecol. Manage.*, **11**, 241–4.

Chrispeels, M. J. and Varner, J. E. (1976) Gibberellic acid enhanced synthesis and release of α-amylase and ribonuclease by isolated barley aleuron layers. *Plant Physiol.*, **42**, 398–486.

Cooper-Driver, G. A. and Le Quesne, P. W. (1987) Diterpenoids as insect antifeedants and growth inhibitors: Role in *Solidago* species, in *Allelochemicals, Role in Agriculture and Forestry* (ed G. R. Waller), ACS Symp. Ser. 330, Amer. Chem. Soc., Washington, DC, pp. 534–50.

Duke, O. Stephen (1988) Glyphosate, in *Herbicides, Chemistry, Degradation, and Mode of Action* (eds P. C. Kearney and D. D. Kaufman), Marcel Dekker, New York and Basel, pp. 1–69.

Duke, O. Stephen, Paul, R. N. and Lee, S. M. (1988) Terpenoids from the genus *Artemisia* as potential pesticides, in *Biologically Active Natural Products, Potential Use in Agriculture* (ed. H. G. Cutler), pp. 318–34.

Duke, O. Stephen, Vaughn, K. C., Croom, E. M. and Elsohly, H. N. (1987) Artemisinin, a constituent of annual wormwood (*Artemisia annua*), is a selective phytotoxin. *Weed Sci.*, **35**, 499–505.

Dzyubenno, N. N. and Petrenko, N. I. (1971) On biochemical interaction of cultivated plants and weeds, in *Physiological-Biochemical Basis of Plant Interactions in Phytocenoses* (ed. A. M. Grodzinsky), Naukova Dumka, Kiev, pp. 16–19.

Einhellig, F. A. (1984) Allelopathy – Natural protection, allelochemicals, in *Hand Book of Natural Pesticide: I. Theory, Practice and Detection* (ed. H. B. Mandava), CRC Press, Florida, pp. 161–200.

Epstein, S. S., Andreae, J., Jaffe, H. *et al.* (1967) Carcinogenicity of the herbicide, maleic hydrazide. *Nature*, **215**, 1388–90.

Evenari, M. (1949) Germination inhibitors. *Bot. Rev.*, **15**, 153–94.

Farkas, G. L. and Kiraly, Z. (1962) Role of phenolic compounds in the physiology of plant diseases and disease resistance. *Phytopathol. Z.*, **44**, 105–50.

Fawcett, C. H. and Spencer, D. M. (1970) Plant chemotherapy with natural products. *Ann. Rev. Phytopath.*, **8**, 403–18.

Green, F. B. and Corcoran, M. R. (1975) Inhibitory action of five tannins on growth induced by several gibberellins. *Plant Physiol.*, **56**, 801–6.

Grollman, A. P. (1967) Structural basis for the inhibition of protein biosynthesis, mode of action of tubulosine. *Science*, **157**, 84–5.

Hedin, P. A. (1986) Developing research trends in the chemistry of plant resistance to pests, in *Natural Resistance of Plants to Pests, Role of Allelochemicals* (eds M. D. Green and P. A. Hedin), ACS Symp. Ser. 296. Amer. Chem. Soc., Washington, DC, pp. 1–14.

Huber, D. M. and Abney, T. S. (1986) Soybean allelopathy and subsequent cropping. *J. Agron. Crop Sci.*, **157**, 73–8.

Huxley, P. A. (1983) *Plant Research and Agroforestry*, International Council for Research in Agroforestry, Nairobi, Kenya.

Javorska, T. (1978) Effect of combined herbicides on the occurrence of morphoses in the spikes of spring barley. *Agrochemia*, **18**, 37–42.

Jimnez, J., Sanchez, L. and Vasquez, D. (1975) Location of resistance to the alkoids, Narciclasine in the 60S ribosomal subunit. *FEBS Lett.*, **55**, 53–6.

Jonson, C. B., Whittington, W. J. and Blackwood, G. C. (1976) Nitrate reductase activity, a possible and predictive test for ultimate crop yield. *Nature*, **261**, 133.

Kil, Bong-Seop and Lee, S. Y. (1987) Allelopathic effects of *Chrysanthemum morifolium* on germination and growth of several herbaceous plants. *J. Chem. Ecol.*, **13**, 299–308.

Kimbler, R. W. L. (1973) Phytotoxicity from plant residues. III. The relative effect of toxins and nitrogen immobilization on the germination and growth of wheat. *Plant and Soil*, **38**, 543–55.

King, K. F. S. and Chandler, M. T. (1978) *The Waste Lands*, ICRAF, Nairobi.

Kubo, I. and Nakanishi, K. (1977) Insect antifeedants and repellents from African plants, in *Host Plant Resistance to Pests* (ed. P. A. Hedin), ACS Symp. Ser. 62, Amer. Chem. Soc., Washington, DC, pp. 165–78.

Kuo, Y. L., Chou, C. H. and Hu, T. W. (1983) Allelopathic potential of *Leucaena leucocephala*, in *Allelochemicals and Pheromones* (eds G. R. Waller and C. H. Chou), Academia Sinica, Monograph Series No. 5, Taipei, ROC, pp. 107–19.

Kuo, C. G., Chou, M. H. and Park, H. G. (1981) Effect of Chinese cabbage residue on mung bean. *Plant and Soil*, **61**, 473–7.

Leon, W. B. (1976) Phytotoxicité inhité par les résidues de recolte de *Sorghum vulgare* dans les sols sableux de Touest African. *Bot. Rev.*, **45**, 15–109.

Lockhart, J. A. and Wiseman, J. A. (1970) *Introduction to Crop Husbandry*. Pergamon Press, Oxford.

Lydon, John and Duke, O. Stephen (1989) Pesticide effects on secondary metabolism of higher plants. *Pestic. Sci.*, **25**, 361–73.

Mahler, H. R. and Baylor, M. B. (1967) Effect of steroidal diamines on DNA duplication and mutagenesis. *Proc. Natl. Acad. Sci. USA*, **51**, 256.

Mathur, S. N., Mukerji, D., Rizvi, S. J. H. and Jaiswal, V. (1982) Plant products as herbicides of future, in *Widening Horizons of Plant Sciences* (ed. C. P. Mallick), Kalyani Publishers, Ludhiana, India, pp. 256–66.

Mathur, R. S., Sharma, K. K. and Ansari, M. Y. (1984) Economics of *Eucalyptus* plantations under agroforestry. *Indian Forester*, **110**, 171–201.

Matsunaka, S. and Kwatsuka, S. (1975) Environmental problems related to herbicidal use. *Environ. Quality and Safety*, **4**, 149–59.

Mikulas, J. (1984) Allelopathy of *Sorghum halepense* (L.) Pers. on weeds and crops. *Acta Agron. Acad. Scient. Hung.*, **33**, 423–7.

Mishra, G. P. (1986) Physiology of allelopathic interactions in tobacco cropping system. MSc Thesis, Rajendra Agricultural Univ. Pusa, India.

Mowbray, D. L. (1986) Pesticide poisoning in Papua New Guinea and the South Pacific. *Papua New Guinea Med. J.*, **29**, 131–41.

Muller, C. H. and del Moral, R. (1966) Soil toxicity induced by terpenes from *Salvia leucophylla*. *Bull. Torrey Bot. Club*, **93**, 130–7.

Netzly, D. H., Riopel, J. L., Ejeta, G. and Butler, L. G. (1988) Germination stimulants for witch weed (*Striga asiatica*) from hydrophobic root exudate of sorghum (*Sorghum bicolor*). *Weed Sci.*, **36**, 441–6.

Perlman, S. and Penman, S. (1970) Mitochondrial protein synthesis: Resistance to emetine and response to DNA synthesis inhibitors. *Biochem. Biophys. Res. Commun.*, **40**, 941–8.

Papas, T. S., Sandhas, L., Chirigos, M. A. and Furusawa, E. (1973) Inhibition of DNA polymerase of avian myoloblastosis virus by an alkaloid extract from *Narcissus tazetta* L. *Biochem. Biophys. Res. Commun.*, **52**, 88–92.

Patrick, Z. A., Toussoun, T. A. and Koch, L. W. (1964) Effect of crop residue decomposition products on plant roots. *Ann. Rev. Phytopathol.*, **2**, 267–92.

Paul, B. S. and Vadlamudi, V. P. (1976) Teratogenic studies of fenitrothion on white leghorn chick embryos. *Bull. Environ. Contam. Toxicol.*, **15**, 223–9.

Putnam, A. R. and DeFrank, J. (1983) Use of phytotoxic plant residues for selective weed control. *Crop Protect.*, **2**, 173–81.

Putnam, A. R. and Tang, C. S. (eds) (1986) *The Science of Allelopathy*, Wiley, New York.

Rakhteenko, I. N., Kaurov, I. A. and Minko, I. T. (1973) Effect of water-soluble metabolites of a series of crops on some physiological processes, in *Physiological-Biochemical Basis of Plant Interactions in Phytocenoses* (ed. A. M. Grodzinsky), Naukova Dumka, Kiev, pp. 16–19.

Ramakrishna, N. and Ramachandran, B. V. (1978) Malathion – fenitrothion synergism. *Indian Journal of Biochemistry and Biophysics*, **16**, 58–61.

Raza, S. M. and Bano, Z. (1980) Effect of sumithion and methyl parathion on the growth of *Phaseolus aureus. Ind. J. Air Poll. Control*, **3**, 56–61.

Rice, E. L. (1984) *Allelopathy*, Academic Press, New York.

Rice, E. L. (1985) Allelopathy – an overview, in *Chemically Mediated Interactions between Plants and other Organisms* (eds G. A. Cooper-Driver, T. Swain and E. E. Conn), Plenum, New York, pp. 81–105.

Ries, Stanley and Houtz, R. (1983) Triacontanol as a plant growth regulator. *Hort. Sci.*, **18**, 654–62.

Ries, S. K., Richanan, T. L. and Wert, V. (1978) Growth and yield of crops treated with triacontanol. *J. Amer. Soc. Hort. Sci.*, **103**, 361–4.

Rizvi, S. J. H., Mukerji, D. and Mathur, S. N. (1980a) A new report on a possible source of natural herbicide. *Ind. J. Exp. Biol.*, **18**, 77–8.

Rizvi, S. J. H., Jaiswal, V., Mukerji, D. and Mathur, S. N. (1980b) A multipurpose pesticide from *Coffea arabica. Ind. J. Mycol. Plant Pathol.*, **10**, 72.

Rizvi, S. J. H., Jaiswal, V., Mukerji, D. and Mathur, S. N. (1980c) Antifungal properties of 1,3,7-trimethylxanthine isolated from *Coffea arabica. Naturwiss.*, **67**, 459–60.

Rizvi, S. J. H., Pandey, S. K., Mukerji, D. and Mathur, S. N. (1980d) 1,3,7-Trimethylxanthine, a new chemosterilant for stored grain pest – *Callosobruchus chinensis. Z. Angew. Entomol.*, **90**, 777–8.

Rizvi, S. J. H., Mukerji, D. and Mathur, S. N. (1981) 1,3,7-Trimethylxanthine, a new natural herbicide. *Agril. Biol. Chem.*, **54**, 1255–6.

Rizvi, S. J. H. and Rizvi, V. (1984) Allelopathy: A new strategy in weed control. *Proc. 1st Trop. Weed Sci. Conf. Haatyai, Thailand*, **2**, 286–300.

Rizvi, S. J. H. and Rizvi, V. (1987) Improving crop productivity in India: Role of allelochemicals, in *Allelochemicals: Role in Agriculture and Forestry* (ed. G. R. Waller), ACS Symp. Ser. 330, Amer. Chem. Soc., Washington, DC, pp. 69–75.

Rizvi, S. J. H., Mishra, G. P. and Rizvi, V. (1989a) Allelopathic effects of nicotine on maize. I: Its potential application in crop rotation. *Plant and Soil*, **116**, 289–91.

Rizvi, S. J. H., Mishra, G. P. and Rizvi, V. (1989b) Allelopathic effects of nicotine on maize. II: Some aspects of its mechanism of action. *Plant and Soil*, **116**, 292–3.

Rizvi, S. J. H., Rizvi, V., Mukerji, D. and Mathur, S. N. (1987) 1,3,7-Trimethylxanthine an allelochemical from seeds of *Coffea arabica*: some aspects of its mode of action as a natural herbicide. *Plant and Soil*, **98**, 81–91.

Rizvi. S. J. H., Singh, V. K., Rizvi, V. and Waller, G. R. (1988) Geraniol, an allelochemical of possible use in integrated pest management. *Plant Protect.*, **3**, 112–14.

Rizvi, S. J. H., Sinha, R. C. and Rizvi, V. (1990) Implications of mimosine allelopathy in agroforestry. *Proc. 19th IUFRO World Cong. Forestry*, Montreal, Canada, **2**, 22–7.

Robinson, T. (1974) Metabolism and function of alkaloids in plant. *Science*, **184**, 430–5.

Samersov, V. F. and Prishchepa, I. A. (1978) Effect of herbicide – insecticide mixture on amount of nitrogenous substances in barley plants. *Khim. Sci. Sk. Khoz.*, **16**, 58–61.

Singh, R. and Bawa, R. (1982) Effect of leaf leachates from *Eucalyptus globulus* Labill. and *Aesculus indica* Colebr. on seed germination of *Glaucium flavum* Grantz. *Ind. J. Ecol.*, **9**, 21–8.

Smith, T. M. and Stratton, G. W. (1986) Effects of synthetic pyrethroid insecticides on nontarget organisms. *Residue Rev.*, **97**, 93–114.

Talwar, K. K., Kumar, Ish and Kalsi, P. S. (1983) A dramatic role of terpenoids in increasing rice production. *Experientia*, **39**, 117–19.

ter-Meulen, V. U. and El-Harith, E. A. (1985) Mimosine, a factor limiting the use of *Leucaena leucocephala* as an animal feed. *Z. Landwirtschaft Tropen Subtropen*, **86**, 109–27.

Thompson, A. C. (1985) *The Chemistry of Allelopathy*, *Biochemical Interactions Among Plants*, ACS Symp. Ser. 268, Amer. Chem. Soc., Washington, DC.

Verma, S. R., Rani Sarita and Dabla, R. C. (1984) Effect of pesticides and their combinations on three serum phosphatase of *Mystus vittatus*. *Water, Air and Soil Pollution*, **21**, 9–14.

Waller, G. R. (ed.) (1987) *Allelochemicals: Role in Agriculture and Forestry*, ACS Symp. Ser. 330, Amer. Chem. Soc., Washington, DC.

Waller, G. R., Friedman, J., Chou, C. H. *et al.* (1983) Hazards, benefits, metabolism and translocation of caffeine in *Coffea arabica* L. plants and surrounding soil, in *Allelochemicals and Pheromones* (eds G. R. Waller, and C. H. Chou), Academia Sinica, Monograph Series No. 5, Taipei, ROC., pp. 239–60.

Waller, G. R., Krenzner, E. G., McPherson, J. K. and McGown, S. R. (1987) Allelopathic compounds in soil from no tillage vs conventional tillage in wheat production. *Plant and Soil*, **98**, 5–15.

Westcott, N. D., Lee, Y. W. and McKinlay, K. S. (1987) Persistence and toxicity of dimethoate on wheat herbage and sweetclover herbage. *J. Environ. Sci. Health*, **B22**, 379–90.

Wild, D. (1975) Mutagenicity studies on organophosphorus insecticide. *Mutation Res.*, **32**, 133.

Wink, Michael (1987) Chemical ecology of quinolizidine alkaloids, in *Allelochemicals: Role in Agriculture and Forestry* (ed. G. R. Waller), ACS Symp. Ser. 330, Amer. Chem. Soc., Washington, DC, pp. 524–33.

Zimmerman, R. H., Lieberman, M. and Broome, O. C. (1977) Inhibitory effect of rhizobitoxine analog on bud growth after release from dormancy. *Plant Physiol.*, **59**, 158–60.

INDEX

Page numbers in *italics* refer to tables